RIVER MECHANICS

SECOND EDITION

The second edition of Julien's textbook presents an analysis of rivers, from mountain streams to estuaries. The book is rooted in fundamental principles to promote sound engineering practice. State-of-the-art methods are presented to underline theory and engineering applications. River mechanics blends the dual concepts of water conveyance and sediment transport. Like the first edition, this textbook contains ample details on river equilibrium, river dynamics, bank stabilization, and river engineering. Complementary chapters also cover the physical and mathematical modeling of rivers. As well as being completely updated throughout, three new chapters have been added on watershed dynamics, hillslope stability, and stream restoration. Throughout the text, hundreds of examples, exercises, problems, and case studies assist the reader in learning the essential concepts of river engineering. The textbook is very well illustrated to enhance advanced student learning, while researchers and practitioners will find the book to be an invaluable reference.

PIERRE Y. JULIEN is Professor of Civil and Environmental Engineering at Colorado State University. He has 35 years of professional engineering experience in the fields of hydraulics and river sedimentation. Julien has authored more than 500 scientific contributions, including two textbooks (the first edition of *River Mechanics*, and *Erosion and Sedimentation* (Cambridge University Press 2010, second edition)), 25 book chapters and manuals, 185 refereed journal articles, and 230 professional presentations and conference papers. He has delivered 20 keynote addresses and guided more than 130 graduate students to completion of their engineering degrees. He is the recipient of the Hans Albert Einstein Award of the American Society of Civil Engineers (ASCE), delivered the Hunter Rouse Lecture of the Environmental and Water Resources Institute (ASCE) in 2015, and is a former editor for the ASCE *Journal of Hydraulic Engineering*.

"This elegantly written book covers the major topics associated with water flow and sediment transport in rivers. It thoughtfully guides readers through descriptions and formulations of key physical processes, and offers many illustrations and worked examples to aid understanding. The book is a comprehensive companion to the author's book *Erosion and Sedimentation*, which focuses on alluvial sediment transport in rivers."
—Robert Ettema, *Colorado State University*

"As an engineering professional facing the challenges of sediment transport, I found the new edition of *River Mechanics* to be a great reference and a very useful resource. Its presentation of material has been substantially revised and expanded, including several new chapters. I especially liked the expanded treatment of watershed processes and new material on stream restoration. *River Mechanics* stands on its own and is even more useful in tandem with Pierre Julien's other book *Erosion and Sedimentation* as its companion. Having been familiar with the first edition from my days in graduate school, this new edition will undoubtedly prove to be an indispensable resource for students and practitioners alike."
—Mark Velleux, *HDR*

"A book in river engineering taking the interested reader from its sources to the estuary, painted with concise problem statements and solved by adequate engineering methods and techniques. Prof. Julien's second edition can be fully recommended to graduate students, researches and practicing engineers in the fields of river basins, river mechanics, river flows, river stability, river equilibrium, river models, and river restoration. Prof. Julien should be praised for his integral approach, his technical formulation and his updated presentation involving both problems in practice and exercises of the complicated topic."
—Willi Hager, *ETH Zurich*

"A rare must-read on modern river mechanics that covers the subject not only comprehensively and rigorously but also inspirationally. The author's philosophy 'from observations to physical understanding to mathematical modelling and numerical simulations' underpins every topic in the book, making it very clear and complete. Undoubtedly, this text will quickly become a benchmark source equally important to students, engineers and researchers. It will also be noteworthy to geoscientists and stream ecologists working at the borders between their disciplines and engineering. A genuine pleasure to read!"
—Vladimir Nikora FRSE, *University of Aberdeen*

RIVER MECHANICS

SECOND EDITION

PIERRE Y. JULIEN
Colorado State University

CAMBRIDGE
UNIVERSITY PRESS

CAMBRIDGE
UNIVERSITY PRESS

University Printing House, Cambridge CB2 8BS, United Kingdom

One Liberty Plaza, 20th Floor, New York, NY 10006, USA

477 Williamstown Road, Port Melbourne, VIC 3207, Australia

314-321, 3rd Floor, Plot 3, Splendor Forum, Jasola District Centre, New Delhi - 110025, India

79 Anson Road, #06-04/06, Singapore 079906

Cambridge University Press is part of the University of Cambridge.

It furthers the University's mission by disseminating knowledge in the pursuit of education, learning and research at the highest international levels of excellence.

www.cambridge.org
Information on this title: www.cambridge.org/9781107462779
DOI: 10.1017/9781316107072

First published 2018

A catalogue record for this publication is available from the British Library

Library of Congress Cataloging in Publication data
Names: Julien, Pierre Y.
Title: River mechanics / Pierre Y. Julien, Colorado State University.
Description: Cambridge, United Kingdom : Cambridge University Press, [2018]
    | Includes bibliographical references and index.
Identifiers: LCCN 2017058314 | ISBN 9781107462779 (pbk.)
Subjects: LCSH: River engineering.
Classification: LCC TC405 .J85 2018 | DDC 627/.12—dc23 LC record
    available at https://lccn.loc.gov/2017058314

ISBN 978-1-107-46277-9 Paperback

Additional resources for this publication at www.cambridge.org/river2

Dedicated to my deceased mother Yolande and my brother Michel

Contents

Preface

Water is essential to sustain life and rivers are truly fascinating. Most prosperous cities are located near river confluences and river engineers must design structures to draw benefits from the fluvial system for developing societies. Ideally, scientists should develop new methods to improve engineering design, while practitioners must understand why certain structures work and others fail. Fundamentally, river mechanics requires understanding of hydrodynamic forces governing the motion of water and sediment in complex river systems. Additionally, the fluvial network must seek equilibrium in its ability to carry water and transport sediment through dynamic river systems. Nowadays, river engineers are concerned not only about urban drainage, flood control, and water supply, but also about water quality, contamination, and aquatic habitat. This textbook broadens this perspective by integrating knowledge of climatology, hydrology, and geomorphology.

This textbook has been prepared for engineers and scientists developing a broad-based technical expertise in river mechanics. It has been specifically designed for graduate students, for scholars actively pursuing scientific research, and for practitioners keeping up with recent developments in river engineering. The prerequisites for reading it and making use of it are simply a basic knowledge of undergraduate fluid mechanics and of partial differential equations. The textbook *Erosion and Sedimentation* from Cambridge University Press serves as prerequisite material for the graduate course, River Mechanics, that I have taught at Colorado State University over the past three decades.

My teaching philosophy has been detailed in my recent Hunter Rouse lecture (Julien, 2017). Sketch I.1 illustrates the key points that I seek to develop among my graduate students and postdoctoral advisees.

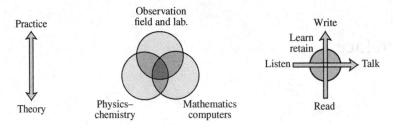

Sketch I.1. Professional development in river engineering

First, the essential complementarity of theory and practice cannot be over-emphasized. Theory can best enhance engineering applications when the fundamental understanding has been grounded in practical observations. Second, there is a need to develop three main poles, where observations from field and laboratories lead to physical understanding, prior to mathematical calculations. Expertise is developed by expanding the overlapping areas of these three poles. Finally, while the processes of listening and reading are essential to the ability to learn and retain new knowledge, my teaching emphasizes also the need to develop verbal and written communication skills. The ability to express dynamic thinking is a tremendous asset for any successful professional career.

Rather than being a voluminous encyclopedia, this textbook scrutinizes selected methods which meet pedagogical objectives. There is sufficient material for a 45-h graduate-level course. Beside basic theory and lecture material, the chapters of this book contain various exercises and problems, data sets and examples, computer problems, and case studies. They illustrate specific aspects of the profession from theoretical derivations, through exercises and problems, to practical solutions with the analysis of case studies. Most problems can be solved with a few algebraic equations; others require the use of computers. No specific computer code or language is required. Instead of promoting the use of commercial software packages, I stimulate students' creativity and originality in developing their own computer programs. Throughout the book, a solid diamond (♦) denotes equations and problems of particular significance; double diamond (♦♦) denotes the most important.

The book covers topics essentially from the mountains to the oceans:

Chapter 1 outlines the physical properties of water and sediment;
Chapter 2 reviews the governing equations of motion and sediment transport;
Chapter 3 describes river basins in terms of the source of water and sediment;
Chapter 4 looks at river basin dynamics;
Chapter 5 treats the steady-flow conditions in canals and rivers;

Chapter 6 delves into flood-wave propagation in rivers;

Chapter 7 introduces some numerical methods used to solve river engineering problems;

Chapter 8 copes with hillslope and riverbank stability;

Chapter 9 deals with riverbank protection measures;

Chapter 10 delineates the hydraulic geometry and equilibrium in alluvial rivers;

Chapter 11 explains the concepts of river dynamics and response;

Chapter 12 focuses on physical modeling techniques;

Chapter 13 provides essential knowledge on stream restoration;

Chapter 14 presents several river engineering techniques; and

Chapter 15 covers waves and tides in river estuaries.

My teaching has been greatly inspired by Drs. Marcel Frenette, Daryl B. Simons, Hunter Rouse, Yvon Ouellet, E. V. Richardson, Jean Louis Verrette, Steven R. Abt, Jose D. Salas, Richard Eykholt, HsiehWen Shen, Jim Ruff, Carl F. Nordin, Jean Rousselle, and Stan Schumm, as well as many others. They have greatly influenced my professional development and university teaching since 1979. I am also thankful to Drs. Phil Combs, Drew Baird, and Patrick O'Brien for sharing their practical expertise in river engineering. This book would not have been the same without contributions and suggestions from a couple of generations of graduate students at Colorado State University. They helped me tailor this textbook to meet their needs under the constraints of quality, concision, and affordability. Jean Parent patiently drafted all the figures. Finally, it has been a renewed pleasure to collaborate with Matt Lloyd, Esther Migueliz, and the Cambridge University Press production staff.

Notation

c_G	group celerity
c_u	undrained cohesion
C	Chézy coefficient
C	sediment concentration
C_a	reference concentration
$\hat{C}$	cropping management factor
C_{fl}	Courant–Friedrichs–Lewy condition
C_k	grid dispersion number
C_{0i}	upstream sediment concentration
C_r	runoff coefficient
$C_u = u\Delta t/\Delta x$	Courant number
C_v, C_w, C_{ppm}, $C_{mg/l}$	sediment concentration by volume, weight, parts per million, and milligrams per liter
d_{10}, d_{50}	particle size distribution, % finer by weight
d_m	effective riprap size
d_s	particle size
d_*	dimensionless particle diameter
D	pipe/culvert diameter
D	headcut height
D_d	degree-days
D_p	drop height of a grade-control structure
D_x	oxygen deficit
DO	dissolved oxygen content
e	void ratio
E	specific energy
E	gross erosion
E_{tons}	expected soil loss in tons
$\hat{E}$	soil loss per unit area
$\tilde{E}$	total energy of a wave
$E(\ )$	exceedance probability
ΔE	specific energy lost in a hydraulic jump
f	Darcy–Weisbach friction factor
f_l	Lacey silt factor
$f(t)$	infiltration rate
F	force
$\tilde{F}$	fetch length of wind waves
F_B	buoyancy force
F_c	centrifugal force
F_D	drag force

F_g	gravitational force
F_h	hydrodynamic force
F_i	inertial force
F_L	lift force
F_M	momentum force
F_p	pressure force
F_s	shear force in a bend
F_S	submerged weight of a particle
$F_{Vf} = V/\sqrt{gL_f}$	fish Froude number
F_w	weight of water
F_W	weight of a particle
$F(\ )$	nonexceedance probability
$F_n(z)$	standard normal distribution
$F(t)$	cumulative infiltration
$F_a(t)$	actual cumulative infiltration
$F_p(t)$	potential cumulative infiltration
Fr	Froude number
g	gravitational acceleration
G	specific gravity of sediment
Gr	gradation coefficient
G_u	universal gravitation constant
h	flow depth
h_c	critical flow depth
h_d	downstream flow depth
h_n	normal flow depth
h_p	pressure head at the wetting front
h_r	rainfall depth
h_s	cumulative snowmelt
h_t	tailwater depth
h_u	upstream flow depth
h_w	partial elevation drop on a watershed
Δh	local change in flow depth
H	Bernoulli sum
ΔH	energy loss over a meander wavelength
H_c	critical hillslope soil thickness
$H_o(\theta_m)$	Struve function
$\tilde{H}_s = 2\tilde{a}$	wave height
H_w	elevation drop on a watershed
i	rainfall intensity

i_b	riverbed infiltration rate
i_e	excess rainfall intensity
i_f	snowmelt rate
i_{30}	maximum 30-min rainfall intensity
j	space index
$J_0(\theta_m)$	zeroth-order Bessel function of the first kind
k	decay coefficient
k_0	resistance parameter for laminar overland flow
k_s	surface roughness
k'_s	grain roughness height
k_t	total resistance to laminar overland flow
$\tilde{k}$	wave number
K	saturated hydraulic conductivity
K	conveyance coefficient
$\hat{K}$	soil erodibility factor
K_1, K_2	coefficients of the pier scour equation
K_b	ratio of maximum shear stress in a bend to a straight channel
K_c	riprap coefficient
K_d	dispersion coefficient
K_d	flood-wave diffusivity
K_d	soil–water partition coefficient or ratio of sorbed to dissolved metals
$\overline{KE}$	average kinetic energy per unit area
$K_G(T)$	frequency factor of the Gumbel distribution
K_{num}	numerical dispersion coefficient
K_{oc}	soil–water partition coefficient normalized to organic carbons
K_{ow}	octanol–water partition coefficient
K_p	plunging jet coefficient
$K_p(\gamma)$	frequency factor of the log-Pearson III distribution
K_S	ratio of the sediment volume
K_{sj}	submerged jet coefficient
$\Delta l = a/R$	mean annual migration rate
l_1 to l_4	moment arms
l_c, l_d	moment arms in radial stability of river bends
L	sinuous river length
L	field runoff length
L_a	abutment length

LC_{50}	lethal concentration resulting in 50% mortality
L_f	depth of the wetting front
L_0	normalized channel length
L_p	pier length
L_r	river length
L_r	length ratio
L_{sb}	settling-basin length
$\hat{L}$	slope-length factor
L_f	fish length
L_M	runoff-model grid-cell size
L_R	grid size of rainfall precipitation
L_S	correlation length of a storm
L_W	length scale of a watershed
L_Δ	length of arrested saline wedge
m	exponent of the resistance equation
m_E	mass of the Earth
m_M	mass of the Moon
m_s	sediment mass eroded from a single storm
M	mass
M	specific momentum
M	snowmelt rate
M_f	melt factor
M_1, M_2	first and second moments of a distribution
M, N	particle-stability coefficients
M/N	ratio of lift to drag moments of force
n	Manning coefficient n
$\vec{n}$	normal vector pointing outside of the control volume
$\tilde{n}$	wave number index
N	number of points per wavelength
N	number of storms
$N_0(\theta_m)$	Neumann function, or the zeroth-order Bessel Y function
$O(\ )$	order of an approximation
p	pressure
$p(\ )$	probability density function
p_{cl}	mean annual percentage lateral migration rate
p_0	porosity
p_{0e}	effective porosity

p_{0i}	initial water content
p_{0r}	residual water content
Δp_c	fraction of material coarser than d_{sc}
Δp_i	sediment size fraction
Δp_0	change in water content at the wetting front
P	wetted perimeter
$P(\ )$	probability
ΔP	power loss in a hydraulic jump
$\hat{P}$	conservation practice factor
$\tilde{P}$	total power of a wave
PCB	polychlorinated biphenyls
$\overline{PE}$	average potential energy per unit surface area
P_0	power loss
P_Δ	grid Peclet number
q	unit discharge
q_{bv}	unit sediment discharge by volume
$q_{bv}^* = q_{bv}/\omega_0 d_s$	dimensionless unit sediment discharge
q_l	lateral unit discharge
q_m	maximum unit discharge
q_s	unit sediment discharge
q_{sj+1}, q_{sj}	upstream and downstream unit sediment discharge
q_t	total unit sediment discharge
Q	river discharge
Q_{bv}	bed sediment discharge by volume
Q_p	peak discharge
Q_s	sediment discharge
r	radial coordinate
r^*	dimensionless radius of curvature
r, θ, z	cylindrical coordinate system r lateral, θ downstream, and z upward
r_Q	discharge ratio
R	risk
R	radius of curvature of a river
$\hat{R}$	rainfall-erosivity factor
ΔR_e	excess rainfall
R_E	radius of the Earth
Re	Reynolds number
$\text{Re}_* = u_* d_s/v$	grain shear Reynolds number
R_h	hydraulic radius

R_m	minimum radius of curvature of a channel
$Ro = \omega/\kappa u_*$	Rouse number
S	slope
$\hat{S}$	slope-steepness factor
S_D	specific degradation
S_{DR}	sediment delivery ratio
S_e	effective saturation
S_0, S_f, S_w	bed, friction, and water-surface slopes
S_{0x}, S_{0y}	bed-slope components in x and y
S_r, S_{wr}	radial water-surface slope
S_r^*	dimensionless radial slope
SF	safety factor
t	time
t	trapezoidal section parameter
Δt	time increment
Δt_s	time increment for sediment
t_a	cumulative time with positive air temperature
t_e	time to equilibrium
t_f	cumulative duration of snowmelt
t_f	fish swimming duration
t_r	rainfall duration
$t_r^* = t_r/\bar{t}_r$	normalized storm duration
t_t	transversal mixing time
t_v	vertical mixing time
T	period of return of extreme events
T	wave period
T°	temperature
T_{50}	time for half the channel-width change
T_E	trap efficiency
T_s	windstorm duration
u, v	velocity along a vertical profile
$\bar{u}$	average flow velocity
u_*	shear velocity
u_{*c}	critical shear velocity
U_f	fish swimming velocity
U_w	wind speed
v_h	migration rate of headcuts
v_s	local velocity against the stone
v_x, v_y, v_z	local velocity components

V	mean flow velocity
V_c	critical velocity
V_x, V_y, V_z	Cartesian mean flow velocities in x, y, and z
V_Δ	densimetric velocity
V_θ	downstream velocity in cylindrical coordinates
$\forall$	volume
$\forall_v, \forall_t$	volume of voids and total volume
W	channel width
W	weight of soil per unit width
W, W_0, W_e	active, initial and equilibrium channel width
W_m	meander width
W_0	overland plane width
x, y, z	coordinates usually x downstream, y lateral, and z upward
x_r, y_r, z_r	length ratios for hydraulic models
X_{max}	downstream distance with the maximum oxygen deficit
Δx	grid spacing
X	runoff length
X_c	reach length
X_e	equilibrium runoff length
X_{max}	maximum endurance fish swimming distance
y_d, y_u	downstream and upstream wave amplitude
Y	sediment yield
z_b	bed elevation
z_w	water-surface elevation
z^*	dimensionless depth
Δz	scour depth

Greek Symbols

α	coefficient of the stage–discharge relationship
α, β	parameters of the gamma distribution
α_b	deflection angle of barges
α_e	Coriolis energy correction factor
$\tilde{\alpha} = 2\pi/N$	phase angle
β	exponent of the stage–discharge relationship
β	bed particle-motion angle
β_m	momentum correction factor

γ	specific weight of water
γ	skewness coefficient
γ_m	specific weight of a water–sediment mixture
γ_{md}	dry specific weight of a water–sediment mixture
γ_s	specific weight of sediment
$\Gamma = \sqrt{1 + 4kK/U^2}$	dimensionless settling parameter
$\Gamma(x)$	gamma function
δ	angle between streamline and particle direction
$\delta_L = \ln(y_d/y_u)$	wave amplification over length L_0
ξ	ratio of exceedance probabilities
$\xi_r = W_r/h_r$	channel width–depth ratio
$\tilde{\xi}$	wave displacement in the x direction
η	sideslope stability number
$\tilde{\eta}$	wave surface elevation
$\zeta_{\tilde{n}}^k$	Fourier coefficients
κ	von Kármán constant
λ	streamline deviation angle
λ	wavelength
λ_f	snowmelt intensity
$\lambda_r = t_r/t_e$	hydrograph equilibrium number
λ_s	significant wavelength
Λ	meander wavelength
μ	dynamic viscosity of water
ν	kinematic viscosity of water
φ	angle of repose of bed material
ϕ	latitude
Φ	potential function for waves
ρ	mass density of water
ρ_m	mass density of a water–sediment mixture
ρ_{md}	dry mass density of a water–sediment mixture
ρ_s	mass density of sediment
ρ_{sea}	mass density of seawater
$\Delta\rho$	mass density difference
$\Pi = \ln[-\ln E(x)]$	double logarithm of exceedance probability
ω	settling velocity
ω_E	angular velocity of the Earth
Ω	sinuosity
Ω_R	ratio of centrifugal force to shear force in bends
θ	downstream orientation of channel flow

θ	angular coordinate
θ_c	critical angle of the failure plane
θ_j	jet angle measured from the horizontal
θ_m	maximum orientation of channel flow
θ_p	flow orientation angle against a pier
θ_r	raindrop angle
θ_0, Θ_0	downstream bed angle
Θ_1	sideslope angle
$\Theta = (t - t_r)/t_e$	dimensionless time
σ	stress components
σ	standard deviation
$\sigma = 2\pi L_0/\lambda$	dimensionless wave number
σ'	effective stress
σ_g	gradation coefficient
$\sigma_x, \sigma_y, \sigma_z$	normal stresses (negative pressure)
$\sigma_{\Delta t}$	standard deviation of dispersed material
σ_θ	normal stress on a plane at an angle θ from the principal stresses
$\tilde{\sigma}$	angular frequency of surface waves
τ	shear stress
τ_0, τ_b	bed shear stress
τ_{0x}, τ_{0y}	downstream and lateral bed shear stresses
τ_{bn}	bed shear stress at a normal depth
τ_c	critical shear stress
τ_f	failure shear strength of the soil
τ_r	radial shear stress
τ_r^*	dimensionless radial shear stress
τ_s	side shear stress
τ_{sc}	critical shear stress on a sideslope
τ_w	wind shear stress
τ_{zx}	shear stress in the x direction in a plane perpendicular to z
τ_*	Shields parameter
τ_{*c}	critical value of the Shields parameter
τ_θ	tangential stress on a plane at an angle θ from the principal stresses
$\psi = q/i_e L$	dimensionless discharge
Ψ	reduced variable

Superscripts and Diacritics

$\tilde{n}$	wave properties
$\hat{C}$	parameters of the universal soil-loss equation
$\bar{e}$	average value
h^k	time index k

Subscripts

a_r, a_θ	cylindrical coordinate components
a_x, a_z	Cartesian components
n_o, n_c	roughness values for overbank and main channel
τ_c	critical shear stress
h_{j+1}	space index at $j + 1$
L_m, Q_m	model value
L_p, Q_p	prototype value
L_r, Q_r	similitude scaling ratio
K_1, K_2, K_3	correction factors of the CSU scour equation
$W_{1/2}, h_{1/2}, S_{1/2}$	width, depth, and slope for half the discharge
t_{63}, X_{63}	time and distance scale for 63% of the sediment to deposit
ρ_m, γ_m	properties of a water–sediment mixture
ρ_{md}, γ_{md}	properties of a dry water–sediment mixture
ρ_s, γ_s	sediment properties

1

Physical Properties

As a natural science, the variability of river processes must be examined through the measurement of physical parameters. This chapter describes dimensions and units (Section 1.1), physical properties of water (Section 1.2), and sediment (Section 1.3).

1.1 Dimensions and Units

Physical properties are usually expressed in terms of the following fundamental dimensions: mass (M), length (L), and time (T). Temperature ($T°$) is also sometimes considered. The fundamental dimension of mass is preferred to the corresponding force.

The fundamental dimensions are measurable in quantifiable units. In the SI system of units, the units for mass, length, time, and temperature are the kilogram (kg), the meter (m), the second (s), and degrees Kelvin (K). The Celsius scale (°C) is commonly preferred in river engineering because it refers to the freezing point of water as 0°C. The abbreviations for cubic meters per second (1 cms = 1 m^3/s) and cubic feet per second (1 cfs = 1 ft^3/s) are commonly used to describe the flow discharge of a river.

A Newton (N) is the force required to accelerate 1 kg at 1 m/s^2, or 1 N = 1 kg m/s^2. The gravitational acceleration at the Earth's surface is $g = 9.81$ m/s^2. The weight of one kilogram is $F =$ mass $\times g = 1$ kg $\times 9.81$ $m/s^2 = 9.81$ N. The pressure is given in pascals from 1 Pa = 1 N/m^2. The unit of work (or energy) is the joule (J), which equals the product of 1 N $\times$ 1 m. The unit of power is a watt (W), which is 1 J/s. Prefixes indicate multiples or fractions of units by powers of 10:

$$\mu(\text{micro}) = 10^{-6}, \quad k(\text{kilo}) = 10^{3},$$
$$m(\text{milli}) = 10^{-3}, \quad M(\text{mega}) = 10^{6},$$
$$c(\text{centi}) = 10^{-2}, \quad G(\text{giga}) = 10^{9}.$$

For example, sand particles are coarser than 62.5 μm or microns; gravels are coarser than 2 mm; and 1 megawatt (MW) equals 1 million watts (1,000,000 or 10^6 W).

In the English system of units, the time unit is a second, the fundamental units of length and mass are, respectively, the foot (ft), equal to 30.48 cm, and the slug, equal to 14.59 kg. The force to accelerate a mass of one slug at 1 ft/s^2 is a pound force (lb). In this text, a pound always refers to a force, not a mass. Temperature in degrees Celsius, $T°C$, is converted to the temperature in degrees Fahrenheit, $T°F$, by $T°F = 32.2°F + 1.8\ T°C$.

Variables are classified as geometric, kinematic, dynamic, and dimensionless variables. As shown in Table 1.1, geometric variables describe the geometry in terms of length, area, and volume. Kinematic variables describe the

Table 1.1. *Geometric, kinematic, dynamic, and dimensionless variables*

Variable	Symbol	Fundamental dimensions	SI units
Geometric (L)			
Length	L, x, h, d_s	L	m
Area	A	L^2	m^2
Volume	$\forall$	L^3	m^3
Kinematic (L, T)			
Velocity	v_x, V, u_*	LT^{-1}	m/s
Acceleration	a, a_x, g	LT^{-2}	m/s^2
Kinematic viscosity	ν	L^2T^{-1}	m^2/s
Unit discharge	q	L^2T^{-1}	m^2/s
Discharge	Q	L^3T^{-1}	m^3/s
Dynamic (M, L, T)			
Mass	m	M	1 kg
Force	$F = ma, mg$	MLT^{-2}	1 kg m/s^2 = 1 N
Pressure	$p = F/A$	$ML^{-1}T^{-2}$	1 N/m^2 = 1 Pa
Shear stress	$\tau_{xy}, \tau_0, \tau_c$	$ML^{-1}T^{-2}$	1 N/m^2 = 1 Pa
Work or energy	$E = Fd$	ML^2T^{-2}	1 Nm = J
Mass density	ρ, ρ_s	ML^{-3}	kg/m^3
Specific weight	$\gamma, \gamma_s = \rho_s g$	$ML^{-2}T^{-2}$	N/m^3
Dynamic viscosity	$\mu = \rho \nu$	$ML^{-1}T^{-1}$	1 kg/m s = 1 Pa s
Dimensionless			
Slope	S_0, S_f	–	–
Specific gravity	$G = \gamma_s/\gamma$	–	–
Reynolds number	Re $= Vh/\nu$	–	–
Grain shear Reynolds number	Re$_* = u_* d_s/\nu$	–	–
Froude number	Fr $= u/(gh)^{0.5}$	–	–
Shields parameter	$\tau_* = \tau/(\gamma_s - \gamma)d_s$	–	–
Concentration	C_v, C_w	–	–

Table 1.2. *Unit conversions*

Unit	kg, m, s	N, Pa, W
1 acre	4,047 m^2	
1 acre-foot (acre-ft)	1,233 m^3	
1 atmosphere (atm)	101,325 kg/m s^2	101.3 kPa
1 bar	100,000 kg/m s^2	100 kPa
1 barrel (US, dry) (bbl)	0.1156 m^3	
1 cubic foot per second (ft^3/s)	0.0283 m^3/s	
(1 m^3/s = 35.32 ft^3/s)		
1 degree Celsius (°C) = $(T°F - 32°)$ 5/9	1 degree Kelvin (K)	
1 degree Fahrenheit (°F) = 32 + 1.8 T°C	0.5556 degree Kelvin	
1 drop	61.6 mm^3	
1 dyne (dyn)	0.00001 kg m/s^2	10 μN
1 dyne per square centimeter (dyn/cm^2)	0.1 kg/m s^2	0.1 Pa
1 fathom (fath)	1.829 m	
1 foot (ft)	0.3048 m	
1 gallon (US gal) (1 US gal = 3.785 l)	0.003785 m^3	
1 horsepower (hp) = 550 lb ft/s	745.7 kg m^2/s^3	745.7 W
1 inch (in.) (1 ft = 12 in.)	0.0254 m	
1 inch of mercury (in. Hg)	3,386 kg/m s^2	3,386 Pa
1 inch of water	248.8 kg/m s^2	248.8 Pa
1 kip (1 kip = 1,000 lb)	4,448 kg m/s^2	4,448 N
1 knot	0.5144 m/s	
1 liter (l) (1 m^3 = 1,000 l)	0.001 m^3	
1 micrometer or micron (μm)	1×10^{-6} m	
1 mile (nautical)	1,852 m	
1 mile (statute) (1 mile = 5,280 ft)	1,609 m	
1 million gallons per day (mgd) = 1.55 ft^3/s	0.04382 m^3/s	
1 ounce (avoirdupois) (oz)	0.02835 kg	
1 fluid ounce (US)	2.957×10^{-5} m^3	
1 pascal (Pa)	1 kg/m s^2	1 N/m^2
1 pint (US pint)	0.0004732 m^3	
1 poise (P)	0.1 kg/m s	0.1 Pa s
1 pound-force (lb) (1 lb = 1 slug × 1 ft/s^2)	4.448 kg m/s^2	4.448 N
1 pound-force per cubic foot (lb/ft^3)	157.1 kg/m^2 s^2	157.1 N/m^3
1 pound-foot (lb-ft)	1.356 kg m^2/s^2	1.356 N m
1 pound per square foot (lb/ft^2 or psf)	47.88 kg/m s^2	47.88 Pa
1 pound per square inch (lb/in.^{2} or psi)	6,895 kg/m s^2	6,895 Pa
1 quart (US) (1 qt = 2 pint)	0.0009463 m^3	
1 slug	14.59 kg	
1 slug per cubic foot (slug/ft^3)	515.4 kg/m^3	
1 stoke (S) = 1 cm^2/s	0.0001 m^2/s	
1 ton (UK long)	1,016 kg	
1 ton (SI metric) (1,000 kg = 1 Mg)	1,000 kg	
1 ton (US short) = 2,000 lb	8,900 kg m/s^2	8.9 kN
1 yard (yd) (1 yd = 3 ft)	0.9144 m	

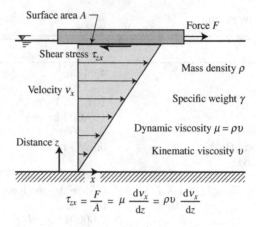

$$\tau_{zx} = \frac{F}{A} = \mu \frac{dv_x}{dz} = \rho v \frac{dv_x}{dz}$$

Figure 1.1. Newtonian fluid properties

motion of fluids and solids, and are depicted by only two fundamental dimensions: L and T. Dynamic variables always include mass M terms. Force, pressure, shear stress, work, energy, power, mass density, specific weight, and dynamic viscosity are common examples of dynamic variables. Several conversion factors are listed in Table 1.2.

1.2 Water Properties

Rivers carry water to the oceans. The properties of water are sketched in Figure 1.1.

Mass density of water ρ. The mass of water per unit volume defines the mass density ρ. The maximum mass density of water is 1,000 kg/m^3 at 4°C and decreases slightly with temperature, as shown in Table 1.3. In comparison, the mass density of sea water is approximately 1,025 kg/m^3, and the mass density of air at sea level is 1.29 kg/m^3 at 0°C. The conversion factor for mass density is 1 slug/ft^3 = 515.4 kg/m^3. The density of ice is approximately 10 percent less than that of water and it increases as the subzero temperature decreases. For instance, the ice cover on lakes can crack during winter nights and expand in daytime. When the cracks fill up with water at night, very large forces will be applied on the banks of rivers, lakes, and reservoirs.

Specific weight of water γ. The weight per unit volume is the specific weight γ. At 10°C, water has a specific weight, $\gamma = 9,810$ N/m^3 or 62.4 lb/ft^3

Table 1.3. *Physical properties of clear water at atmospheric pressure*

Temperature (°C)	Density ρ (kg/m³)	Specific weight γ (N/m³)	Dynamic viscosity μ (N s/m² or kg/m s)	Kinematic viscosity ν (m²/s)
−30	921	9,035	Ice	Ice
−20	919	9,015	Ice	Ice
−10	918	9,005	Ice	Ice
0	999.9	9,809	1.79×10^{-3}	1.79×10^{-6}
4	**1,000**	**9,810**	1.56×10^{-3}	1.56×10^{-6}
5	999.9	9,809	1.51×10^{-3}	1.51×10^{-6}
10	999.7	9,807	1.31×10^{-3}	1.31×10^{-6}
15	999	9,800	1.14×10^{-3}	1.14×10^{-6}
20	998	9,790	$\mathbf{1.0 \times 10^{-3}}$	$\mathbf{1.0 \times 10^{-6}}$
25	997	9,781	8.91×10^{-4}	8.94×10^{-7}
30	996	9,771	7.97×10^{-4}	8.00×10^{-7}
35	994	9,751	7.20×10^{-4}	7.25×10^{-7}
40	992	9,732	6.53×10^{-4}	6.58×10^{-7}
50	988	9,693	5.47×10^{-4}	5.53×10^{-7}
60	983	9,643	4.66×10^{-4}	4.74×10^{-7}
70	978	9,594	4.04×10^{-4}	4.13×10^{-7}
80	972	9,535	3.54×10^{-4}	3.64×10^{-7}
90	965	9,467	3.15×10^{-4}	3.26×10^{-7}
100	958	9,398	2.82×10^{-4}	2.94×10^{-7}
°F	slug/ft³	lb/ft³	lb s/ft²	ft²/s
0	1.7844	57.40	Ice	Ice
10	1.7839	57.34	Ice	Ice
20	1.7816	57.31	Ice	Ice
30	1.7787	57.25	Ice	Ice
32	1.938	62.40	3.75×10^{-5}	1.93×10^{-5}
40	**1.94**	62.43	3.23×10^{-5}	1.66×10^{-5}
50	1.938	**62.4**	2.73×10^{-5}	1.41×10^{-5}
60	1.936	62.37	2.36×10^{-5}	1.22×10^{-5}
70	1.935	62.30	$\mathbf{2.0 \times 10^{-5}}$	$\mathbf{1.0 \times 10^{-5}}$
80	1.93	62.22	1.80×10^{-5}	0.930×10^{-5}
100	1.93	62.00	1.42×10^{-5}	0.739×10^{-5}
120	1.92	61.72	1.17×10^{-5}	0.609×10^{-5}
140	1.91	61.38	0.981×10^{-5}	0.514×10^{-5}
160	1.90	61.00	0.838×10^{-5}	0.442×10^{-5}
180	1.88	60.58	0.726×10^{-5}	0.385×10^{-5}
200	1.87	60.12	0.637×10^{-5}	0.341×10^{-5}
212	1.86	59.83	0.593×10^{-5}	0.319×10^{-5}

(1 lb/ft³ = 157.09 N/m³). Specific weight varies with temperature as given in Table 1.3. The specific weight γ equals the product of mass density ρ and gravitational acceleration $g = 32.2$ ft/s² = 9.81 m/s²:

$$\gamma = \rho g. \tag{1.1}$$

Dynamic viscosity μ. As a fluid is brought into low rates of deformation, the velocity of the fluid at any boundary equals the velocity of the boundary. The ensuing rate of fluid deformation causes a shear stress τ_{zx} that is proportional to the dynamic viscosity μ and the rate of deformation of the fluid, dv_x/dz:

$$\tau_{zx} = \mu \frac{dv_x}{dz}. \tag{1.2}$$

The fundamental dimension of the dynamic viscosity μ is M/LT. In Table 1.3, the dynamic viscosity of water decreases with temperature. The dynamic viscosity of clear water at 20°C is 1 centipoise: 1 cP = 0.01 P = 0.001 kg/m s = 0.001 N s/m^2 = 0.001 Pa s. The conversion factor for the dynamic viscosity is 1 lb s/ft^2 = 47.88 N s/m^2 = 47.88 Pa s.

Kinematic viscosity of water v. The kinematic viscosity is obtained when the dynamic viscosity of a fluid μ is divided by its mass density ρ. The kinematic viscosity v of water in L^2/T is shown in Table 1.3 to decrease with temperature. The viscosity of clear water at 20°C is 1 centistokes = 1 cS = 0.01 cm^2/s = 1 × 10^{-6} m^2/s. The conversion factor is 1 ft^2/s = 0.0929 m^2/s. The change in kinematic viscosity of water v with temperature $T°$ in degrees Celsius can be roughly estimated from

$$v = \frac{\mu}{\rho} = (1 + 0.0337 T_C° + 0.0002217 T_C°^2)^{-1} \times 1.78 \times 10^{-6} \text{m}^2/\text{s} \tag{1.3}$$

1.3 Sediment Properties

Rivers also carry vast amounts of sediment that helps shape their own morphology. The physical properties of sediment are classified into single particles (Section 1.3.1), sediment mixtures (Section 1.3.2), sediment suspensions (Section 1.3.3), and sediment deposits (Section 1.3.4).

1.3.1 Sediment Particles

The physical properties of a single solid particle of volume $\forall_s$ are sketched in Figure 1.2. The mass density of a solid particle ρ_s describes the solid mass per volume of solids. The mass density of quartz particles is ρ_s = 2,650 kg/m^3 (1 slug/ft^3 = 515.4 kg/m^3).

Specific weight of sediment γ_s. The particle specific weight γ_s corresponds to the solid weight per unit volume of solid. Typical values of γ_s are 26.0 kN/m^3 or 165.4 lb/ft^3 and the conversion factor is 1 lb/ft^3 = 157.09 N/m^3. The specific

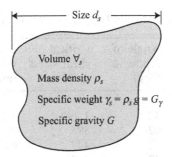

Figure 1.2. Physical properties of a single solid particle

weight of a solid γ_s is the product of the mass density of a solid particle ρ_s and gravitational acceleration g

$$\gamma_s = \rho_s g. \tag{1.4}$$

Specific gravity of sediment G. The ratio of the specific weight of a solid particle to the specific weight of fluid at a standard reference temperature defines the specific gravity G. With reference to water at 4°C, the specific gravity of quartz particles is $G = 2.65$

$$G = \frac{\gamma_s}{\gamma} = \frac{\rho_s}{\rho} = 2.65. \tag{1.5}$$

The specific gravity is a dimensionless parameter independent of the system of units.

Submerged specific weight of sediment $\tilde{\gamma}_s$. The specific weight of a solid particle γ_s submerged in a fluid of specific weight γ equals the difference

$$\tilde{\gamma}_s = \gamma_s - \gamma = (G - 1)\gamma. \tag{1.6}$$

Sediment size d_s. The most important physical property of a sediment particle is its size. The first two columns of Table 1.4 show the grade scale commonly used in sedimentation (1 in. = 25.4 mm). The size of sediment particles can be determined in a number of ways described in detail in Julien (2010). Some of the main characteristics associated with particle size are listed in Table 1.4.

1.3.2 Sediment Mixtures

The properties of a sediment mixture are sketched in Figure 1.3. The total volume, $\forall_t$, is the total of the volume of solids $\forall_s$ and the volume of voids $\forall_v$.

Table 1.4. *Sediment grade scale and approximate properties*

Class name	Particle diameter d_s (mm)	Angle of repose ϕ (deg)	Critical shear stress τ_c (N/m^2)	Critical shear velocity u_{*c} (m/s)	Settling velocity ω_0 (mm/s)
Boulder					
Very large	>2,048	42	1,790	1.33	5,430
Large	>1,024	42	895	0.94	3,839
Medium	>512	42	447	0.67	2,715
Small	>256	42	223	0.47	1,919
Cobble					
Large	>128	42	111	0.33	1,357
Small	>64	41	53	0.23	959
Gravel					
Very coarse	>32	40	26	0.16	678
Coarse	>16	38	12	0.11	479
Medium	>8	36	5.7	0.074	338
Fine	>4	35	2.71	0.052	237
Very fine	>2	33	1.26	0.036	164
Sand					
Very coarse	>1.000	32	0.47	0.0216	109
Coarse	>0.500	31	0.27	0.0164	66.4
Medium	>0.250	30	0.194	0.0139	31.3
Fine	>0.125	30	0.145	0.0120	10.1
Very fine	>0.062	30	0.110	0.0105	2.66
Silt					
Coarse	>0.031	30	0.083	0.0091	0.67
Medium	>0.016	30	0.065	0.0080	0.167[a]
Fine	>0.008				0.042[a]
Very fine	>0.004				0.010[a]
Clay					
Coarse	>0.0020				2.6×10^{-3}[a]
Medium	>0.0010		Cohesive		6.5×10^{-4}[a]
Fine	>0.0005		material		1.63×10^{-4}[a]
Very fine	>0.00024				4.1×10^{-5}[a]

[a]Possible flocculation.

Sediment particle-size distribution. The example of particle-size distribution in Figure 1.3 also shows the percentage by weight of material finer than a given sediment size. The median grain diameter is the size d_{50} for which 50 percent of the material is finer. Likewise d_{90} and d_{10} are sediment sizes for which 90 percent and 10 percent of the material are finer.

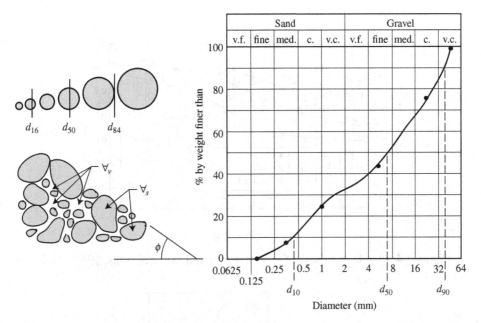

Figure 1.3. Particle-size distribution

Gradation coefficients σ_g and Gr. The gradation of the sediment mixture is a measure of the particle-size distribution of a sediment mixture. It can be described as

$$\sigma_g = \left(\frac{d_{84}}{d_{16}}\right)^{1/2} \tag{1.7a}$$

or

$$\mathrm{Gr} = \frac{1}{2}\left(\frac{d_{84}}{d_{50}} + \frac{d_{50}}{d_{16}}\right). \tag{1.7b}$$

Both gradation coefficients reduce to unity for uniform sediment mixtures, i.e. when $d_{84} = d_{50} = d_{16}$. The gradation coefficient increases with nonuniformity, and high gradation coefficients describe well-graded mixtures.

Angle of repose ϕ. Typical values of the angle of repose ϕ of granular material are shown in Figure 1.4. The angle of repose varies with grain size and angularity of the material. Typical values of the angle of repose are also given in column three of Table 1.4 for material coarser than medium silt.

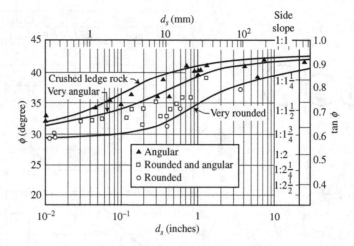

Figure 1.4. Angle of repose of granular material (after Simons, 1957)

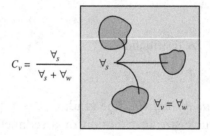

Figure 1.5. Properties of a sediment suspension

1.3.3 Sediment Suspensions

The properties of a sediment suspension are sketched in Figure 1.5, with the volume of void $\forall_v$ completely filled with water $\forall_w$, i.e. without air in the mixture.

Sediment concentration. The volumetric sediment concentration C_v is the ratio of the volume of solids $\forall_s$ to the total volume $\forall_t = \forall_s + \forall_w$.

$$C_v = \frac{\text{sediment volume}}{\text{total volume}} = \frac{\forall_s}{\forall_s + \forall_w}. \tag{1.8a}$$

The sediment concentration is commonly measured in milligrams per liter $C_{\text{mg/l}}$ while other measurements include the concentration by weight C_w and the concentration in parts per million (ppm). The following conversions from Julien (2010) are

$$C_w = \frac{\text{sediment weight}}{\text{total weight}} = \frac{C_v G}{1 + (G-1)C_v}, \tag{1.8b}$$

Table 1.5. *Equivalent concentrations for* C_v, C_w, C_{ppm}, $C_{mg/l}$, p_0, *and* γ_{md}

C_v	C_w	C_{ppm}	$C_{mg/l}$	p_0	γ_{md} lb/ft^3	γ_{md} kN/m^3
Suspension						
0.0001	0.00026	265	265			
0.001	0.00264	2,645	2,650			
0.0025	0.00659	6,598	6,625			
0.005	0.01314	13,141	13,250			
0.0075	0.01963	19,632	19,875			
0.01	0.02607	26,069	26,500			
0.025	0.06363	63,625	66,250			
Hyperconcentrations						
0.05	0.12240	122,401	132,500	0.95		
0.075	0.17700	176,863	198,750	0.925		
0.1	0.22750	227,467	265,000	0.9	16.5	2.6
0.20	0.39850	398,496	530,000	0.8	33.0	5.2
0.25	0.46900	469,027	662,500	0.75	41.3	6.5
0.30	0.53200	531,772	795,000	0.70	49.6	7.8
0.40	0.63900	638,554	1,060,000	0.60	66.2	10.4
Sediment deposits						
0.5	0.72600	726,027	1,325,000	0.50	82.7	13
0.6	0.79900	798,995	1,590,000	0.40	99.2	15.6
0.7	0.86080	860,788	1,855,000	0.30	116	18.2
0.75	0.88800	888,268	1,987,500	0.25	124	19.5

Note: Calculations based on $G = 2.65$.

$$C_{ppm} = 10^6 C_w, \tag{1.8c}$$

$$C_{mg/l} = \frac{\text{sediment mass}}{\text{total volume}} = \rho G C_v. \tag{1.8d}$$

Equivalent sediment concentrations are listed in columns one to four of Table 1.5. In practice, there is a negligible difference (<10 percent) between C_{ppm} and $C_{mg/l}$ when $C_{ppm} < 145,000$.

Specific weight of a mixture γ_m. The specific weight of a submerged mixture is the total weight of solid and water in the voids-per-unit total volume. The specific weight of a mixture γ_m is a function of the volumetric concentration C_v as

$$\gamma_m = \frac{\text{total weight}}{\text{total volume}} = \gamma_s C_v + \gamma(1 - C_v). \tag{1.9}$$

The specific mass ρ_m of a submerged mixture is the total mass of solid and water in the voids-per-unit total volume. The specific mass is given by $\rho_m = \gamma_m/g$.

1.3.4 Sediment Deposits

The properties of sediment deposits usually include the porosity p_0 and the dry specific weight γ_{md} of sediment mixtures.

Porosity p_0. The porosity p_0 is the volume of void $\forall_v$ per total volume $\forall_t$.

$$p_0 = \frac{\forall_v}{\forall_t}. \tag{1.10}$$

The values of porosity are listed in column five of Table 1.5.

Dry specific weight of a mixture γ_{md}. The dry specific weight of a mixture is the weight of solid-per-unit total volume, including the volume of solids and voids. The dry specific weight of a mixture γ_{md} is a function of porosity p_0 as

$$\gamma_{md} = \frac{\text{sediment weight}}{\text{total volume}} = \gamma G C_v = \gamma_s (1 - p_0). \tag{1.11}$$

The dry specific weight of sand deposits is approximately ~ 14.75 kN/m^3 or 93 lb/ft^3. The dry specific mass of a mixture is the mass of solid-per-unit total volume. The dry specific mass of a mixture is $\rho_{md} = \gamma_{md}/g$.

Problem 1.1

Determine the mass density, specific weight, dynamic viscosity, and kinematic viscosity of clear water at 20°C (a) in SI units and (b) in the English system of units.

[*Answers*: (a) $\rho = 998$ kg/m^3, $\gamma = 9{,}790$ N/m^3, $\mu = 1.0 \times 10^{-3}$ N s/m^2, $v = 1 \times 10^{-6}$ m^2/s, (b) $\rho = 1.94$ slug/ft^3, $\gamma = 62.3$ lb/ft^3, $\mu = 2.1 \times 10^{-5}$ lb/ft^2, $v = 1.1 \times 10^{-5}$ ft^2/s]

♦Problem 1.2

Determine the sediment size, mass density, specific weight, submerged specific weight, and angle of repose of small quartz cobbles (a) in SI units and (b) in English units.

♦♦Problem 1.3

The volumetric sediment concentration of a sample is $C_v = 0.05$. Determine the corresponding: (a) concentration by weight C_w, (b) concentration in parts per million C_{ppm}, (c) concentration in milligrams per liter $C_{mg/l}$, (d) porosity p_0, and (e) void ratio e.

[*Answers*: The answers are found in Table 1.5.]

◆*Problem 1.4*

The porosity of a sandy loam is 0.45. Determine the corresponding soil properties: (a) volumetric concentration, (b) void ratio e, (c) specific weight γ_m, (d) specific mass ρ_m, (e) dry specific weight γ_{md}, and (f) dry specific mass ρ_{md}.

◆*Problem 1.5*

Calculate the gradation coefficients σ_g and Gr from the particle distribution in Figure 1.3.

[*Answer: It is a well-graded sediment mixture* $\sigma_g = 8$, *and* Gr $= 10$.]

2

Mechanics of Rivers

This chapter describes the fundamental relationships that govern the transport of water and sediment in fluvial systems. Most of these relationships were developed in detail in the companion text on Erosion and Sedimentation (Julien, 2010). A summary of the main relationships is presented here. This chapter is divided into three parts: governing equations for river flows (Section 2.1); governing equations for sediment motion (Section 2.2); and a discussion on why rivers form (Section 2.3).

2.1 Equations Governing River Flows

The governing equations describing river flows include those for velocity and acceleration (Section 2.1.1), for conservation of mass or continuity (Section 2.1.2), for fluid motion (Section 2.1.3), for conservation of momentum (Section 2.1.4), for specific momentum (Section 2.1.5), covering the Saint-Venant principle (Section 2.1.6), and for specific energy (Section 2.1.7).

2.1.1 River Flow Velocity and Acceleration

Flow kinematics describes fluid motion in terms of velocity and acceleration. As shown in Figure 2.1, two orthogonal coordinate systems following the "right-hand" rule are commonly used in rivers: (1) Cartesian (x, y, z) coordinates with x in the main downstream direction, y in the lateral direction to the left bank, and z upward and (2) cylindrical (r, θ, z) coordinates where r is the river radius of curvature in a horizontal plane, θ in the downstream direction, and z upward (coming out of the page). Cartesian coordinates describe most rivers, while the cylindrical coordinates are useful for delineating single meander bends. For one-dimensional analyses, a natural coordinate system follows the center line of the channel.

14

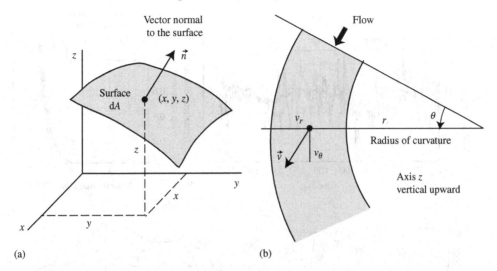

Figure 2.1. (a) Cartesian and (b) cylindrical coordinates

The rate of change in the position of a fluid element is the measure of its velocity. Velocity is defined as the ratio between the displacement ds and the corresponding increment of time dt. Velocity is a vector quantity $\vec{v}\,(t, x, y, z)$ that varies both with time t and space (x, y, z). The velocity magnitude v at a given time equals the square root of the sum of squares of its orthogonal components $v = \sqrt{v_x^2 + v_y^2 + v_z^2}$, where $v_x = dx/dt$, $v_y = dy/dt$, and $v_z = dz/dt$. Examples of river velocity measurements along a vertical and along a cross-section are shown in Figure 2.2a and b.

The notion of a volumetric flux passing through a given area is defined by the scalar product of the velocity vector $\vec{V}$ times the cross-section area vector $\vec{A}$ oriented in the direction $\vec{n}$ shown in Figure 2.1. The flux is obtained from the scalar product, or dot product, of two vectors. For instance, the volumetric flux of water in a river is the flow discharge obtained from $Q = \vec{A} \cdot \vec{V} = AV\cos\theta$. This flow discharge is equal to the product of the scalar magnitude of the flow velocity, the cross-section area, and the cosine of the angle between the two vectors. For instance, as shown in Figure 2.2c, the projection of the flow-velocity vector in the direction perpendicular to the surface area contributes to the downstream discharge. Examples of cross-sectional profiles of flow depth and depth-averaged velocity are shown in Figure 2.2d.

A line tangent to the velocity vector at every point at a given instant is known as a streamline. The pathline of a fluid element is the successive locations of the fluid element through time, e.g. the path followed by a single

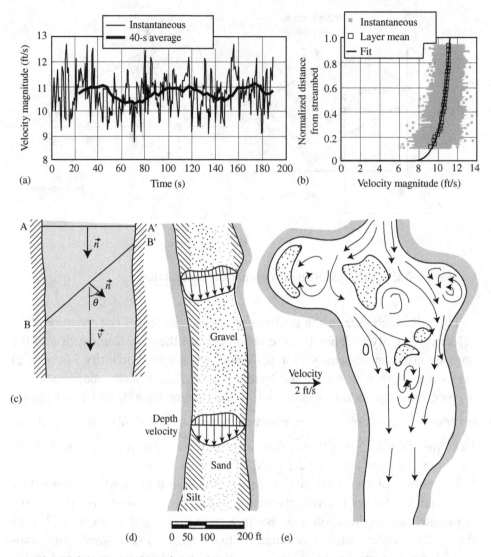

Figure 2.2. Examples of velocity measurements: (a) time series of point velocity measurements; (b) vertical velocity profile using ADCP measurements (after P. O'Brien personal communication); (c) flux definition sketch; (d) cross-sectional depth and velocity profiles; and (e) surface streamlines of the Matamek River (after Frenette and Julien, 1980)

buoy on a river. A streakline is defined as the line connecting all fluid elements that have passed successively at a given point in space, e.g. an instantaneous position of all buoys released over time from a single point on a river. The distinction between streamline, pathline, and streakline becomes important in the case of flows that change direction or magnitude through time, e.g. tidal flows in estuaries. For instance, in the case of contamination from a point source, such as an oil spill, the streakline can be viewed from a photograph, while the pathlines are helpful in finding out which areas have been contaminated. All lines are identical in steady flow, as shown in the example in Figure 2.2e.

The mathematical description of the change of velocity in a Cartesian coordinate system is now considered. The differential velocity components over an infinitesimal distance ds (dx, dy, dz) and time increment dt at a point (x, y, z) are

$$\mathrm{d}v_x = \frac{\partial v_x}{\partial t}\mathrm{d}t + \frac{\partial v_x}{\partial x}\mathrm{d}x + \frac{\partial v_x}{\partial y}\mathrm{d}y + \frac{\partial v_x}{\partial z}\mathrm{d}z \tag{2.1a}$$

$$\mathrm{d}v_y = \frac{\partial v_y}{\partial t}\mathrm{d}t + \frac{\partial v_y}{\partial x}\mathrm{d}x + \frac{\partial v_y}{\partial y}\mathrm{d}y + \frac{\partial v_y}{\partial z}\mathrm{d}z \tag{2.1b}$$

$$\mathrm{d}v_z = \frac{\partial v_z}{\partial t}\mathrm{d}t + \frac{\partial v_z}{\partial x}\mathrm{d}x + \frac{\partial v_z}{\partial y}\mathrm{d}y + \frac{\partial v_z}{\partial z}\mathrm{d}z . \tag{2.1c}$$

$$\underbrace{\qquad}_{\text{local}} \underbrace{\qquad\qquad\qquad}_{\text{convective}}$$

The Cartesian acceleration components are obtained directly after the velocity equations are divided by the time increment dt,

$$a_x = \frac{\mathrm{d}v_x}{\mathrm{d}t} = \frac{\partial v_x}{\partial t} + v_x\frac{\partial v_x}{\partial x} + v_y\frac{\partial v_x}{\partial y} + v_z\frac{\partial v_x}{\partial z}, \tag{2.2a}$$

$$a_y = \frac{\mathrm{d}v_y}{\mathrm{d}t} = \frac{\partial v_y}{\partial t} + v_x\frac{\partial v_y}{\partial x} + v_y\frac{\partial v_y}{\partial y} + v_z\frac{\partial v_y}{\partial z}, \tag{2.2b}$$

$$a_z = \frac{\mathrm{d}v_z}{\mathrm{d}t} = \frac{\partial v_z}{\partial t} + v_x\frac{\partial v_z}{\partial x} + v_y\frac{\partial v_z}{\partial y} + v_z\frac{\partial v_z}{\partial z}. \tag{2.2c}$$

$$\underbrace{\qquad}_{\text{local}} \underbrace{\qquad\qquad\qquad}_{\text{convective}}$$

In cylindrical coordinates with $v_r = \mathrm{d}r/\mathrm{d}t$, $v_\theta = r\,\mathrm{d}\theta/\mathrm{d}t$, and $v_z = \mathrm{d}z/\mathrm{d}t$, additional convective terms are obtained because of the change in orientation of the unit velocity vectors. In curvilinear coordinates, the two additional convective terms are the centrifugal acceleration v_θ^2/r in Equation (2.3a) and the Coriolis acceleration $v_r v_\theta/r$ in Equation (2.3b):

$$a_r = \frac{\mathrm{d}v_r}{\mathrm{d}t} = \frac{\partial v_r}{\partial t} + v_r\frac{\partial v_r}{\partial r} + \frac{v_\theta}{r}\frac{\partial v_r}{\partial \theta} - \frac{v_\theta^2}{r} + v_z\frac{\partial v_r}{\partial z} \tag{2.3a}$$

$$a_\theta = \frac{\mathrm{d}v_\theta}{\mathrm{d}t} = \frac{\partial v_\theta}{\partial t} + v_r\frac{\partial v_\theta}{\partial r} + \frac{v_\theta}{r}\frac{\partial v_\theta}{\partial \theta} + \frac{v_r v_\theta}{r} + v_z\frac{\partial v_\theta}{\partial z} \tag{2.3b}$$

$$a_z = \frac{\mathrm{d}v_z}{\mathrm{d}t} = \frac{\partial v_z}{\partial t} + v_r\frac{\partial v_z}{\partial r} + \frac{v_\theta}{r}\frac{\partial v_z}{\partial \theta} + v_z\frac{\partial v_z}{\partial z} . \tag{2.3c}$$

$$\underbrace{\qquad}_{\text{local}} \underbrace{\qquad\qquad\qquad}_{\text{convective}}$$

It is shown in Equations (2.2) and (2.3) that the total acceleration can be separated into local acceleration terms for the change with time and local acceleration terms for the changes in space. Steady flows do not change with time and the local acceleration terms vanish at any point. River flows are uniform when all convective acceleration terms vanish. Flows in river bends are considered nonuniform, even when the cross section does not change in the downstream direction, because the convective acceleration terms are not negligible. Steady-uniform flows describe the particular case of motion without any acceleration component. It is noted that the Coriolis acceleration terms included here are caused by the change in velocity vector orientation, and not by the rotation of the Earth. The effects of the Earth rotation will be discussed in Chapter 15.

2.1.2 Conservation of Mass for River Flows

The equation of continuity, or law of conservation of mass, states that mass cannot be created or destroyed. It is assumed in most river studies that water remains in the liquid phase and the changes from liquid to solid/gas phases are negligible. This applies to most river conditions and the volumetric conversions from water to ice and the water losses to evaporation and sublimation are relatively small. The continuity equation can be written either in differential form or in integral form; let's review a two-dimensional formulation and a one-dimensional simplification.

Differential form of continuity. We shall consider the infinitesimal control volume in Figure 2.3 filled with a fluid of mass density ρ_m.

The difference between the mass fluxes entering the differential control volume equals the rate of increase of internal mass. For instance, in the x direction, the net mass flux leaving the control volume by advection is $\partial \rho_m v_x / \partial x$ times the cross-section area $dy\,dz$. The change in internal mass is $\partial \rho_m / \partial t$ times the volume $dx\,dy\,dz$. Repeating the process in the y and the z directions yields the following differential relationship in Cartesian coordinates (x, y, z) and in cylindrical coordinates (without derivation here) as

$$\frac{\partial \rho_m}{\partial t} + \frac{\partial}{\partial x}\left(\rho_m v_x\right) + \frac{\partial}{\partial y}\left(\rho_m v_y\right) + \frac{\partial}{\partial z}\left(\rho_m v_z\right) = 0 \qquad (2.4a) \blacklozenge$$

$$\frac{\partial \rho_m}{\partial t} + \frac{1}{r}\frac{\partial}{\partial r}\left(\rho_m r v_r\right) + \frac{1}{r}\frac{\partial}{\partial \theta}\left(\rho_m v_\theta\right) + \frac{\partial}{\partial z}\left(\rho_m v_z\right) = 0. \qquad (2.4b) \blacklozenge$$

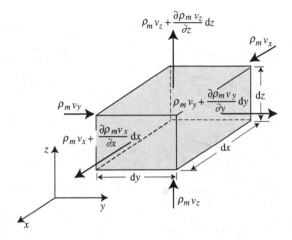

Figure 2.3. Infinitesimal element of fluid

A similar approach can be used for the conservation of solid mass or contaminants. When advection is the main transport mechanism, conservation of solid mass can also be defined after replacing ρ_m in Equation (2.4) with the sediment concentration C_v. However, in sediment/contaminant transport problems where turbulent diffusion and dispersion are significant, additional fluxes due to turbulent mixing must be included, as discussed in Section 10.2 of Julien (2010). For homogeneous incompressible suspensions without settling, the mass density is independent of space and time (ρ_s, ρ, ρ_m = const). Consequently, the divergence of the velocity vector must be zero, which reduces Equation (2.4a) to

$$\frac{\partial v_x}{\partial x} + \frac{\partial v_y}{\partial y} + \frac{\partial v_z}{\partial z} = 0. \tag{2.4c}$$

In most rivers, the concentrations are sufficiently low that compressibility effects can be discounted, and we find that Equation (2.4c) is applicable.

Integral form of continuity. The general form of the continuity equation is simply the integral over a control volume $\forall$ of the differential form in Equation (2.4):

$$\int_\forall \frac{\partial \rho}{\partial t}\, d\forall + \int_\forall \left(\frac{\partial \rho v_x}{\partial x} + \frac{\partial \rho v_y}{\partial y} + \frac{\partial \rho v_z}{\partial z} \right) d\forall = 0. \tag{2.5}$$

This volume integral of partial derivatives can be transformed into surface integrals owing to the divergence theorem, e.g. with application to a vector $\vec{F}$ such that

$$\int_\forall \frac{\partial \vec{F}}{\partial x}\, d\forall = \int_A \vec{F} \frac{\partial x}{\partial n}\, dA, \tag{2.6}$$

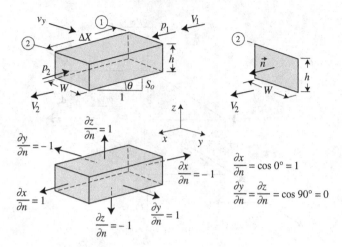

Figure 2.4. Integral form of the continuity equation

in which $\partial x/\partial n$ is the cosine of the angle between the coordinate x and the normal vector $\vec{n}$ pointing outside the control volume, as sketched in Figure 2.4.

Depth-integrated 2-D continuity for rivers. The integral form of the continuity equation states simply that the difference between inflow and outflow results in volumetric storage. Depth-integrated formulations of continuity are helpful in two-dimensional river models. The surface integration is carried out over a control volume of fixed grid dx and dy with bed elevation z_b and water-surface elevation z_w that vary in the horizontal space, as sketched in Figure 2.5.

The system of coordinates typically sets the x axis in the main downstream direction at a very small average bed slope θ_0 from the horizontal axis. The depth-integrated continuity relationship is obtained for homogeneous suspensions (constant ρ_m) from the integration of Equation (2.4a) along the upward z axis (this z axis is also deflected θ_0 from the vertical). Given the flow depth $h = z_w - z_b$, the depth-integrated velocity components V_x and V_y are

$$V_x = \frac{1}{h} \int_{z_b}^{z_w} v_x \, dz \quad \text{and} \quad V_y = \frac{1}{h} \int_{z_b}^{z_w} v_y \, dz. \qquad (2.7)$$

The net volume flux leaving the control volume in the x direction is

$$dy \left[hV_x + \frac{\partial(hV_x)}{\partial x} dx \right] - dy\, hV_x = \frac{\partial(hV_x)}{\partial x} dx\, dy. \qquad (2.8)$$

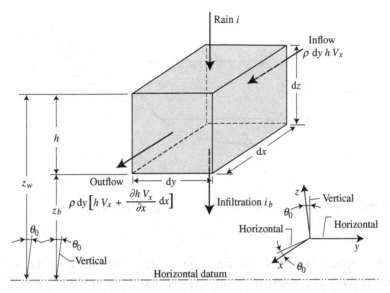

Figure 2.5. Definition sketch for the 2-D continuity formulation

Repeating the process in the y direction gives $\partial(hV_y)/\partial y\,dx\,dy$. Also considering the infiltration flux $i_b\,dx\,dy$ on the channel bed and the rainfall flux $i\,dx\,dy$ at the free surface, the net volume change within the control volume is

$$dx\,dy\frac{\partial h}{\partial t}+dx\,dy\frac{\partial hV_x}{\partial x}+dx\,dy\frac{\partial hV_y}{\partial y}+(i_b-i)dx\,dy=0. \qquad (2.9)$$

The depth-integrated form of continuity is obtained after Equation (2.9) is divided by $dx\,dy$:

$$\frac{\partial h}{\partial t}+\frac{\partial hV_x}{\partial x}+\frac{\partial hV_y}{\partial y}+i_b-i=0. \qquad (2.10)\blacklozenge\blacklozenge$$

The rate of change in flow depth $\partial h/\partial t$ describes volumetric storage and is obtained from the net downstream flux $\partial(hV_x)/\partial x$, the net lateral flux $\partial(hV_y)/\partial y$, the downward vertical infiltration rate i_b is positive (outward) and the downward rainfall intensity i is negative (inward). All parameters, h, V_x, V_y, i_b, and i, can vary in space x, y, and time t.

One-dimensional continuity for rivers. The definition sketch in Figure 2.6 describes a river reach with a top width W, cross-section area A, and wetted perimeter P. The flow discharge Q equals the product of the mean flow velocity V perpendicular to the cross-section area A; the unit discharge or flow rate per unit width of the lateral flow is q_l. The rainfall intensity i at the free surface and the infiltration rate i_b through the wetted perimeter are also considered.

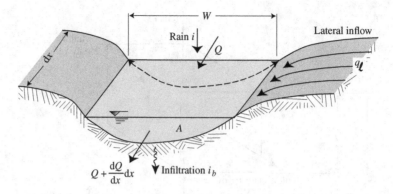

Figure 2.6. Definition sketch for the 1-D continuity formulation

The net volumetric flux entering the control volume is $q_l\,\mathrm{d}x + iW\,\mathrm{d}x$. The flux leaving the control volume is $(\partial Q/\partial x)\,\mathrm{d}x + i_b P\,\mathrm{d}x$. The difference corresponds to volumetric storage $\partial A\,\mathrm{d}x$ per unit time ∂t. After dividing by $\mathrm{d}x$, we easily demonstrate that

$$\frac{\partial A}{\partial t} + \frac{\partial Q}{\partial x} + i_b P - iW - q_l = 0. \tag{2.11}$$

For an impervious channel ($i_b = 0$) without rainfall ($i = 0$) and lateral inflow ($q_l = 0$), the one-dimensional (1-D) equation of continuity simply reduces to

$$\frac{\partial Q}{\partial x} + \frac{\partial A}{\partial t} = 0. \tag{2.12} \blacklozenge\blacklozenge$$

This differential equation describes conservation of mass and will be used extensively to describe the propagation of flood waves in Chapter 6.

2.1.3 *Equations of Fluid Motion*

Fluid motion results from the application of forces on the fluid mass inside a control volume $\forall$. Given that the force F equals the product of mass m and acceleration a, the approach for fluids of mass density $\rho = m/\forall$ stems from $a = F/m = F/\rho\forall$. The forces acting on a Cartesian element of fluid ($\mathrm{d}x$, $\mathrm{d}y$, $\mathrm{d}z$) are classified as either internal forces or external forces. The internal acceleration, or body force per unit mass, is the gravitational acceleration g with its components g_x, g_y, and g_z, as shown in Figure 2.7.

The external forces per unit area define stresses. Two types of external stresses are applied on each face of the element: normal and tangential

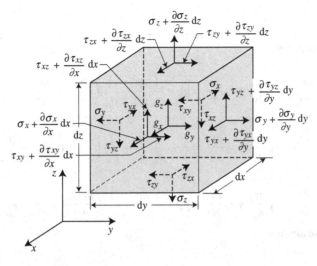

Figure 2.7. Definition sketch for the equations of motion

stress components. The normal stresses σ_x, σ_y, and σ_z are perpendicular to the surface and designated as positive for tension. There are six shear stresses, τ_{xy}, τ_{yx}, τ_{xz}, τ_{zx}, τ_{yz}, and τ_{zy}, with two orthogonal components applied on each face. The first subscript indicates the direction normal to the face, and the second subscript designates the direction in which the stress is applied.

The acceleration components of a cubic element of fluid in Figure 2.7 are equal to the sum of the forces per unit mass in each direction, x, y, and z:

$$a_x = g_x + \frac{1}{\rho_m} \frac{\partial \sigma_x}{\partial x} + \frac{1}{\rho_m} \frac{\partial \tau_{yx}}{\partial y} + \frac{1}{\rho_m} \frac{\partial \tau_{zx}}{\partial z} \tag{2.13a}$$

$$a_y = g_y + \frac{1}{\rho_m} \frac{\partial \sigma_y}{\partial y} + \frac{1}{\rho_m} \frac{\partial \tau_{xy}}{\partial x} + \frac{1}{\rho_m} \frac{\partial \tau_{zy}}{\partial z} \tag{2.13b}$$

$$a_z = g_z + \frac{1}{\rho_m} \frac{\partial \sigma_z}{\partial z} + \frac{1}{\rho_m} \frac{\partial \tau_{xz}}{\partial x} + \frac{1}{\rho_m} \frac{\partial \tau_{yz}}{\partial y}. \tag{2.13c}$$

These equations of motion are general, without any restriction as to compressibility, viscous shear, turbulence, or other effects. The normal stresses can be rewritten as a function of the pressure p and the additional normal stresses, τ_{xx}, τ_{yy}, and τ_{zz}, such that $\sigma_x = -p + \tau_{xx}$, $\sigma_y = -p + \tau_{yy}$, and $\sigma_z = -p + \tau_{zz}$. After considering the acceleration components a_x, a_y, and a_z from Equation (2.2), the equations of motion in Cartesian and cylindrical coordinates can be written as shown in Table 2.1.

Table 2.1. *Equations of motion*

Cartesian coordinates
x-component

$$a_x = \frac{\partial v_x}{\partial t} + v_x \frac{\partial v_x}{\partial x} + v_y \frac{\partial v_x}{\partial y} + v_z \frac{\partial v_x}{\partial z} = g_x - \frac{1}{\rho_m}\frac{\partial p}{\partial x} + \frac{1}{\rho_m}\left(\frac{\partial \tau_{xx}}{\partial x} + \frac{\partial \tau_{yx}}{\partial y} + \frac{\partial \tau_{zx}}{\partial z}\right) \quad (2.14a)\blacklozenge$$

y-component

$$a_y = \frac{\partial v_y}{\partial t} + v_x \frac{\partial v_y}{\partial x} + v_y \frac{\partial v_y}{\partial y} + v_z \frac{\partial v_y}{\partial z} = g_y - \frac{1}{\rho_m}\frac{\partial p}{\partial y} + \frac{1}{\rho_m}\left(\frac{\partial \tau_{xy}}{\partial x} + \frac{\partial \tau_{yy}}{\partial y} + \frac{\partial \tau_{zy}}{\partial z}\right) \quad (2.14b)\blacklozenge$$

z-component

$$a_z = \frac{\partial v_z}{\partial t} + v_x \frac{\partial v_z}{\partial x} + v_y \frac{\partial v_z}{\partial y} + v_z \frac{\partial v_z}{\partial z} = g_z - \frac{1}{\rho_m}\frac{\partial p}{\partial z} + \frac{1}{\rho_m}\left(\frac{\partial \tau_{xz}}{\partial x} + \frac{\partial \tau_{yz}}{\partial y} + \frac{\partial \tau_{zz}}{\partial z}\right) \quad (2.14c)\blacklozenge$$

Cylindrical coordinates
r-component

$$\frac{\partial v_r}{\partial t} + v_r \frac{\partial v_r}{\partial r} + \frac{v_\theta}{r}\frac{\partial v_r}{\partial \theta} - \frac{v_\theta^2}{r} + v_z \frac{\partial v_r}{\partial z} = g_r - \frac{1}{\rho_m}\frac{\partial p}{\partial r} + \frac{1}{\rho_m}\left[\frac{1}{r}\frac{\partial}{\partial r}(r\tau_{rr}) + \frac{1}{r}\frac{\partial \tau_{\theta r}}{\partial \theta} - \frac{\tau_{\theta\theta}}{r} + \frac{\partial \tau_{zr}}{\partial z}\right]$$

$$(2.15a)\blacklozenge$$

θ-component

$$\frac{\partial v_\theta}{\partial t} + v_r \frac{\partial v_\theta}{\partial r} + \frac{v_\theta}{r}\frac{\partial v_\theta}{\partial \theta} + \frac{v_r v_\theta}{r} + v_z \frac{\partial v_\theta}{\partial z} = g_\theta - \frac{1}{\rho_m r}\frac{\partial p}{\partial \theta} + \frac{1}{\rho_m}\left[\frac{1}{r^2}\frac{\partial}{\partial r}(r^2\tau_{r\theta}) + \frac{1}{r}\frac{\partial \tau_{\theta\theta}}{\partial \theta} + \frac{\partial \tau_{z\theta}}{\partial z}\right]$$

$$(2.15b)\blacklozenge$$

z-component

$$\frac{\partial v_z}{\partial t} + v_r \frac{\partial v_z}{\partial r} + \frac{v_\theta}{r}\frac{\partial v_z}{\partial \theta} + v_z \frac{\partial v_z}{\partial z} = g_z - \frac{1}{\rho_m}\frac{\partial p}{\partial z} + \frac{1}{\rho_m}\left[\frac{1}{r}\frac{\partial(r\tau_{rz})}{\partial r} + \frac{1}{r}\frac{\partial \tau_{\theta z}}{\partial \theta} + \frac{\partial \tau_{zz}}{\partial z}\right] \quad (2.15c)\blacklozenge$$

2.1.4 Conservation of Momentum for River Flows

Momentum equations define the hydrodynamic forces exerted by surface flows. In fluids, the element of force is given from $dF = dm\,a = \rho\,a\,d\forall$. Since acceleration varies in space and time, the hydrodynamic forces are obtained by the integral over the control volume $\forall$ of the product of mass density times the equations of motion, or $F = \int \rho\,a\,d\forall$. The acceleration terms on the left-hand side of this integral describe motion and represent the rate of momentum change per unit volume. The terms of the right-hand side result from the applied stresses and define the impulse per unit volume. The force balance in rivers is also sometimes called the impulse–momentum equation or simply the momentum equation. For example, the *x* component in the Cartesian coordinates for steady flow is obtained from Equation (2.14):

$$\int_\forall \rho_m \left(\frac{\partial v_x}{\partial t} + v_x \frac{\partial v_x}{\partial x} + v_y \frac{\partial v_x}{\partial y} + v_z \frac{\partial v_x}{\partial z} \right) d\forall = \int_\forall \rho_m g_x \, d\forall - \int_\forall \frac{\partial p}{\partial x} \, d\forall$$

$$+ \int_\forall \left(\frac{\partial \tau_{xx}}{\partial x} + \frac{\partial \tau_{yx}}{\partial y} + \frac{\partial \tau_{zx}}{\partial z} \right) d\forall. \tag{2.16}$$

The convective acceleration terms on the left-hand side can be rewritten as follows from the identity $a\partial b/\partial x = (\partial ab/\partial x) - b\partial a/\partial x$.

$$\frac{\partial \rho v_x^2}{\partial x} + \frac{\partial \rho v_x v_y}{\partial y} + \frac{\partial \rho v_x v_z}{\partial z} - v_x \left(\frac{\partial \rho v_x}{\partial x} + \frac{\partial \rho v_y}{\partial y} + \frac{\partial \rho v_z}{\partial z} \right). \tag{2.17}$$

From the continuity Equation (2.4c), the terms in parentheses of Equation (2.17) can be dropped. The volume integrals of the remaining momentum and stress terms can be transformed into surface integrals by means of the divergence theorem, Equation (2.6). The resulting impulse–momentum relationship is a vector with three components:

x-component

$$\frac{d}{dt} \int_\forall \rho_m v_x d\forall + \int_A \rho_m v_x \left(v_x \frac{\partial x}{\partial n} + v_y \frac{\partial y}{\partial n} + v_z \frac{\partial z}{\partial n} \right) dA$$

$$= \int_\forall \rho_m g_x d\forall - \int_A p \frac{\partial x}{\partial n} dA + \int_A \left(\tau_{xx} \frac{\partial x}{\partial n} + \tau_{yx} \frac{\partial y}{\partial n} + \tau_{zx} \frac{\partial z}{\partial n} \right) dA, \tag{2.18a}$$

y-component

$$\frac{d}{dt} \int_\forall \rho_m v_y d\forall + \int_A \rho_m v_y \left(v_x \frac{\partial x}{\partial n} + v_y \frac{\partial y}{\partial n} + v_z \frac{\partial z}{\partial n} \right) dA$$

$$= \int_\forall \rho_m g_y d\forall - \int_A p \frac{\partial y}{\partial n} dA + \int_A \left(\tau_{xy} \frac{\partial x}{\partial n} + \tau_{yy} \frac{\partial y}{\partial n} + \tau_{zy} \frac{\partial z}{\partial n} \right) dA, \text{ and} \tag{2.18b}$$

z-component

$$\frac{d}{dt} \int_\forall \rho_m v_z d\forall + \int_A \rho_m v_z \left(v_x \frac{\partial x}{\partial n} + v_y \frac{\partial y}{\partial n} + v_z \frac{\partial z}{\partial n} \right) dA$$

$$= \int_\forall \rho_m g_z d\forall - \int_A p \frac{\partial z}{\partial n} dA + \int_A \left(\tau_{xz} \frac{\partial x}{\partial n} + \tau_{yz} \frac{\partial y}{\partial n} + \tau_{zz} \frac{\partial z}{\partial n} \right) dA. \tag{2.18c}$$

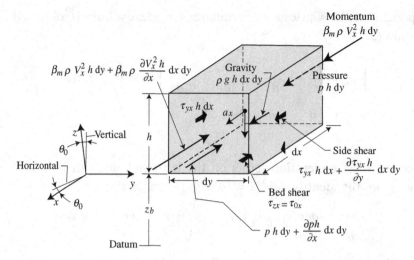

Figure 2.8. Definition sketch for depth-integrated 2-D river momentum

This formulation is very general and sounds quite complicated, and it can be in some cases. The solution to the most complex river engineering problems does require a thorough understanding of the general formulation. However, simplifications to these equations are sufficiently accurate for the vast majority of river-engineering applications. Let's now consider the most practical 2-D and 1-D simplifications that are available.

Depth-integrated 2-D momentum for rivers. For most rivers, depth-integrated formulations of the equation of motion are sufficiently accurate. The hydrostatic approximation is valid as long as the accelerations in the near-vertical z direction are negligible. The sketch in Figure 2.8 is used to define the 2-D depth-integrated formulation for homogeneous suspensions (constant $\rho_m = \rho$, or low sediment concentration).

It is assumed that the wind and rainfall forces applied at the free surface are negligible. The pressure is hydrostatic such that the depth-averaged pressure is given by $p = 0.5\rho gh$ and the hydrostatic pressure force is $F_p = 0.5\rho gh^2$. The bed shear stress $\tau_{zx} = \tau_{0x}$, and the sum of forces, using Equation (2.18a), gives

$$\rho a_x h \mathrm{d}x \mathrm{d}y = \rho g h \mathrm{d}x \mathrm{d}y \sin\theta - \frac{\rho g}{2}\frac{\partial h^2}{\partial x}\mathrm{d}x\mathrm{d}y + \frac{\partial \tau_{yx}h}{\partial y}\mathrm{d}x\mathrm{d}y - \tau_{0x}\mathrm{d}x\mathrm{d}y. \quad (2.19)$$

Strictly speaking, when performing depth-integrals of the momentum terms on the left-hand side of Equation (2.18), we obtain a Boussinesq momentum correction factor β_m, given the cross-sectional averaged velocity V_x:

$$\beta_m = \frac{1}{AV_x^2}\int_A v_x^2\, \mathrm{d}A. \quad (2.20)$$

The value of β_m in rivers is generally close to unity and β_m can often be dropped. Dividing all terms of Equation (2.19) by the mass $\rho h \, dx dy$ and considering a small angle θ_0, such that $\sin \theta_0 \cong \tan \theta_0 = \overline{S}_{0x}$ (and locally $\sin \theta = \overline{S}_{0x} - \partial z_b / \partial x$) results in an integrated form of the equation of motion

$$a_x = \underbrace{\frac{\partial V_x}{\partial t}}_{\text{local}} + \underbrace{V_x \frac{\partial V_x}{\partial x} + V_y \frac{\partial V_x}{\partial y} + V_z \frac{\partial V_x}{\partial z}}_{\text{convective}}$$

$$= \underbrace{g \overline{S}_{0x}}_{\text{gravity}} - \underbrace{g \frac{\partial z_b}{\partial x}}_{\text{bed elevation}} - \underbrace{g \frac{\partial h}{\partial x}}_{\text{pressure}} - \underbrace{\frac{\tau_{0x}}{\rho h}}_{\text{bed shear}} + \underbrace{\frac{\partial \tau_{yx}}{\rho \partial y}}_{\text{bank shear}}. \qquad (2.21)$$

In most rivers, it can be further assumed that V_z is small. Also, the riverbank shear force in wide rivers is small compared to the bed shear force. The following two-dimensional approximation for the depth-integrated equation of motion in rivers becomes

$$\frac{\partial V_x}{\partial t} + V_x \frac{\partial V_x}{\partial x} + V_y \frac{\partial V_x}{\partial y} = g \overline{S}_{0x} + g S_{0x} - g \frac{\partial h}{\partial x} - \frac{\tau_{0x}}{\rho h}. \qquad (2.22)$$

With a similar analysis applied in the y direction, we obtain

$$\frac{\partial V_y}{\partial t} + V_x \frac{\partial V_y}{\partial x} + V_y \frac{\partial V_y}{\partial y} = g S_{0y} - g \frac{\partial h}{\partial y} - \frac{\tau_{0y}}{\rho h}, \qquad (2.23)$$

where $\overline{S}_{0x}$ is the average downstream bed slope and the average lateral slope is $\overline{S}_{0y} = 0$. The local bed slopes are $S_{0x} = -\partial z_b / \partial x$ and $S_{0y} = -\partial z_b / \partial y$, respectively. These formulations are used for 2-D raster-based river models with high width–depth ratios.

Momentum in a river bend. For flow in bends, the relative magnitude of radial acceleration terms in cylindrical coordinates from Equation (2.15a) indicates that the centrifugal acceleration is primarily counterbalanced by the pressure gradient and the radial shear stress, as suggested by Rozovskii (1957):

$$\frac{V_\theta^2}{r} = g S_r - \frac{1}{\rho} \frac{\partial \tau_r}{\partial z}, \qquad (2.24)$$

where the local downstream velocity V_θ, the radial shear stress τ_r, and the radial water-surface slope S_r vary with the radius of curvature r, as sketched in Figure 2.9. It is interesting to note that the shear stress applied onto the fluid control volume is pointing to the outer bank. Of course, the fluid will exert a radial shear stress on the riverbed material in the opposite direction,

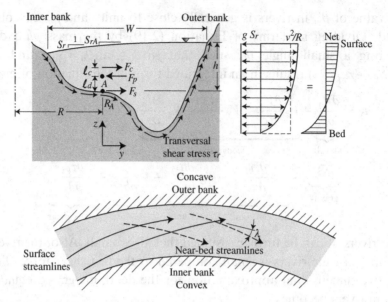

Figure 2.9. Flow momentum in a river bend

i.e. toward the inner bank. We can already expect from this analysis that sediment particles will tend to move toward the inner bank while the river bend will migrate toward the outer bank.

Scaling factors define dimensionless parameters for radial surface slope $S_r^* = S_r/S_{rA}$, channel width $w^* = w/W$, flow depth $z^* = z/h$, radius of curvature $r^* = r/R$, velocity $v^* = v/\overline{V}$, and radial shear stress $\tau_r^* = \tau_r/\tau_{rR}$. The element of fluid or control volume $d\forall = dx\,dy\,dz$ for the portion of a river bend of length $dx = R\,d\theta$ is reduced to a dimensionless volume $d\forall^* = d\forall/WRh$.

The radial equation of motion (Equation (2.24)) is multiplied by ρ and $d\forall$, reduced in dimensionless form, and then integrated over the dimensionless volume $\forall^*$ of the reach. The resulting dimensionless momentum equation in the radial direction is

$$\underbrace{\rho Wh\overline{V}^2 \int_{\forall^*} \frac{v^{*2}}{r^*}\,d\forall^*}_{\text{centripetal force Fc}} = \underbrace{\rho gWRhS_{rA} \int_{\forall^*} S_r^*\,d\forall^*}_{\text{pressure force Fp}} \ - \ \underbrace{WR\tau_{rR} \int_{\forall^*} \frac{\partial \tau_r^*}{\partial z^*}\,d\forall^*}_{\text{shear force Fs}}. \qquad (2.25)$$

The integrals of dimensionless terms in Equation (2.25) result in pure numbers and each term represents a force. The corresponding force diagram is sketched in Figure 2.9. The inward pressure force is found to balance the sum of the outward centrifugal force exerted at a distance l_c above point A and the outward shear force exerted at a distance l_d below A. It is interesting

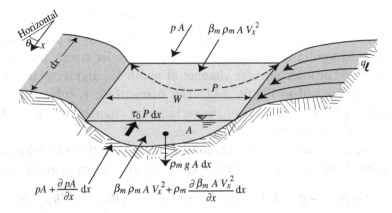

Figure 2.10. One-dimensional river-flow momentum

to use the sum of moments about A to find $\tau_{rR} \sim \rho \overline{V}^2 h / R$. This relationship will be used extensively to describe river meandering in Chapter 10.

One-dimensional river momentum. One-dimensional approximations of the momentum equation are useful when the channel width and cross-sectional area do not vary much with x and time, as sketched in Figure 2.10.

The one-dimensional momentum equation is obtained from Equation (2.18a). On the left-hand side, the volume integral of the local acceleration term is $\rho(\partial A V_x / \partial t) dx$, the net momentum from the convective acceleration terms is $\rho(\partial \beta_m Q V_x / \partial x) dx$. On the right-hand side, the gravitational force component is $\rho g S_o A dx$, the force from the pressure gradient is $\rho(g \partial h / \partial x) A dx$, and the bed shear stress force is $\tau_0 P dx$. The sum of all components, after dividing by ρdx, gives

$$\frac{\partial Q}{\partial t} + \frac{\partial \beta_m Q V}{\partial x} = g A S_0 - g \frac{A \partial h}{\partial x} - \frac{\tau_0 P}{\rho}. \tag{2.26}$$

We can further assume $\beta_m \cong 1$, and further simplifications arise in wide rivers such that $P \cong W$ and $R_h \cong h$. Also, when the channel width and cross-section area remain fairly constant, the one-dimensional velocity reduces to $V = V_x$, the bed slope to $S = S_{0x}$, and the bed shear stress to $\tau_0 = \tau_{0x}$:

$$\frac{\partial V_x}{\partial t} + \frac{V_x \partial V_x}{\partial x} = g \overline{S}_{0x} + g S_{0x} - g \frac{\partial h}{\partial x} - \frac{\tau_{0x}}{\rho_m h}, \tag{2.27a}$$

where $\overline{S}_{0x}$ is the reach-averaged bed slope and the local slope $S_{0x} = -\partial z_b / \partial x$.

$$\frac{\partial v_x}{\partial t} + v_x \frac{\partial v_x}{\partial x} \cong g S_0 - \frac{1}{\rho} \frac{\partial p}{\partial x} + \frac{1}{\rho} \frac{\partial \tau_{zx}}{\partial z}. \tag{2.27b}$$

2.1.5 Specific Momentum Equations

An application example of the momentum concept leads to the definition of specific momentum in a rectangular channel. Consider steady flow at a discharge $Q = AV$ in a rectangular channel of width W, and cross-section area $A = Wh$, as sketched in Figure 2.11. The free surface is subjected to wind shear equal to τ_w. On the sides of the rectangular channel, the bank shear is $\tau_{yx} = \tau_s$, and the bed shear stress is $\tau_{zx} = \tau_0$.

The downstream momentum relationship from Equation (2.18a) is applied to this control volume $\forall$ of length X_c, width W, and height h, and results in

$$\beta_m \rho A_2 V_2^2 + p_2 A_2 - \beta_m \rho A_1 V_1^2 - p_1 A_1 = \gamma \forall \sin\theta - \tau_0 W X_c - \tau_s 2 h X_c + \tau_w W X_c.$$

(2.28)

Further simplification arises when the bed and bank shear stresses are equal, $\tau_0 = \tau_s$, in rivers with small bed slope θ, $\sin\theta \cong S_0$, without wind shear and of with $\beta_m \cong 1$, as

$$p_1 A_1 + \rho A_1 V_1^2 = p_2 A_2 + \rho A_2 V_2^2 - \gamma \forall S_0 + \tau_0 (W + 2h) X_c.$$

(2.29)

The hydrodynamic force F_h on the left-hand side can be written as

$$F_h = pA + \rho A V^2 = 0.5 \rho g h W h + \rho W h \left(\frac{Q}{Wh}\right)^2.$$

(2.30)

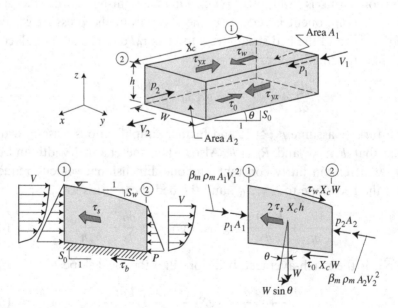

Figure 2.11. Definition sketch for specific momentum

The specific-momentum function M is then simply obtained, after dividing the hydrodynamic force F_h by the channel width W and the specific weight γ

$$M = \frac{F_h}{\rho g W} = \frac{h^2}{2} + \frac{q^2}{gh}, \qquad (2.31)\blacklozenge$$

where the unit discharge $q = Q/W$. The concept of specific momentum will be used in the analysis of rapidly varied river flows in Chapter 5.

2.1.6 Saint-Venant Equation

The Saint-Venant equation is a simplified version of the momentum equation for a wide rectangular channel at a bed slope S_0, as shown in Figure 2.12. The flow is one dimensional (1-D) in the x direction; thus $v = v_x$ and $v_y = v_z = 0$.

The equation of motion, Equation (2.14a), is applied to describe the motion of this element of fluid. Three approximations can be made for wide-rectangular channels: (i) the wetted perimeter $P = (W + 2h)/h$ is approximately equal to the channel width W; (ii) the hydraulic radius $R_h = Wh/P \cong h$; and (iii) the bank surface is very small compared to the bed surface and the bank shear force is small compared to the bed shear force. We also neglect the shear stress due to element stretching $\tau_{xx} = 0$. At small bed-slope angles, $\sin\theta \cong \tan\theta$ and $g_x = g\sin\theta \cong gS_0$, and thus, Equation (2.14a) reduces to

$$\frac{\partial v_x}{\partial t} + v_x\frac{\partial v_x}{\partial x} \cong gS_0 - \frac{1}{\rho}\frac{\partial p}{\partial x} + \frac{1}{\rho}\frac{\partial \tau_{zx}}{\partial z}. \qquad (2.32)$$

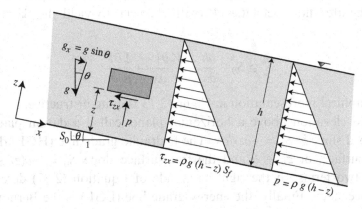

Figure 2.12. Shear stress and pressure distribution

We obtain the pressure distribution by integrating Equation (2.14c), given that $a_z = 0$ and that all shear-stress variations in the vertical direction are small. The resulting hydrostatic-pressure distribution is obtained from the integral of $g_z = g \cos \theta \cong -g$ along the z direction:

$$\int_p^0 dp = \rho \int_z^h -g \, dz \quad \text{or} \quad p = \rho g(h - z). \tag{2.33}$$

Accordingly, the relative pressure vanishes at the free surface, $p = 0$ at $z = h$, and the pressure at the bed $z = 0$ is $p_0 = \rho g h$.

In most river flows, the shear stress vanishes at the free surface (unless winds are quite strong) and varies linearly over the depth to reach the bed shear stress $\tau_0 = \rho g h S_f$. The shear-stress distribution in the water column is thus

$$\tau_{zx} = \rho g(h - z) S_f. \tag{2.34} \blacklozenge$$

When comparing p and τ_{zx} from Equations (2.33) and (2.34), it becomes interesting to think of the friction slope as a ratio of bed shear stress to pressure. Also, substituting p and τ_{zx} into the equation of motion Equation (2.32) leads to the following simplified version of the equation of motion because $\partial S_f / \partial z = \partial h / \partial z = 0$;

$$\frac{\partial v_x}{\partial t} + v_x \frac{\partial v_x}{\partial x} \cong g S_0 - g \frac{\partial h}{\partial x} - g S_f. \tag{2.35}$$

We can solve the equation of motion (Equation (2.35)) in dimensionless form after dividing by g:

$$S_f \cong S_0 - \frac{\partial h}{\partial x} - \frac{v_x \partial v_x}{g \partial x} - \frac{\partial v_x}{g \partial t}. \tag{2.36}$$

The practical significance is that point velocities v_x can be replaced with the depth-integrated flow velocities V (with $\beta_m \cong 1$), to yield the Saint-Venant equation:

$$S_f \cong S_0 - \frac{\partial h}{\partial x} - \frac{V}{g} \frac{\partial V}{\partial x} - \frac{1}{g} \frac{\partial V}{\partial t}. \tag{2.37} \blacklozenge\blacklozenge$$

The graphical representation in Figure 2.13 is quite instructive.

The bed elevation above a horizontal plane, called a datum plane, is z_b and the bed slope is $S_0 = -\partial z_b / \partial x$. The hydraulic grade line (HGL) describes the free surface, or $z_b + h$ and the free surface slope $S_0 = -\partial(z_b + h)/\partial x$. The first two terms on the right-hand side of Equation (2.37) describe the free-surface slope. Finally, the energy grade line (EGL) is the Bernoulli sum equal to $H = z_b + h + V^2/2g$.

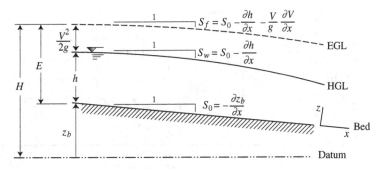

Figure 2.13. Hydraulic and energy grade lines for steady flow

2.1.7 Specific Energy Equation

When considering flow in the wide-rectangular channel sketched in Figure 2.13, the specific energy E is defined as the sum of the flow depth and the velocity head. When steady flow is considered ($\partial V/\partial t = 0$) and $\partial V^2/\partial x = 2V\partial V/\partial x$, the remaining four terms of the Saint-Venant equation, Equation (2.37), can be rearranged in the following manner:

$$\frac{dE}{dx} \cong \frac{d}{dx}\left(h + \frac{V^2}{2g}\right) = S_0 - S_f. \tag{2.38}$$

At a constant unit discharge $q = Vh$, the specific energy E corresponds to

$$E = h + \frac{V^2}{2g} = h + \frac{q^2}{2gh^2} = h + \frac{Q^2}{2gA^2}. \tag{2.39}\blacklozenge$$

The specific-energy function E and Equation (2.38) will be helpful to describe rapidly varied river flows in Chapter 5.

2.2 Equations Governing Sediment Motion

This section presents a brief digest of governing equations describing sediment transport in rivers. This section first covers some useful relationships describing incipient motion, sediment transport and settling (Section 2.2.1), before a description of the conservation of sediment mass in rivers (Section 2.2.2).

2.2.1 Incipient Motion, Sediment Transport and Settling

First, the dimensionless particle diameter d_* is defined from the specific gravity G of sediment, the kinematic viscosity of the fluid υ, and the gravitational acceleration g as

$$d_* = d_s\left[\frac{(G-1)g}{\upsilon^2}\right]^{1/3}. \tag{2.40}$$

34 *Mechanics of Rivers*

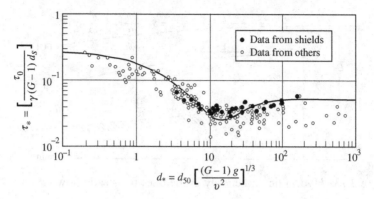

$$d_* = d_{50} \left[\frac{(G-1)g}{v^2} \right]^{1/3}$$

Figure 2.14. Modified Shields diagram for incipient motion (from Julien, 2010)

The dimensionless particle diameter is very helpful in defining other sedimentation characteristics such as the settling velocity and incipient motion. Values of d_* for $G = 2.65$ and $v = 1.0 \times 10^{-6}$ m²/s are listed in Table 1.4.

Critical shear stress τ_c and shear velocity u_{*c}. The ratio of the bed shear force to the submerged particle weight defines the Shields parameter τ_*

$$\tau_* = \frac{\tau_0}{(\gamma_s - \gamma)d_s} = \frac{u_*^2}{(G-1)gd_s} \cong \frac{hS_f}{(G-1)d_s} \qquad (2.41) \blacklozenge\blacklozenge$$

where $\tau_0 = \rho u_*^2$ is the bed shear stress, u_* is the shear velocity, γ_s is the specific weight of a sediment particle, γ is the specific weight of water, d_s is the particle size, and g is the gravitational acceleration. The bed shear stress can also be calculated as $\tau_0 = \gamma R_h S_f \cong \gamma h S_f$, where R_h is the hydraulic radius, h is the flow depth, and S_f is the friction slope.

The critical value of the Shields parameter τ_{*c} matches the beginning of motion ($\tau_0 = \tau_c$) and depends on d_*, as shown in Figure 2.14. Critical values of the Shields parameter τ_{*c} and critical shear stress τ_c for different particle sizes on plane horizontal surfaces are listed in Table 1.4. The corresponding critical shear velocity u_{*c} is defined as $u_{*c} = \sqrt{\tau_c/\rho}$. Note that both τ_c and u_{*c} do not change significantly for silts.

Figure 2.15 shows the critical shear stress value for coarse sediment. As a practical approximation, a shear-stress value of $\tau = 0.1$ Pa is sufficient to move silts but not sands, and $\tau_c = 1$ Pa is sufficient to move sands but not gravels. For materials coarser than sands, the critical shear stress in Pa is approximately equal to the particle diameter in mm (i.e. a 10-mm particle will be near incipient motion under a shear stress of 10 Pa).

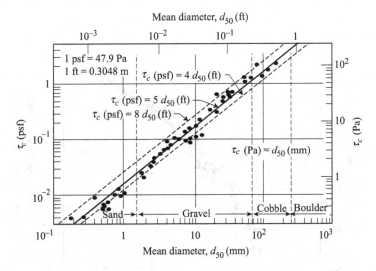

Figure 2.15. Critical shear stress for coarse particles (from Julien, 2010)

Settling velocity ω. The settling velocity ω of sediment particles in clear water at 10°C is calculated from

$$\omega \cong \frac{8\upsilon}{d_s}\left[\left(1+\frac{d_*^3}{72}\right)^{0.5}-1\right] \qquad (2.42)\blacklozenge$$

where d_s is the particle diameter, υ is the kinematic viscosity, G is the specific gravity, and g is the gravitational acceleration. The values of the settling velocity in clear water at 20°C and $G = 2.65$ are also given in Table 1.4.

Sediment transport. Several sediment transport formulas are described in Julien (2010). For gravel-bed streams, the unit sediment discharge by volume q_{bv} [L^2/T] can be defined from the Meyer–Peter and the Müller formulas as

$$q_{bv} = 8(\tau_* - \tau_{*c})^{1.5}[(G-1)gd_s^3]^{0.5}. \qquad (2.43)$$

An alternative formulation uses the dimensionless transport parameter $q_{bv}/\omega d_s$ as a function of the Shields parameter and the settling velocity as shown in Figure 2.16. The strength of this formulation will be demonstrated when defining the river similitude for sediment transport in Chapter 12.

Also, a very crude approximation for sands, where $0.1 < \tau^* < 1$, is

$$q_{bv} \approx 18\sqrt{g}\, d_s^{3/2}\tau_*^2. \qquad (2.44)$$

This relationship will be useful in defining the downstream hydraulic geometry of alluvial channels in Chapter 11.

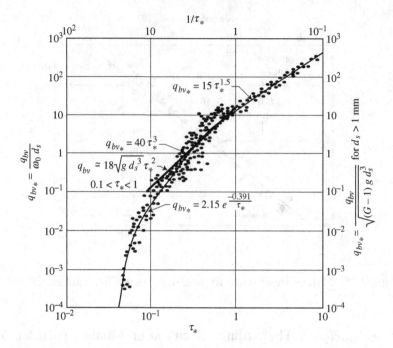

Figure 2.16. Sediment transport versus Shields parameter (from Julien, 2010)

Sediment concentration profiles. The sediment concentration C at an elevation z above the bed can be calculated from the Rouse equation as

$$C = C_a \left[\left(\frac{h-z}{z} \right) \left(\frac{a}{h-a} \right) \right]^{\frac{\omega}{\kappa u_*}} \qquad (2.45)$$

where C_a is the concentration at an elevation a above the bed, h is the flow depth, κ is the von Kármán constant ($\kappa \cong 0.4$), ω is the settling velocity, and u_* is the shear velocity. The concentration profiles in Figure 2.17 illustrate how the Rouse number $Ro = \omega/\kappa u_*$ can be evaluated from a vertical sediment concentration profile.

Sediment-rating curve. A sediment-rating curve defines the relationship between sediment transport and the flow discharge. Typically, the sediment discharge Q_s can be calculated from the surface integral of concentration times flow velocity.

$$Q_s = W q_t = C_f Q = \int C v \, dA \qquad (2.46)$$

where C_f is the flux-averaged concentration and Q is the flow discharge. Also, the unit sediment discharge q_t equals the total sediment discharge per unit

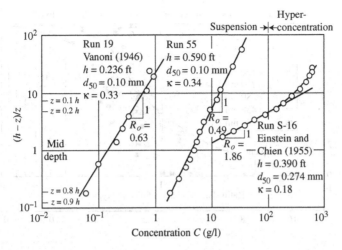

Figure 2.17. Examples of sediment concentration profiles (after Woo et al., 1988)

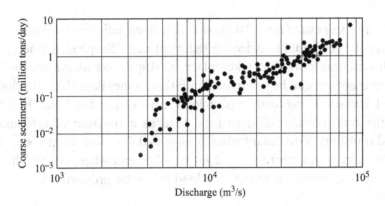

Figure 2.18. Sediment-rating curve for the Jamuna River (after FAP24, 1996)

channel width W. For instance, the sediment load in metric tons per day can be obtained from the product of the sediment concentration $C_{mg/l}$ in milligrams per liter times the discharge Q_{cms} in cubic meters per second (or Q_{cfs} in cubic feet per second) as

$$Q_{s\ metric\ tons\ per\ day} = 0.0864 C_{mg/l} Q_{cms} = 0.002446 C_{mg/l} Q_{cfs}. \qquad (2.47)$$

As an example, Figure 2.18 shows the sediment-rating curve of the Jamuna River. Sediment concentration usually increases with discharge, i.e. $Q_s = CQ = AQ^B$, and $B > 1$.

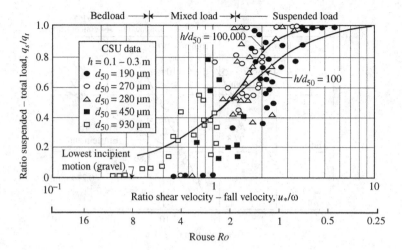

Figure 2.19. Ratio of suspended to total load versus u_*/ω (after Shah-Fairbank et al., 2011)

Bedload or suspended load. "Bedload" describes sediment transport in a thin layer near the bed equal to twice the sediment size. "Suspended load" refers to the sediment transport that takes place in suspension above the bed layer. Sediment transport can be subdivided into three zones describing the dominant mode of transport: bedload, mixed load, and suspended load. Figure 2.19 shows the ratio of suspended to total load as a function of u_*/ω and h/d_s. Bedload transport is dominant when $u_*/\omega < \sim 0.4$. A transition zone, called a mixed load, is found where $0.4 < u_*/\omega < \sim 2.5$, and where both bedload and suspended load contribute to the total load in similar proportions.

2.2.2 Conservation of Sediment Mass

The term "conservation of sediment mass" denotes the internal change in volumetric sediment concentration C_v as a function of the changes in volumetric sediment fluxes. In differential form, the sediment continuity relationship is

$$\frac{\partial C_v}{\partial t} + \frac{\partial q_{tx}}{\partial x} + \frac{\partial q_{ty}}{\partial y} + \frac{\partial q_{tz}}{\partial z} = 0, \qquad (2.48)\blacklozenge$$

where the unit sediment fluxes q_{tx}, q_{ty}, and q_{tz} account for the total unit sediment discharge by volume in the x, y, and z directions, respectively.

Settling conditions can be examined for a steady supply of sediment ($\partial C_v/\partial t = 0$) in a wide channel without lateral sediment inflow ($\partial q_{ty}/\partial y = 0$).

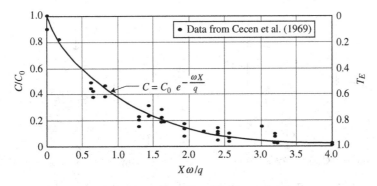

Figure 2.20. Downstream increase in trap efficiency

In this case, Equation (2.48) reduces to two terms where the advective fluxes $q_{tx} = v_x C_v$ and $q_{tz} = -\omega C_v$ are dominant. As shown in Figure 2.20, the solution at a constant unit discharge $q = Vh$ is a function of the upstream sediment concentration C_0

$$C = C_0 e^{-\frac{X\omega}{hV}}. \tag{2.49}$$

The concentration left in suspension is negligible ($C/C_0 = 0.01$) at a distance X_C:

$$X_C \cong 4.6 \frac{hV}{\omega}. \tag{2.50}$$

This relationship is very useful in the design of settling basins and for the analysis of reservoir sedimentation. The percentage of sediment that settles within a given distance X defines the trap efficiency T_E as:

$$T_E = \frac{C_0 - C}{C_0} = 1 - e^{\frac{X\omega}{hV}} = 1 - e^{\frac{-WX\omega}{Q}}. \tag{2.51}\blacklozenge$$

It is interesting to note that at a constant flow discharge $Q = WhV$, the trap efficiency of a given particle size simply increases with the surface area of the settling basin, described by the basin width W times length X. In numerical models, when the grid size ΔX is larger than Xc, the trap efficiency is essentially unity and aggradation responds directly to changes in the sediment transport capacity. For ΔX smaller than Xc, only part of the sediment load in suspension will settle within the given reach.

In the vertical direction, there is an interaction between the sediment concentration in suspension and the sediment deposition on the river bed. As sketched in Figure 2.21, the net settling flux $\omega C_v W \Delta X$ must be equal to the rate of change in volumetric storage on the bed $(1 - p_0) W \Delta X \Delta z_b / \Delta t$.

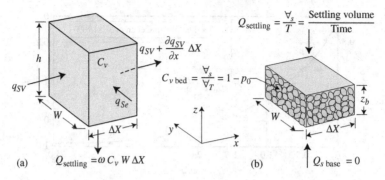

Figure 2.21. (a) Sediment settling and (b) bed elevation changes (from Julien, 2010)

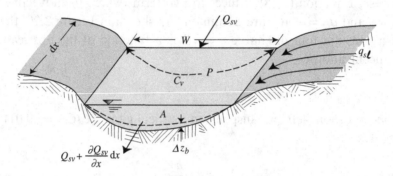

Figure 2.22. Bed elevation changes in a one-dimensional channel

The 1-D formulation of sediment continuity can be defined from Equation (2.48) for the one-dimensional river sketched in Figure 2.22,

$$\frac{\partial A C_v}{\partial t} + (1-p_0)\frac{\partial P z_b}{\partial t} + \frac{\partial Q_{sv}}{\partial x} = q_{sl}, \qquad (2.52)$$

where A is the cross-section area, P is the wetted perimeter, z_b is the bed elevation, Q_{sv} is the volumetric sediment discharge, W is the channel width, and q_{sl} is the unit sediment discharge by volume of the lateral inflow.

The bed elevation changes in rivers are determined from the sediment continuity relationship. In most cases, the internal change of suspended sediment concentration and the lateral inflow in Equation (2.52) are usually negligible. The bed elevation changes as a function of time can be calculated from

$$\frac{\partial z_b}{\partial t} = -\frac{T_E}{(1-p_0)}\frac{\partial Q_{tx}}{W \partial x}. \qquad (2.53)$$

In sediment transport models, we should ensure that $\Delta X > Xc$ from Equation (2.50) in order to ensure that all sediment can settle within the reach ΔX. In this case, the resulting formulation is called the Exner equation, when the trap efficiency $T_E = 1$. For sand-and gravel-bed streams, it is common to assume $T_E = 1$ and $p_0 = 0.43$. For different grain sizes, the values of porosity from Table 1.5 are suggested as a first approximation.

2.3 Why Do Rivers Form?

After all this hard work, have we learned anything? The reader has already gathered enough understanding of river mechanics to answer the fundamental question: why do rivers form? This can lead to an interesting in-class discussion! The concepts of gravity and the hydrologic cycle may come up first. The need for erodible material or alluvium should also quickly emerge. However, all of this does not explain why rivers form. Could we live on a planet where water spreads out everywhere on the land surface, as in the everglades? Can the reader apply the concepts discussed in this chapter to explain the fundamental mechanics behind the formation of alluvial rivers? Well, for a river to form, flow convergence must be preferable to flow divergence. That is to say that if flow convergence would result in aggradation, then the sedimentation process would force the water away from the point of convergence and rivers would never form. Rivers can therefore only form when flow convergence results in riverbed degradation, as sketched in Figure 2.23.

From the governing equations discussed in this chapter, those for conservation of water and sediment must be applied. Aggradation and degradation result from the principle of conservation of sediment. The relevant question then becomes: does converging flow tend to cause aggradation or degradation? Can we formulate an intuitive understanding by using a simple sediment-rating curve of the type $q_s = Cq = aq^b$, where q_s is the unit sediment

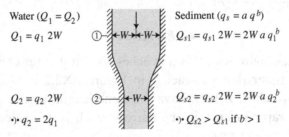

Figure 2.23. Flow convergence and sediment continuity

discharge, C is the sediment concentration, and q is the unit discharge. As sketched in Figure 2.23, the total sediment load at the upstream section 1 is $2Waq^b$, while the sediment load in the narrow section 2 is $Wa(2q)^b$. It is then clear that convergence results in a greater sediment-transport capacity at section 2 than at section 1 when $b > 1$. Converging flows would thus result in channel degradation when $b > 1$ and aggradation when $b < 1$. Is there any reason, inferable from our understanding of erosion and sedimentation, that supports the possibility that $b > 1$? Simply put, with $q_s = Cq$ where C is the sediment concentration, if concentration increases with discharge, then we do have favorable conditions for rivers to form. I guess our improved understanding of river mechanics has helped us answer our fundamental question!

Now this discussion highlighted another fundamental aspect of river mechanics. The need for a hydrologic cycle and stability of erodible material merits further study. The reader should also clearly understand that the flow of water is not the only fluvial process. There are two main components at work in rivers: water and sediment. The importance of sediment transport is often neglected, but we now understand that it cannot be overstated. Let's pursue our investigation with a closer look at the hydrologic cycle in the Chapter 3, after solving a few exercises and problems.

♦♦*Exercise 2.1*

With reference to Figure 2.6, apply the law of conservation of mass through a 1-D control volume of discharge $Q = \int_0^A v_x \, dA$, cross-sectional area A, and top width W. Consider the lateral inflow of unit discharge q_l, rainfall intensity i, and infiltration rate i_b leaving through the wetted perimeter P. [*Hint*: Calculate the volume fluxes entering the control volume $\forall = A \, dx$ and compare to the rate of change in control volume $\partial \forall / \partial t = dx \, \partial A / \partial t$.]

[*Answer*: Equation (2.11).]

♦♦*Exercise 2.2*

Can you explain why wider river reaches have a greater capacity to trap sediment? Consider that subreach 1 in Figure EX2.2 is twice as wide as the lower subreach 2. If both have the same length, discharge, and grain size, and the trap efficiency in the narrow reach is 0.5, calculate the trap efficiency in the wider reach.

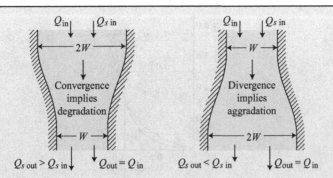

Figure EX2.2. Flow convergence

◆Exercise 2.3

Demonstrate the 1-D formulation of the momentum equation, Equation (2.18a), for the application to the flow situation in Figure 2.10. Consider the momentum flux $F = \beta_m \rho A V_x^2 = \rho \int_0^A v_x^2 \, dA$ passing through a cross-sectional area A and top width W. Ignore momentum contributions from rainfall, lateral inflow, and infiltration. The shear stress τ_0 is applied over the wetted perimeter P and the bed slope is $S_0 = \tan \theta$.
[*Answer:* The answer is Equation (2.26).]

Exercise 2.4

At an amusement park, water is flowing north at a velocity of 1 m/s. The flow is steady for 1 second and suddenly changes direction. Every second, the flow direction rotates clockwise by 30° while the magnitude remains constant at 1 m/s, e.g. flowing east after 3 s and south after 6 s. If one buoy is released every second at the origin of the coordinate system, track the position of all the buoys over time and define the streamline, the pathline and the streakline of the buoys after 5 and 10 s.

◆Problem 2.1

From the velocity profile shown in Figure P2.1 with a flow depth of 3.7 ft, determine the depth-averaged flow velocity, the Boussinesq momentum correction factor $\beta_m = \frac{1}{A V_x^2} \int_A v_x^2 \, dA$, and the Coriolis energy correction factor $\alpha_e = \frac{1}{A V_x^3} \int_A v_x^3 \, dA$.

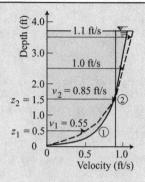

Figure P2.1. Example of velocity profile

[*Answer*: Calculations per unit channel width.

$$\overline{V} \cong \frac{1}{3.7 \, \text{ft}} \, [0.55 + 0.85 + 1.0 + (1.1 \times 0.7)] \frac{\text{ft}^2}{\text{s}} = 0.857 \frac{\text{ft}}{\text{s}},$$

$\beta_m = 1.057$, and

$$\alpha_e \cong \frac{1}{3.7 \, \text{ft}} \frac{\text{s}^3}{(0.857)^3 \, \text{ft}^3} \, [0.55^3 + 0.85^3 + 1.0^3 + (1.1^3 \times 0.7)] \frac{\text{ft}^4}{\text{s}^3} = 1.166.]$$

◆◆*Problem 2.2*

Table P2.2 provides the measurements of flow depth h, depth-averaged velocity v, and flow direction θ (i.e. $\theta = 0°$ downstream) across the Matamek River. Fill in the blanks in the table, plot the profiles, and find the cross-section average velocity V.

Table P2.2. *Cross-section data of the Matamek River*

Distance from the left bank y_i (ft)	Flow depth h_i (ft)	Depth-averaged velocity v_i (ft/s)	Velocity angle θ_i (°)	Incremental cross-section area a_i (ft²)	Incremental discharge ΔQ_i (ft³/s)
0	0	0	–	0	
10	1.3	+0.6	170		−7.68
20	1.7	+0.2	190	17	
30	1.8	0.4	20		6.76
40	2.1	1.1	0	31.5	
60	2.4	1.4	−1		67.19
80	2.9	2.1	−4	58	
100	3.6	2.6	−5		186.48
120	4.1	1.9	−2	82	
140	2.2	0.8	0		35.2
160	0	0	–	0	

[*Answer:* $a_i = w_i h_i = 0.5(y_{i+1} + y_{i-1})h_i$, $A = \sum a_i = 383.5 \text{ ft}^2$, $Q = AV = \Sigma \Delta Q_i = \Sigma a_i v_i \cos \theta_i = 596.5 \text{ ft}^3/\text{s}$, and $V = Q/A = 1.55 \text{ ft/s}$.]

Problem 2.3

Evaluate the Boussinesq momentum correction factor $\beta_m = \frac{1}{AV_x^2} \int_A v_x^2 \, dA$ for the velocity profile of the Mississippi River in Figure 2.2 if the flow depth is 60 ft.

◆Problem 2.4

Strictly speaking, the depth integral of the velocity head in the specific energy Equation (2.28) is

$$\frac{d\tilde{E}}{dx} = \frac{d}{dx}\left(\frac{p}{\gamma_m} + \alpha_e \frac{V^2}{2g}\right) = \frac{-dz_b}{dx} - S_f = S_0 - S_f \cong 0.$$

Consider the velocity profile example of the Mississippi River in Figure 2.2. Calculate the Coriolis energy correction factor $\alpha_e = \frac{1}{AV_x^3} \int_A v_x^3 \, dA$. Is there any good reason to consider the approximation $\alpha_e \cong 1$ in Equation (2.38)?
[*Answer:* Yes, $\alpha_e = 1.015$ for this example from the Mississippi River.]

◆Problem 2.5

Consider an impervious plane surface under a constant rainfall intensity i. Apply the continuity equation to link the unit discharge q and flow depth h as a function of x and t.
[*Answer:* From Equation (2.20), $\partial h/\partial t + \partial q/\partial x = i$.]

◆◆Problem 2.6

From a look at Figure 2.9, explain why the shear force applied on the control volume is acting toward the concave bank. Are the streamlines near the riverbed not acting toward the convex bank? So, why is the shear force applied on the fluid in the opposite direction?

◆Problem 2.7

In the river bend sketched in Figure 2.9, use the sum of moments about A with the forces defined in Equation (2.25) to find a relationship for the transversal shear stress.
[*Answer:* $l_c \rho Wh\overline{V}^2 \int_{\forall^*} \frac{v^{*2}}{r^*} d\forall^* = l_d WR\tau_{rR} \int_{\forall^*} \frac{\partial \tau_{rz}^*}{\partial z^*} d\forall^*$, thus $\tau_{rR} = \text{const} \times \rho \overline{V}^2 h/R$.]

◆◆*Problem 2.8*

Examine the discharge of 1 cubic meter per second of slurry into a tailing pond. If the volumetric sediment concentration is 35 percent and the pond area is 2 km^2, what is the sedimentation rate in the pond? Also, if the particle size distribution is 40 percent fine sand, 46 percent silt, and 13 percent clay with $d_{50} = 50$ μm and $d_{30} = 10$ microns, what is the expected sediment concentration in mg/l and the size distribution at the downstream end of the pond?

◆◆*Problem 2.9*

Consider the sediment transport of the Jamuna River in Bangladesh, as shown in Figure 2.18. Can you plot lines of constant concentration in mg/l on this figure? Does the sediment concentration increase or decrease with discharge? Can you define B in $Q_s = CQ = AQ^B$?

3

River Basins

It is important to figure out where the river water and sediment come from. The terms river basin, watershed, and catchment are used interchangeably to describe the land area where surface waters drain to a river, or a stream. It is particularly important to examine river basins in terms of surface runoff and sediment yield. The chapter focuses on the spatial variability in river-basin characteristics (Section 3.1), followed by an analysis of excess rainfall precipitation (Section 3.2), and then one of surface runoff (Section 3.3). Methods to determine sediment sources and sediment yield for river basins can be found in Section 3.4.

3.1 River Basin Characteristics

River-basin characteristics include topography and physiography, geology and pedology, forestry and climatology. Watershed boundaries are separated by drainage divides located at high points between watersheds, as shown in Figure 3.1. Lines of constant elevation are called contour lines. Hypsometric curves give the relative basin area higher than a given elevation, which helps determine the type of precipitation, in terms of rain or snow.

Geographic Information Systems (GIS) are helpful for watershed delineation, topography and mapping slope, soil type, and land use. The surface slope is a dominant parameter in the calculation of surface runoff and sediment transport. Raster-based GIS display watershed information in a square grid format, typically at a 30-m resolution. Digital Elevation Models (DEM) provide a raster-based topographic description of a watershed. Figure 3.2 shows characteristic watershed and rainstorm scales.

Rivers follow the low points along the watershed topographic profiles. With very few exceptions, in arid areas and in perched rivers, the lowest point of a watershed is located at the river outlet. The drainage network of a

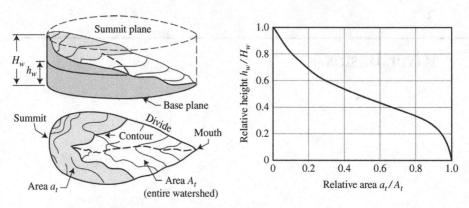

Figure 3.1. River-basin delineation and hypsometric curve

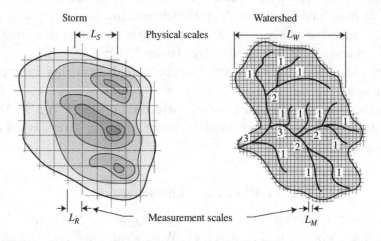

Figure 3.2. Characteristics of watersheds and rainstorms

watershed implies a discontinuous increase in drainage area and discharge at river confluences. The Strahler channel numbering system, starting from the upland areas, is also illustrated in Figure 3.2. The river length is measured from the remotest point on the watershed to the outlet and varies with drainage area, as shown in Figure 3.3. River length increases approximately with the square root of the drainage area. Very small watersheds cover less than 10 km^2, small watersheds less than 100 km^2, and large watersheds exceed 1,000 km^2.

Geologic information demonstrates the overall soil erodibility of watersheds. For instance, the Yellow River in China carries a very large sediment load from the Loess plateau, while the St. Lawrence River drains the high-grade metamorphic rocks of the Canadian Shield at very low sediment concentrations ($C < 20$ mg/l). Knowing the location of faults assists in

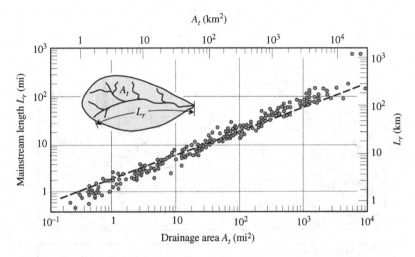

Figure 3.3. River length versus drainage area

detecting bed elevation changes caused by tectonic activity and also in tracking possible lateral shifting of rivers through geologic times.

Climatic conditions are known to vary over very large watersheds, and may also change over long time periods, e.g. desertification. For instance, the climate of the Niger River watershed ranges from arid to humid, as shown in Figure 3.4a. Vegetation changes from a desert to a rain forest in the south, depending on mean annual rainfall precipitation and potential evapotranspiration, as shown in Figure 3.4b and c.

3.2 Excess Rainfall Precipitation

This section introduces the concept of the hydrologic cycle with more details on rainfall precipitation (Section 3.2.1), infiltration (Section 3.2.2), and excess rainfall (Section 3.2.3). The depiction of the hydrologic cycle in Figure 3.5 illustrates processes contributing to the source and yield of both water and sediment from upland areas to the fluvial system.

Figure 3.5 shows the hydrologic processes of condensation, precipitation, interception, evaporation, transpiration, infiltration, subsurface flow, exfiltration, deep percolation, groundwater flow, surface runoff, and surface-detention storage. All of these processes play a role in hydrology; however, precipitation, infiltration, surface runoff, and upland erosion are most important in the analysis of rivers.

Floods and surface runoff are calculated from the excess rainfall. The excess rainfall precipitation is calculated by subtracting the interception,

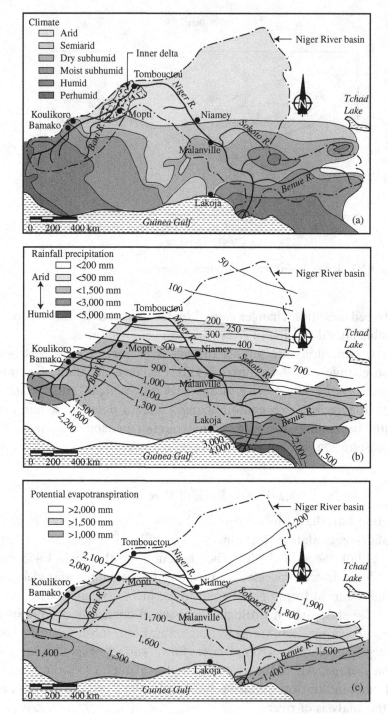

Figure 3.4. Niger River basin: (a) climate, (b) mean annual rainfall precipitation, and (c) potential evapotranspiration

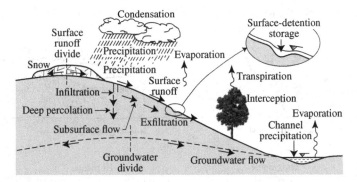

Figure 3.5. The hydrologic cycle

evapotranspiration, surface storage, and infiltration losses from the rainfall precipitation. Some rainfall is intercepted by vegetation before it reaches the ground. The amount of interception varies with the type, density, and stage of vegetation growth, rainfall intensity, and wind speed. On a single storm basis, the interception storage capacity is generally less than 4 mm, and will typically reach ~1 mm for coniferous and broadleaf forests. Interception storage becomes negligible under heavy and extreme rainfall events.

Evapotranspiration is the combination of evaporation and transpiration. "Evaporation" refers to the phase change of water from wet surfaces, from liquid to vapor. Water evaporation from plant surfaces is called "transpiration." The evaporation from a leaf surface includes both transpiration and intercepted water. Evapotranspiration generally returns a large fraction of the total precipitation to the atmosphere, e.g. Figure 3.4. As regards single storms, evapotranspiration is usually negligible during severe rainstorms.

Surface storage is the volume of water required for filling land depressions before surface runoff begins. It is estimated at 0.2–0.6 in. (5–15 mm) in pervious areas such as open fields, woodlands, and grasses. Values of 0.05–0.3 in. (1–8 mm) are typical for paved surfaces and roofs in urban areas.

3.2.1 Rainfall Precipitation

Rainfall precipitation depends primarily on the size distribution of raindrops and their fall velocity. Raindrops with diameter $d < 1$ mm remain spherical but their shape resembles an oblate spheroid for larger sizes, up to 8 mm (Smith 1992). Atlas and Ulbrich (1977) measured the raindrop fall velocity v in m/s based on raindrop diameter in mm from $v = 3.9d^{0.67}$, when $0.8 < d < 4$ mm. The typical range of raindrop sizes passing through vegetation is 1–6 mm (Calder 1992). "Light rain" denotes rainfall with a precipitation

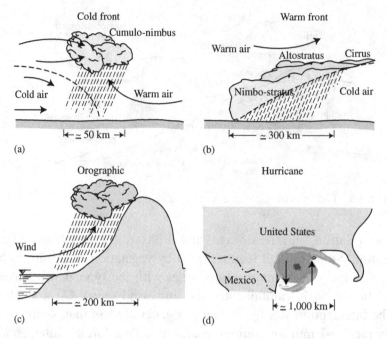

Figure 3.6. Typical rainfall precipitation patterns for cold and warm fronts, orographic precipitation, and hurricanes: (a) short and heavy rainstorms, (b) long and moderate rainstorms, (c) long and light rainstorms, and (d) long and extreme rainstorms

rate of less than 1 mm/h, and there is also moderate rain (1–4 mm/h), heavy rain (4–16 mm/h), very heavy (16–50 mm/h), and extreme rain (>50 mm/h).

Rainstorms are the result of the cooling of saturated air masses. Some examples sketched in Figure 3.6 illustrate the cases of cold and warm fronts, orographic precipitation, and hurricanes. Cold fronts will typically produce heavy but short thunderstorms, while warm fronts will produce long and light to moderate rainstorms. Orographic precipitation is generated from winds pushing saturated air masses up a mountain range; the resulting precipitation is generally light and a steady drizzle. Hurricanes and typhoons generate heavy to extreme rainstorms on large surface areas with strong winds and of long durations. Most devastating floods result from hurricanes and tropical storm precipitation.

The mean annual rainfall precipitation in the United States increases southward from 50 mm in the north to over 4,000 mm near the Mississippi River delta, as shown in Figure 3.7a. Such spatial variability in climate is typical for very large rivers. This contrasts with the relative homogeneity of most small watersheds. The corresponding spatial distribution of a 3-h rainstorm with a period of return of 100 years is shown in Figure 3.7b. Large precipitation events in the United States are typically determined by the magnitudes

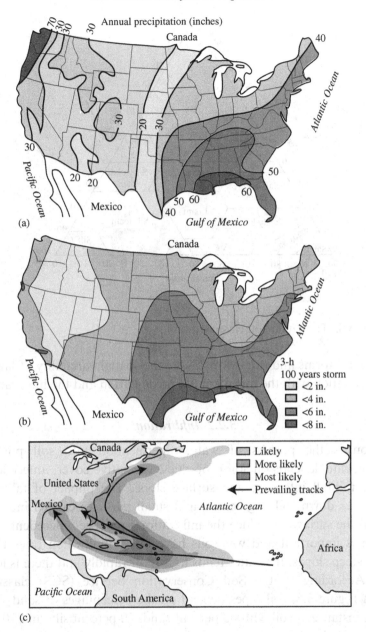

Figure 3.7. (a) Mean annual rainfall precipitation in the United States [1 in. = 25.4 mm]; (b) prevailing hurricane tracks [after NOAA]; and (c) 3 h rainstorm with 100-year return period in the United States

and paths of hurricanes. Hurricanes form when sea surface temperatures are higher than 26°C (>79°F). Figure 3.7c illustrates the prevailing path line of the hurricanes near the Gulf of Mexico. Formed at sea, they are most devastating when they move inland, where they lose their strength but cause

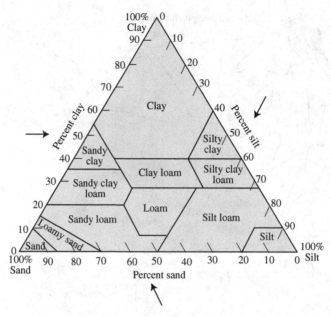

Figure 3.8. Triangular soil classification

extreme windstorms and rainstorms. Note the spatial variability of large rainstorms in relation with the mean annual precipitation and the hurricane path.

3.2.2 Infiltration

Infiltration is the process of water's permeation into soil pores. The
infiltration rate depends on soil properties, vegetative cover, antecedent soil-
moisture, rainfall intensity, and surface slope. The impact of falling raindrops breaks down soil aggregates, and small particles carried into the soil
pores seal the surface to reduce the infiltration rate. The antecedent moisture
condition is important and wet soils have lower infiltration rates than dry
soils. On steep slopes, the water tends to run off rapidly and there is less infiltration. According to the Soil Conservation Service (SCS) classification
shown in Figure 3.8, soil types depend on the percentages of sand, silt, and
clay. For instance, a soil with 60 percent sand, 30 percent silt, and 10 percent
clay is a sandy loam.

Green and Ampt (1911) developed an approximate infiltration model based
on the saturated hydraulic conductivity K from Darcy's law. As sketched in
Figure 3.9a, the thoroughly drained (or residual) water content is p_{0r} and the
initial water content is assumed uniform at p_{0i}. The effective porosity is given
by $p_{0e} = p_0 - p_{0r}$ and the effective saturation is $S_e = p_{0i}/p_{0e}$. The change in
water content across the wetting front Δp_0 is obtained from $\Delta p_0 = p_{0e} - p_{0i} =
(1 - S_e)p_{0e}$. The values for p_0, p_{0e}, h_p, and K from Rawls et al. (1983) are

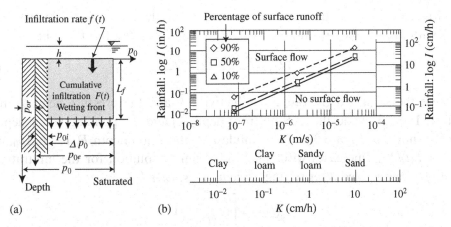

Figure 3.9. (a) Piston-type infiltration approximation and (b) percentage of surface runoff (after Sharma, 2000)

Table 3.1. *Green–Ampt infiltration parameters and cumulative infiltration depths for dry soils*

Soil texture	p_0[a]	p_{0e}[a]	h_p[a] (cm)	K[a] (cm/h)	F_{1h}[b] (cm)	F_{3h}[b] (cm)	F_{24h}[b] (cm)
Sand	0.437	0.417	4.95	11.78	16	42	295
Loamy sand	0.437	0.401	6.13	2.99	6.0	13.5	80
Sandy loam	0.453	0.412	11.0	1.09	3.9	7.8	37
Silt loam	0.501	0.486	16.7	0.65	3.7	7.0	28
Loam	0.463	0.434	8.89	0.34	1.9	3.5	14
Sandy clay loam	0.398	0.330	21.9	0.15	1.6	2.9	10
Clay loam	0.464	0.309	20.9	0.10	1.2	2.16	7.3
Silty clay loam	0.471	0.432	27.3	0.10	1.6	2.86	9.2
Sandy clay	0.430	0.321	23.9	0.06	1	1.8	5.7
Silty clay	0.479	0.423	29.2	0.05	1.15	2.05	6.3
Clay	0.475	0.385	31.6	0.03	0.9	1.55	4.7

[a]From Rawls et al. (1983), which contains more information on these parameters, including their standard deviations and values for various soil horizons.
[b]Cumulative infiltration calculated assuming dry initial condition.

given for different soil types in Table 3.1. An approximation of how much of the rainfall infiltrates and how much flows on the ground surface is shown in Figure 3.9b. The results were obtained from two-dimensional surface runoff simulations with the model CASC2D by Sharma (2000). For instance, a rainfall intensity of 1 in./h (2.54 cm/h) on a soil with $K = 2$ μm/s will generate significant surface runoff (~90 percent surface and 10 percent subsurface flow). This graph shows that very high intensities are required to generate surface runoff on sandy soils and the opposite is true for clayey soils.

More detailed calculations are possible where the infiltration rate $f(t)$ varies with time as the piston-type wetting front propagates into the soil:

$$f(t) = \frac{K(h + L_f + h_p)}{L_f}, \qquad (3.1)$$

where the saturated hydraulic conductivity K and the pressure head h_p are from Table 3.1, and the wetting front depth L_f and the ponding water depth h are shown in Figure 3.9. The cumulative infiltration depth $F(t) = L_f \Delta p_0$, or $L_f = F(t)/\Delta p_0$, increases with time and a simple solution for the infiltration rate $f(t)$ is obtained after setting aside h ($h << L_f + h_p$)

$$f(t) = K\left[\frac{h_p \Delta p_0}{F(t)} + 1\right]. \qquad (3.2)$$

The cumulative infiltration $F(t)$ is then found by integration of Equation (3.2) as

$$F(t) = Kt + h_p \Delta p_0 \ln\left[\frac{F(t)}{h_p \Delta p_0} + 1\right]. \qquad (3.3)$$

Equations (3.2) and (3.3) are applicable when the ponded depth h is negligible (the error is less than 10 percent when the ponded depth is less than 10 mm). When this is not the case, h_p should be replaced with $h_p + h$. The practical way to solve this implicit equation is to first solve Equation (3.3) for time t at various values of $F(t)$. The infiltration rate $f(t)$ can then be calculated using Equation (3.2) with the same values of $F(t)$. Example 3.1 illustrates how to calculate infiltration using the Green–Ampt method.

Example 3.1 Calculation of the infiltration potential

Calculate infiltration for a dry soil containing 20 percent sand, 60 percent silt, and 20 percent clay.

Solution: This is a silt loam with $S_e = 0$. Table 3.1 provides the following characteristics: $p_{0e} = 0.486$, $h_p = 16.7$ cm, and $K = 0.65$ cm/h. We then calculate $\Delta p_0 = (1 - S_e)p_{0e} = (1 - 0)0.486 = 0.486$. For instance, we compute the time t at which the cumulative infiltration $F(t) = 7$ cm by rearranging Equation (3.3):

$$t = \frac{F(t) - h_p \Delta p_0 \ln\left[1 + \dfrac{F(t)}{h_p \Delta p_0}\right]}{K} = \frac{7 - 16.7(0.486)\ln\left[1 + \dfrac{7}{16.7(0.486)}\right]}{0.65} = 3 \text{ h.}$$

From Equation (3.2), the infiltration rate at $t = 3$ h is

$$f(3h) = K\left[\frac{h_p \Delta p_0}{F(t)} + 1\right] = 0.65\left[\frac{16.7(0.486)}{7} + 1\right] = 1.40 \text{ cm/h.}$$

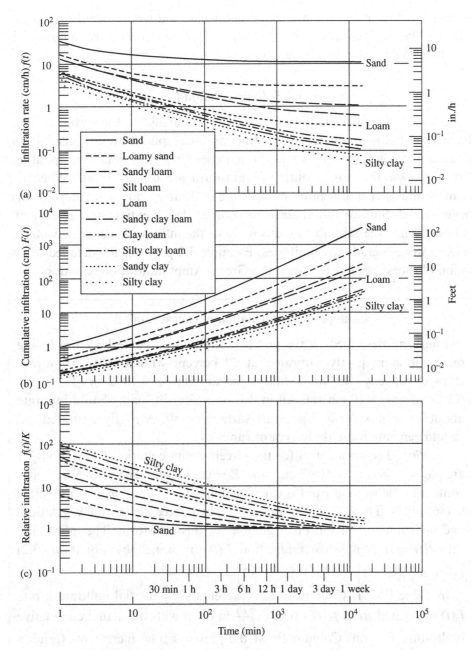

Figure 3.10. Infiltration curves for dry soils

Values of the cumulative infiltration potential for dry soils are listed in the last columns of Table 3.1 for durations of 1, 3, and 24 h, respectively. The potential infiltration curves for initially dry soils in terms of $f(t)$, $F(t)$, and $f(t)/K$ are also plotted in Figure 3.10a–c, respectively. In general, soils that contain more than 90 percent sand are highly permeable and will not generate surface runoff. When soils contain at least 20 percent clay, the cumulative infiltration is less than 2.5 cm (1 in.) in 3 h and 10 cm (4 in.) in a day.

3.2.3 Excess Rainfall

Excess rainfall represents the amount of water available for surface runoff. Excess rainfall is calculated by subtracting interception, evapotranspiration, detention storage losses, and infiltration losses from the rainfall precipitation. Evapotranspiration losses during a rainstorm are usually small. When the rainfall rate after interception exceeds the infiltration rate, detention storage begins to fill. Surface runoff starts as soon as the detention storage capacity is filled. When the rainfall rate drops below the infiltration rate, the water in surface storage gradually infiltrates. Example 3.2 provides the detailed calculation of excess rainfall by use of the Green–Ampt infiltration equation.

Example 3.2 Calculation of excess rainfall

A rainstorm described by the first two columns of Table E3.2.1 is falling on a silt loam partly saturated at 30 percent with a 0.75 cm surface storage capacity. From a first look at Figure 3.10b, or Table 3.1 (Column six), a 10-cm rainfall in 3 h on a dry silt loam should infiltrate about 7 cm of water and generate surface runoff. A partly saturated soil should generate more than 3 cm of runoff.

Solution: The parameters for the Green–Ampt equation from Table 3.1 are $p_{0e} = 0.486$, $h_p = 16.68$ cm, and $K = 0.65$ cm/h. The change in water content at the wetting front is calculated as $\Delta p_0 = (1 - S_e)p_{0e} = (1 - 0.3) \times 0.486 = 0.34$. The first mm of rain would be intercepted by the vegetation, and this amount (~ 1 mm) is negligible for this rainstorm. The infiltration rate $f(t)$ and cumulative infiltration $F(t)$ are linked by Equation (3.2)

$$f(t) = K\left[\frac{h_p \Delta p_0}{F(t)} + 1\right] = 0.65\left[\frac{16.7(0.34)}{F(t)} + 1\right] = 0.65 + \frac{3.69}{F(t)}.$$

In Table E3.2.1, Column three represents the potential infiltration rate $f_p(t)$ calculated from $f_p(t) = 0.65 + \frac{3.69}{F_a(t)}$ in cm/h with the actual cumulative infiltration F_a from Column six at the previous time increment. Column four gives the potential infiltration volume $\Delta F_p(t)$ calculated as $f_p(t)\Delta t$. The actual infiltration volume $\Delta F_a(t)$ in Column five is the smaller of the value in Column four and (the sum of the values in Column two and Column seven) for the given time increment. Note that during the first hour, all the rainfall infiltrates because the potential infiltration rate exceeds the rainfall rate and the actual infiltration volume is limited by the rainfall amount. In Column six, the cumulative infiltration $F_a(t)$ corresponds to the previous time increment plus $\Delta F_a(t)$ for the current time increment. At time $t = 75$ min, the potential infiltration rate falls below the rainfall rate and some detention storage and rainfall excess are generated. Because at this time the surface storage is empty, the rainfall first satisfies infiltration with 0.73 cm, then 0.75 cm in detention storage, thus leaving 0.07 cm in rainfall excess for surface runoff. This process is continued until the end of the storm. At time $t = 135$ min, $\Delta F_a(t)$ is 0.40 cm, while the rainfall is only 0.36, cm so that 0.04 cm of water taken from the detention storage infiltrates. During the last two time increments, the rain ceases and the ponded water depth infiltrates into the ground. Mass balance must be preserved at all times, thus the cumulative sum of excess rainfall, Column eight of Table E3.2.1, plus the values of infiltration and detention storage in Columns six and seven must equal the cumulative rainfall precipitation in Column two.

Table E3.2.1. *Excess rainfall calculations*

t^a (min)	Rainfall[a] $i\Delta t$ (cm)	Potential $f_p(t)$[b] (cm/h)	Potential $\Delta F_p(t)$[c] (cm)	Actual infiltration[d] $\Delta F_a(t)$ (cm)	Actual $F_a(t)$[e] (cm)	Detention storage[f] (cm)	Excess rainfall[g] (cm)
0	0.00				0.00	0.00	0.00
15	0.25	∞	∞	0.25	0.25	0.00	0.00
30	0.30	15.40	3.85	0.30	0.55	0.00	0.00
45	0.43	7.36	1.84	0.43	0.98	0.00	0.00
60	0.66	4.42	1.10	0.66	1.64	0.00	0.00
75	1.55	2.90	0.73	0.73	2.37	0.75	0.07
90	2.85	2.21	0.55	0.55	2.92	0.75	2.30
105	1.91	1.91	0.48	0.48	3.40	0.75	1.43
120	1.51	1.74	0.43	0.43	3.83	0.75	1.08
135	0.36	1.61	0.40	0.40	4.23	0.71	0.00
150	0.28	1.52	0.38	0.38	4.61	0.61	0.00
165	0.00	1.45	0.36	0.36	4.97	0.25	0.00
180	0.00	1.39	0.33	0.25	5.22	0.00	0.00
Totals	$\Sigma = 10.1$		=		5.22	0.00	$+ \Sigma 4.88$

[a] The hyetograph in the first two columns is based on the net rainfall obtained by subtracting the interception (less than 0.4 cm) and negligible evapotranspiration losses from the rainfall.
[b] Potential infiltration rate from the actual cumulative infiltration F_a at the previous time increment.
[c] Potential infiltration incremental volume $\Delta F_p = f_p \Delta t$ with $\Delta t = 0.25$ h.
[d] The actual infiltration volume is the smaller of Column four and (Column two + Column seven).
[e] Actual cumulative infiltration $F_a(t) = F_a(t-1) + \Delta F_a(t)$.
[f] Detention storage on the soil surface, equal to 7.5 mm in this example.
[g] Excess rainfall available for runoff.

3.3 Surface Runoff

Runoff is the surface flow occurring during and immediately after precipitation events. "Base flow" refers to seepage and groundwater flow between precipitation events. Surface runoff is added to the base flow to determine the total flow. The base flow of ephemeral streams and small watersheds can often be neglected in the computation of surface runoff from large rainstorms. Excess rainfall generates surface runoff as overland flow and channel flow. To capture the essential features of the rainfall–runoff relationship, this section aims at defining the corresponding runoff hydrograph, which requires knowledge of resistance to overland flow (Section 3.3.1) and overland flow hydraulics (Section 3.3.2). Surface runoff is then calculated for overland flow (Section 3.3.3), with a brief discussion of snowmelt runoff (Section 3.3.4).

3.3.1 Resistance to Overland Flow

Resistance to flow defines the relationship between flow depth h and depth-averaged flow velocity $\bar{u}$. Resistance to flow can be written in terms of the Darcy–Weisbach friction factor f, Manning coefficient n, or Chézy coefficient C. The corresponding definitions of C, n, and f are, respectively,

$$\bar{u} = C\, h^{1/2} S^{1/2},\tag{3.4}$$

$$\bar{u} = \frac{1}{n}\, h^{2/3} S^{1/2} \quad \text{in SI units},\tag{3.5}$$

$$\bar{u} = \sqrt{\frac{8g}{f}}\, h^{1/2} S^{1/2},\tag{3.6}\blacklozenge$$

Note that only the Darcy–Weisbach coefficient f is dimensionless; the Manning coefficient n has the dimensions $T/L^{1/3}$ and the numerator is replaced with 1.49 in the English system of units. The Chézy coefficient C has the dimensions $L^{1/2}/T$. In the case of overland flow, Table 3.2 from Woolhiser (1975) lists typical values of the Manning coefficient n, the Darcy–Weisbach friction factor f, and the Chézy coefficient C given the identity $C = \sqrt{8g/f}$. In the case of laminar flow, the laminar resistance coefficient k_0 remains constant for given surface conditions.

For smooth impervious surfaces, Figure 3.11 shows the Darcy–Weisbach friction factor f as a function of the Reynolds number $\mathrm{Re} = \bar{u}\, h/\upsilon = q/\upsilon$ given the flow depth h, the mean flow velocity $\bar{u}$ or unit discharge q, and kinematic viscosity υ. It is important to note that two flow regimes are observed: (1) laminar flow with $f = k_0/\mathrm{Re}$ with a constant roughness value for k_0 when

Table 3.2. *Resistance to overland flow (after Woolhiser, 1975)*

Surface type	Laminar flow k_0	Turbulent flow		
		Manning n	Chézy C ($ft^{1/2}/s$)	Darcy–Weisbach f
Concrete or asphalt	24–108	0.01–0.013	73–38	0.03–0.4
Bare sand	30–120	0.01–0.016	65–33	0.04–0.5
Graveled surface	90–400	0.012–0.03	38–18	–
Bare clay–loam soil	100–500	0.012–0.033	36–16	–
Sparse vegetation	1,000–4,000	0.053–0.13	11–5	0.1–1,000
Short grass prairie	3,000–10,000	0.10–0.20	6.5–3.6	0.5–13,000
Bluegrass sod	7,000–100,000	0.17–0.48	4.2–1.8	1–10,000

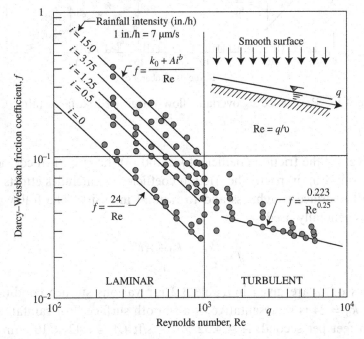

Figure 3.11. Resistance to overland flow on a smooth surface (after Li and Shen, 1973)

Re < 1,000 and (2) turbulent flow approximated by the Blasius equation $f \cong 0.223/Re^{0.25}$ when Re > 1,000. It is also interesting that raindrop impact only increases resistance to flow in the laminar-flow regime. The effects of rainfall intensity are negligible in turbulent flows.

Laminar flows with raindrop impact can be described by the Darcy–Weisbach equation, in which the friction factor f relates to (1) the Reynolds

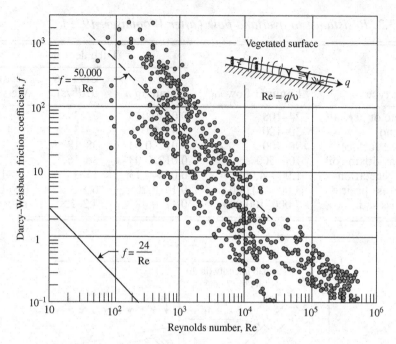

Figure 3.12. Resistance to overland flow on grassed surfaces (after Chen, 1976)

number Re, (2) the friction coefficient k_0, and (3) the empirical coefficients A and b for raindrop impact. The friction coefficient k_t includes effects that are due both to surface roughness and to rainfall intensity. The following relationship is generally used:

$$f = \frac{k_t}{\text{Re}} = \frac{k_0 + Ai^b}{\text{Re}}. \tag{3.7}$$

The values of k_0 are given in Table 3.2 for various surface conditions and the value $k_0 = 24$ is representative of a smooth surface. For rainfall intensity i given in feet per second, $A \cong 4.32 \times 10^5$ s/ft ($A \cong 1.42 \times 10^6$ s/m for i in meters per second) and $b \cong 1$.

In the case of densely vegetated surfaces, the Darcy–Weisbach friction factor f is shown in Figure 3.12. The experiments of Chen (1976) showed that resistance to flow on grassed surfaces is much higher than it is on smooth impervious surfaces. The laminar regime extends up to Re $< 10^5$, and turbulent flows are found when Re $> 10^5$. The values of k_0 for grassed surfaces in Table 3.2 indicate that k_0 can be several orders of magnitude larger than for a smooth surface. Resistance from raindrop impact becomes negligible compared to the effects of vegetation.

Table 3.3. *Resistance relationships $q = \alpha h^\beta$ for overland flow*

Flow type	Resistance coefficient	α	β	t_e
Laminar	$k_t = $ constant	$\dfrac{8gS}{k_t v}$	3	$\left(\dfrac{k_t v L}{8gSi^2}\right)^{1/3}$
Turbulent Darcy–Weisbach	f	$\sqrt{\dfrac{8gS}{f}}$	1.5	$\left(\dfrac{fL^2}{8gSi}\right)^{1/3}$
Chézy	C	$CS^{1/2}$	1.5	$\left(\dfrac{L^2}{C^2 Si}\right)^{1/3}$
Manning (SI)	n	$S^{1/2}/n$	1.67	$\left(\dfrac{nL}{S^{1/2}i^{0.667}}\right)^{0.6}$

3.3.2 Overland Flow Hydraulics

Since the flow velocity $\bar{u}$ varies with flow depth h, the unit discharge $q = \bar{u}h$ also varies with flow depth. In general, resistance to flow can be written as a depth–discharge relationship where the unit discharge q is a power function of flow depth h, or

$$q = \alpha h^\beta. \tag{3.8}\blacklozenge$$

The values for the resistance coefficient α and exponent β for overland flow on rectangular planes are given in Table 3.3 as a function of the parameters k_t, f, C, and n listed in Table 3.2. The other parameters are gravitational acceleration g, surface slope S, kinematic viscosity v, and excess rainfall intensity i.

Similarly, flow depth h, flow velocity $\bar{u}$, bed shear stress τ_0, and Froude number Fr can be determined as functions of discharge from

$$h = \left(\frac{q}{\alpha}\right)^{1/\beta}, \tag{3.9}$$

$$\bar{u} = \alpha \left(\frac{q}{\alpha}\right)^{\frac{\beta-1}{\beta}}, \tag{3.10}$$

$$\tau_0 = \gamma h S = \gamma S \left(\frac{q}{\alpha}\right)^{1/\beta}, \text{ and} \tag{3.11}$$

$$\mathrm{Fr} = \frac{\bar{u}}{\sqrt{gh}} = \alpha^{\frac{3}{2\beta}} q^{\frac{2\beta-3}{2\beta}} g^{-1/2}. \tag{3.12}$$

It is important to understand that constant values of n do not correspond to constant values of k_t, C, or f. For instance, after combining Equations (3.5–3.7), we obtain

$$n = \left(\frac{k_t}{8g}\right)^{5/9} \frac{v^{1/9}}{S^{1/18}} \mathrm{Re}^{-4/9} \text{ and} \qquad (3.13)$$

$$f = \frac{8gS^{0.1}n^{1.8}}{v^{0.2}\mathrm{Re}^{0.2}}. \qquad (3.14)$$

For laminar flow, k_t is constant and Manning n decreases with discharge, or Re. At very low discharges, Manning n can become extremely high, e.g. greater than one. For turbulent flow, a constant Manning n corresponds to $f \sim \mathrm{Re}^{-0.2}$, which is approximately equivalent to the Blasius relationship in Figure 3.11.

3.3.3 Overland Flow Hydrographs

Analytical expressions for overland-flow hydrographs are derived for a plane-rectangular surface of length L and width W at a constant slope S_0 under a constant excess-rainfall intensity i_e. It is assumed that the overland-flow plane is initially dry ($h = 0$ and $q = 0$) before the beginning of precipitation at time $t = 0$. The flow depth increases linearly with time $h = i_e t$ during the rising limb until it reaches the equilibrium discharge ($q_m = i_e L$). As sketched in Figure 3.13, at a given time t, the upstream portion of the

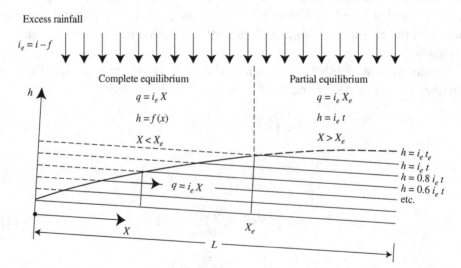

Figure 3.13. Sketch of overland-flow depth

plane $X < X_e$ reached complete equilibrium with unit discharge $q = i_e x$. The discharge for the complete equilibrium portion does not change with time and the flow is steady and nonuniform. The downstream portion of the plane $X > X_e$ reached partial equilibrium, where the flow is unsteady and uniform. In the downstream portion, the unit discharge at any point is constant in space at $q = i_e X_e$ but changes with time at a corresponding flow depth $h = i_e t$.

The continuity relationship [see Equation (2.10)] describing the conservation of fluid mass is applied to wide-rectangular flow:

$$\frac{\partial h}{\partial t} + \frac{\partial q}{\partial x} = i_e. \qquad (3.15)\blacklozenge$$

Note that this continuity relationship describes a steady flow in the complete-equilibrium domain ($X < X_e$) when $\partial h/\partial t = 0$ and a nonuniform flow for partial equilibrium results from $\partial q/\partial x = i_e$. As rainfall duration increases, the length X_e increases and the complete-equilibrium domain becomes larger with time. From combining $h = it$, $q = iX_e$, and $q = \alpha h^\beta$, we can define the position X_e as a function of time as

$$X_e = \frac{\alpha}{i_e}(i_e t)^\beta. \qquad (3.16)$$

The time to equilibrium t_e is obtained when X_e reaches the downstream end of the plane and the entire plane thus becomes under steady-flow; this state is called complete equilibrium. The equilibrium time t_e is obtained when the equilibrium discharge $q_m = i_e L = \alpha(i_e t_e)^\beta$, which is solved for t_e to give

$$t_e = \left(\frac{L}{\alpha}\right)^{1/\beta} i_e^{\frac{1-\beta}{\beta}} = \frac{1}{i_e}\left(\frac{i_e L}{\alpha}\right)^{1/\beta}. \qquad (3.17)\blacklozenge\blacklozenge$$

The time to equilibrium can be calculated as a function of the excess-rainfall intensity i_e, the length of the plane L, and α and β from Table 3.3. Values of t_e for different resistance relationships are also determined from the channel length in Figure 3.3 and the mean flow velocity estimate, a rough approximation for the time to equilibrium of small to large watersheds is shown in Figure 3.14. It essentially shows that hourly rainfall data are suitable for small and medium watersheds, while daily rainfall precipitation data will be suitable for large watersheds.

The discharge at the beginning of rainfall is $q = \alpha(i_e t)^\beta$, and the maximum unit discharge is $q = i_e L$. Surface runoff hydrographs for different

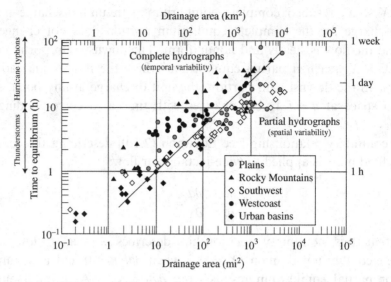

Figure 3.14. Time to equilibrium for small and large watersheds

flow-resistance equations can be written in dimensionless form. Complete-equilibrium hydrographs are those for which the rainfall duration t_r exceeds the time to equilibrium t_e; hence the dimensionless time $\lambda_r = t_r/t_e$. The surface-runoff hydrograph can be subdivided into three parts: the rising limb, equilibrium, and the falling limb. In general terms, surface runoff can be written in dimensionless form as $\psi = q/(i_e\,L)$ as a function of dimensionless time $\Theta = (t - t_r)/t_e$. Figure 3.15 illustrates the dimensionless surface-runoff hydrographs for complete ($\lambda_r \geq 1$) and partial ($\lambda_r < 1$) hydrographs. The rising limb of both cases is given by $\psi = (\Theta + \lambda_r)^{\beta}$. The maximum discharge equals $\psi = 1$ for complete hydrographs and $\psi = \lambda_r^{\beta}$ for partial hydrographs. The falling limb for both cases is given by

$$\Theta = \frac{1 - \psi}{\beta \psi^{\frac{\beta-1}{\beta}}} \tag{3.18}$$

as shown in Figure 3.15b. Example 3.3 presents detailed calculations for surface runoff hydrographs and for overland-flow hydraulics.

3.3.4 Snowmelt Runoff

Energy budgets can help determine snowmelt conditions (see, e.g. Gray and Prowse, 1992; Follum et al., 2015). The density of snow crystals typically ranges from 50 to 120 kg/m^3. The snow water equivalent (SWE) of fresh snow is approximately 100 kg/m^3, which is only 10 percent of the mass

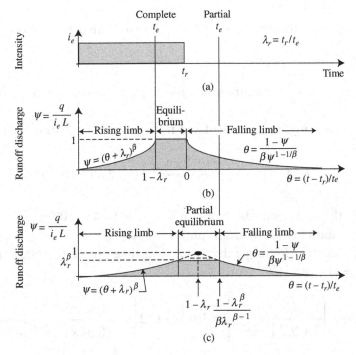

Figure 3.15. Dimensionless overland-flow hydrographs: (a) Hyetograph, (b) Complete hydrograph $(\lambda_r = t_r/t_e > 1)$, and (c) Partial hydrograph $(\lambda_r = t_r/t_e < 1)$

Example 3.3 Calculation of surface-runoff hydrographs

Calculate the surface-runoff hydrograph for a 1-h rainstorm of constant excess-rainfall intensity of 1 in./h on a rectangular plane 100 ft wide, 400 ft long, on a 20 percent slope. Also calculate the maximum flow depth, velocity, and shear stress at the corresponding Froude number. The plane is sparsely vegetated.

In SI units, $i_e = 7 \times 10^{-6}$ m/s, $L = 122$ m, and $t_r = 3,600$ s. Considering the kinematic viscosity $\upsilon = 1 \times 10^{-6}$ m²/s, the maximum Reynolds number is $\mathrm{Re} = i_e L/\upsilon = 7 \times 10^{-6} \times 122/1 \times 10^{-6} = 860$ and the flow is laminar.

For a sparsely vegetated field, a value of $k_0 \cong 2,000$ is selected from Table 3.2. From Table 3.3, $\beta = 3$ for laminar flow and

$$\alpha = \frac{8gS}{k\upsilon} = \frac{8 \times 9.81 \times 0.2}{2,000 \times 1 \times 10^{-6}} = 7,848 \text{ m}^{-1}\text{s}^{-1}.$$

The time to equilibrium from Table 3.3 is calculated from $t_e = \left(\frac{k\upsilon L}{8gS i_e^2}\right) =$

$$\left[\frac{2{,}000\times 1\times 10^{-6}\times 122}{8\times 9.81\times 0.2\times (7\times 10^{-6})^2}\right]^{1/3} = 682 \text{ s.}$$

The complete-equilibrium hydrograph $(t_r > t_e)$ is calculated in three parts:

Part 1: the rising limb is calculated from $\psi = (\Theta + \lambda_r)^\beta$ or $q = i_e L (t/t_e)^3$

$$q = i_e L \left(\frac{t}{t_e}\right)^3 = \left(\frac{7\times 10^{-6}\times 122}{682^3}\right) t^3 = 2.69 \times 10^{-12} t^3.$$

Part 2: the maximum discharge, $q = i_e L = 8.54 \times 10^{-4} \text{m}^2/\text{s}$ is reached after 682 s and remains constant until $t = 3{,}600$ s.

Part 3: the unit discharge decreases with time after $t > 3{,}600$ s, according to $\Theta = (1 - \Psi)/\beta\Psi^{(\beta-1)\beta}$ or

$$t = t_r + t_e \left[\frac{i_e L - q}{3 \times (i_e L)^{1/3} q^{2/3}}\right] = 3{,}600 + 682 \left[\frac{8.54 \times 10^{-4} - q}{3 \times (8.54\times 10^{-4})^{1/3} q^{2/3}}\right] \text{s.}$$

Part 4: the falling limb is calculated by substituting discharge values, $0 < q < 8.54 \times 10^{-4} \text{ m}^2/\text{s}$, into this equation to calculate the time t at which the discharge will occur. For instance, half the peak discharge ($q = 4.27 \times 10^{-4} \text{ m}^2/\text{s}$) will be observed at $t = 3{,}780$ s.

During complete equilibrium, the maximum flow depth, velocity, shear stress, and Froude number are obtained from Equations (3.9–3.12):

$$h = \left(\frac{q}{\alpha}\right)^{1/\beta} = \left(\frac{8.54 \times 10^{-4}}{7{,}848}\right)^{1/3} = 4.8 \text{ mm,}$$

$$\bar{u} = \alpha\left(\frac{q}{\alpha}\right)^{\frac{\beta-1}{\beta}} = 7{,}848 \left(\frac{8.54 \times 10^{-4}}{7{,}848}\right)^{2/3} = 0.18 \text{ m/s,}$$

$$\tau_0 = \gamma h S = 9{,}810 \times 4.8 \times 10^{-3} \times 0.2 = 9.4 \text{ Pa,}$$

$$\text{Fr} = \frac{\bar{u}}{\sqrt{gh}} = \frac{0.18}{\sqrt{9.81 \times 4.8 \times 10^{-3}}} = 0.83.$$

The flow is subcritical because the roughness is fairly high and the shear stress is could remove small gravel particles and break the bonds between clay particles.

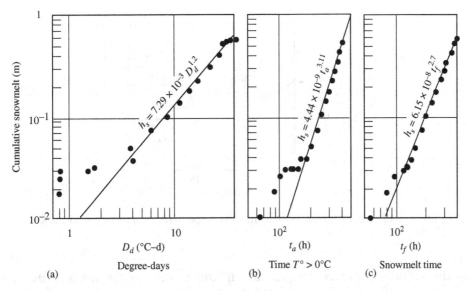

Figure 3.16. Cumulative snowmelt (after Julien and Frenette, 1986)

density of clear water. During winter, the density of a snowpack increases up to 200–400 kg/m^3. Snowmelt rates M in mm of water can be estimated from the simple degree-days method; $M \cong M_f T_{\circ C}$ where $T_{\circ C}$ is the mean daily air temperature. The melt factor M_f typically ranges from 3 to 6 mm/°C. As an example, Figure 3.16 shows hourly snowmelt-runoff discharge data on a small experimental plot in Canada from Rousseau (1979). The cumulative snowmelt h_s in meters of SWE was successfully correlated with three factors: (a) the cumulative number of degree days D_d in degrees Celsius times days; (b) the cumulative time t_a in hours when the air temperature is above 0°C; and (c) the cumulative snowmelt duration t_f in hours measured from the experimental plot.

3.4 Sediment Sources and Sediment Yield

A brief description of the sediment sources on a watershed is followed with an analysis of the Universal Soil-loss Equation (USLE) to obtain the mean annual erosion losses (Section 3.4.1). This is followed with a brief description of new methods for dynamic modeling of watersheds (Section 3.4.2) and an analysis of sediment yield and specific degradation (Section 3.4.3).

The multiple sediment sources on river basins include upland erosion originating from sheet and rills, while channel erosion is associated with gullies and river banks. In some particular cases, wind and ice can also contribute

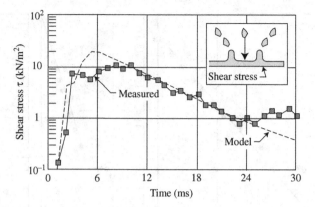

Figure 3.17. Shear stress from raindrop impact (after Hartley and Julien, 1992)

to the sediment sources. In general, upland erosion is the main source of sediment at the watershed scale. Upland erosion by water can be classified into sheet erosion, also called inter-rill erosion, and rill erosion. Sheet erosion involves: (1) the detachment of material by raindrop impact, or thawing of frozen ground and (2) its subsequent removal by sheet flow.

The surface-erosion process begins when raindrops hit the ground and detach soil particles by splash. The characteristics of raindrop splash depend on raindrop size and sheet-flow depth, and a crown-shaped crater forms a few milliseconds after impact. As shown in Figure 3.17, Hartley and Julien (1992) experimentally measured the shear stress exerted by the impact of single raindrops on shallow sheet flows. Peak shear stress measurements reached 10 Pa, which far exceeds the critical shear stress of 2.5 Pa for cohesive soils. In general, raindrop impact can be ignored when the flow depth exceeds three times the raindrop diameter.

Sheet flow naturally converges into micro channels called rills, which are small enough to be removed by normal tillage. In general, the upland erosion equations can be written as a power function of surface slope S and unit discharge q. Julien and Simons (1985) found that the exponent of slope is typically between 1.2 and 1.9 and the exponent of discharge ranges from 1.4 to 2.4. For bare sandy soils, Kilinc (1972) found $q_s \cong 25{,}500 S^{1.66} q^{2.035}$, where q_s is the unit sediment discharge from sheet and rill erosion in metric ton/ms, S is the slope, and q is the unit discharge in square meters per second. The following modification of the Kilinc–Richardson (1973) equation has a much broader range of applicability:

$$q_t \text{ (tons/ms)} = 1.7 \times 10^5 S_0^{1.66} q^{2.035} \hat{K}\hat{C}\hat{P}. \qquad (3.19)\blacklozenge$$

Where the surface slope S_0 is in m/m, the unit discharge q is in m²/s, $\hat{K}$ is the soil erodibility factor, $\hat{C}$ is the cropping management factor, and $\hat{P}$ is the conservation practice factor. The last three factors are from the USLE, discussed in the next section.

3.4.1 Universal Soil Loss Equation

The USLE from Wischmeier and Smith (1978) determines upland erosion losses. Renard et al. (1997) later proposed a revised version, RUSLE, based on the same concept. The mean annual soil loss $\hat{E}$ at a given site is the product of six major factors:

$$\hat{E} = \hat{R}\hat{K}\hat{L}\hat{S}\hat{C}\hat{P} \qquad (3.20)\blacklozenge\blacklozenge$$

where $\hat{E}$ is the soil loss per unit area, $\hat{R}$ is the rainfall-erosivity factor, $\hat{K}$ is the soil-erodibility factor, $\hat{L}$ and $\hat{S}$ are the slope-length and slope-steepness factors, $\hat{C}$ is the cropping-management, and $\hat{P}$ is the conservation-practice factor.

The rainfall erosivity factor $\hat{R}$ describes rainstorm properties. Soil-erosion losses from single rainstorms strongly correlate with the maximum 30-min rainfall intensity i_{30} in inches per hour (1 in./h = 25.4 mm/h). The annual sum of individual storms $\hat{R} = 0.01\sum i_{30}(916 + 331\log i_{30})$ gives the mean annual rainfall-erosivity factor, which ranges from 0 to 600 in the United States. The mean annual $\hat{R}$ values also correlate well with the mean annual precipitation as shown in Figure 3.18. From the analysis of US data by Cooper (2011), regions prone to frequent low-intensity rainstorms have lower $\hat{R}$ values than regions affected by frequent thunderstorms and tropical cyclones. The relationships $\hat{R} = 0.8P_{\text{in}}^{1.5} \cong 0.0064P_{\text{mm}}^{1.5}$, with P_{in} in inches or P_{mm} in mm, are close approximations to those of Bols (1978), Renard and Freimund (1994), Yu and Rosewell (1996), and Teh (2011).

The soil-erodibility factor $\hat{K}$ describes the inherent erodibility of the soils expressed in the same units as the annual erosion losses, tons per acre (1 ton/ acre = 225 metric tons per km²). The erodibility of cohesive soils varies with grain-size distribution, texture, permeability, and organic content. Based on the triangular soil classification (Figure 3.8), typical values for $\hat{K}$ are listed in Table 3.4.

The slope length-steepness factor $\hat{L}\hat{S}$ is a topographic factor relating erosion losses to a given field slope and length. The factor $\hat{L}\hat{S}$ is normalized to a standard plot 72.6 ft (22 m) long at a 9 percent slope. The length-steepness factor is obtained from the field runoff length L in m and slope s_0 in m/m as $\hat{L}\hat{S} = \sqrt{3.28L}(0.0076 + 0.53s_0 + 7.6s_0^2)$. The values of $\hat{L}\hat{S}$ are plotted in Figure 3.19 as a function of the field length and slope.

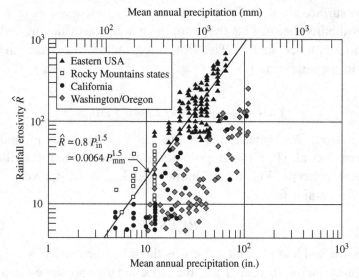

Figure 3.18. Mean annual rainfall erosivity factor $\hat{R}$ vs mean annual precipitation (after Cooper, 2011)

Table 3.4. *Soil erodibility factor $\hat{K}$ in tons/acre (after Schwab et al., 1981)*

Textural class	Organic matter content (percent)	
	0.5	2
Fine sand	0.16	0.14
Very fine sand	0.42	0.36
Loamy sand	0.12	0.10
Loamy very fine sand	0.44	0.38
Sandy loam	0.27	0.24
Very fine sandy loam	0.47	0.41
Silt loam	0.48	0.42
Clay loam	0.28	0.25
Silty clay loam	0.37	0.32
Silty clay	0.25	0.23

The cropping–management factor $\hat{C}$ accounts for soils under different cropping and management combinations such as different vegetation, crop residues, and mulching. The $\hat{C}$ factor is normalized to a standard value equal to unity for a tilled soil with continuous fallow. Typical values of $\hat{C}$ are given in Table 3.5.

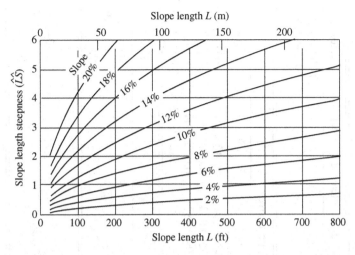

Figure 3.19. Slope-length-steepness factor of the USLE (after Wischmeier and Smith, 1978)

The conservation-practice factor $\hat{P}$ describes the soil-loss reduction from sound soil conservation measures. Factor $\hat{P}$ is normalized to a standard value equal to unity for straight-row farming up and down the slope. Typical values for contouring, strip-cropping, and terracing are given in Table 3.6.

The term "gross erosion losses" refers to the total amount of sediment eroded from the watershed surface. When using GIS at grid-cell sizes less than 100 m, the gross erosion is determined as the sum of all erosion losses on all pixels of a given watershed. Coarser grids can also be used with a correction factor introduced by Julien (1979) and developed by Julien and Frenette (1985–7) and Julien and del Tanago (1991). A calculation example is given in Example 3.4 with detailed soil erosion maps using GIS and the USLE (Molnar and Julien, 1998) or the RUSLE (Kim and Julien, 2006; Teh, 2011; Sahaar, 2013). From the six factors of the USLE, the topographic factors $\hat{S}$ and the land-use factor $\hat{C}$ typically determine most of the spatial variability in soil-erosion losses.

3.4.2 Dynamic Watershed Modeling with TREX

Dynamic watershed modeling is becoming popular for examining the runoff and sediment transport characteristics over entire watersheds. These simulations provide information that is complementary to that concerning mean annual averages derived using the USLE/RUSLE approach. The fully

Table 3.5. *Cropping-management factor Ĉ values (after Wischmeier and Smith, 1978)*

Percentage of area covered by trees and undergrowth	Percentage of area covered by duff at least 2 in. deep	Factor Ĉ
100–75	100–90	0.0001–0.001
70–45	85–75	0.002–0.004
40–20	70–40	0.003–0.009

Canopy (height)	Percentage of ground cover					
	0	20	40	60	80	95+
No canopy	0.45	0.20–0.24	0.10–0.15	0.042–0.09	0.013–0.043	0.003–0.01
Tall grasses (20 in. or 50 cm)	0.17–0.36	0.10–0.20	0.06–0.13	0.032–0.083	0.011–0.041	0.003–0.011
Brush or bushes (6 1/2 ft or 2 m)	0.28–0.40	0.14–0.22	0.08–0.14	0.036–0.087	0.012–0.042	0.003–0.011
Trees (13 ft or 4 m)	0.36–0.42	0.17–0.23	0.09–0.14	0.039–0.089	0.012–0.042	0.003–0.011

Type of mulch	Mulch rate (tons/acre)	Factor Ĉ
Straw	1.0–2.0	0.06–0.20
Crushed stone 0.25 to 1.5 in.	7–240	0.02–0.08
Wood chips	12–25	0.02–0.05

	Factor Ĉ
Tilled continuous fallow	1.0
Rough fallow	0.30–0.80
Residues left on the field	0.10–0.50
No tillage	0.05–0.25

Example 3.4 Upland erosion losses on a small watershed

Consider a watershed area covering 600 acres (247 acres = 1 km^2) as sketched in Figure E3.4.1. Compute the annual soil-erosion loss given a mean annual rainfall of 37 in. The rainfall erosivity has been determined as $\hat{R} = 185$, and all soils are a silt loam with a soil-erodibility factor $\hat{K} \simeq 0.46$ from Table 3.4.

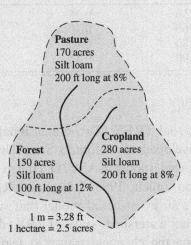

Pasture
170 acres
Silt loam
200 ft long at 8%

Cropland
280 acres
Silt loam
200 ft long at 8%

Forest
150 acres
Silt loam
100 ft long at 12%

1 m = 3.28 ft
1 hectare = 2.5 acres

Figure E3.4.1. Upland erosion example

Step 1: the cropland area covers 280 acres of continuous corn cultivated up and down a 200-ft-long hill at 8 percent slope with residues left on the field. The topography factor is $\hat{LS} = 1.4$ from Figure 3.19, the cropping-management factor $\hat{C} \cong 0.4$ for continuous cultivation, and $\hat{P} = 1$. The annual soil loss is $\hat{E} \approx 185 \times 0.46 \times 1.4 \times 0.4 = 47$ tons/acre.

Step 2: the Pasture area covers 170 acres, half of which is covered with short brush (0.5-m fall height), and half by grasses, and the 8 percent slopes are 200 ft long. The parameters are $\hat{R} = 185$, $\hat{K} = 0.46$, $\hat{LS} = 1.4$, $\hat{C} = 0.012$ from Table 3.5, and $\hat{P} = 1$. The annual soil loss is 1.4 tons per acre.

Step 3: finally, 150 acres are densely forested. Slopes are 100 ft long with a 12 percent incline. The parameters are $\hat{R} = 185$, $\hat{K} = 0.46$, $\hat{LS} = 1.8$ from Figure 3.19, and $\hat{C} = 0.009$ from Table 3.5. The annual soil loss is 1.37 tons/acre.

Step 4: The total soil-erosion loss on this watershed equals 13,600 tons/year (47 tons/acre × 280 acres + 1.4 tons/acre × 170 acres + 1.37 tons/acre × 150 acres).

Table 3.6. *Conservation-practice factor* $\hat{P}$ *for contouring, strip-cropping, and terracing (after Wischmeier and Smith, 1978)*

| | $\hat{P}$ value | | | |
| | | | Terracing | |
Land slope (percent)	Farming on contour	Contour strip-crop	farmland	off-field
2–7	0.50	0.25	0.50	0.10
8–12	0.60	0.30	0.60	0.12
13–18	0.80	0.40	0.80	0.16
19–24	0.90	0.45	0.90	0.18

distributed two-dimensional TREX (Two-dimensional Runoff, Erosion, and Export) model was developed at Colorado State University, based on the physically based watershed model CASC2D (Julien et al., 1995; Johnson et al., 2000; Ogden and Julien, 1993; Julien and Rojas, 2002). The basic CASC2D framework is an event-based model that simulates overland flow and soil erosion at the watershed scale. TREX development expanded CASC2D to incorporate the chemical transport features from the WASP/IPX series of stream water-quality models (Ambrose et al., 1993; Velleux et al., 2001). A conceptual diagram of chemical model processes is presented in Figure 3.20. More detailed descriptions of all processes in TREX are presented by Velleux et al. (2006, 2008).

TREX calculates the excess rainfall precipitation from the Green-Ampt infiltration equation, uses the Manning resistance equation, and solves the diffusive wave approximation of the Saint-Venant equations for surface runoff. Input data are prepared using ArcGIS 9.3 and the surface topography of a watershed is discretized at a 30×30 m grid scale. Coarser grid sizes (not exceeding 100×100 m per Rojas et al., 2008) can also be used for large watersheds. Calibrated model parameters include the saturated hydraulic conductivity, K_h, and Manning n, which are typically the most sensitive parameters during calibration. The parameter evaluation is straightforward, given the infiltration parameters from Table 3.1 and surface roughness conditions from Table 3.2.

The calibration and validation procedure focuses on the simulated peak discharge, the total runoff volume, and time to peak. To evaluate the model performance, the Nash-Sutcliffe Efficiency Coefficient (NSEC), the Percent

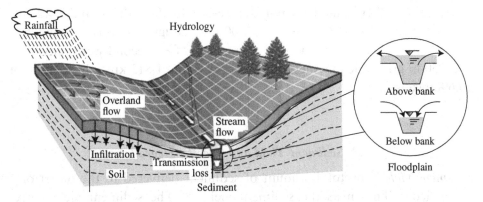

Figure 3.20. Schematic of the TREX model (after Velleux et al., 2008)

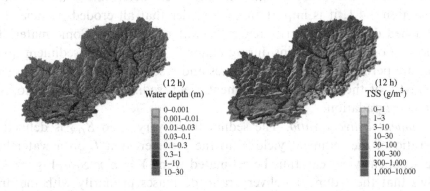

Figure 3.21. TREX simulation of flow depth and TSS at Naesung stream (after Ji et al., 2014)

BIAS (PBIAS), and the Relative Percent Difference (RPD) parameters can be used. Values of NSEC between 0.0 and 1.0 are acceptable, with 1.0 being optimal. PBIAS measures the tendency to underestimate or overestimate and the optimal value of PBIAS is 0.0. RPD is also useful for evaluating the total volume, peak discharge, and time to peak.

Several applications can be found on small mountain watersheds (Velleux, 2005; Halgren, 2012; Steininger, 2014), on a large 12,000 km^2 watershed (England, 2006), as well as applications in South Korea (Kim, 2012; Ji et al., 2014) and under tropical monsoons (Abdullah, 2013). Dynamic video simulations of storm events are usual for this type of model. Figure 3.21

illustrates the TREX model simulation results for flow depth and total suspended solids (TSS) at 12 h for a 300-mm design rainstorm lasting 6 h
(i.e. 50 mm/h). The calculations at a time step of ~1 second were performed
at a 150-m grid scale on Naesung stream, covering 1,815 km^2, thus a total of
80,690 pixels.

3.4.3 Sediment Yield and Specific Degradation

Sediment yield. The total amount of sediment delivered to the outlet of a
watershed is known as the sediment yield Y. The sediment yield corresponds to the mean annual sediment load of a river, typically determined
from the flow-duration/sediment-rating curve method described in Julien
(2010). Once the gross erosion E has been determined, as shown in
Subsection 3.4.1, it is important to consider that all eroded particles in a
watershed do not necessarily reach the watershed outlet. Some material is
deposited on the floodplains during major floods and large sediment quantities are permanently trapped in lakes and reservoirs. Figure 3.22 shows a
schematic of the concepts of sediment yield, sediment–delivery ratio, and
specific degradation.

Sediment–delivery ratio. The sediment–delivery ratio S_{DR} is defined as
the ratio of the sediment yield Y to the gross erosion E on a watershed.
The sediment yield can thus be estimated from $Y = E \times S_{DR}$. Figure 3.23
shows that the sediment–delivery ratio decreases primarily with the drainage area.

Specific degradation. The specific degradation S_D of a watershed is
obtained from the sediment yield divided by the drainage area. Typical
values from many reservoirs in the United States are shown in
Figure 3.24.

A study of water and sediment discharge to the oceans from the major rivers around the world shows a total sediment yield of 13×10^9 metric tons
per year. Examination of some of the most important rivers listed in
Table 3.7 indicates that approximately half of the global sediment discharge
to the oceans originates from rivers in Southeast Asia. In comparison, the
total freshwater flow to the oceans from all rivers of the world combined
adds up to 1.2×10^6 m^3/s. The average sediment concentration of river flows
to oceans is ~360 mg/l.

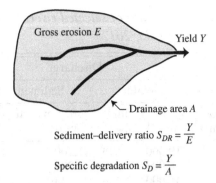

Figure 3.22. Schematic of gross erosion and sediment yield

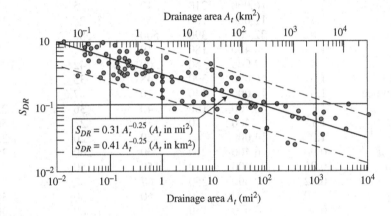

Figure 3.23. Sediment–delivery ratio

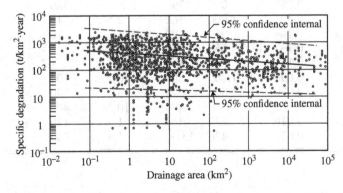

Figure 3.24. Specific degradation versus drainage area for US reservoirs (after Kane and Julien, 2007)

Table 3.7. *Water and sediment loads of selected rivers (revised after Jansen et al., 1979; Milliman and Syvitski, 2017)*

		Discharge				
		Water		Sediment S_D		
River	Drainage (10^6 km^2)	m^3 s^{-1}	mm year^{-1}	10^6 tons year^{-1}	tons/ km^2 year	Concentration (mg/l)
Amazon	6.1	19,000	100	1,200	190	2,000
Mississippi	3.9	18,000	150	200	50	350
Congo	3.7	44,000	370	43	11	30
Parana	3.0	19,000	200	90	30	150
Ob	3.0	12,000	130	16	5.3	40
Nile	2.9	3,000	30	80	27	850
Yenisei	2.6	17,000	210	11	4.2	20
Lena	2.4	16,000	210	12	5	25
Amur	2.1	11,000	160	52	25	150
Yangtze	1.8	22,000	390	500	280	750
Volga	1.5	8,400	180	19	13	70
Missouri	1.4	2,000	50	200	142	3,200
Zambezi	1.3	16,000	390	48	35	100
St-Lawrence	1.3	14,000	340	10	7.6	20
Niger	1.1	5,700	160	40	36	220
Murray	1.1	400	10	30	27	2,500
Ganges	1.0	14,000	440	520	520	1,200
Indus	0.96	6,400	210	250	260	1,200
Orinoco	0.95	25,000	830	150	150	190
Danube	0.82	6,400	250	67	82	330
Mekong	0.80	15,000	590	160	200	340
Yellow	0.77	4,000	160	1,400	1,800	11,000
Brahmaputra	0.64	19,000	940	540	850	900
Dnieper	0.46	1,600	110	2.1	4.6	42
Irrawaddy	0.41	13,000	1,000	260	630	650
Rhine	0.36	2,200	190	0.72	2	10
Magdalena	0.28	7,000	790	220	780	1,000
Vistula	0.19	1,000	160	2.5	13	80
Chao Phraya	0.16	960	190	11	70	360
Elbe	0.13	650	160	0.84	6	40
Oder	0.11	530	150	0.13	1.2	10
Rhone	0.096	1,700	560	10	104	190
Po	0.070	1,500	670	13	190	280
Nakdong	0.024	380	490	10	400	840
Ishikari	0.016	230	450	1.7	110	240
Tiber	0.013	420	1,000	1.8	140	140

♦*Exercise 3.1*

Combine the equations $h = it_e$, $q = iL$, and $q = \alpha h^\beta$ to derive the expressions for the time to equilibrium in Table 3.3.

♦♦*Exercise 3.2*

Combine the relationships between the Darcy–Weisbach friction factor f, Manning n, and the Reynolds number Re, to derive Equation (3.14).

♦♦*Exercise 3.3*

Demonstrate that in Table 3.3, $\alpha = 1.49 \, S^{1/2}/n$ and $\beta = 5/3$ for Manning's equation in English units, and show that $q = \alpha h^\beta$.

♦*Problem 3.1*

Repeat the infiltration calculations of Example 3.2 for a silty clay soil. Plot the hyetograph, infiltration rate, detention storage, and excess rainfall as functions of time from Table E3.2.1. Compare the results for silty loam with those for silty clay.

♦*Problem 3.2*

Estimate the time to equilibrium for a 3 h rainstorm with a period of return of 100 years in Tennessee on a rectangular plot of farmland covering 150 acres. Estimate the infiltration on bare soil for a silty clay loam. Can you estimate the maximum discharge for this storm?
[*Answer*: The total rainfall of 5 in. in 3 h corresponds to $\cong 42$ mm/h. The drainage area corresponds to 60 ha $= 0.6$ km^2 and the time to equilibrium is much shorter than 1 h. A silty clay loam is relatively impervious and the infiltration rate will decrease to 3–4 mm/h after 100 min. The $Q_{max} \cong iA_w = 39$ mm/h $\times 0.6$ km$^2 = 6.5$ m^3/s $\cong 230$ ft^3/s.]

♦♦*Problem 3.3*

Consider a 1 h storm with intensity of 1 in./h on a 100 m plot at a 5 percent slope. Determine the maximum flow depth, flow velocity, Froude number, and shear stress if the surface is rough and impervious. Also calculate and plot the surface-runoff hydrograph. [*Hint*: Determine the appropriate resistance equation.]

Problem 3.4

From the data in Figure 3.16, find the equivalent rainfall intensity in mm/h from the average snowmelt rate. What is the average daily snowmelt rate in mm/°C?

◆Problem 3.5

From the data in Table 3.7, plot on a log–log scale the sediment discharge as a function of water discharge, and concentration as functions of drainage area. Also plot the specific degradation for these large rivers as an extension of Figure 3.24.

◆◆Problem 3.6

Duksan Creek covers 33.1 km^2 and the channel length is 12 km. This steep mountain watershed has an average slope of 25 percent and Manning n is approximately 0.03. Estimate the time to equilibrium for a 3 h rainfall intensity of 60 mm/h. Determine the maximum discharge for two conditions: (1) the surface is impervious and (2) the soils are primarily a sandy loam.

[*Answer*: (1) $T_e \cong 1.2$ h, $Q_{max} \cong 550$ m^3/s and (2) the infiltration rate of the third hour ~20 mm/h and with infiltration, $Q_{max} \cong 365$ m^3/s for a quick and rough estimate. A detailed simulation with the TREX model by Kim (2012) gave $Q_{max} \cong 450$ m^3/s.]

4

River Basin Dynamics

This chapter examines the dynamics of river basins in terms of time variability in rainfall, runoff, and sediment transport. The time variability in rainfall precipitation is discussed in Section 4.1, followed by analyses of duration curves (Section 4.2), of flood frequency (Section 4.3), and of extreme floods (Section 4.4).

4.1 Rainfall Precipitation

This section discusses rainfall precipitation with weather radars. Point rainfall precipitation is covered in Section 4.1.1 and daily and annual rainfall precipitation in Section 4.1.2. Polarimetric weather radar techniques offer an opportunity to track the development and motion of rainstorms with unprecedented resolution. The National Weather Service WSR-88D weather radar data system has been in service for several decades. High-resolution data are also available from the Colorado State University radar facility (CSU-CHILL), near Greeley, Colorado. The CSU-CHILL radar is a linearly polarized coherent radar that operates at 2.75 GHz, which corresponds to a wavelength of 10.7 cm in the S band. Attenuation does not affect S-band weather radars, even in the heaviest rainfall, making them ideal for weather observation. The example in Figure 4.1 shows rainfall rates of up to 60 mm/h for storms at a 1×1 km grid size within the 80×80 km data domain at 12:48 MDT and 13:56 MDT.

This example illustrates the rapid development and motion of summer thunderstorms. The size of individual storm cells was approximately 4–10 km, with an average storm cell size of approximately 8 km. The rainfall data from successive radar scans were processed, and the correlation length 2.3 km in this case means that the rainfall precipitation from rain gauges 2.3 km apart

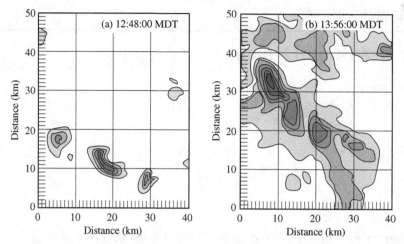

CSU-CHILL Rainfall (mm/h); June 3, 1991; Elevation angle 1.6°; contour interval 10 mm/h

Figure 4.1. Colorado rainfall rates from the CSU-CHILL radar (after Ogden and Julien, 1994)

would be essentially uncorrelated. This means that a very large number of rain gauges would be required to cover large watersheds when each rain gauge appropriately describes thunderstorms covering $\sim$5–10 km^2. Many thunderstorm cells travel at average velocities exceeding 100 km/h, which means that rainstorm events may reach rainfall intensities up to 60 mm/h for durations smaller than 6 min. While radars provide tremendous spatial coverage of rainstorms, ground calibration remains an active research area for improved accuracy of radar measurements.

May and Julien (1998) examined rainfall events measured by a network of 46 rain gauges with an average spacing of 10.6 km in the Denver area. The example shown in Figure 4.2 has a small mesoscale rainstorm with a diameter of 20–30 km. The spatial cross correlation between 2-min rainfall data in a frame of reference moving with the rainstorm is shown in the same figure. The correlation coefficient becomes close to zero at a rain gauge separation distance of $\sim$16 km. This means that, for this particular storm, 2-min rainfall measurements with an average spacing of $\sim$15 km would be essentially uncorrelated. This shows the importance of using a system of coordinates that moves with the center of mass of the storm cell (also called Lagrangian) in order to have physically representative correlation coefficients. Radar measurements have been linked to spatially distributed runoff models by Ogden and Julien (1994) and Jorgeson and Julien (2005).

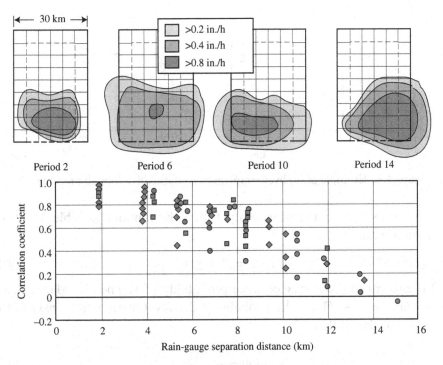

Figure 4.2. Rainfall correlation coefficients versus rain gauge distance (after May and Julien, 1998)

4.1.1 Point Rainfall Precipitation

Point rainfall precipitation can be measured at a fixed location on the ground with the use of rain gauges which collect and measure the amount of rainfall precipitation. Tipping-bucket rain gauges are commonly used to obtain rainfall precipitation during short time intervals, e.g. $\Delta t = 2$ min. The amount of rain in mm within Δt defines the rainfall intensity. The variation of rainfall intensity i with time defines the hyetograph and the cumulative rainfall intensity with time defines the total rainfall precipitation. The example in Figure 4.3 shows three discrete rainfall events, each having a finite duration and constant intensity. The two principal variables of rainfall precipitation are the storm duration t_r and the storm intensity i. The rainstorm depth h_r is obtained from $h_r = it_r$.

Probability density functions. Let X denotes a random variable and x denote a possible value of X. The cumulative distribution function (CDF) of

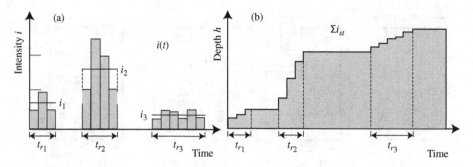

Figure 4.3. Example of (a) hyetograph and (b) cumulative rainfall precipitation

X, denoted as $F(x)$, is the probability P that the random variable X will be less than or equal to x:

$$F(x) = P(X \leq x). \tag{4.1}$$

$F(x)$ is also called the nonexceedance probability of x. The exceedance probability $E(x) = 1 - F(x)$. The probability density function (pdf) $p(x)$ is the derivative of the CDF, or

$$p(x) = \frac{\mathrm{d}F(x)}{\mathrm{d}x}. \tag{4.2}$$

The following discussion will present practical distribution functions for the analysis of rainfall precipitation. The exponential distribution is often used to characterize rainfall intensity and duration. The gamma function is well suited to defining the distribution of daily rainfall depth, while the measure of normal distribution is usually appropriate for defining the distribution of annual rainfall precipitation.

Rainfall duration. The average storm duration is simply $\bar{t}_r = \frac{1}{N}\sum_1^N t_r$, where N is the number of storms. The dimensionless storm duration $t_r^* = t_r/\bar{t}_r$ denotes rainfall duration in comparison with the mean value. The probability of very long storm durations is much smaller than the probability of a short duration. The probability density function $p(\ )$ decreases exponentially with rainfall duration t_r^* as

$$p(t_r^*) = e^{-t_r^*}. \tag{4.3}$$

The corresponding cumulative distribution function $F(t_r^*)$, or nonexceedance probability, is obtained from

$$F(t_r^*) = \int_0^{t_r^*} p(t_r^*)\mathrm{d}t_r^* = \int_0^{t_r^*} e^{-t_r^*}\mathrm{d}t_r^* = 1 - e^{-t_r^*}. \tag{4.4}$$

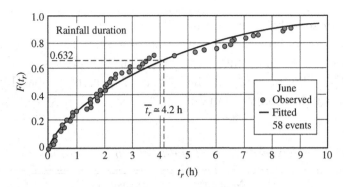

Figure 4.4. Cumulative distribution function of rainfall duration

This distribution is quite interesting because the probability density function $p(\ )$ and the exceedance probability function $E(\ ) = 1 - F(\ )$ are identical

$$E(t_r^*) = p(t_r^*) = e^{-t_r^*}. \tag{4.5}$$

Figure 4.4 provides an example of exponential fit to rainfall duration measurements. Example 4.1 also shows how to compare the rainfall distribution to an exponential distribution.

Example 4.1 Exponential distribution of rainfall duration

A total of 10 rainstorms were recorded during the month of July. The storm durations were 45, 150, 10, 90, 20, 60, 40, 25, 5, and 30 min. Compare the distribution of these events with the exponential distribution, and determine the probability of getting a rainstorm longer than 3 h during that month. In Table E4.1.1, the measurements are listed in Column one and ranked in increasing order in Column two. Using the Weibull plotting position, their ranks in Column three are each divided by $N + 1 = 11$ in Column four to determine the measured nonexceedance probability F_m. The measured exceedance probability $E_m = 1 - F_m$ is given in Column five. For comparisons with the exponential distribution, the rainfall duration t_r is normalized as $t_r^* = t_r/\bar{t}_r$ in Column six. The calculated nonexceedance probability $F_c = 1 - e^{-t_r^*}$ is listed in Column seven. Figure E4.1.1 shows the cumulative distribution functions from the comparison between measured and calculated nonexceedance probabilities. The dimensionless duration of a 3 h = 180 min rainstorm is $t_r^* = 180/47.5 = 3.8$ and the exceedance probability is $E(t_r^*) = e^{-t_r^*} = e^{-3.8} = 0.022$. This means that there is a 2.2 percent probability that a July storm will last longer than 3 h.

Table E4.1.1. *Nonexceedance probability of rainfall duration*

Duration t_r (min)	Ranked duration	Rank m	Measured nonexceedance probability $F_m = m/(N+1)$	Measured exceedance probability $E_m = 1 - F_m$	$t_r^* = \frac{t_r}{\bar{t}_r}$	Calculated nonexceedance probability $F_c = 1 - e^{-t_r^*}$
45	5	1	0.09	0.91	0.10	0.10
150	10	2	0.18	0.82	0.21	0.19
10	20	3	0.27	0.73	0.42	0.34
90	25	4	0.36	0.64	0.52	0.41
20	30	5	0.45	0.55	0.63	0.47
60	40	6	0.54	0.46	0.84	0.57
40	45	7	0.64	0.36	0.94	0.61
25	60	8	0.73	0.27	1.26	0.72
5	90	9	0.82	0.18	1.89	0.85
30	150	10	0.91	0.09	3.15	0.96
sum	475					

$N = 10$, and the average duration $\bar{t}_r = 475 \text{ min}/10 = 47.5$ min.

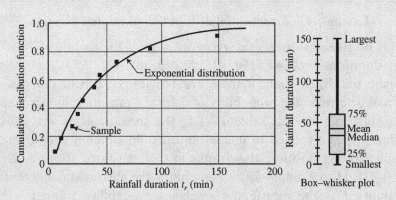

Figure E4.1.1. CDF and box–whisker plot of rainfall duration

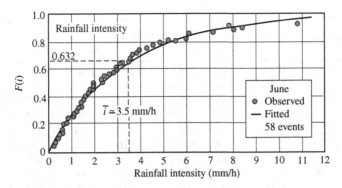

Figure 4.5. Example of exponential CDF of rainfall intensity

Rainfall intensity. Similarly, rainstorm intensity i can be normalized after considering the average intensity $\bar{i} = \frac{1}{N}\sum_1^N i$ over N storms. The dimensionless intensity $i_* = i/\bar{i}$ is also exponentially distributed with a probability density function $p(\ )$ equal to the exceedance probability $E(\ )$ as

$$E(i_*) = p(i_*) = e^{-i_*},\qquad (4.6)$$

$$F(i) = F(i_*) = 1 - E(i_*).\qquad (4.7)$$

For instance, the observed and the fitted exponential distribution functions for rainstorm intensity, $F(i) = F(i_*)$, are shown in Figure 4.5.

The examples shown in Figures 4.4 and 4.5 (from Julien and Frenette, 1985) included 6 years of daily rainfall data. Monthly periods proved sufficiently long to assume homogeneity and to calculate average rainfall duration and intensity. Statistical analyses also demonstrated the mutual independence of duration and intensity of rainstorms.

4.1.2 Daily and Annual Rainfall Precipitation

Daily rainfall depth. The amount of daily rainfall precipitation x is often close to an exponential distribution but may be better represented by a gamma distribution. The probability density function $f(x)$ of a two-parameter gamma function is given by

$$f(x) = \frac{1}{|\alpha|\Gamma(\beta)} \left(\frac{x}{\alpha}\right)^{\beta-1} e^{-\frac{x}{\alpha}}.\qquad (4.8)$$

From a total of N events, the mean value $\bar{x} = \sum_1^N x/N$ and the variance $\sigma^2 = \sum_1^N (x_i - \bar{x})^2/(N-1)$ are used to calculate the parameters $\beta = (\bar{x}/\sigma)^2$

Table 4.1. *Gamma, factorial, normal, hyperbolic, and error functions*

z	$\overline{x^2}/\overline{x}^2$	$\Gamma(z)$	$!(z)$	$F_n(z)$	sinh	cosh	tanh	erf	erfc
0.1	184,756	9.51	0.951	0.540	0.100	1.005	0.099	0.113	0.887
0.15	2,213	6.22	0.933	0.560	0.150	1.011	0.148	0.168	0.832
0.2	252	4.59	0.918	0.580	0.201	1.020	0.197	0.222	0.778
0.25	70	3.62	0.906	0.599	0.252	1.031	0.244	0.276	0.724
0.3	30.2	2.99	0.897	0.618	0.304	1.045	0.291	0.329	0.671
0.4	10.86	2.21	0.887	0.655	0.410	1.081	0.379	0.428	0.572
0.5	6	1.77	0.886	0.692	0.521	1.127	0.462	0.521	0.479
0.6	4.090	1.49	0.893	0.726	0.636	1.185	0.537	0.604	0.396
0.7	3.138	1.30	0.908	0.758	0.758	1.255	0.604	0.678	0.322
0.8	2.588	1.16	0.931	0.790	0.888	1.337	0.664	0.742	0.258
0.9	2.238	1.07	0.962	0.816	1.026	1.433	0.716	0.797	0.203
1	2	1.00	1.00	0.841	1.175	1.543	0.761	0.843	0.157
1.2	1.700	0.918	1.10	0.885	1.509	1.810	0.833	0.910	0.090
1.4	1.523	0.887	1.24	0.919	1.904	2.150	0.885	0.952	0.048
1.6	1.409	0.893	1.43	0.945	2.375	2.577	0.921	0.976	0.024
1.8	1.330	0.931	1.68	0.964	2.942	3.107	0.946	0.989	0.011
2	1.273	1.00	2.0	0.977	3.626	3.762	0.964	0.995	0.005
2.5	1.183	1.33	3.32	0.994	6.505	6.132	0.986	0.999	0.001
3	1.132	2.0	6.0	0.999	10.01	10.06	0.995	0.999	0.000
3.5	1.100	3.32	11.6	0.999	16.54	16.57	0.998	0.999	0.000
4	1.078	6.0	24.0	0.999	27.28	27.30	0.999	0.999	0.000
4.5	1.063	11.6	52.3	0.999	45.00	45.01	1.00	1.00	0.000

and $\alpha = \overline{x}/\beta$. It is remembered that the factorial function $!z$ applies to integers such that $!z = z(z-1)(z-2)\cdots1$, with for instance $!4 = 4 \times 3 \times 2 \times 1 = 24$. The new function in Equation (4.8) is the gamma function $\Gamma(z)$, which is a generalization of the factorial function to nonintegers. The relationship between the factorial and gamma functions is $!(z) = \Gamma(z + 1) = z\Gamma(z)$ and practical values are listed in Table 4.1. For instance, $!(3) = \Gamma(4) = 6$, and $\Gamma(2.5) = 1.5 \times \Gamma(1.5) = 1.5 \times 0.886 = 1.33$. The reader finds $\Gamma(3.75) = 4.40$ from Table 4.1. For large values of the argument z, the Stirling asymptotic series can be used:

$$\Gamma(z+1) \simeq \sqrt{2\pi z}z^z e^{-z}\left\{1 + \frac{1}{12z} + \frac{1}{288z^2} - \frac{139}{51,840z^3}\cdots\right\}. \qquad (4.9)$$

Example 4.2 shows how to fit a gamma distribution to a rainfall depth sample.

Annual rainfall precipitation. The annual rainfall precipitation consists of the sum of numerous individual rainfall precipitation events. The distribution of annual rainfall precipitation tends to be normally distributed. From a set

Example 4.2 Gamma distribution of daily rainfall depth

As an example, an application of this method to the daily rainfall precipitation for tropical rainstorms during the monsoons of Malaysia is shown in Figure E4.2.1. For the daily precipitation at Subang Airport from 1960 to 2011, Muhammad (2013) found that the two-parameter gamma function is most suitable for representing the distribution of daily rainfall depth. The mean daily rainfall precipitation is $\bar{x} = 12.77$ mm and the standard deviation $\sigma = 17.24$ mm. These values lead to the estimation of shape parameters $\beta = (\bar{x}/\sigma)^2 = 0.55$ and $\alpha = \bar{x}/\beta = 23.2$. The gamma distribution of the rainfall amount x from a single day is

$$f(x) = \frac{1}{|23.2|\Gamma(0.55)} \left(\frac{x}{23.2}\right)^{0.55-1} e^{-\frac{x}{23.2}}.$$

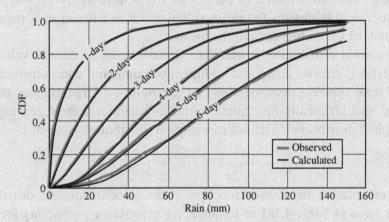

Figure E4.2.1. Example of rainfall depth distribution from one to six consecutive rainy days (after Muhammad et al., 2015)

One of the properties of the gamma distribution is that the distribution of a sum of gamma distributions is also a gamma distribution. The parameter $\beta \simeq 0.6t$ increases linearly with time. It then becomes possible to describe the distribution of multiday rainfall precipitation events from the bivariate probability distribution of rainfall amount and duration

$$f(x, t) \cong \frac{1}{|24|\Gamma(0.6t)} \left(\frac{x}{24}\right)^{0.6t-1} e^{-\frac{x}{24}}.$$

where x is the total rainfall amount in mm for t consecutive rainy days. For a single rainy day, 60 percent of the rainfall amounts are less than 10 mm. However, there is a 25 percent chance that the total rainfall amount from five consecutive rainy days will exceed 100 mm.

of observations of parameter x, estimators for the mean $\bar{x}$, variance σ^2, and skewness coefficient γ are obtained from

$$\bar{x} = \sum_1^n \frac{x_i}{n}, \tag{4.10a}$$

$$\sigma^2 = \frac{\sum_1^n (x_i - \bar{x})^2}{n-1}, \tag{4.10b}$$

$$\gamma = \frac{n \sum_1^n (x_i - \bar{x})^3}{(n-1)(n-2)\sigma^3}. \tag{4.10c}$$

The standard deviation σ is simply the square root of the variance and the coefficient of variation is defined as $C_v = \sigma/\bar{x}$. In order to compare how the value of a parameter compares to the mean and standard deviation, the following normalized value z of parameter x is defined as $z_i = (x_i - \bar{x})/\sigma$. For instance, $z_i = 0$ represents the mean value, and $z_i = 2$ describes a parameter two standard values higher than the mean.

The normal distribution is useful to describe average annual values like precipitation, stream flow, and sediment/contaminant concentration. The central limit theorem demonstrates that the sum of a large number of independent and identically distributed variables becomes normal regardless of the original distribution. The pdf of a normal distribution is given by

$$f_n(x) = \frac{1}{\sigma\sqrt{2\pi}} e^{-\frac{1}{2}\frac{(x-\bar{x})^2}{\sigma^2}} = \frac{1}{\sigma\sqrt{2\pi}} e^{-\frac{z^2}{2}}. \tag{4.11}$$

The cumulative distribution function of the standard normal distribution $F_n(z)$ is given in Table 4.1. For instance, the exceedance probability for $z = 2$ is $1 - F_n(2) = 0.023$ is less than 2.5 percent. Some practical approximations include

$$F_n(z) = 1 - \frac{1}{2} e^{-\left[\frac{562+351z+83z^2}{165+(703/z)}\right]}, \tag{4.12a}$$

and

$$z = \frac{F_n^{0.135} - (1-F_n)^{0.135}}{0.1975}. \tag{4.12b}$$

For instance, the value of z exceeded 10 percent of the time would be $z = 1.28$, obtained from Equation (4.12b) with $F_n = 0.9$.

Lognormal distributions of variable y can be handled with the same normal distribution after considering $x = \log y$. The results can be back-transformed with $y = 10^x$.

Finally, when considering the sum of n independent values from the same population of standard deviation σ, the variance of the sum is equal to the sum of variances. Thus, the standard deviation σ_s of the sum of n values is equal to $\sigma_s = \sigma/\sqrt{n}$. Case Study 4.1 shows an example of a devastating flood on the Big Thompson River in 1976.

Case Study 4.1 The Big Thompson River flood, Colorado, United States

The two reports of Grozier et al. (1976) and McCain et al. (1979) document the Big Thompson River flood of July 31–August 1, 1976. On July 31st, several violent thunderstorms (Figure CS4.1.1) released large volumes of rain from Estes Park to the Wyoming border. As much as 12 in. (305 mm) of rain fell on the Big Thompson River basin, causing a devastating flood between Estes Park and Loveland, Colorado. Larimer County officials reported 139 lives lost and property damage at a cost of $16.5 million.

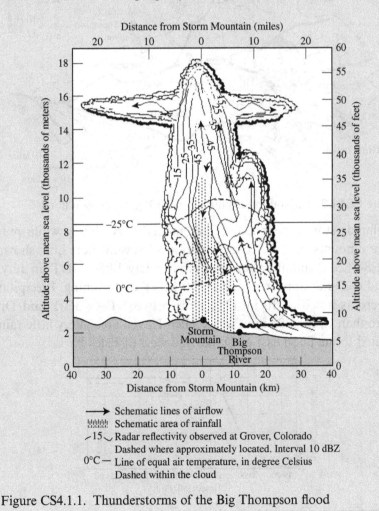

→ Schematic lines of airflow
▦ Schematic area of rainfall
⌐15⌐ Radar reflectivity observed at Grover, Colorado
      Dashed where approximately located. Interval 10 dBZ
0°C — Line of equal air temperature, in degree Celsius
      Dashed within the cloud

Figure CS4.1.1. Thunderstorms of the Big Thompson flood

The isohyetal map of the total precipitation from July 31 to August 2, 1976, is shown in Figure CS4.1.2. Conditions in Eastern Colorado on July 31, 1976 were favorable for heavy rain for a number of reasons. The surface map of that morning showed a slowly moving cold front in the state. Such fronts display lines of convergence that lift air to form thunderstorms. Also favorable was the easterly wind just north of the front, moving air upslope and aiding the frontal lifting. The low-level air was very moist and the moisture aloft was also unusually high. Thunderstorms move with the speed and the direction of the winds ~ 5 knots and were not expected to travel much during the day.

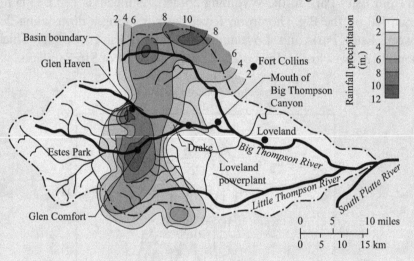

Figure CS4.1.2. Rainfall precipitation of the Big Thompson flood

The thunderstorms near Estes Park moved very slowly while putting out large amounts of water over a period of several hours, as shown in Figure CS4.1.3. Rainfall began at approximately 18:30 MDT on July 31, 1976, and ended at approximately 23:30 MDT that evening. Precipitation totals were as much as 10 in. (254 mm) between Estes Park and Drake and more than 12 in. (305 mm) in the Glen Haven area. Very little rainfall contributed to the flood east of Drake and west of Estes Park.

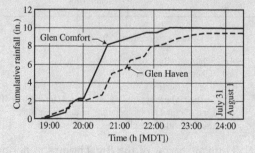

Figure CS4.1.3. Cumulative rainfall precipitation

Flood runoff in the Big Thompson basin derived from an area of approximately 60 square miles (155 km^2) centered on the Big Thompson River from Lake Estes to Drake. The topography of the area is characterized by steep north- and south-facing slopes with rugged rock faces and a thin soil mantle. Because of the steep slopes and small storage capacity of the soils, the storm runoff quickly reached nearby surface channels. The flood lasted only a few hours.

The reported peak stages on the Big Thompson River occurred as follows: 20:00 at Glen Comfort, 21:00 at Drake, 21:30 at the Loveland power plant, and approximately 23:00 at the mouth of the canyon ~8 miles (13 km) west of Loveland. The peak on the Big Thompson River just downstream from Drake occurred before the peak from the North Fork arrived at Drake. The flood peak moved through the 7.3-mile (11.7-km) reach between Drake and the canyon mouth in ~2 h with no apparent reduction in discharge. East of the canyon mouth, the Big Thompson River valley widens rapidly and the flood discharge was quickly reduced by valley storage. The peak discharge at the mouth of the Big Thompson River near LaSalle was only ~2,500 ft^3/s (71 m^3/s), occurring at noon on August 1, as compared with 31,200 ft^3/s (883 m^3/s) ~35 miles (56 km) upstream at the mouth of the canyon. The peak discharges at various locations are shown as a function of drainage area in Figure CS4.1.4.

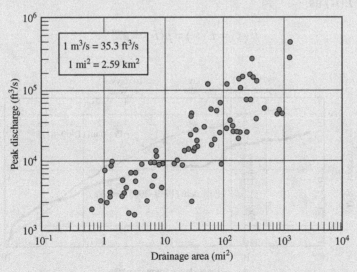

Figure CS4.1.4. Big Thompson peak discharge versus drainage area

4.2 River Flow Duration Curves

Two interesting river applications are presented for duration curves in Section 4.2.1 and the mean annual sediment load in Section 4.2.2.

4.2.1 Duration Curves

Duration curves describe the percentage of time during which a certain level of water, or sediment, discharge is exceeded. The example in Figure 4.6 illustrates the value of discharge as a function of the exceedance probability from the procedure discussed in Example 4.1 applied to daily discharge data. A log-log plot usually highlights the period with dominant flows that are observed 1 day per 2–5 years.

A deeper analysis of flow duration curves considers a relationship with the exponential distribution of rainfall intensity where $E(i) = E(i_*) = 1 - F(i_*) = e^{-i^*}$ from Equations (4.6 and 4.7). Variable x is assumed to be a power function of the rainfall intensity i_*

$$x = \hat{a} i_*^{\hat{b}} \quad \text{or} \quad i_* = ax^b \tag{4.13}$$

where a and b are the coefficient and exponent, and $\hat{a}$ and $\hat{b}$ are the equivalent parameters such that $\hat{a} = (1/a)^{1/b}$ and $\hat{b} = (1/b)$, and $a = \hat{a}^{-1/\hat{b}} = \hat{a}^{-b}$ and $b = (1/\hat{b})$. For instance, if x denotes flow discharge it becomes interesting to consider that large floods should be caused by large rainstorms. From Equations (4.6, 4.7, and 4.13), the exceedance probability E_x should be linked to $E(i_*)$ as

$$E(x) = E(i_*) = p(i_*) = e^{-ax^b}. \tag{4.14}$$

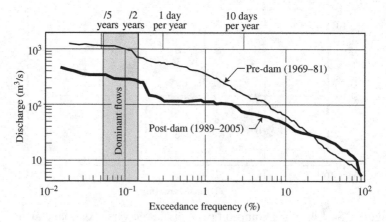

Figure 4.6. Typical flow duration curve

From taking the logarithm of Equation (4.14) twice, it follows that

$$-\ln E(x) = ax^b, \tag{4.15a}\blacklozenge$$

$$\Pi(x) = \ln[-\ln E(x)] = \ln a + b \ln x, \tag{4.15b}\blacklozenge$$

and a and b are determined from a linear plot of $\Pi(x)$ vs $\ln x$, as shown in Example 4.3.

The exponent $\hat{b}$ measures the nonlinearity between rainfall intensity and parameter x. A typical range of values of $\hat{b}$ for various processes is summarized in Table 4.2. From practical applications, Julien (1996) showed that the nonlinearity typically increases from point rainfall $(0.8 < \hat{b} < 1.3)$ to surface runoff and sediment concentration $(1.3 < \hat{b} < 1.8)$, and to sediment yield $(1.7 < \hat{b} < 2.9)$. Also, the expected value $\bar{x}$ can be simply evaluated from a simple gamma function, $\bar{x} = \hat{a}!\hat{b}$, where ! designates the factorial function of the argument $\hat{b}$.

Let us examine the daily measurements of discharge and sediment load on the Rio Grande in Figure 4.7. The exceedance probability $E(x)$ of a variable x can be calculated from

$$E(x) = e^{-ax^b}, \tag{4.15c}\blacklozenge$$

given parameters a and b. For instance, the probability of exceeding a discharge $Q = 5,000$ ft^3/s on the Rio Grande for the data shown in Figure 4.7 with $a = 0.018$ and $b = 0.613$ is simply $E(5,000) = e^{-0.018(5,000)^{0.613}} = 0.0357$, corresponding to 13 days per year. This method also enables the user to estimate what value of x will be exceeded a certain fraction of the time directly from

$$x = [-\ln E(x)]^{1/b} \left(\frac{1}{a}\right)^{1/b} = \hat{a}[-\ln E(x)]\hat{b}. \tag{4.16}$$

For the daily sediment load of the Rio Grande with $\hat{a} = 2,641$ and $\hat{b} = 2$ in Figure 4.7b, the value exceeded 1 percent of the time is $Q_s = 2,641(-\ln 0.01)^2 = 56,000$ tons per day.

One interesting property of this analysis is the ability to determine the relative magnitude of large events. From Equation (4.16), the ratio of two events is simply given as

$$\frac{x_1}{x_2} = \left[\frac{\ln E(x_1)}{\ln E(x_2)}\right]^{\hat{b}}. \tag{4.17}$$

Example 4.3 Flow duration curve analysis

Consider the following sample of variable x: 4.5, 1.0, 7.0, 2.0, 9.0, 0.5, 6.0, 11.0, 3.5. Note that all values must be positive.

Step 1. Rank the $N = 9$ values in decreasing order as shown in Column two of Table E4.3.1; the natural logarithms of the ranked values are in Column three.

Table E4.3.1. *Duration curve example*

Sample x	Ranked x	$\ln(x)$	Rank m	$E(x)$ $m/(N+1)$	$\Pi(x)$ $\ln[-\ln E(x)]$
4.5	11.0	2.3979	1	0.1	0.8340
1.0	9.0	2.1972	2	0.2	0.4759
7.0	7.0	1.9459	3	0.3	0.1856
2.0	6.0	1.7918	4	0.4	−0.0874
9.0	4.5	1.5041	5	0.5	−0.3665
0.5	3.5	1.2528	6	0.6	−0.6717
6.0	2.0	0.6931	7	0.7	−1.0309
11.0	1.0	0.0000	8	0.8	−1.4999
3.5	0.5	−0.6931	9	0.9	−2.2504

There are $N = 9$ values, and note that all x values must be positive.

Step 2. The numbers are ranked in decreasing order from 1 to N in Column four. Using the Weibull plotting position, the exceedance probability $E(x)$ in Column five is the rank divided by $(N + 1)$. Values of $\Pi(x) = \ln[-\ln E(x)]$ are then calculated in Column six.

Step 3. The values of $\Pi(x)$ are plotted versus $\ln x$ in Figure E4.3.1. A straight line $\Pi(x) = -2.3 + 1.3 \ln(x)$ is fitted in the zone of interest, i.e. the higher values of $\ln x$. The values $a = e^{-2.3} = 0.1$ and $b = 1.3$ correspond to $\hat{a} = (1/a)^{(1/b)} = (1/0.1)^{(1/1.3)} = 5.89$ and $\hat{b} = 1/b = 0.77$.

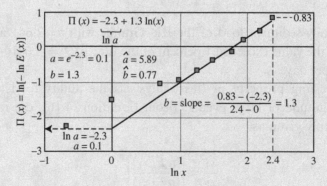

Figure E4.3.1. Duration curve example

Table 4.2. *Multiplication factor k for duration curves* $x = kx_{0.01}$

				Inverse transform coefficient $\hat{b}$				
				Rain fall $\longleftrightarrow$		Runoff $\longleftrightarrow$		Sediment
Exceeded	$E(x)$	$-\log E(x)$	$\Pi(x)$	1	1.5	2	2.5	3
10 percent of the time	0.10	1	0.83	0.5	0.35	0.25	0.18	0.125
5 percent of the time	0.05	1.3	1.09	0.65	0.52	0.42	0.34	0.275
1 percent of the time	0.01	2.0	1.53	1	1	1	1	1
1 day per year	2.74×10^{-3}	2.56	1.77	1.28	1.45	1.64	1.86	2.1
1 day per 1.5 years	1.83×10^{-3}	2.74	1.84	1.37	1.60	1.87	2.19	2.6
1 day per 2 years	$\mathbf{1.37 \times 10^{-3}}$	**2.86**	**1.88**	**1.43**	**1.71**	**2.05**	**2.45**	**2.9**
1 in 1,000 days	0.001	3.0	1.93	1.50	1.83	2.25	2.76	3.4
1 day per 5 years	5.47×10^{-4}	3.26	2.02	1.63	2.08	2.66	3.40	4.3
1 day per 10 years	2.73×10^{-4}	3.56	2.10	1.78	2.37	3.17	4.24	5.6
1 in 10,000 days	0.0001	4.0	2.22	2.50	3.95	6.25	9.88	15.6

For practical applications, the ln can be replaced with log and with the value of $x_{0.01}$ exceeded 1 percent of the time as a reference, the ratio k simply becomes

$$k = \frac{x}{x_{0.01}} = [-0.5 \log E(x)]\hat{b}. \tag{4.18}$$

Values of the multiplication coefficient in Equation (4.18) are given in Table 4.2.

The distribution of daily flows of the Hwang River before the construction of Hapcheon Dam is shown in Figure 4.8. The discharge exceeded 10 percent of the time is $Q_{0.01} = 350 \text{ m}^3/\text{s}$, and the discharge exceeded 0.1 percent of the time is $Q_{0.001} = 950 \text{ m}^3/\text{s}$. The corresponding ratio, $k_{0.001} = Q_{0.001}/Q_{0.01} = 2.71$, gives $\hat{b} \cong 2.5$ in Table 4.2. One of the practical river-engineering applications of this method is the determination of the 2-year flood, which usually corresponds to the dominant discharge of alluvial rivers. For the Hwang River, $\hat{b} \cong 2.5$ and $k_{2 \text{ years}} = 2.45$ from Table 4.2, the dominant discharge is $Q_{2 \text{ years}} \cong 2.45 \times 350 = 860 \text{ m}^3/\text{s}$.

4.2.2 Mean Annual Sediment Load

An interesting application of the flow duration curves in Section 4.2.1 emerges when the sediment-rating curve is known. From Equation (4.13)

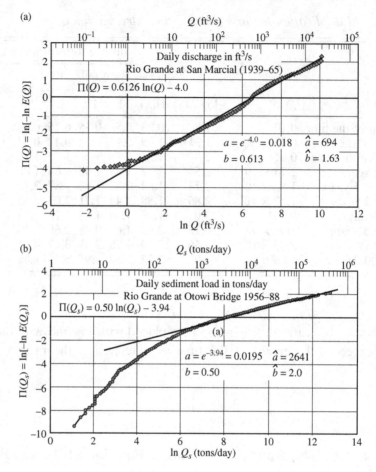

Figure 4.7. Examples of (a) flow and (b) sediment duration curves on the Rio Grande

with $Q = \hat{a} i_*^{\hat{b}}$ and Equation (4.6) we can estimate the mean annual sediment load of a river as follows:

$$\overline{Q}_s = A \int_0^\infty Q^B p(i_*) di_* = A \hat{a}^B \int_0^\infty i_*^{B\hat{b}} e^{-i_*} di_* = A \hat{a}^B \Gamma(1 + B\hat{b}). \qquad (4.19a)$$

For instance, consider the sediment-rating curve $Q_s = A Q^B$ where Q is the flow discharge in ft³/s and Q_s is the daily sediment load in tons per day. From the values of $\hat{a}$ and $\hat{b}$ for the duration curve of the daily flow discharges in ft³/s, the mean annual sediment load in ktons per year is simply estimated from

$$\overline{Q}_s \simeq 0.365 A \hat{a}^B \Gamma(1 + B\hat{b}). \qquad (4.19b)$$

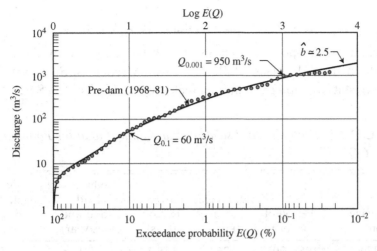

Figure 4.8. Hwang River flow duration curve (after Shin and Julien, 2010)

For example, with $A = 0.0286$, $B = 1.80$ from the sediment-rating curve, and $\hat{a} = 4.93$, $\hat{b} = 2.61$ from the flow duration curve, we estimate the mean annual sediment load at 13.7 thousand tons per year. It is interesting to note that this equation can also approximate the mean annual discharge when $A = B = 1$; however, the low flows can deviate significantly from the flow-duration curve equation. Therefore, the predictions of the sediment load are generally better. Case Study 4.2 shows an example of application of flow-duration curves and sediment load for Fountain Creek in Colorado.

4.3 Flood Frequency Analysis

The concept of flood frequency analysis is introduced with focus on the log-Pearson III distribution in Section 4.3.1, the Gumbel distribution in Section 4.3.2, and risk analysis in Section 4.3.3. A frequency analysis describes the distribution of events measured over certain time intervals. For instance, a flood frequency analysis will emphasize the magnitude of floods likely to occur over certain time intervals. Figure 4.9 sketches the comparison of the data series used for flood frequency compared to flow duration. While all circles shown are used in the flow-duration curve analysis, only the highest peak on an annual basis will be used in the flood frequency analysis.

When daily flows are available for a sufficiently long period (10–100 years), the annual peak values are listed out for the analysis. Two distributions are particularly useful for flood frequency analysis: the Gumbel distribution and the log-Pearson III distribution.

Case Study 4.2 Fountain Creek, Colorado

The sediment transport in Fountain Creek was analyzed by the graduate students in the 2017 class CIVE 717 River Mechanics. Summary results from five stations along Fountain Creek are listed in Table CS4.2.1.

Table CS4.2.1. *Summary of discharge and sediment load for Fountain Creek, CO*

USGS gage location	Period of record	Drainage area (km^2)	Mean discharge (ft^3/s)	Sediment load (thousand tons/year)	Specific degradation (tons/km^2 year)
A-7103700 FC near Colo. Springs	1958–2017	264	18.3	14	50
B-7103700 Monument Creek	1997–2015	465	28.2	12	24
C-7105500 FC at Colo. Springs	1976–2017	1,014	101	170	151
D-7105800 FC at Security	1964–2017	1,294	120	85	59
E-7106500 FC at Pueblo	1950–2017	2,394	153	275	105

Slightly lower values of specific degradation are observed in the mountains (Stations A and B) and where Fountain Creek develops a floodplain (Station D). Figure CS4.2.1b shows that the sediment-rating curves of all these stations are very similar. There is a tremendous variability in river flow discharges up to $\sim 20,000$ cfs during the major floods near Pueblo. The flow-duration curves in Figure CS4.2.1c are remarkably similar, with values of $\hat{b} \cong 2.5$ observed at all stations. This demonstrates the similarity in the rainfall–runoff relationship from mountain watersheds. Figure CS4.2.1c plots the cumulative distribution function of the sediment load at these stations. More than half the annual sediment load is transported during floods at discharges greater than five times the mean discharge.

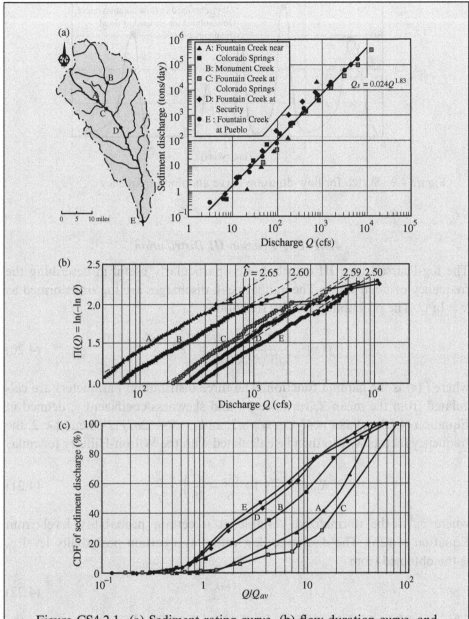

Figure CS4.2.1. (a) Sediment-rating curve, (b) flow duration curve, and (c) CDF of sediment load at Fountain Creek, Colorado

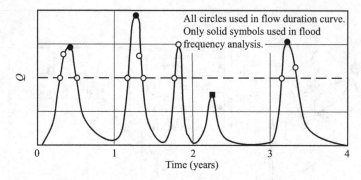

Figure 4.9. Sketch for flow-duration curve and flood frequency

4.3.1 Log-Pearson III Distribution

The log-Pearson type III distribution is particularly useful in describing the frequency of flood flows. The annual peak discharges are log transformed as $x = \ln Q$. The probability density function is

$$f(x) = \frac{1}{|\alpha|\Gamma(\kappa)} \left(\frac{x-\xi}{\alpha}\right)^{\kappa-1} e^{-\frac{(x-\xi)}{\alpha}} \qquad (4.20)$$

where $\Gamma(\kappa)$ is the gamma function. The three distribution parameters are calculated from the mean $\bar{x}$, variance σ^2, and skewness coefficient γ, defined in Equation (4.10a–c) as $\kappa = 4/\gamma^2$, $\alpha = \gamma\,\sigma/2$, and $\xi = \bar{x} - (2\sigma/\gamma)$. When $|\gamma| < 2$, the frequency factors K_p can then be calculated with the Wilson-Hilferty formula,

$$K_p(\gamma) = \frac{2}{\gamma}\left(1 + \frac{\gamma z_p}{6} - \frac{\gamma^2}{36}\right)^3 - \frac{2}{\gamma}, \qquad (4.21)$$

where z_p is the normalized variable at a certain probability level from Equation (4.12b). The discharge (log value) at a certain probability level z_p is the obtained from

$$x_p = e^{(\bar{x} + \sigma K_p)}. \qquad (4.22)$$

This sounds complicated, but Example 4.4 shows how to perform the calculations.

4.3.2 Gumbel Distribution

The Gumbel distribution simply requires the mean $\bar{x}$ and standard deviation σ from Equation (4.10a–b) to calculate parameters $\alpha = 0.7797\sigma$ and

$\xi = \bar{x} - 0.5772\alpha$. The Gumbel distribution has a fixed skewness coefficient $\gamma = 1.14$, which is similar to the log normal distribution $\gamma = 1.13$. The Gumbel probability density function is

$$f(x) = \frac{1}{\alpha} \exp\left[-\frac{x - \xi}{\alpha} - \exp\left(-\frac{x - \xi}{\alpha} \right) \right]. \tag{4.23}$$

The frequency factor for a period of return T is obtained from

$$K_G(T) = -\frac{\sqrt{6}}{\pi} \left\{ 0.5772 + \ln\left[\ln\left(\frac{T}{T-1} \right) \right] \right\}, \tag{4.24}$$

which is used to calculate the discharge from

$$x_p = \bar{x} + \sigma K_G(T). \tag{4.25}$$

An application example of the Gumbel method is provided in Example 4.4. The reader is referred to hydrology textbooks (like Stedinger et al., 1992; Anctil et al., 2012) for parameter estimation using L-moments and for the Weibull, Pareto, and generalized extreme value GEV distributions.

Example 4.4 Flood frequency analysis

Consider the annual peak discharges of the Muda River given in Julien et al. (2010). The ranked peak daily discharges from the annual series from 1961–2005 are: Q = 1340, 1225, 1200, 1100, 980, 912, 861, 789, 781, 706, 661, 640, 626, 626, 612, 602, 572, 572, 565, 565, 549, 546, 542, 539, 516, 500, 480, 450, 449, 436, 433, 399, 393, 382, 377, 375, 374, 340, 332, 326, 319, 315, 268, and 264 m^3/s.

Solution: The mean $\bar{x}$ = 587 and standard deviation σ = 263 are used for the Gumbel method where α = 205. After logarithmic transformation with $x = \ln Q$, the three parameters from Equation 4.10(a–c) are the mean $\bar{x} = 6.29$, standard deviation σ = 0.408, and skewness coefficient γ = 0.418. From these values, the three parameters of the log-Pearson III are ξ = 4.338, α = 0.085, and κ = 22.89. The log normal values in Table E4.4.1 are simply obtained from $x_p = e^{(\bar{x} + \sigma z_p)}$ while the log-Pearson III values are obtained from $x_p = e^{(\bar{x} + \sigma K_p)}$. Notice that the Gumbel distribution uses $x_p = \bar{x} + \sigma K_G$, and the results in Figure E4.4.1 are quite similar to the log-normal distribution. The differences become increasingly pronounced for long return periods.

Table E4.4.1. *Muda River flood frequency calculations*

Return period T in years	$F(Q)$ $1-1/T$	z_p	Q_{LN} m³/s	K_G	Q_G m³/s	K_p	Q_{LPIII} m³/s
			$\bar{x}=6.29$ $\sigma = 0.408$		$\bar{x}=587$ $\sigma = 263$		$\bar{x}=6.29$ $\sigma = 0.408$
2	0.5	0.0	539	−0.164	544	−0.069	524
5	0.8	0.842	760	0.719	776	0.814	751
10	0.9	1.282	910	1.304	929	1.317	923
20	0.95	1.645	1,055	1.866	1,077	1.754	1,103
25	0.96	1.751	1,101	2.044	1,124	1.886	1,164
50	0.98	2.054	1,247	2.592	1,268	2.271	1,362
100	0.99	2.326	1,394	3.137	1,410	2.630	1,577
200	1.995	2.576	1,543	3.679	1,553	2.969	1,811

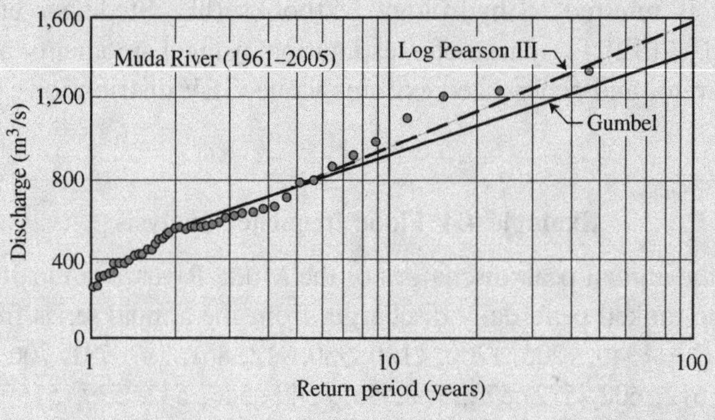

Figure E4.4.1. Flood frequency of the Muda River

4.3.3 Risk Analysis

The exceedance probability $E(x) = 1 - F(x)$ of an event describes how often a certain value will be exceeded during a given time interval. The return period (sometimes called the recurrence interval) is usually used in preference to the exceedance probability. The return period is the inverse of $E(x)$, or $T = 1/E(x)$. For instance, if a flood has a 50-year period of return, every year, a flood exceeding the 50-year flood has a 2 percent probability.

Table 4.3. *Period of return T in years corresponding to the risk level R over a useful life n*

Risk R percent	Project useful life n (in years)						
	2	5	10	20	25	50	100
50	3.43	7.74	14.9	29.4	36.6	72.6	145
25	7.46	17.9	35.3	70.0	87.4	174	348
10	19.5	48.1	95.4	190	238	475	950
5	39.5	98.0	196	390	488	976	1,950
2	99.5	248	496	990	1,250	2,480	4,950
1	199	498	996	1,990	2,490	4,980	9,950

Given the probability of an event p, and thus nonoccurence $1 - p$, the binomial distribution defines the probability P of having k occurrences in n independent trials is

$$P_k = \left[\frac{!n}{!k!(n-k)} \right] p^k (1-p)^{n-k}. \tag{4.26}$$

For instance, the probability of having two floods exceeding the 10-year flood in the next 5 years would be calculated from Equation (4.26) with $p = 0.1$, $n = 5$, and $k = 2$, thus

$$P_2 = \left[\frac{!5}{!2!3} \right] 0.1^2 (0.9)^3 = \left(\frac{5 \times 4}{2} \right) 0.01 \times 0.729 = 0.0729. \tag{4.27}$$

This concept can be extended to determine the probability not to have a single large flood in the next n years as $P_0 = (1 - p)^n$. The probability that at least one event with a period of return T may occur in the next n years is called the risk R and can be calculated from

$$R = 1 - \left(1 - \frac{1}{T} \right)^n. \tag{4.28}$$

For instance, consider a dam designed to withstand a 100-year flood with a useful life of 30 years. The risk that the 100-year flood may be exceeded at least once in 30 years is $R = 1 - 0.99^{30} = 0.26$. It is incorrect to think that the 100-year flood will reoccur periodically every 100 years. In reality, there is a 36.6 percent chance (i.e. 0.99^{100}) that the 100-year flood may not be exceeded in a 100-year period. Note that the term "risk" sometimes includes the damages incurred when the discharge is exceeded. Table 4.3 illustrates the relationship between risk and period of return over the useful life of a project, from Equation (4.28).

4.4 Extreme Floods

Rare events have a period of return of between 100 and 1,000 years and extreme events have a period of return larger than 1,000 years. Figure 4.10a provides a schematic of flow discharge as a function of the Annual Exceedance Probability (AEP). A guideline for the design of dams and levees as proposed by the Corps of Engineers and the Bureau of Reclamation is also shown in Figure 4.10b.

Two main types of engineering methods can be used for the analysis of rare and extreme events: (1) an analysis of discharge and (2) an analysis of rainfall precipitation in conjunction with a rainfall runoff model. For large and rare floods, a flood frequency analysis and a regional analysis can be conducted. A review of the latest standards of practice for the analysis of hydrologic hazard curves at Reclamation can be found in Swain et al. (2006). Salas et al. (2014) also provide a recent review of the Gradex

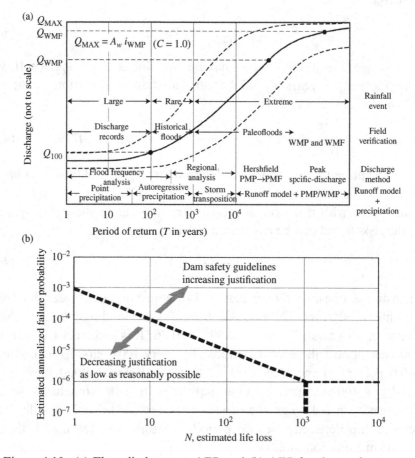

Figure 4.10. (a) Flow discharge vs AEP and (b) AEP for dam safety

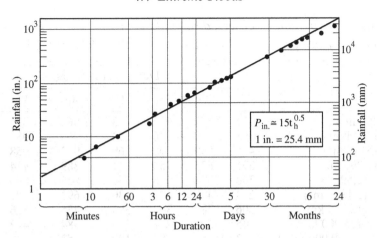

Figure 4.11. World maximum precipitation

method, stochastic event-based surface runoff models, and the Hershfield method. Application examples with the TREX model include results on: (1) a 12,000-km^2 semi-arid watershed in the United States, by England (2006); (2) a small mountain stream devastated by debris flows, by Kim (2012); and (3) tropical monsoon watersheds up to 1,300 km^2, by Abdullah et al. (2016). Novelties include a stochastic storm transposition approach by England et al. (2014), a field application of autoregressive precipitation models like DARMA by Muhammad et al. (2015), field verification with paleo-flood records by England et al. (2010), and simulations including the Probable Maximum Precipitation (PMP) and Probable Maximum Flood (PMF) by England et al. (2007).

The distribution of extreme rainfall precipitation is shown in Figure 4.11. The maximum rainfall depth depends on rainfall duration and on the surface area. In the United States, the maximum observed rainfall depths are listed in Table 4.4. For instance, the daily maximum precipitation is 983 mm, which is somewhat less than the world maximum daily point precipitation of 1,825 mm.

It is very interesting to examine the variability in maximum discharges on watersheds. Figure 4.12 illustrates an example of variability in runoff coefficient for a small watershed in the tropical rainforest of Malaysia. In this example from Abdullah (2013), the model TREX was used to simulate scenarios with high, medium, and low values of the two main parameters: surface roughness and soil infiltration. While keeping all other parameters constant, the peak discharge divided by the drainage area and rainfall intensity gives the peak runoff coefficient. These simulation results illustrate the

Table 4.4. *World maximum observed rainfall depth (in mm) as function of*
duration and surface area for the United States (after WMO, 1986)

Area Mile^2	Duration (in hours)						
	6	12	18	24	36	48	72
10 mi^2 (26 km^2)	627	757	922	983	1,062	1,095	1,148
100 mi^2 (260 km^2)	498	668	826	894	963	988	1,031
200 mi^2 (518 km^2)	455	650	798	869	932	958	996
500 mi^2 (1,295 km^2)	391	625	754	831	889	914	947
1,000 mi^2 (2,590 km^2)	340	574	696	767	836	856	886
2,000 mi^2 (5,180 km^2)	284	450	572	630	693	721	754
5,000 mi^2 (12,950 km^2)	206	282	358	394	475	526	620
10,000 mi^2 (25,900 km^2)	145	201	257	307	384	442	541
100,000 mi^2 (259,000 km^2)	43	64	89	109	152	170	226

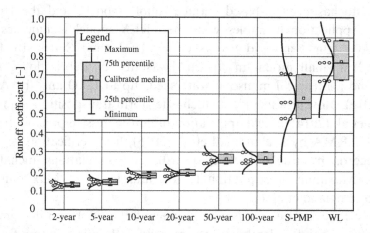

Figure 4.12. Example of variability in runoff coefficient under large and
extreme precipitation (after Abdullah and Julien, 2014)

nonlinearity in the rainfall–runoff relationship. It is observed that the runoff
coefficient of the 100-year flood is about twice the runoff coefficient for the
2-year flood. When considering that the 100-year rainfall intensity is 2.5 times
that of the 2-year rainstorm, the 100-year flood at that site should be about
five times the magnitude of the 2-year flood. Now looking at extreme events
(WL in Figure 4.12), the runoff coefficient of the most densely vegetated
watersheds in the world will result in a runoff coefficient of about 0.75,
which is about three times the value for the 100-year floods. If the world
maximum precipitation is three times the 100-year precipitation and the

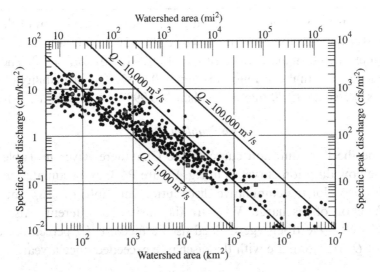

Figure 4.13. Specific discharge of the largest measured floods as a function of drainage area (after Creager et al., 1945; Abdullah, 2013)

runoff coefficient also increases three times, the maximum flood at that site should be about nine times larger than the 100-year flood.

Finally, let's look at the largest measured floods, shown in Figure 4.13. Specific discharge values are obtained by dividing the peak discharge values Q_p by the drainage area of a river basin A_t.

Problem 4.1

From the data for the Big Thompson flood in Case Study 4.1, compare these extreme flood discharges with those of Figure 4.13.

◆Problem 4.2

The mean annual precipitation readings from 10 rain gauges in the Ndjili river basin in Democratic Republic of the Congo are respectively: 1,497, 1,468, 1,348, 1,472, 1,491, 1,486, 1,462, 1,389, 1,491, and 1,394 mm. Determine the mean value, the standard deviation, the coefficient of variation, and the skewness coefficient. What is the standard deviation of the average rainfall precipitation on the basin? What is the probability that the mean annual value would exceed 1,500 mm?
[*Answer*: The parameters of the distribution are $\bar{x} = 1,450$ mm, $\sigma = 52.8$ mm, $C_v = 0.036$, and $\gamma = -1.08$. The variance of the sum of

10 stations is the sum of variances, thus the standard deviation of the river basin average precipitation is $\sigma_s = \sigma/\sqrt{10} = 16.7$ mm. To exceed $x = 1,500$ mm, the normalized value of $z = (1,500-1,450)/16.7 = 3.0$. The CDF of a standard normal distribution is $F(z) = 0.9987$. The probability of distribution's exceeding 1,500 mm is less than $1-0.9987$, i.e. less than 1 percent.]

◆◆*Problem 4.3*

Examine the flow duration curve of the Chaudiere River in Table P4.3. Plot the flow duration curve, as in e.g. Figure P4.3, and estimate the exponent $\hat{b}$. Also plot the cumulative distribution functions of $Q\Delta p$ and $Q_s\Delta p$ as a function of discharge. Why are the curves so different? Also determine the effective discharge defined as the median (50 percentile) of the CDF of $Q_s\Delta p$. Compare with the discharge exceeded once a year.

Table P4.3. *Long-term sediment yield of the Chaudière River from flow-duration-sediment-rating curve method (from Julien, 2010)*

Time intervals (percent) (1)	Interval midpoint (percent) (2)	Interval Δp (percent) (3)	Discharge Q_{cfs} (ft³/s) (4)	Concentration C (mg/l)[a] (5)	$Q_{cfs}\Delta p$ (ft³/s) (6)	Sediment load $Q_s\Delta p$[*] (tons/year)[b] (7)
0.00–0.02	0.01	0.02	58,000	607	12	6,280
0.02–0.1	0.06	0.08	52,000	526	42	19,539
0.1–0.5	0.3	0.4	43,000	411	172	63,102
0.5–1.5	1	1	33,000	292	330	85,820
1.5–5.0	3.25	3.5	21,000	162	737	106,213
5–15	10	10	10,640	67	1,064	63,529
15–25	20	10	5,475	29	548	13,782
25–35	30	10	3,484	16	348	4,873
35–45	40	10	2,435	10	244	2,138
45–55	50	10	1,839	7	184	1,121
55–65	60	10	1,375	4.7	138	575
65–75	70	10	1,030	3.2	103	296
75–85	80	10	763	2.1	76	149
85–95	90	10	547	1.4	55	69
95–100	97.5	5	397	0.9	14	12
Total		100			4,067	367,500

[a]The concentration C in mg/l is calculated from $C_{mg/l} = 0.04Q^{1.3}_{cms} = 3.89 \times 10^{-4}Q^{1.3}_{cfs}$.
[b]The annual sediment yield in metric tons/year from $Q_s\Delta p = 0.893CQ\Delta p$ with C in mg/l and Q in ft³/s.
Notes: Columns two and four define the flow-duration curve. Columns four and five define the sediment-rating curve. The product of Columns three and four is given in Column six.

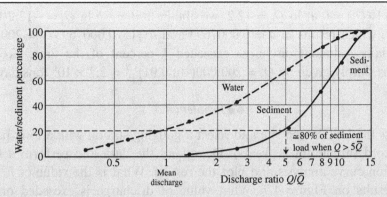

Figure P4.3. Percentage of water and sediment versus discharge for the Chaudière River

◆*Problem 4.4*

The duration curve for the daily sediment discharge of the Colorado River at Lee's Ferry is shown in Figure P4.4 (from Julien, 1996). Graphically estimate the parameters and estimate the sediment discharge that is exceeded 1 percent of the time.

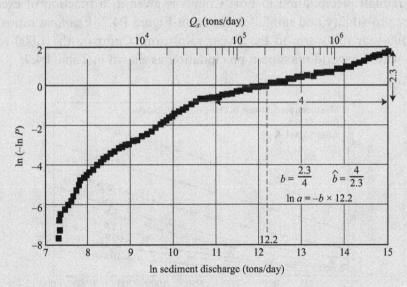

Figure P4.4. Sediment duration curve of the Colorado River

[*Answer*: From the upper part of the graph, the slope of the measurements is 2.3 as $\ln x$ increases from 11 to 15, which gives $\hat{b} = 1/b = 4/2.3 = 1.7$.

From $\Pi(Q_s) = 0$ at $\ln Q_s = 12.2$, we obtain $\ln a = -b \ln Q_s = -12.2/1.7 = -7.17$, which results in $a = 0.0007693$ and $\hat{a} = (1/0.0007693)^{1.7} = 200,000$. The daily sediment discharge exceeded 1 percent of the time, $E(Q_s) = 0.01$, is simply given by $Q_s = 200,000(-\ln 0.01)^{1.7} = 2.7 \times 10^6$ tons/day.]

◆◆*Problem 4.5*

Access the USGS web site and select a gauging station with a long period of record (minimum 30 years). Download the data and perform a flow-duration curve analysis and plot the results. What is the value of $\hat{b}$? Plot the results on Figure 4.7. What value of discharge is exceeded once a year?

◆◆*Problem 4.6*

For the data collected in Problem 4.5, carry out a flood frequency analysis. Compare with the log-Pearson III, the log-normal, and the Gumbel distributions. How much difference is there between these three methods?

◆*Problem 4.7*

The rainfall precipitation in Fort Collins is given as a function of excee-dance probability and rainfall duration in Figure P4.7. Examine ratios of the 100-year rainstorm to the 2-year rainstorm? Compare the 1,000-year data with the world maximum precipitation as shown in Table P4.7.

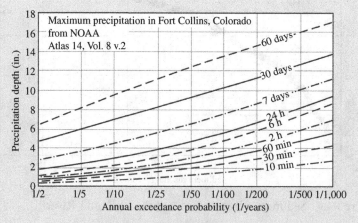

Figure P4.7. Maximum rainfall precipitation in Fort Collins, CO

Table P4.7. *World maximum rainfall precipitation in mm*

Rainfall duration	Precip. in mm	Location	Date
1 min	38	Guadeloupe	11/26/1970
15 min	198	Jamaica	12/5/1916
30 min	280	China	3/7/1974
1 h	401	China	3/7/1975
3 h	724	USA (PA)	18/7/1942
6 h	840	China	1/8/1977
1 day	1,825	Reunion	3/15/1952
7 days	5,400	Reunion	24/2/2007
1 month	9,300	India	1/7/1861
1 year	26,461	India	1/8/1860

◆◆*Problem 4.8*

With reference to Fountain Creek in Case Study 4.2, estimate the mean annual sediment load from Equation (4.19) based on the individual flow duration and sediment-rating curves listed in Table P4.8. Also recalculate $\overline{Q}_s$ from the rating curve in Figure CS4.2.1a.

Table P4.8. *Sediment yield estimates from Equation (4.19)*

Station	Sediment curve A	B	Flow duration $\hat{a}$	$\hat{b}$	Measured $\overline{Q}_s$ (ktons/year)	Estimated $\overline{Q}_s$ (ktons/year)	Figure CS4.2.1a $\overline{Q}_s$ (ktons/year)
A	0.0003	2.64	2.51	2.65	14		
B	0.0286	1.80	4.93	2.61	12	14	13.6
C	0.037	1.84	13.1	2.59	170	125	75.7
D	0.017	1.82	18.6	2.49	85	69	
E	0.024	1.83	26.9	2.50	275		214

5

Steady Flow in Rivers

"Steady flow" refers to conditions that do not change with time. Mathematically, steady flow implies that $(\partial h/\partial t) = 0$, $(\partial V/\partial t) = 0$, and $(\partial Q/\partial t) = 0$. "Uniform flow" refers to conditions that do not change with space. Nonuniform flow is possible when the mean flow velocity, channel width, and flow depth change in the downstream direction, or $(\partial V/\partial x) \neq 0$, $(\partial W/\partial x) \neq 0$, and $(\partial h/\partial x) \neq 0$. Section 5.1 describes steady and uniform river flows and Section 5.2 focuses on steady nonuniform flow conditions.

5.1 Steady-Uniform River Flow

Several properties of steady river flows are covered in this section, starting with longitudinal profiles and at-a-station geometry (Section 5.1.1), flow-rating curves and specific gage records (Section 5.1.2), resistance to flow relationships (Section 5.1.3), resistance to flow parameters (Section 5.1.4), flow conveyance and compound channels (Section 5.1.5), and normal depth (Section 5.1.6).

Rivers follow the low points along the watershed topographic profiles. With very few exceptions in arid areas, the lowest point of a watershed is located at the river outlet. The drainage network of a watershed implies a discontinuous increase in drainage area and discharge at river confluences. Except for river captures and man-made changes like transmountain water diversions, the drainage network does not change with time. The river length is usually measured in the downstream direction from the uppermost river elevation, from a confluence, or from a landmark. Accordingly, left and right banks are usually determined in a downstream-looking direction. River mileage is often measured downstream from a major river confluence or upstream from the mouth

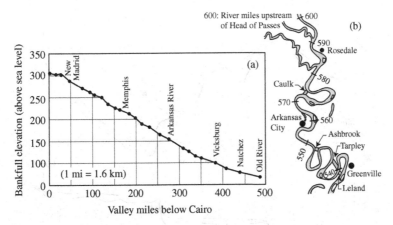

Figure 5.1. (a) Longitudinal profile and (b) planform geometry of an alluvial river, e.g. the Mississippi River

of a river. It is important to consider that the exact location of these reference points usually changes over the years. The straightening of river reaches through meander cutoffs can change the lengths of rivers and their corresponding sinuosity and slope.

A longitudinal profile is quite useful in detecting bedrock controls, headcuts and falls, nickpoints and rapids, and alluvial reaches. The approximate slope of alluvial reaches can be estimated from gradual changes in elevation over long distances. The valley slope corresponds to the floodplain elevation drop over the valley length. The channel slope corresponds to the slope of the energy grade line. Channel and valley slopes are defined as positive although the elevation decreases in the downstream direction. The river sinuosity is then defined as the ratio of the channel length to the valley length between two points located on the river. Alluvial rivers form in the material that they transport. Examples of river reaches and longitudinal profiles are shown in Figure 5.1 for an alluvial river, and Figure 5.2 for a semialluvial river. The longitudinal profile of an alluvial river reach is gradual. Bedrock control in semialluvial river reaches causes discontinuities in longitudinal profiles and bed-material sizes.

At engineering time scales, channel characteristics can change significantly during extreme floods or after active tectonic and volcanic periods. River reaches can be examined in terms of their physical characteristics, such as length, width, depth, sinuosity, surface roughness, and hydraulic-resistance factor. The following discussion focuses on at-a-station hydraulic geometry (Section 5.1.1) and resistance to flow in rivers (Section 5.1.2).

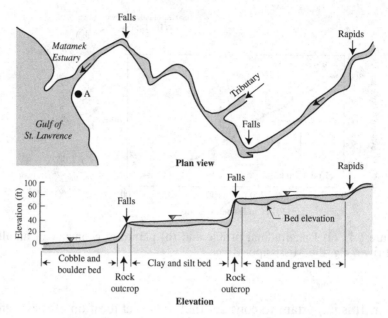

Figure 5.2. Longitudinal profile and planform geometry of a semialluvial river, e.g. the Matamek River (after Frenette and Julien, 1980)

5.1.1 At-a-station Hydraulic Geometry

It is important to reference the elevation to a benchmark that provides an exact horizontal-reference-plane elevation such as the mean sea level. Elevation scales, also called river stage, are usually linked to the geodetic reference elevation (above mean sea level, or ASL) or the World Geodetic System (WDS84), or the North American Vertical Datum of 1988 (NAVD88), which superseded the National Geodetic Vertical Datum (NGVD of 1929). For many navigable waterways, the elevation can be given with respect to a Low-Water Reference Plane (LWRP), which is the water-surface elevation in a river that is exceeded 97 percent of the time. Note that the LWRP is not horizontal, but slopes with the river.

At a given station along the river, a cross-section profile can be drawn in the direction perpendicular to the main flow direction. Cross sections that are not measured perpendicular to the flow direction will appear wider than they are in reality. The cross-section profile shows the flow-depth distribution across the river as well as the elevation of the banks including the floodplain. From a cross-sectional profile, the following geometrical parameters can be determined at a given water level: (1) cross-section area A; (2) wetted perimeter P; (3) hydraulic radius $R_h = A/P$; (4) top-channel width W; and (5)

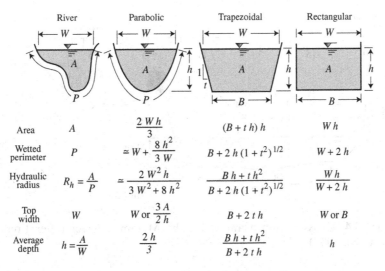

Figure 5.3. Geometric parameters of typical cross-sections

mean flow depth $\overline{h} = A/W$. These parameters describe the geometry of a cross section. It is important to consider that the mean flow depth is different from the stage. An increase/decrease in mean flow depth does not necessarily correspond to the same increase/decrease in stage. Examples of typical cross sections are shown in Figure 5.3. The wetted perimeter approximation is valid when $B > 4h$, which is usually the case for open-channel flows. The parameter t of a trapezoidal section defines the horizontal distance for a unit vertical distance. A rectangular section is the particular case of a trapezoidal section where the side slope parameter $t = 0$ and the base channel width B equals the top width, or $B = W$. A triangular cross section would also simply be the particular case where $B = 0$.

Figure 5.4 shows an example of a river cross section with indications of the substrate material and floodplain vegetation types in terms of deciduous and coniferous trees, shrubs, and grasses. Riverbank information may describe aquatic habitat and floodplain roughness for runoff simulations during floods with large overbank flows.

As discussed in Chapter 2, the depth-averaged velocity $\overline{v}$ is obtained from a measured velocity profile, as shown in Figure 5.5a. The total cross-sectional area A is the sum of incremental areas a_i, $A = \sum_i a_i = \sum_i \Delta W_i h_i$, where ΔW is the spacing between verticals of a cross section. The total discharge is $Q = \sum_i a_i \overline{v}_i$, where $\overline{v}_i$ is the depth-averaged flow velocity normal to the incremental area. The cross-sectional average velocity is $V = Q/A$. Figure 5.5b shows cross-sectional flow depth and velocity profiles. For most

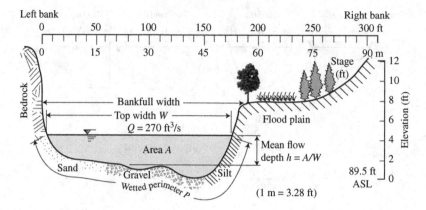

Figure 5.4. Cross-section example for the Matamek River (after Frenette and Julien, 1980)

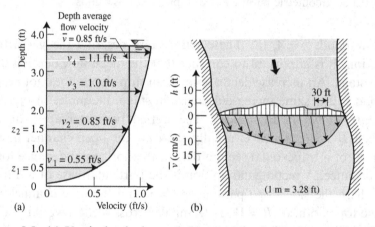

Figure 5.5. (a) Vertical velocity and (b) transversal velocity profiles of the Matamek River (after Frenette and Julien, 1980)

rivers, the wetted perimeter is approximately equal to the top width, $P \cong W$, and the hydraulic radius is close to the mean flow depth, $R_h \cong \overline{h}$.

The analysis of at-a-station hydraulic-geometry relationships becomes simple once the channel width W, cross-section area A, hydraulic radius $R_h = A/P$, and mean flow depth $h = A/W$ are determined at a given stage, as shown in Figure 5.6. Beyond bankfull flows, the channel width suddenly increases, as sketched in Figure 5.6c, and this will significantly reduce the mean flow depth and hydraulic radius, as sketched in Figure 5.6d.

This process is repeated at various discharges to define the at-a-station hydraulic-geometry relationships. Figure 5.7 shows cross sections over a 1.4-km reach of the Matamek River at flow conditions less than bankfull discharge.

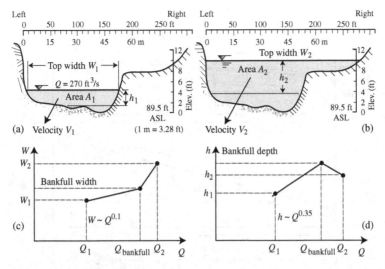

Figure 5.6. At-a-station hydraulic geometry of Matamek River: (a) low flow, (b) high flow, (c) width, and (d) flow depth

The hydraulic geometry parameters of this semialluvial river are plotted versus discharge on a log–log scale. For instance, the wetted perimeter in Figure 5.7a can be approximated by $P \cong 126Q^{0.1}$, with Q in cubic feet per second and P in feet. Note that the exponent is very small and the use of a rectangular cross section is a fairly good practical approximation for many rivers. The variability in mean flow depth and cross-section average velocity increases approximately as $\bar{h} = 0.36Q^{0.36}$ and $V \cong 0.022Q^{0.54}$, with Q in cubic feet per second and $\bar{h}$ in feet. The discharge must equal the product of width, depth, and velocity, thus $Q = W\bar{h}V \cong 126Q^{0.1} \times 0.36Q^{0.36} \times 0.022Q^{0.54}$. Accordingly, both the product of coefficients ($126 \times 0.36 \times 0.022 = 1$) and the sum of exponents ($0.1 + 0.36 + 0.54 = 1$) must equal unity.

Typical at-a-station hydraulic geometry relationships below bankfull conditions show that the changes in discharge primarily affect flow velocity and, to a lesser extent, flow depth. This can be quite important for aquatic species, since the periods of droughts may result in very shallow depths and much-reduced flow velocities. It is also particularly important in the case of salmonids, where the reduced flow velocity may imply less oxygenation of the eggs in the spawning grounds, or redds. In the case of the Matamek River, the low winter flow conditions always correspond to the coldest periods and a deep freeze of the spawning grounds is always a major concern. Finally, it is interesting to notice in Figure 5.7d that the water-surface slope of a river may also change slightly with discharge. This is particularly true upstream of falls on nonalluvial rivers. It is also the case for meandering channels, which tend to flow in a "straightened" manner and cut through chute and neck cutoffs during major floods.

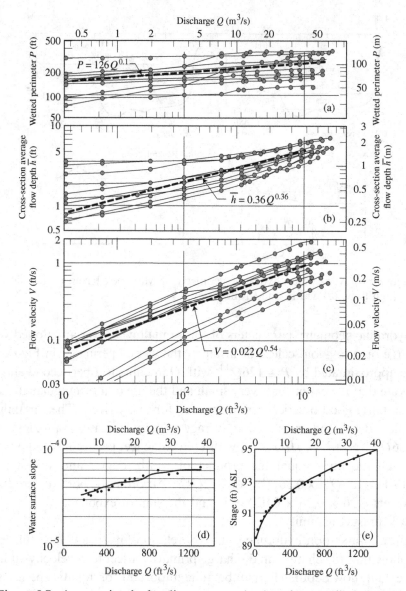

Figure 5.7. At-a-station hydraulic geometry (a–c) and stage–discharge relationship (d–e) for the Matamek River (after Frenette and Julien, 1980)

5.1.2 *Flow-Rating Curves and Specific Gage Records*

A flow-rating curve, or stage–discharge relationship, displays the change in stage, or water level, with discharge. Stage–discharge relationships are particularly important in the design of levees to protect living communities from major floods. In channels with bedrock control, the stage–discharge relationship is

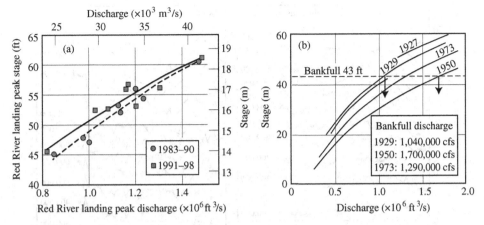

Figure 5.8. Stage–discharge relationships for the Mississippi River (a) at Red River Landing, Louisiana (after USACE, 1999) and (b) at Vicksburg, Mississippi (after Winkley, 1977)

unique and well defined, e.g. the Matamek River in Figure 5.7e. In alluvial rivers, the stage–discharge relationship may shift over time because of a combination of processes, including: (1) bed aggradation or degradation, causing a deformation of the cross-section geometry; (2) changes in bedform configuration; or (3) loop-rating effects that are due to dynamic flood routing; this will be discussed in greater detail in the next chapter. The example of the Mississippi River is shown in Figure 5.8. The shifts in stage–discharge relationships are quite significant, and Figure 5.8b particularly illustrates that the bankfull discharge at a stage of 43 ft varied from 1,040,000 cfs in 1929 to 1,700,000 cfs in 1950.

Specific gage records indicate the changes in water level or stage at a given flow discharge as a function of time. For instance, Figure 5.9 shows the aggradation–degradation trends of the Mississippi River at Columbus and Vicksburg from 1920 to 1980.

When specific gage records are conducted at numerous stations along a river, the successive specific gage records at a given discharge can be highly informative with respect to the morphological changes of an alluvial river. For instance, the reach of the Atchafalaya River between Simmesport and Morgan City from 1950 to 1997 is presented at a discharge of 240,000 cfs, in Figure 5.10. It is inferred from this graph that the changes in stage at this discharge correspond to changes in bed elevation. It is important to consider that without tributary inflow, rivers will tend to maintain a constant slope. The Atchafalaya River has been gradually degrading between Simmesport and Chicot Pass and aggrading downstream of Chicot Pass.

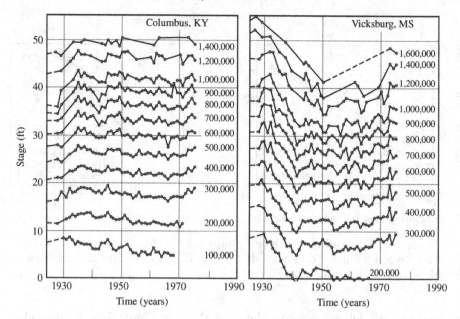

Figure 5.9. Specific gage records of the Mississippi River at Columbus, KY, and Vicksburg, MS, from 1930 to 1975 (after Winkley, 1977)

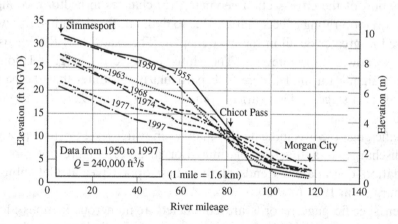

Figure 5.10. Specific gage records of the Atchafalaya River between Simmesport and Morgan City from 1950 to 1997 at a discharge of 240,000 cfs (after US Army Corps of Engineers, 1999)

In some alluvial rivers, water temperature can also affect the specific gage records. For the example of the Mississippi River at Union Point and Tarbert Landing in Figure 5.11, the water temperature at a fixed discharge of 10^6 cfs can alter the stage by as much as 5 ft. Higher stages correspond to warm water temperature. In his analysis of water-temperature effects,

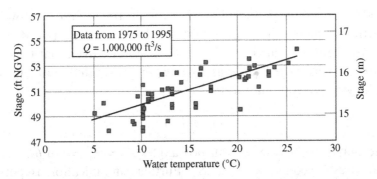

Figure 5.11. Water temperature effects on alluvial river stage, e.g. the Mississippi River (after US Army Corps of Engineers, 1999)

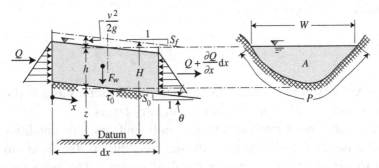

Figure 5.12. Average bed shear stress for steady-uniform flows

Akalin (2002) showed that at a given discharge, warm waters result in slower flow velocities. Warmer temperatures also decrease the settling velocity of suspended sediment. For instance, at the same discharge, the average concentration of fine sands can typically decrease from 100 to 40 mg/l as the water temperature increases from 9 to 30°C. The fact that water viscosity decreases with water temperature can result in a drastic change in bedform configuration from plane bed to large dunes, as discussed in Julien (2010, p. 181). This is due to the change in the laminar sublayer thickness, which is in the transition regime between hydraulically smooth and hydraulically rough.

5.1.3 Resistance to Flow Relationships

Resistance to flow describes the relationship between velocity (or discharge) and flow depth (or stage) and is determined empirically under steady-uniform flow conditions. A typical cross section is sketched in Figure 5.12 for the determination of the average bed shear stress.

With identical pressure distributions at the upstream and the downstream cross sections, the weight component in the downstream x direction is balanced by the bed shear force $F_w \sin \theta = \tau_0 W \, dx$. Given that the weight is $F_w = \gamma A \, dx$, the average bed shear stress τ_0 on the wetted perimeter reduces to the following general relationship:

$$\tau_0 = \gamma R_h S_f. \tag{5.1}$$

For steady-uniform flow and small bed angles, $S_f = S_0 = \tan \theta \cong \sin \theta$ and the shear stress becomes $\tau_0 = \gamma R_h S_0$. Further simplification is possible in wide-rectangular channels when $B \gg h$, as the hydraulic radius $R_h \cong h$ and the shear stress become $\tau_0 \cong \gamma h S_0$.

Shear velocity u_* defines a kinematic substitute for the bed shear stress τ_0. The identity stems from $\tau_0 = \rho u_*^2$ or

$$u_* = \sqrt{\frac{\tau_0}{\rho}} = \sqrt{g R_h S_f}. \tag{5.2}\blacklozenge$$

The shear velocity is not a measurable quantity, but serves as a scaling parameter for velocity profiles in turbulent boundary layers.

Resistance to flow is evaluated from steady-uniform flow conditions, $S_f = S_0$. Because no shear exists under hydrostatic conditions, the shear stress τ_0 is assumed to vary with the square of the fluid velocity. This assumption $\tau_0 = f\rho V^2/8$ leads to the dimensionless Darcy–Weisbach friction factor f, defined as

$$f = \frac{8\tau_0}{\rho V^2} = \frac{8g R_h S_f}{V^2}. \tag{5.3}\blacklozenge\blacklozenge$$

Resistance to flow can thus be described from the Darcy–Weisbach friction factor f and it is described from Equations (5.2) and (5.3) as $\sqrt{8/f} = V/u_*$. The relationship between mean flow velocity and the hydraulic radius is then simply obtained as

$$V = \sqrt{\frac{8g}{f}} R_h^{1/2} S_f^{1/2} = C R_h^{1/2} S_f^{1/2}, \tag{5.4}\blacklozenge$$

where C is the Chézy coefficient. The identity $C^2 = 8g/f$ is always valid, and it is important to note that f describes flow resistance, whereas C describes flow conveyance. The Chézy coefficient C is a constant as long as f is a constant. Also, the Chézy coefficient C depends on the system of units, while the Darcy–Weisbach friction factor f is dimensionless. It has been observed over the years that the Chézy coefficient C or the Darcy–Weisbach factor f varies

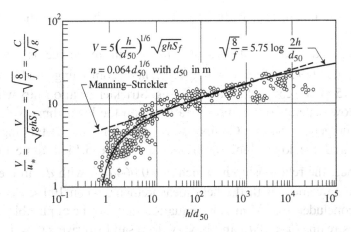

Figure 5.13. Resistance to flow for hydraulically rough plane-bed rivers (from Julien, 2002)

with flow depth and boundary roughness. Manning's equation is a convenient empirical relationship for rivers and can be written as

$$V = \frac{1}{n} R_h^{2/3} S_f^{1/2} \text{ (in SI units)} = \frac{1.49}{n} R_h^{2/3} S_f^{1/2} \text{ (in English units)}, \quad (5.5)\blacklozenge\blacklozenge$$

where SI units provide velocities in m/s from the hydraulic radius R_h in m. In English units, the velocity in ft/s is obtained from R_h in ft, and the unit conversion factor 1.49. There are three equivalent flow-resistance formulations commonly used in river engineering practice: (1) the Chézy coefficient C; (2) the Manning coefficient n; and (3) the Darcy–Weisbach friction factor f. Both Chézy and Manning coefficients are dimensional, and their equivalence is simply obtained from Equations (5.3–5.5) as

$$C \equiv \sqrt{\frac{8g}{f}} \equiv \frac{R_h^{1/6}}{n} \text{ (SI units)} \equiv \frac{1.49\, R_h^{1/6}}{n} \text{ (English units)}. \quad (5.6)\blacklozenge\blacklozenge$$

It is interesting to consider that both f and n are resistance parameters and are inversely proportional to Chézy C, which is a conveyance parameter.

With the developments of the boundary-layer theory by Prandtl and von Karman, logarithmic relationships became available to describe resistance to turbulent flows with hydraulically rough boundaries. For practical purposes, it is clear from Figure 5.13 that resistance to flow can be approximated by

$$\frac{V}{u_*} = \frac{C}{\sqrt{g}} = \sqrt{\frac{8}{f}} = 5.75 \log\left(\frac{2h}{d_{50}}\right), \quad (5.7)\blacklozenge\blacklozenge$$

where h is the mean flow depth and d_{50} is the median grain diameter of the bed material.

In gravel-bed and cobble-bed streams, grain resistance is predominant and Strickler (1923) found $n \sim d_s^{1/6}$. When combining with Manning's Equation (5.6) we can define a Manning–Strickler relationship where resistance to flow depends on the power rather than the logarithmic value of relative submergence h/d_{50}. Comparisons between the Manning–Strickler approach and the logarithmic approach in Figure 5.13 is quite instructive. For instance, the relationship in which $n = 0.064 d_{50}^{1/6}$, with d_{50} in meters, is in reasonable agreement with the field measurements when $100 < d_{50} < 10,000$. It is also concluded that Manning's equation may not be applicable in shallow mountain streams ($h < 10 d_s$) and in very deep sandbed rivers ($h > 10,000 d_s$).

The logarithmic form of grain resistance in Equation (5.7) can be transformed into an equivalent power form in which the exponent m varies with relative submergence h/d_{50}:

$$\frac{V}{u_*} = \sqrt{\frac{8}{f}} \cong 5.75 \log\left(\frac{2h}{d_{50}}\right) \equiv \frac{2.5}{m}\left(\frac{h}{d_{50}}\right)^m. \qquad (5.8)\blacklozenge$$

It is possible to meet the criteria that both the value and the first derivative stay identical when

$$m = \frac{1}{2.3 \log(2h/d_{50})}. \qquad (5.9)$$

The power relationships are useful in coarse-bed channels like cobble- and boulder-bed streams, as shown in Figure 5.14.

5.1.4 Resistance to Flow Parameters

A database including 2,604 river measurements has been examined by Lee and Julien (2012a). Natural channels (1,865 measurements) are primarily described in terms of their substrate or bed material, and four main types are recognized: sands, gravels, cobbles, and boulders. In the case of 739 vegetated channels, the main types include grasses, shrubs, and trees. The range of channel parameters is summarized in Table 5.1, and the resistance parameters in terms of the Darcy–Weisbach friction factor are shown for natural channels in Figure 5.15, and very similar results are observed for vegetated channels.

The analysis of the Darcy–Weisbach roughness coefficients is presented in Figure 5.16. The variability is high, but definite trends show that the friction coefficient decreased with discharge and increased with friction slope.

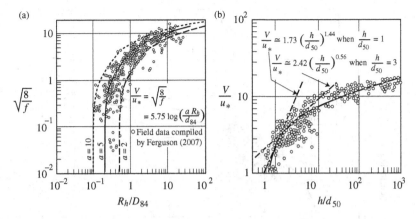

Figure 5.14. Resistance relationships for (a) mountain streams and (b) rivers

Table 5.1. *River database for resistance to flow (after Lee and Julien, 2017b)*

Type	Number	Discharge Q (m³/s)	Friction slope S_f (–)	Mean diameter d_{50} (mm)	Velocity V (m/s)	Flow depth h (m)
Natural channels						
Sand	172	0.14–26,560	0.00010–0.02860	0.01–1.64	0.02–3.64	0.10–15.67
Gravel	989	0.01–14,998	0.00009–0.08100	2–63.6	0.04–4.70	0.04–11.15
Cobble	651	0.02–3,820	0.00001–0.05080	64–253	0.07–4.29	0.10–6.94
Boulder	53	2.00–1,700	0.02060–0.03730	263–945	0.32–5.11	0.28–4.09
Vegetated						
Grass	281	0.01–750	0.00007–0.01790	0.33–305	0.06–3.66	0.16–3.96
Shrub	150	0.38–542	0.00001–0.03400	16–893	0.1–3.64	0.04–3.08
Tree	308	0.02–3,220	0.00010–0.04050	0.17–397	0.07–5.11	0.10–9.17

Very similar trends are observed for natural and vegetated channels. Overall, the roughness coefficient increased with bed material particle size. The friction factor for shrubs is slightly higher than those for grasses and trees. Finally, it is interesting to notice in Figure 5.16b that the flow in natural and vegetated channels is almost always subcritical (0.1 < Fr < 1.0). The box–whisker plots for f show similar variability for natural and vegetated channels. The variability around the mean is about the same regardless of the types of bed roughness or vegetation. The variability of boulders and shrubs is slightly higher than the others.

The analysis of Manning roughness coefficients is summarized in Figure 5.17. In it, Manning roughness coefficients are shown to vary slightly with discharge Q and friction slope S_f, but the trends are less pronounced than those observed for Darcy–Weisbach f.

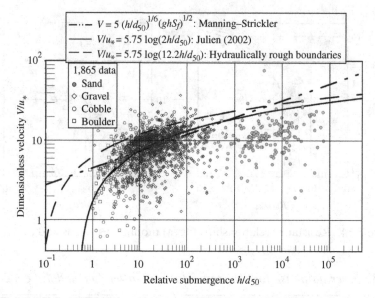

Figure 5.15. Resistance parameters for natural channels (after Lee and Julien, 2017b)

Recommended values for the resistance to flow of near-bankfull channels are listed in Table 5.2. It is observed that the variability in Manning n is less than the variability in Darcy–Weisbach f. In general, the average values of Manning n are fairly comparable but somewhat higher than suggested by Chow (1959). For natural channels, Manning n increases with grain diameter and is the highest for boulder-bed streams, as expected. Of the three vegetation types, shrubs show higher resistance to flow. Example 5.1 shows how to determine resistance to flow parameters from a velocity profile.

5.1.5 Flow Conveyance and Compound Channels

The flow discharge is simply obtained from $Q = AV$, which from the Manning resistance relationship described in Section 5.1.4 can be written as

$$Q = AV = \frac{1}{n} A R_h^{2/3} S_f^{1/2} \text{ (SI units)} = \frac{1.49}{n} A R_h^{2/3} S_f^{1/2} \text{ (British units)} \quad (5.10)$$

Flow conveyance describes the ability of a given cross section to deliver a certain flow rate. Flow conveyance is the relationship between discharge Q and friction slope S_f,

$$Q = K S_f^{1/2}, \quad (5.11)$$

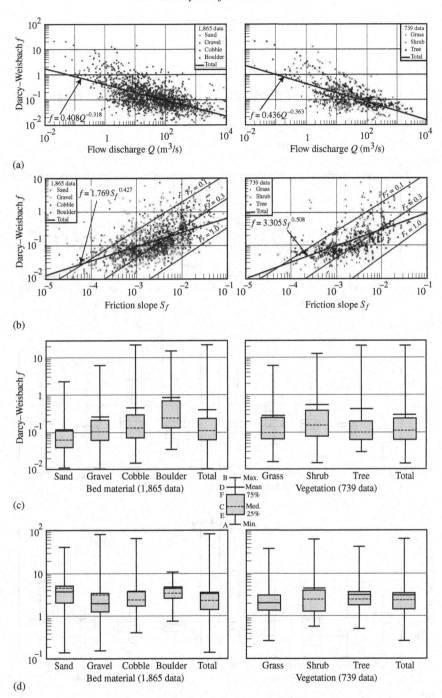

Figure 5.16. Darcy–Weisbach f roughness coefficients for rivers: (a) Regression between $\log f$ and $\log Q$, (b) Regression between $\log f$ and $\log S_f$, (c) Box–whisker plots for f, and (d) Box–whisker plots for $f Q^{1/4}/S_f^{1/3}$ (after Lee and Julien, 2017b)

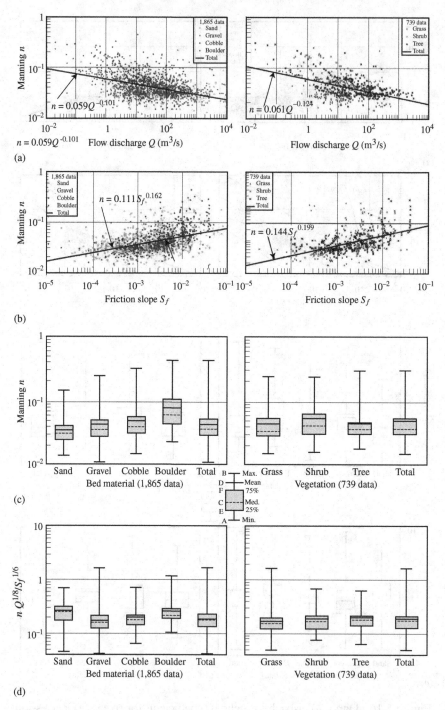

Figure 5.17. Manning n roughness coefficients for rivers: (a) Regression between $\log n$ and $\log Q$, (b) Regression between $\log n$ and $\log S_f$, (c) Box–whisker plots for n, and (d) Box–whisker plots for $nQ^{1/8}/S_f^{1/6}$ (after Lee and Julien, 2017b)

Table 5.2. *Distribution range of Darcy–Weisbach f and Manning n for natural and vegetated channels*

Roughness type	Name	Number	Darcy–Weisbach f			Manning n		
			Min.	Mean	Max.	Min.	Mean	Max.
Natural channels	Sand	172	0.011	**0.115**	2.188	0.014	**0.036**	0.151
(1,865 data)	Gravel	989	0.010	**0.251**	6.121	0.011	**0.045**	0.250
	Cobble	651	0.015	**0.465**	21.462	0.015	**0.051**	0.327
	Boulder	53	0.034	**0.794**	14.592	0.023	**0.080**	0.444
Vegetated channels	Grass	281	0.016	**0.271**	6.121	0.015	**0.045**	0.250
(739 data)	Shrub	150	0.015	**0.580**	12.910	0.016	**0.057**	0.250
	Tree	308	0.030	**0.434**	21.462	0.018	**0.047**	0.310

Other typical Manning n roughness values
Plexiglass	0.010–0.014
Timber and concrete	0.011–0.020
Riveted metal	0.015–0.020
Corrugated metal	0.022–0.030
Straight earth channel	0.013–0.025
Sand dunes and ripples	0.018–0.040
Antidunes	0.013–0.020
Rocky irregular channel	0.026–0.060
Overbank short grass	0.025–0.035
Tall grasses and reeds	0.030–0.050
Floodplain crops	0.020–0.050
Sparse vegetation	0.035–0.050
Medium to dense brush	0.070–0.016

where K is the conveyance coefficient, which is a function of the cross-section hydraulic geometry and roughness coefficient. For instance, the conveyance coefficient for the Manning resistance formula is given by

$$K = \frac{1}{n} A R_h^{2/3} \text{ (SI units)} = \frac{1.49}{n} A R_h^{2/3} \text{ (English units)}. \quad (5.12)$$

The friction slope can be obtained from the discharge and the conveyance coefficient from

$$S_f = \left(\frac{nQ}{A R_h^{2/3}}\right)^2 = \left(\frac{nV}{R_h^{2/3}}\right)^2 \text{ (SI units)} = \left(\frac{nV}{1.49 R_h^{2/3}}\right)^2 \text{ (English units)}. \quad (5.13)$$

Example 5.1 Resistance to flow parameters from a velocity profile

Consider the given measured velocity profile in a 200-ft-wide river, in Figure E5.1. When plotting on a semilog scale, a straight line is commonly observed and a logarithmic velocity profile can be fitted through the data $v = \frac{u_*}{\kappa} \ln \frac{z}{k_s'} = 2.3 \frac{u_*}{\kappa} \log \frac{z}{k_s'}$, where v is the flow velocity at an elevation z above the bed, u_* is the shear velocity, and k_s' is the Nikuradse roughness height.

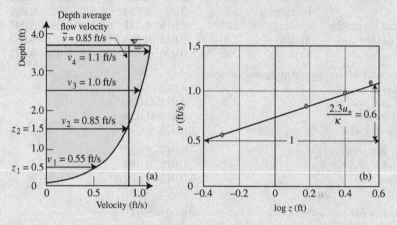

Figure E5.1. Velocity profile: (a) linear, and (b) semi-logarithmic

(a) depth-averaged flow velocity:
  Note that $dh_i = 1$ ft, except for the uppermost velocity measurement ($dh_4 = 0.7$ ft), and the flow depth is 3.7 ft.

$$\overline{V} = \frac{1}{A} \int_A v_i \, dA \cong \frac{1}{h} \sum_i v_i \, dh_i$$

$$= \frac{1}{3.7 \text{ ft}} [0.55 + 0.85 + 1.0 + (1.1 \times 0.7)] \frac{\text{ft}^2}{\text{s}} = 0.85 \frac{\text{ft}}{\text{s}} = 0.26 \frac{\text{m}}{\text{s}}$$

The flow velocity at 40 percent of the flow depth, i.e. $z_{0.4} = 0.4 \times 3.7$ ft $= 1.48$ ft, is 0.852 ft/s, which is also very close to the depth-averaged flow velocity.

(b) the momentum correction factor:

$$\beta_m = \frac{1}{A V_x^2} \int_A v_x^2 \, dA \cong \frac{1}{h V_x^2} \sum_i v_{xi}^2 \, dh_i,$$

$$\beta_m \cong \frac{1}{3.7 \text{ ft}} \frac{\text{s}^2}{(0.85)^2 \text{ ft}^2} [0.55^2 + 0.85^2 + 1.0^2 + (1.1^2 \times 0.7)] \frac{\text{ft}^3}{\text{s}^2} = 1.074.$$

(c) energy correction factor:

$$\alpha_e = \frac{1}{AV_x^3} \int_A v_x^3 \, dA \cong \frac{1}{hV_x^3} \sum_i v_{xi}^3 \, dh_i = 1.194.$$

Note that $\alpha_e > \beta_m$ and both are greater than unity.

(d) unit and total discharge:

$$q = Vh = 0.85 \text{ ft/s} \times 3.7 \text{ ft} = 3.1 \text{ ft}^2/\text{s} = 0.29 \text{ m}^2/\text{s},$$
$$Q = Wq = 200 \text{ ft} \times 3.1 \text{ ft}^2/\text{s} = 630 \text{ ft}^3/\text{s} = 17.8 \text{ m}^3/\text{s}.$$

(e) hydraulic radius:

$$R_h = \frac{A}{P} = \frac{Wh}{W + 2h} = \frac{200 \times 3.7 \text{ ft}^2}{200 + 2 \times 3.7 \text{ ft}} = 3.57 \text{ ft} = 1.09 \text{ m};$$

Note that the hydraulic radius $R_h = 3.57$ ft is very close to the flow depth $h = 3.7$ ft, because the width–depth ratio is large (i.e. $W/h = 200/3.7 = 54$).

(f) Froude number:

$$\text{Fr} = \frac{V}{\sqrt{gR_h}} \cong \frac{V}{\sqrt{gh}} = \frac{0.85 \text{ ft/s}}{\sqrt{32.2 \times 3.7 \frac{\text{ft}^2}{\text{s}^2}}} = 0.078.$$

The following parameters can be determined from two points located on the fitted semilogarithmic velocity profile, i.e. v_1 and v_2 are the flow velocities at z_1 and z_2, respectively.

(g) shear velocity assuming $\kappa = 0.4$:

$$u_* = \frac{\kappa(v_2 - v_1)}{\ln\left(\frac{z_2}{z_1}\right)} = \frac{0.4(0.85 - 0.55) \text{ ft/s}}{\ln\left(\frac{1.5}{0.5}\right)} = 0.11 \text{ ft/s} = 0.033 \text{ m/s}.$$

A graphical approximation is obtained from the semilog plot where the coefficient $2.3\frac{u_*}{\kappa} = 0.6$, and $u_* = 0.104$ ft/s $= 0.032$ m/s.

(h) boundary shear stress:

$$\tau_0 = \rho u_*^2 = \frac{1.92 \text{ slug}}{\text{ft}^3} (0.11)^2 \frac{\text{ft}^2}{\text{s}^2} = 0.023 \frac{\text{lb}}{\text{ft}^2} = 1.1 \text{ Pa}.$$

(i) friction slope:

$$S_f = \frac{\tau_0}{\gamma R_h} \cong \frac{\tau_0}{\gamma h} = \frac{0.023 \text{ lb}}{\text{ft}^2} \frac{\text{ft}^3}{62.4 \text{ lb} \times 3.57 \text{ ft}} = 1 \times 10^{-4} \cong \frac{10 \text{ cm}}{\text{km}}.$$

(j) Darcy–Weisbach friction factor (dimensionless):

$$f = \frac{8u_*^2}{V^2} = \frac{8 \times 0.11^2}{0.85^2} = 0.13,$$

or

$$f = \frac{8S_f}{\mathrm{Fr}^2} = \frac{8 \times 1 \times 10^{-4}}{0.078^2} = 0.13.$$

(k) Manning coefficient:

$$n = \frac{1.49}{V} R_h^{2/3} S_f^{1/2} = \frac{1.49\ \mathrm{ft}^{1/3}\mathrm{s}}{0.85\ \mathrm{ft\ m}^{1/3}}(3.57\ \mathrm{ft})^{2/3}(1 \times 10^{-4})^{1/2} = 0.041$$

$$n = \frac{1}{V} R_h^{2/3} S_f^{1/2} = \frac{1\ \mathrm{s}}{0.26\ \mathrm{m}}(1.09\ \mathrm{m})^{2/3}(10 \times 10^{-5})^{1/2} = 0.041.$$

Note that because of the conversion factor $(1.49\ \mathrm{ft}^{1/3}/\mathrm{m}^{1/3})$, the value of n is the same in both SI and English units.

(l) Chézy coefficient:

$$C = \sqrt{\frac{8g}{f}} = \sqrt{\frac{8 \times 32.2\ \mathrm{ft}}{0.13\ \mathrm{s}^2}} = 44.5\ \mathrm{ft}^{1/2}/\mathrm{s},$$

or

$$C = \sqrt{\frac{8g}{f}} = \sqrt{\frac{8 \times 9.81\ \mathrm{m}}{0.13\ \mathrm{s}^2}} = 24.6\ \mathrm{m}^{1/2}/\mathrm{s}.$$

In English units, the value of C is 1.81 times larger than it is in SI units.

"Compound sections" refers to cross sections with distinct features such as flow depth and roughness. For instance, as sketched in Figure 5.18, a channel may have two main stems with different degrees of roughness around a vegetated island. By assuming a constant energy slope through the reach, the sum of conveyances of each segment denoted with a subscript i, gives the total channel conveyance. Accordingly, the total discharge $Q = \sum_i Q_i$, the total cross-section area $A = \sum_i A_i$, the total wetted perimeter $P = \sum_i P_i$, and the total conveyance $K = \sum_i K_i$ yield the following relationships for composite hydraulic radius $\overline{R_h} = A/P$, and composite roughness $\bar{n} = A\overline{R_h}^{2/3}/K$ in SI and $\bar{n} = 1.49A\overline{R_h}^{2/3}/K$ in English units. For the example in Figure 5.18 at a friction

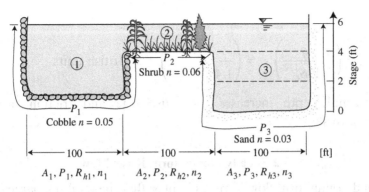

Figure 5.18. Analysis of compound channels

slope of 100 cm/km, the total discharge at a flow depth of 6 ft is $Q = \sum_i Q_i = 4{,}480 \text{ ft}^3/\text{s}$, and the total conveyance is $K = \sum_i K_i = 142{,}000 \text{ ft}^3/\text{s}$, thus resulting in a hydraulic radius $\overline{R_h} = A/P = 1{,}300/318 = 4.1$ ft and a composite Manning $\overline{n} = 1.49 A \overline{R_h}^{-2/3}/K = 0.035$.

5.1.6 Normal Depth

When river flow is steady and uniform, the energy grade line EGL becomes parallel to the channel bed and the friction slope becomes identical to the bed slope, i.e. $S_f = S_0$. The corresponding flow depth in the channel is called the normal depth h_n. The normal depth physically represents the flow depth that rivers strive to attain. For any cross section, the normal depth can be calculated from Equation (5.11) when $S_f = S_0$, or

$$K(h_n) = Q/S_o^{1/2} \tag{5.14}$$

For this general case, the normal depth can always be obtained with equation solvers or numerically by trial and error after changing the flow depth until the conveyance equals the value from the right-hand side of Equation (5.14).

In rivers with a large width–depth ratio, $R_h = h_n$, and Manning's resistance equation is considered. In this case, the normal depth h_n can be simply obtained after $Q = Wq = WVh_n$ is substituted into Equation (5.13):

$$h_n = \left(\frac{nq}{S_0^{1/2}}\right)^{3/5} = \left(\frac{nQ}{WS_0^{1/2}}\right)^{3/5} \quad \text{in SI} \tag{5.15a}\blacklozenge$$

or

$$h_n = \left(\frac{nq}{1.49S_0^{1/2}}\right)^{3/5} = \left(\frac{nQ}{1.49WS_0^{1/2}}\right)^{3/5} \text{ in British units} \qquad (5.15b)\blacklozenge$$

The normal depth increases with discharge and friction factor but decreases with slope.

5.2 Steady Nonuniform River Flow

Steady and nonuniform flow in rivers implies that the total discharge does not change with time, but varies in the downstream direction. Two types of nonuniform flow condition are recognized: (1) rapidly varied flow, in which conditions change within a very short distance and (2) gradually varied flows, in which changes take place over long distances. Rapidly varied flow also branches into two types, depending on whether the flow is converging or diverging. In the case of converging rapidly varied flow, energy is conserved as discussed in Section 5.2.1. In the case of diverging rapidly varied flows, conservation of momentum becomes very important, as reviewed in Section 5.2.2. Gradually varied flows and backwater curves are discussed in Section 5.2.3.

5.2.1 Energy and Rapidly Varied Converging Flows

"Rapidly varied flow" refers to large spatial derivatives over a short channel reach. In converging flows, the energy level is usually constant because the friction losses tend to be negligible over a short distance. The specific energy E derived in Section 2.1.7 can be applied to converging river flows in this section. It is interesting that Equation (2.39) must yield a minimum value of the specific energy E. The minimum when Q and g are constant is obtained when the following condition is satisfied:

$$\frac{dE}{dh} = 1 - \frac{2Q^2}{2gA^3}\frac{dA}{dh}. \qquad (5.16)\blacklozenge$$

Given $dA = W\,dh$, the minimum energy $dE/dh = 0$ defines

$$\frac{Q^2 W}{gA^3} = \frac{V^2 W}{gA} = \frac{V_c^2}{gh_c} = \text{Fr}^2 = 1. \qquad (5.17)\blacklozenge$$

In wide-rectangular channels, $A = Wh$ and $Q = Wq$, the specific energy becomes

$$E = h + \frac{q^2}{2gh^2} = h\left(1 + \frac{\text{Fr}^2}{2}\right). \qquad (5.18)\blacklozenge$$

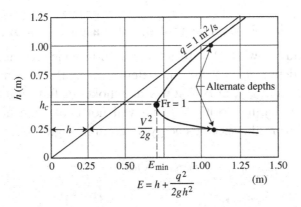

Figure 5.19. Specific-energy diagram and two alternate depths

As plotted in Figure 5.19, the specific energy diagram shows two possible depths at a given energy level, called alternate depths. The diagram also produces a minimum value at the critical depth h_c obtained from $dE/dh = 0$, and

$$h_c = \left(\frac{q^2}{g}\right)^{1/3}. \tag{5.19}\blacklozenge$$

The reader can easily demonstrate that the minimum specific energy $E_{\min} = 1.5h_c$.

Once the critical depth of flow is known, the Froude number can be directly calculated from $V = q/h$ and $q^2 = gh_c^3$ as

$$\mathrm{Fr} = \frac{q}{h\sqrt{gh}} = \left(\frac{h_c}{h}\right)^{3/2}, \quad \text{or} \quad h_c = h\mathrm{Fr}^{2/3}. \tag{5.20}$$

Applications of rapidly varied flows are shown for river-flow contraction in Example 5.2.

Some interesting applications of rapidly converging flows include weirs, gates and spillways, as sketched in Figure 5.20. Weirs can be divided into two groups, depending on the weir thickness, and two types of gates (sluice gates and Tainter gates) are sketched on the same figure. In all cases, the relationship for the flow discharge includes a cross-section area and a flow velocity proportional to the critical velocity characterized by the square root of gravitational acceleration times a flow depth. Approximate relationships for Ogee spillways are also presented in Figure 5.21, with a lot more design information on such structures in Reclamation (1974, 1976, 1977).

Example 5.2 Maximum sill elevation and river contraction

Consider steady flow at $Q = 10$ m³/s in a 10-m wide-rectangular channel at a normal depth, $h_n = 0.788$ m. Determine: (a) the maximum possible elevation of a sill Δz_{max} at A that will not choke the flow and (b) the minimum possible channel width W_2 that will not choke the flow. Answers can be found from the specific-energy diagram in Figure E5.2.

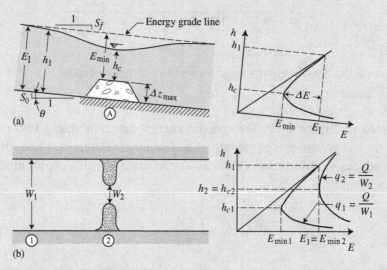

Figure E5.2. Specific energy diagrams for: (a) river sill, and (b) river contraction

Solution: In this example, the unit discharge $q = (Q/W) = [(10$ m³/s$)/10$ m$] = 1$ m²/s, the mean flow velocity $V = q/h_n = 1.27$ m/s, Fr $= V/\sqrt{gh_n} = 0.457$, and $E_1 = h_n(1 + 0.5\text{Fr}^2) = 0.87$ m. The critical depth of flow is $h_c = (q^2/g)^{1/3} = 0.467$ m and the minimum energy $E_{min} = 3/2\, h_c = 0.70$ m.

(a) The maximum elevation of the sill Δz_{max} at A corresponds to critical flow on top of the sill $\Delta z_{max} + E_{min} = E_1$, and thus $\Delta z_{max} = E_1 - E_{min} = 0.870$ m $-$ 0.70 m $= 0.17$ m.

   For a wide-rectangular channel, the conjugate depth can be calculated from the upstream flow depth and Froude number as

$$\Delta z_{max} = h\left(1 + \frac{\text{Fr}^2}{2} - \frac{3\text{Fr}^{2/3}}{2}\right) = 0.788(1 + 0.104 - 0.89) = 0.17 \text{ m.}$$

(b) The minimum channel width $W_{min} = W_2$ at section A that does not cause backwater is such that the total discharge remains constant, $Q = W_1 q_1 = W_2 q_2$.

The flow at section A_2 is critical, or $h_{c2} = 0.667E_{min2} = 0.667E_1$, with $E_1 = h_n(1 + 0.5Fr^2)$ and $1 = q_2^2/gh_{c2}^3$, or

$$W_{min} = \frac{Q}{\sqrt{g\left[0.667h_n\left(1 + 0.5Fr^2\right)\right]^3}}$$

$$= \frac{10 \text{ m}^3 \text{ s}}{\text{s}\sqrt{9.81 \text{ m} \left(0.667 \times 0.788 \text{ m} \times 1.1\right)^3}} = 7.26 \text{ m}.$$

The maximum lateral contraction $\Delta W_{max} = W_1 - W_2 = 10 \text{ m} - 7.22 \text{ m} = 2.78 \text{ m}$.

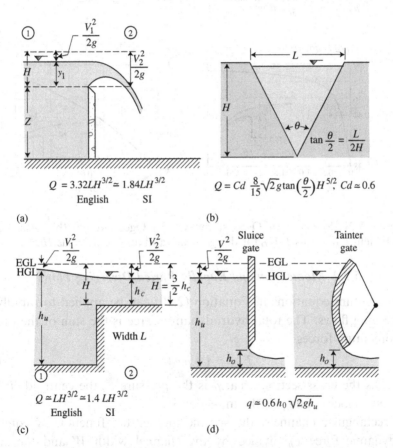

$$Q = 3.32LH^{3/2} \simeq 1.84LH^{3/2}$$
English   SI

$$Q = Cd\,\frac{8}{15}\sqrt{2}g\tan\left(\frac{\theta}{2}\right)H^{5/2},\ Cd \simeq 0.6$$

(a)   (b)

$$Q \simeq LH^{3/2} \simeq 1.4\,LH^{3/2}$$
English   SI

$$q \simeq 0.6h_0\sqrt{2gh_u}$$

(c)   (d)

Figure 5.20. Sketches of weirs and gates: (a) Rectangular sharp-crested weir, (b) Triangular sharp-crested weir, (c) Broad-crested weirs, and (d) Sluice and tainter gates

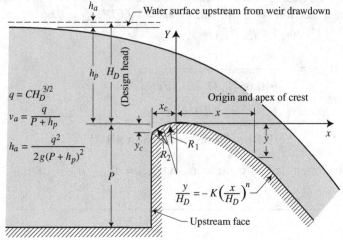

For vertical upstream face and $h_a = 0$; $K = 0.50$; $n = 1.872$; $R_1 = 0.53H_D$; $R_2 = 0.235H_D$; $x_c = 0.283H_D$; $y_c = 0.127H_D$

(a)

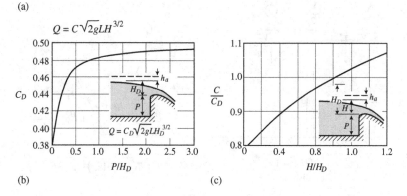

(b)   (c)

Figure 5.21. Sketches of Ogee spillways: (a) Ogee profile, (b) Discharge coefficient C_D versus P/H_D for design head, and (c) C/C_D versus H/H_D

5.2.2 *Momentum and Rapidly Varied Diverging Flows*

The momentum equations in Equation (2.30) can be applied to rapidly varied diverging flows. The total hydrodynamic force is the sum of the pressure and momentum forces

$$F_h = \overline{p}A + \rho A V^2 ,\qquad (5.21)$$

where A is the cross-section area, $\overline{p}$ is the pressure at the centroid, V is the mean flow velocity, and ρ is the mass density.

For rectangular channels, the specific momentum function M equals the hydrodynamic force F_h divided by the channel width W and the specific weight ρg, or

$$M = \frac{F_h}{\rho g W} = \frac{h^2}{2} + \frac{q^2}{gh} = h^2 \left(0.5 + \mathrm{Fr}^2\right),\qquad (5.22)$$

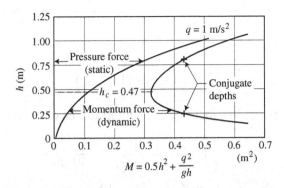

Figure 5.22. Specific-momentum diagram and two conjugate depths

where the unit discharge $q = Vh$. This specific-momentum function M can be plotted as a function of flow depth at a given unit discharge q, e.g. Figure 5.22 for $q = 1$ m²/s. The two flow depths with the same specific momentum are called conjugate depths.

Examples 5.3 and 5.4 show applications of specific momentum and specific energy.

Example 5.3 Momentum application to a hydraulic jump

Consider a 10-m-wide channel with a steady flow $Q = 10$ m³/s. The upstream flow velocity $V = 4$ m/s decreases rapidly in a hydraulic jump. Let's determine the downstream flow depth, velocity, head loss, and the total force. The turbulence level in rapidly decelerating flows far exceeds the energy losses from bed resistance equations. The bed friction force on a smooth surface without baffle blocks is small compared with the pressure and momentum forces. Equilibrium is obtained when the upstream and downstream forces are equal, thus solving for the conjugate depths in Figure 5.22, or

$$\frac{q^2}{g}\left(\frac{1}{h_1} - \frac{1}{h_2}\right) = \frac{1}{2}\left(h_2^2 - h_1^2\right). \tag{E5.3.1}$$

The relationship $\mathrm{Fr}_1^2 = q^2/gh_1^3$ allows Equation (E5.3.1) to be rewritten as

$$\mathrm{Fr}_1^2 = \frac{1}{2}\frac{h_2}{h_1}\left(\frac{h_2}{h_1} + 1\right). \tag{E5.3.2}$$

The solution of this quadratic equation in h_2/h_1 is called the Bélanger equation:

$$\frac{h_2}{h_1} = \frac{1}{2}\left(\sqrt{1 + 8\mathrm{Fr}_1^2} - 1\right). \qquad \text{(E5.3.3)}\blacklozenge$$

The specific energy lost in a hydraulic jump can be calculated directly from

$$\Delta E = \left(h_2 + \frac{q^2}{2gh_2^2}\right) - \left(h_1 + \frac{q^2}{2gh_1^2}\right) = \frac{(h_2 - h_1)^3}{4h_2 h_1}. \qquad \text{(E5.3.4)}\blacklozenge$$

Numerically, for $Q = 10$ m³/s and $W = 10$ m, or $q = Q/W = 1$ m²/s, the upstream flow depth is $h_1 = q/V_1 = 0.25$ m and $\mathrm{Fr}_1^2 = \frac{q^2}{gh_1^3} = 6.52$, or $\mathrm{Fr}_1 = 2.55$. The corresponding downstream flow depth from the Bélanger equation is $h_2 = (0.25/2)(\sqrt{1 + 8 \times 6.52} - 1) = 0.79$ m and the downstream velocity $V_2 = 1.27$ m/s.

The energy loss is obtained from Equation (E5.3.4) as

$$\Delta E = \frac{(0.79 - 0.25)^3}{4 \times 0.79 \times 0.25} = 0.2 \text{ m}.$$

The length of hydraulic jumps is approximately six times the downstream flow depth and would be about 5 m in this case. (See Figure E5.3)

The hydrodynamic force is calculated by multiplying the specific-momentum value by $\rho g W$, or from Equation (5.22), the hydrodynamic force is

$$F_{\text{total}} = \rho g W h^2 (0.5 + \mathrm{Fr}_1^2) = \frac{9.81 \text{ kN}}{\text{m}^3} \times 10 \text{ m} \times (0.25 \text{ m})^2 (0.5 + 6.52) = 43.1 \text{ kN}.$$

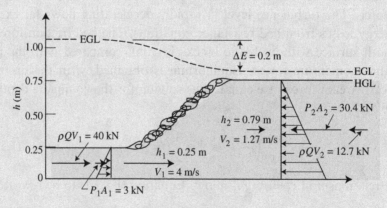

Figure E5.3. Momentum and hydraulic jump

5.2.3 Gradually Varied River Flows

Gradually varied flows are those where changes in width, depth, and velocity take place over reasonably long distances. From the analysis of the Saint-Venant equations in Section 2.1.6, the gradual changes in flow depth h in the downstream x direction for steady flows can be described from Equation (2.38) as:

$$\frac{dE}{dx} \cong \frac{d}{dx}\left(h + \frac{Q^2}{2gA^2}\right) = \frac{d}{dh}\left(h + \frac{Q^2}{2gA^2}\right)\frac{dh}{dx} = S_0 - S_f. \qquad (2.38)$$

In the case of constant discharge, the cross-section area A varies with depth and $dA/dh = W$, such that the derivative of the specific energy gives the following:

$$\frac{dE}{dh} = \frac{d}{dh}\left\{h + \frac{Q^2}{2g[A(h)]^2}\right\} = \left[1 + \frac{Q^2(-2)}{2gA^3}\frac{dA}{dh}\right] = \left(1 - \frac{Q^2 W}{gA^3}\right) = 1 - \mathrm{Fr}^2.$$

$$(5.23)$$

The changes in flow depth with distance x can therefore be determined by combining these two equations and solving for dh/dx as

$$\frac{dh}{dx} = \frac{S_0 - S_f}{1 - \mathrm{Fr}^2}. \qquad (5.24)\blacklozenge\blacklozenge$$

Resistance to flow in gradually varied flow is calculated as for steady-uniform flow. From Equation (5.13), the general form of the backwater equation in SI units is obtained as

$$\frac{dh}{dx} = \frac{S_0 - \left(\dfrac{nQ}{AR_h^{2/3}}\right)^2}{\left(1 - \dfrac{Q^2 W}{gA^3}\right)}. \qquad (5.25)$$

This equation is applicable to any cross-section geometry. In the case of English units, Manning n is replaced by $n/1.49$.

In wide-rectangular channels, simplifications arise from $h = R_h$, and from Equations (5.15) and (5.20), the gradually varied flow Equation (5.24) for constant unit discharge q; and Manning n can be written in dimensionless form as

$$\frac{dh}{dx} = \frac{S_0\left[1 - \left(\dfrac{h_n}{h}\right)^{10/3}\right]}{\left[1 - \left(\dfrac{h_c}{h}\right)^3\right]} \qquad (5.26)$$

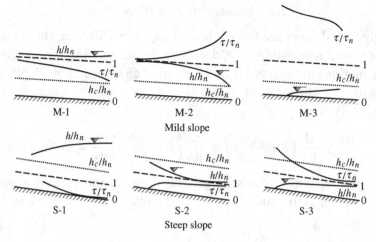

Figure 5.23. Typical backwater profiles for mild and steep slopes

where S_0 is the channel bed slope, the normal depth h_n is calculated from Equation (5.15), and the critical depth h_c from Equation (5.19). From Equation (5.20), the ratio of normal and critical depths is only a function of the Froude number, $h_n/h_c = \mathrm{Fr}^{-2/3}$.

Backwater profiles define the shape of the water-surface elevation changes in the downstream direction. Five types of backwater profiles are possible:

1. S profiles for steep slopes when $h_n < h_c$ or $S_0 > f/8$,
2. C profiles for critical slopes when $h_n = h_c$ or $S_0 = f/8$,
3. M profiles for mild slopes when $h_n > h_c$ or $S_0 < f/8$,
4. H profiles for horizontal surfaces with $h_n \to \infty$,
5. A profiles for adverse slopes when $S_0 < 0$.

The ratio of friction to bed slope $S_f/S_0 = (h_n/h)^{10/3}$ and the ratio τ/τ_n of the shear stress at flow depth h to the shear stress at normal depth h_n for wide-rectangular channels becomes

$$\frac{\tau}{\tau_n} \cong \frac{\gamma h S_f}{\gamma h_n S_0} = \frac{h}{h_n}\left(\frac{h_n}{h}\right)^{10/3} = \left(\frac{h_n}{h}\right)^{7/3}. \qquad (5.27)\blacklozenge$$

Typical water-surface profiles in open channels are shown in Figure 5.23. For supercritical flows, the flow depth is controlled upstream and for subcritical flows, the flow-depth control is downstream. Backwater profiles tend to reach the normal depth asymptotically, while flows become rapidly varied near critical depth.

The bed shear stress increases ($\tau > \tau_n$) at flow depths less than normal depth ($h < h_n$). The bed shear stress increases in the downstream direction

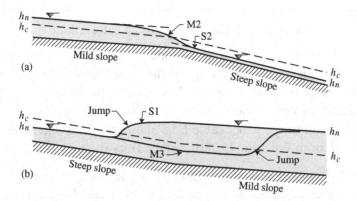

Figure 5.24. Slope transition from: (a) mild to steep, and (b) steep to mild

for converging flows (M-2 and S-2 backwater curves), and decreases for diverging flows (M-1, M-3, S-1, and S-3 backwater curves). Consequently, bed sediment transport increases in the downstream direction for converging flows and decreases in the downstream direction for diverging flows. Transitions from steep to mild slopes are also sketched in Figure 5.24.

There are two basic methods for calculating backwater profiles: (1) the direct step method and (2) the standard step method. Numerical calculations proceed from a given flow depth h_1. The direct step method determines Δx at which $h_2 = h_1 + \Delta h$:

$$\Delta x \cong \frac{\Delta h \left[1 - \left(\dfrac{h_c}{h_1}\right)^3\right]}{S_0 \left[1 - \left(\dfrac{h_n}{h_1}\right)^{10/3}\right]}. \tag{5.28a}$$

The standard step method finds the change in flow depth Δh given the fixed length Δx, or

$$\Delta h \cong \frac{S_0 \Delta x \left[1 - \left(\dfrac{h_n}{h_1}\right)^{10/3}\right]}{\left[1 - \left(\dfrac{h_c}{h_1}\right)^3\right]}. \tag{5.28b}$$

Note that the backwater calculation process is moving upstream when $\Delta x < 0$. An application is presented in Example 5.4.

Example 5.4 Backwater profile calculations

Consider the 10-m-wide canal at a slope of 170 cm/km from Examples 5.2–5.3. Given the normal depth $h_n = 0.788$ m and critical depth $h_c = 0.467$ m at a discharge of 10 m³/s, the M-1 backwater profile can be calculated with the standard step method with $\Delta x = -100$ m starting at a flow depth of 1.45 m upstream of the gate. Table E5.5 shows the backwater calculations in SI with Δh from Equation (5.28b), and bed elevation $z = -xS_0$.

Table E5.5. *Standard step M-1 backwater calculations*

x m	h m	Δh m	z m	HGL m	EGL m
0	1.45	−0.153	0	1.45	1.474
−100	1.297	−0.144	0.17	1.467	1.497
−200	1.153	−0.131	0.34	1.493	1.531
−300	1.022	−0.109	0.51	1.532	1.581
−400	0.913	−0.076	0.68	1.593	1.654
−500	0.837	−0.037	0.85	1.687	1.760
−600	0.799	−0.01	1.02	1.819	1.899
−700	0.789	−0.001	1.19	1.979	2.061

Exercise 5.1

Determine the cross-section area, mean flow velocity, wetted perimeter, mean flow depth, and hydraulic radius of the river sketched below. Find the slope if Manning $n = 0.025$. (See Figure EX5.1)

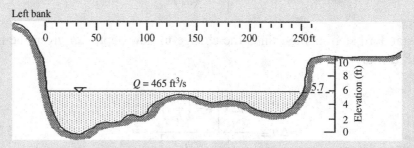

Figure EX5.1. Cross-section profile of the Matamek River

Exercise 5.2

Bray (1979) proposed a relationship for Manning n as $n = 0.048 d_{50}^{0.179}$ where d_{50} is the median diameter of the bed material in ft. Compare with $n = 0.064 d_{50}^{1/6}$ with d_{50} in m.

♦**Problem 5.1 Channel Geometry**

Consider the cross section of a river with a flood plain in the sketch below. With all lengths in feet, develop a spreadsheet to calculate the channel width, cross-section area, wetted perimeter, hydraulic radius, and mean flow depth as a function of maximum flow depth, or stage. Calculate every 4″ up to a maximum flow depth of 10 ft and plot all parameters. Discuss the effects of the discontinuity at a flow depth of 5 ft. (See Figure P5.1)

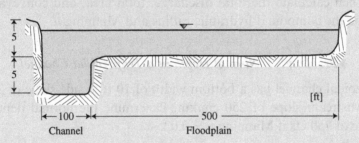

Figure P5.1. Geometry of a floodplain section

♦♦**Problem 5.2 At-a-station Hydraulic Geometry**

A cross-section of the Missouri River near Omaha is sketched below with contours describing sediment concentration in mg/l. Determine the cross-section area, mean flow velocity, mean flow depth, hydraulic radius, and Manning n. Find the coefficients and exponents of at-a-station hydraulic-geometry equations. (See Figure P5.2)

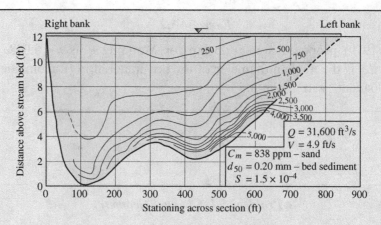

Figure P5.2. Cross section of the Missouri River

◆Problem 5.3 Flow Conveyance

Consider the compound cross section presented in Figure 5.18 at a friction slope $S_f = 0.001$. Use a spreadsheet to calculate the cross-section area, wetted perimeter, hydraulic radius, flow velocity, conveyance, and discharge for each segment of the cross section at incremental depths of 0.5 ft. Then calculate the total discharge, total area, and conveyance to determine the compound hydraulic radius and Manning n.

◆Problem 5.4 Normal Depth in a Trapezoidal Channel

A trapezoidal channel has a bottom width of 10 ft, a side slope of 2H:1V, and downstream slope of 250 cm/km. Determine the normal depth at a discharge of 750 cfs if Manning $n = 0.02$.

[*Answer*: Equation (5.13) gives $AR_h^{2/3} = \frac{nQ}{1.49S_0^{1/2}} = 201$. The area A and hydraulic radius R_h for trapezoidal channels (Figure 5.3) result in $(Bh_n + th_n^2)\left[\frac{Bh_n + th_n^2}{B + 2h_n(1+t^2)^{1/2}}\right]^{2/3} = 201$. With an equation solver, the normal depth for $B = 10$ and $t = 2$ is $h_n = 4.9$ ft.]

◆◆Problem 5.5 Floodplain Flow Characteristics

With reference to the spreadsheet developed for Problem 5.1, consider Manning $n = 0.02$ and a slope of 0.005. Calculate the flow velocity, discharge, and dQ/dA and plot the results as a function of stage from 0 to 10 ft.

Consider a second case after river restoration where the flood plain will be vegetated and the floodplain roughness is increased to $n = 0.1$ while the main channel remains at $n = 0.02$. In this second case, determine the composite Manning n and compare the results. What is the difference in discharge at the maximum stage? What percentage of the total discharge flows in the main channel when the flood plain is with/without vegetation? What is the difference in stage for the two cases during flood conditions? Discuss the results.

◆Problem 5.6 Normal and Alternate Depths in a Trapezoidal Channel

A trapezoidal channel has a base width of 5 m and side slopes at 2H:1V and a downstream slope of 50 cm/km. If Manning n is 0.021, determine the following at a discharge of 50 cms: (a) the normal depth; (b) the Froude number; (c) the alternate depth; and (d) the critical depth.
[*Answers*: Fr = 0.3, h_c = 1.72 m.]

◆◆Problem 5.7 Maximum Sill Height in a Trapezoidal Channel

A trapezoidal channel has a base width of 6 ft and side slopes at 3H:2V with a flow depth of 4.5 ft and a mean flow velocity of 3.5 ft/s. If a 9″ sill is built over a short reach, determine the flow depth at that raised section. Calculate the maximum possible change in bed elevation without causing backwater.
[*Answers*: y_2 = 3.55 ft and Δz_{max} = 1.15 ft.]

◆◆Problem 5.8 Maximum River Contraction

Fotherby (2009) reported on the Platte River in Nebraska. The river width varies widely from 200 to 600 m and the bankfull depth is about 1.0 m. The bed material is slightly coarser than 1.0 mm and the bed slope is 0.0012. Consider a 1.5-year discharge of 100 cms and assume Manning n for dunes for this braided channel and estimate the flow velocity, normal depth, and Froude number. From these values, determine the maximum sill height and minimum channel width that would not cause backwater?
[*Answers*: With $n = 0.03$ and $W = 200$ m, one obtains $h_n = 0.6$ m, $V = 0.83$ m/s, and Fr = 0.34. The critical conditions are $W_{min} = 134$ m, and $\Delta z_{max} = 0.2$ m.]

♦♦♦*Computer Problem 5.1 Backwater Behind a Reservoir*

Consider steady flow $q = 3.72$ m^2/s in the impervious channel sketched below. Assume a wide-rectangular channel with $n = 0.03$. Calculate the backwater profile and determine the following parameters along the 25-km reach of the channel when the water-surface elevation at the dam is 10 m above the bed elevation: (a) flow depth in meters; (b) mean flow velocity in meters per second; and (c) bed shear stress in Pascals. (See Figure CP5.1)

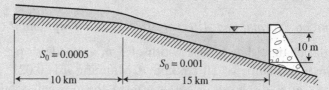

Figure CP5.1. River reach for backwater calculations

6

Unsteady Flow in Rivers

This chapter is concerned with unsteady flow in rivers. The propagation of a solitary wave without friction is discussed in Section 6.1. Unsteady flow equations in open channels with friction define kinematic and dynamic waves in Section 6.2. The concept of flood wave celerity is then introduced, in Section 6.3. The amplification and attenuation of surface waves are then covered, for long waves in Section 6.4 and for short waves in Section 6.5. The effects of flow pulses on sediment transport are introduced in Section 6.6. The chapter ends with a description of river hysteresis and loop-rating curves, in Section 6.7.

6.1 Solitary Wave Propagation

Consider a solitary wave produced by a sudden horizontal displacement of a vertical gate in a laboratory flume. Assume a frictionless channel and a wave traveling without changing shape or velocity. As sketched in Figure 6.1, the solitary wave travels to the right with celerity c in a stationary fluid. An observer moving along the wave crest at a velocity equal to the celerity will perceive steady flow where the wave appears to stand still with the flow moving to the left at a velocity equal to c.

When the continuity relationship is applied to the steady-flow case, constant discharge implies that the relative velocity under the crest is $ch/(h + \Delta h)$. The equation of motion is then applied to the steady relative motion. When friction is neglected and a small slope is assumed, the energy equation between the normal section and the wave crest simply describes conservation of specific energy, or

$$h + \frac{c^2}{2g} = h + \Delta h + \frac{c^2}{2g}\left(\frac{h}{h + \Delta h}\right)^2 \tag{6.1}$$

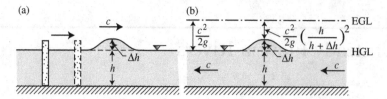

Figure 6.1. Solitary wave without friction: (a) fixed, and (b) moving frame of reference

which can be solved for c to give the formulation from Saint-Venant (1870),

$$c = \sqrt{\frac{2g(h + \Delta h)^2}{2h + \Delta h}}.$$

(6.2)

This description of the 1-D propagation of small waves in still water can be reduced further for waves of small amplitude, $\Delta h < h$, and the wave celerity simply becomes

$$c = \sqrt{gh}.$$

(6.3)

This is known as the Lagrange (1788) celerity relationship, describing the celerity of small amplitude waves in a rectangular channel without friction. The Froude number is often described as $\mathrm{Fr} = V/c$ given by the ratio of the mean flow velocity to the celerity of small perturbations. It is important to remember that this relationship is obtained without resistance to flow and for small-amplitude waves. The effects of resistance to flow are considered in the next section.

6.2 Kinematic and Dynamic Waves

The governing equations for unsteady flow in rivers were derived in Chapter 2. The main relationships include conservation of mass and momentum. The definition sketch in Figure 6.2 portrays a river reach with a top width W, cross-section area A, wetted perimeter P, hydraulic radius $R_h = A/P$, and mean flow depth $h = A/W$. The flow discharge $Q = AV$ where V is the mean flow velocity perpendicular to the area A.

For an impervious channel without rainfall and without lateral inflow, the one-dimensional (1-D) continuity equation, Equation (2.12), simply reduces to

$$\frac{\partial A}{\partial t} + \frac{\partial Q}{\partial x} = 0.$$

(6.4)◆◆

The continuity relationship links space and time and will be used to describe the main characteristics of flood-wave propagation.

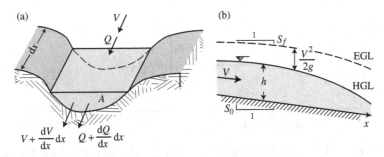

Figure 6.2. River flow: (a) definition sketch, and (b) longitudinal profile

In Section 2.1.6, the momentum equation for one-dimensional channels has been reduced to the Saint-Venant equation, which can be written as

$$S_f \cong S_0 - \frac{\partial h}{\partial x} - \frac{V \partial V}{g \partial x} - \frac{1}{g}\frac{\partial V}{\partial t}$$

$$(1) \quad (2) \quad (3) \quad (4) \qquad (5)$$

$$\underbrace{}_{\text{kinematic}}$$

$$\underbrace{}_{\text{diffusive}}$$

$$\underbrace{}_{\text{quasi-steady}}$$

$$\underbrace{}_{\text{full dynamic}}$$

$$(6.5)\blacklozenge\blacklozenge$$

Different flood-wave propagation types can be identified depending on the various approximations of Equation (6.5). The dynamic-wave approximation of the Saint-Venant equation describes flows where all terms of Equation (6.5) are significant. For all steady flows, the quasi-steady dynamic-wave approximation includes the first four terms of Equation (6.5); this also corresponds to the backwater equations for gradually varied flows used in Chapter 5. The diffusive-wave approximation, $S_f = S_0 - \partial h/\partial x$, is most commonly used to analyze flood-wave propagation in rivers. Finally, the kinematic-wave approximation is obtained when all but the first two terms vanish. This is typically the case when the channel slope is very large, e.g. in some mountain streams.

6.3 Flood-Wave Celerity

Flood-wave celerity involves a relationship between cross-section area A and discharge Q, or simply a resistance to flow equation. Let us illustrate the concept by using a wide-rectangular channel for which resistance to flow can

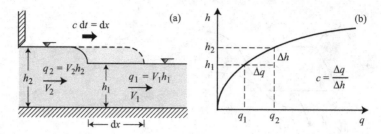

Figure 6.3. Channel celerity: (a) definition sketch, and (b) stage-discharge relationship

be described in here for Manning's relationship, $\alpha = S_f^{1/2}/n$ and $\beta = 5/3$. Figure 6.3 illustrates a simple rectangular channel of width W with steady flow q_1 at a flow depth h_1. Now if the upstream gate is suddenly opened to allow a steady flow q_2 at a flow depth h_2, how fast will the wave front propagate in the downstream direction. Quite simply, by continuity, the difference in flow discharge $W(q_2 - q_1)$ equals the difference in area times the celerity $cW(h_2 - h_1)$. In other words, the speed at which the wave front propagates is called the celerity $c = W(q_2 - q_1)/W(h_2 - h_1)$ or simply $c = \Delta Q/\Delta A = \Delta q/\Delta h$.

In a more rigorous mathematical sense, unsteady flow describes changes in flow discharge Q with time t, and downstream distance x. Therefore the exact differential is

$$dQ = \frac{\partial Q \, dx}{\partial x} + \frac{\partial Q \, dt}{\partial t}. \tag{6.6}$$

By definition, the celerity can also be described as $c = dx/dt$, which defines the location point where the flow is made steady, i.e. $dQ = 0$. The celerity c at which space–time changes take place is simply the solution of Equation (6.6) for dx/dt when $dQ = 0$, or

$$c \equiv \frac{dx}{dt} \equiv \frac{-\partial Q}{\partial t} \frac{\partial x}{\partial Q}. \tag{6.7}$$

If we now consider conservation of mass in a one-dimensional impervious channel, we obtain from Equation (6.4) that $\partial Q/\partial x = -\partial A/\partial t$. Substituting into Equation (6.7) gives

$$c = \frac{\partial Q}{\partial A}. \tag{6.8} ◆◆$$

This relationship for flood-wave celerity is referred to as the Kleitz–Seddon law, from the work of Kleitz (1877) and Seddon (1900). The flood-wave celerity in rivers increases with discharge as long as the river discharge

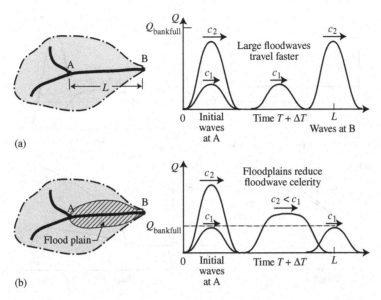

Figure 6.4. Flood-wave propagation characteristics: (a) $Q < Q_{bankfull}$ and (b) $Q > Q_{bankfull}$

level remains below bankfull discharge level. Within the main channel, larger flood waves (larger discharge) propagate faster than small flood waves, as sketched in Figure 6.4.

From Section 5.1.1, the at-a-station hydraulic geometry relationship between mean flow velocity and discharge can be written as $V = aQ^b$, and the celerity then becomes $c = \partial Q/\partial A = [1/(1 - b)]V$. As an example, the at-a-station hydraulic geometry of the Matamek River was $V \cong 0.022Q^{0.54}$, and yields $A = Q/V \cong 45.4Q^{0.46}$ with a celerity $c = \partial Q/\partial A = 0.048Q^{0.54}$, which does increase with discharge. In fact, with $b = 0.54$, the celerity is exactly $(1/0.46) = 2.17$ times greater than the mean flow velocity. Let's remember that flood waves travel faster than the fluid.

When the flow exceeds the bankfull discharge, the situation can be different. Typically, the increase in cross-section area may become very large for flow over vegetated floodplains while the discharge increase is minimal. This slows down the propagation of flood waves when the flow exceeds the bankfull discharge. Thus, floodplains have a particular effect on flood-wave propagation characteristics. Figure 6.5 illustrates the effects of a main channel with and without a floodplain on the propagation characteristics of flood hydrographs. Two values of Manning roughness are considered, n_c and n_f for the main channel and the floodplain, respectively. Figure 6.5b shows that

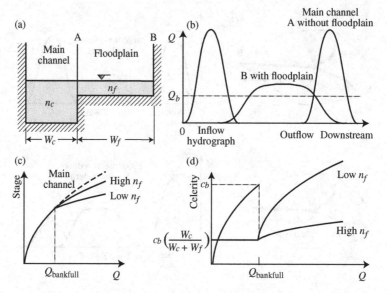

Figure 6.5. Flood-wave propagation characteristics with and without a floodplain: (a) floodplain sketch, (b) effect of hydrographs, (c) stage, and (d) celerity

the channel widening as the stage exceeds bankfull effectively increases the conveyance of the channel and increases the discharge at a given stage. It is clear that dense floodplain vegetation will increase n_f, such that the water depth required for a given discharge will be higher than for the case of low n_f. The effects of floodplains on flood-wave celerity are also sketched in Figure 6.5c and the discontinuity in the celerity relationship near bankfull stage is most interesting. The discontinuity can be analytically examined using the Kleitz–Seddon law, where the increase in discharge ΔQ near bankfull is simply that of the main channel while the cross-section area suddenly increases from $W_c \Delta h$ to $(W_c + W_f)\Delta h$, where W_c is the main channel width and W_f is the width of the floodplain. The flood-wave propagation celerity thus decreases suddenly from the celerity c_b at a stage slightly below bankfull to a celerity $c_b W_c / (W_c + W_f)$ at a stage slightly above bankfull. Surface roughness also affects the celerity with lower flood-wave celerity obtained when the floodplain roughness is high (vegetated floodplains). The effects of floodplains on flood hydrographs propagation and on stage–discharge relationships are sketched in Figure 6.5d. In a nutshell, floodplains reduce both the flood-wave propagation celerity and the peak discharge in the downstream direction. In comparison with Case B for a river with a floodplain, the effects of channelization (sketched as Case A) and floodplain removal would quickly drain surface waters while increasing the downstream peak discharge.

These changes in flood-wave celerity with discharge cause non-linearity in the downstream amplification or attenuation. Techniques based on the linear superposition of different hydrographs, e.g. the unit hydrograph, fail to adequately simulate the flood-wave propagation in river channels. While linear methods may be better suited for small watersheds, as discussed in Section 3.3.3, nonlinear flood wave propagation methods are better suited for partial hydrographs (Saghafian et al., 2002).

6.4 Flood-Wave Propagation

For a wide-rectangular channel, the unit discharge q varies with depth h according to resistance relationships, such as $q = Vh = \alpha h^{\beta}$. This is a very good approximation for long waves like flood waves. The equations describing wave propagation and wave celerity, respectively, reduce to

$$\frac{\partial h}{\partial t} + \frac{\partial q}{\partial x} = \frac{\partial h}{\partial t} + \frac{\partial q}{\partial h}\frac{\partial h}{\partial x} = \frac{\partial h}{\partial t} + c\frac{\partial h}{\partial x} = 0 \tag{6.9}$$

$$c \equiv \frac{dx}{dt} = \frac{\partial q}{\partial h} = \beta \alpha h^{\beta-1} = \beta V. \tag{6.10}\blacklozenge$$

The flood-wave celerity c travels faster than the flow velocity because $\beta > 1$, e.g. $\beta = 3$ for laminar flow and $\beta = 5/3$ for turbulent flow with constant Manning n (see Table 3.3).

Let us include the resistance relationship $V = \alpha h^{\beta-1}$, which yields the following derivatives, $\partial V = (\beta - 1)(V/h)\partial h$. Accordingly, the reader can easily demonstrate that the equation of motion Equation (6.5) can be rewritten as

$$S_f = S_0 - \left[1 + (\beta - 1)\mathrm{Fr}^2\right]\frac{\partial h}{\partial x} - \frac{1}{g}\frac{\partial V}{\partial t}. \tag{6.11a}$$

Note that for steady flow, this equation is slightly different from the gradually varied flow equations in Section 5.2.3. Well, this is because q is not constant here. Now, the continuity relationship Equation (6.4) provides a link between space and time derivatives from $\partial h/\partial t = -\partial q/\partial x = -V\partial h/\partial x - h\partial V/\partial x$. This can be combined with $\partial V = (\beta - 1)(V/h)\partial h$ to reduce the unsteady term in Equation (6.11a) into a space derivative, or

$$S_f = S_0 - \left[1 - (\beta-1)^2\mathrm{Fr}^2\right]\frac{\partial h}{\partial x} = S_0 - K_d\frac{\partial h}{\partial x} \tag{6.11b}\blacklozenge$$

Table 6.1. *Conversion factors for flood-wave propagation*

$\dfrac{\partial h}{\partial x}$	$\dfrac{\partial h}{\partial t}$	$\dfrac{\partial V}{\partial x}$	$\dfrac{\partial V}{\partial t}$	$\dfrac{\partial Q}{\partial x}$	$\dfrac{\partial Q}{\partial t}$
$\dfrac{\partial h}{\partial x} =$	$\dfrac{-1}{\beta V}\dfrac{\partial h}{\partial t}$	$\dfrac{h}{(\beta-1)V}\dfrac{\partial V}{\partial x}$	$\dfrac{-h}{\beta(\beta-1)V^2}\dfrac{\partial V}{\partial t}$	$\dfrac{h}{\beta Q}\dfrac{\partial Q}{\partial x}$	$\dfrac{-Wh^2}{\beta^2 Q^2}\dfrac{\partial Q}{\partial t}$

Note: Calculated from $Q = WhV = W a h^\beta$, $\partial h/\partial t = -\partial Q/W\partial x$, and $c = dx/dt = \beta V$.

where $K_d = 1 - (\beta - 1)^2 \mathrm{Fr}^2$ is the flood-wave diffusivity, which depends on the values of β and Fr. For instance, Manning's equation is applicable in most rivers ($\beta = 5/3$), and K_d is positive when Fr < 1.5, and negative when Fr > 1.5. In most rivers, the Froude number is quite low and K_d often becomes close to unity.

It is interesting to rewrite the diffusivity factor K_d in terms of Manning's equation. Thus from Equation (3.12),

$$K_d = 1 - (\beta-1)^2 \mathrm{Fr}^2 \simeq 1 - \frac{4}{9}\frac{Q^{0.2}S^{0.9}}{gn^{1.8}W^{0.2}}. \qquad (6.12)$$

The reader will notice that K_d becomes increasingly close to one when the channel width and roughness increase. Diffusivity also decreases at high discharges on steep slopes. Following similar reasoning, continuity and resistance to flow can allow us to rewrite $\partial h/\partial x$ as a function of other terms. A list of conversion factors is presented in Table 6.1.

For instance, the Saint-Venant equation can be rewritten as a function of discharge as

$$S_f = S_0 - K_d \frac{\partial h}{\partial x} \cong S_0 + K_d \frac{9Wh^2}{25Q^2}\frac{\partial Q}{\partial t}. \qquad (6.13)\blacklozenge$$

It is clear from this relationship that the friction slope exceeds the bed slope ($S_f > S_0$) during the rising limb ($\partial Q/\partial t > 0$) of a hydrograph, and vice versa, slope ($S_f < S_0$) during the falling limb ($\partial Q/\partial t < 0$). Flashy hydrographs (large $\partial Q/\partial t$) will also see greater deviations from the kinematic wave approximation of the Saint-Venant equations.

The propagation dynamics of flood waves can be studied after the friction slope is brought back into the Manning relationship such that the flow velocity becomes a function of the dynamic changes in flow depth and discharge in the downstream direction. The flow velocity V is used alongside celerity $c = \beta V$ and discharge $Q = AV$:

$$V = \frac{1}{n}R_h^{2/3}\left(S_0 - K_d\frac{\partial h}{\partial x}\right)^{1/2}. \qquad (6.14)$$

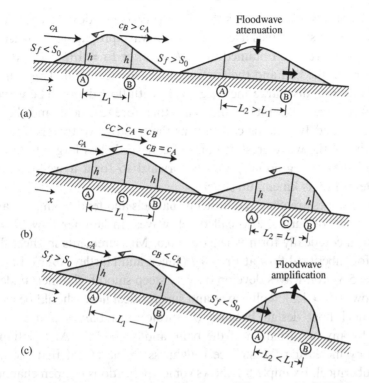

Figure 6.6. Flood-wave attenuation and amplification: (a) Diffusive wave $K_d > 0$ (low Fr), (b) Kinematic wave $K_d = 0$ and $S_f = S_0$, and (c) Dynamic wave $K_d < 0$ (high Fr)

Let's now consider a flood wave in a wide-rectangular channel, as sketched in Figure 6.6. Two points at equal flow depth (same n and R_h) are, respectively, located upstream (A) and downstream (B) of the wave crest. The difference between these two points is that $\partial h/\partial x < 0$ on the downstream side and $\partial h/\partial x > 0$ on the upstream side of the wave crest. The reference discharge Q_r corresponds to the steady discharge from the kinematic wave approximation, i.e. $S_0 = S_f$, or $Q_r = (A/n) R_n^{2/3} S_0^{1/2}$. Three types of flood wave are recognized depending on the value of the diffusivity factor K_d: (1) a diffusive wave when $K_d > 0$, or Fr $< 1/(\beta - 1)$; (2) a kinematic wave when $K_d = 0$, or Fr $= 1/(\beta - 1)$; and (3) a dynamic wave when $K_d < 0$, or Fr $> 1/(\beta - 1)$.

Diffusive waves are most common in rivers, the flow is subcritical and $K_d > 0$. Flood routing is adequately described by the diffusive-wave approximation of the Saint-Venant equation. As sketched in Figure 6.6a, $S_f < S_0$ upstream and $S_f > S_0$ downstream of the wave crest of diffusive waves. This implies that the two points, A and B, at identical flow depth will propagate at different celerities where the downstream point B moves faster downstream than point A. This will result in flood-wave attenuation, or stretching of the distance separating A and B, as the flood propagates downstream. The peak discharge of diffusive waves decreases as the wave propagates downstream.

Kinematic waves are obtained when $K_d > 0$, or $Fr = 1/(\beta - 1)$. As sketched in Figure 6.6b, the bed and the friction slopes are identical. This implies that wave celerity and discharge increase solely with flow depth. The wave celerities and discharges at points A and B are therefore identical, and the distance separating A and B remains constant as the flood wave travels downstream. The celerity of the wave crest at point C is nevertheless larger than it is at A or B, and the crest gradually moves forward to form a well-defined wave front, referred to as kinematic shock.

Dynamic waves are observed in steep channels and tend to amplify and form pulsating flows or surges, also called roll waves. In laminar flow ($\beta = 3$), roll waves can theoretically form when $Fr > 0.5$. Measurements in sheet flows are possible for subcritical flow at $Fr > 0.7$ (Julien and Hartley, 1986). In turbulent flows ($\beta = 5/3$), roll waves develop on very steep smooth channels under supercritical flows ($Fr > 1.5$). Roll waves and supercritical flows should be avoided in open-channel flow design because of surface instabilities and cross waves incurred by any perturbation of the bank and/or the bed. Attenuation can be achieved by increasing the surface roughness to the extent that the flow will remain subcritical. Example 6.1 shows some applications to open channels.

Example 6.1 Application of the Saint-Venant equation

An observer measures the flow depth in a 50-m wide-rectangular channel inclined at $S_0 = 0.003$ with Manning coefficient $n = 0.03$. Initially, the flow depth is 1.0 m and the water level rises at a rate of 1 m/h. Calculate: (1) the initial discharge at a distance of 1 km downstream; (2) the relative magnitude of the acceleration terms in the Saint-Venant equation; and (3) whether the flood wave attenuates as it propagates downstream.

Step 1: Assuming first $S_f = S_0$, the initial upstream discharge is

$$Q_u = \frac{W}{n} h_u^{5/3} S_0^{1/2} = \frac{50}{0.03} 1^{5/3} \sqrt{3 \times 10^{-3}} = 91.3 \text{ m}^3/\text{s}.$$

From the continuity equation,

$$\Delta Q = \frac{-W \Delta h \Delta X}{\Delta t} = -50 \times \frac{1}{3,600 \text{ s}} \times 1,000 \text{ m}^3 = -13.9 \text{ m}^3/\text{s}$$

and the downstream discharge Q_d is $91.3 - 13.9 = 77.4$ m^3/s.

Step 2: The downstream flow depth is

$$h_d = \left(\frac{nQ}{WS^{1/2}}\right)^{3/5} = \left(\frac{0.03 \times 77.4}{50 \times \sqrt{3 \times 10^{-3}}}\right)^{0.6} = 0.906 \text{ m}.$$

The upstream and the downstream velocities are, respectively,

$$V_u = \frac{Q_u}{Wh_u} = \frac{91.3}{50 \times 1} = 1.826 \text{ m/s}, \quad V_d = \frac{Q_d}{Wh_d} = \frac{77.4}{50 \times 0.906} = 1.708 \text{ m/s}.$$

Over a distance of 1 km, the flow depth changes by −0.094 m and $\Delta V = -0.118$ m/s. Note that the flow depth decreases downstream as the stage rises. The terms of the Saint-Venant equation [Equation (6.2)] are

$$(2) \quad S_0 = 0.003 = 3 \times 10^{-3},$$

$$(3) \frac{\partial h}{\partial x} = \frac{-0.094 \text{ m}}{1,000 \text{ m}} = -9.4 \times 10^{-5},$$

$$(4) \frac{V}{g}\frac{\partial V}{\partial x} = \frac{1.77}{9.81}\frac{(-0.118)}{1,000} = -2.13 \times 10^{-5}.$$

Given the flood-wave celerity $c = \beta V = 5/3 \times 1.77$ m/s $= 2.95$ m/s, the 1-km distance is traveled in 340 s ($\sim$5.6 min). At one location, the velocity would increase by 0.118 m/s in 340 s, or

$$(5) \quad \frac{1}{g}\frac{\partial V}{\partial t} = \frac{\text{s}^2}{9.81 \text{ m}}\frac{0.118 \text{ m}}{340 \text{ s}^2} = 3.54 \times 10^{-5}.$$

Notice that the time derivative is positive at one point as the flow depth increases in time, while decreasing downstream. The friction slope is $S_f = S_0 - (\partial h/\partial x) - (V\partial V/g\partial x) - (1/g)(\partial V/\partial t) = 0.00308$. Note that the Froude number,

$$\text{Fr} = \frac{V_u}{\sqrt{gh_u}} = \frac{1.826}{\sqrt{9.81}} = 0.583,$$

is less than unity and term (4) < term (3), while term (5) > term (4).

Step 3: We can use Equation (6.12b) to calculate the flood-wave diffusivity term as $1 - (\beta-1)^2\text{Fr}^2 = 1 - \frac{4}{9}(0.583)^2 = 0.85$. The flood wave is diffusive and attenuates as it propagates downstream.

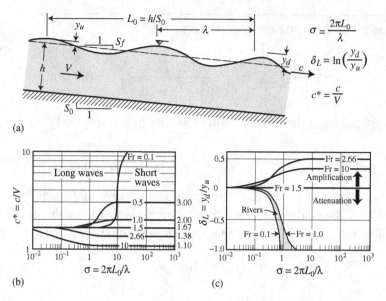

Figure 6.7. Surface waves: (a) definition sketch, (b) celerity, and (c) attenuation

6.5 Short-Wave Propagation

The following discussion presents theoretical developments on the one-dimensional propagation of short and long waves in open channels. The detailed analysis stems from Tsai and Yen (2004), with subsequent applications to steep open channels by Friesen (2007) and Julien et al. (2010). Figure 6.7 shows water surface deformations around the normal depth h in a channel of bed slope S_0, Manning n, and depth-averaged flow velocity V. A normalized channel length L_0 is defined as $L_0 = h/S_0$. The characteristics of the surface wave include a constant wavelength λ and celerity c, and the upstream wave amplitude y_u increases (or decreases) to y_d as the wave travels the distance L_0.

Dimensionless parameters are defined as the Froude number $\mathrm{Fr} = V/\sqrt{gh}$, the dimensionless celerity is $c^* = c/V$, and a dimensionless wave number $\sigma = 2\pi L_0/\lambda$. The wave amplification over length L_0 is defined as $\delta_L = \ln(y_d/y_u)$. The amplification over a single wavelength λ is $\delta_1 = \delta_L/\sigma = \ln(y_2/y_1)$. The applications from Julien et al. (2010) are presented here without the lengthy derivation from Tsai and Yen (2004):

$$c^* = 1 + (1/\sigma\mathrm{Fr}^2)C\sin(\theta/2) \tag{6.15}$$

$$\delta_L = 2\pi\sigma\left[\frac{-1 + C\cos(\theta/2)}{\sigma\mathrm{Fr}^2 + C\sin(\theta/2)}\right], \tag{6.16}$$

where $A = 1 - \sigma^2\mathrm{Fr}^2$, $B = (-4/3)\sigma\mathrm{Fr}^2$, $C = (A^2 + B^2)^{1/4}$, and $\theta = \cos^{-1}(A/C^2)$.

The main characteristics of surface-wave propagation are plotted in Figure 6.7 as a function of Froude number Fr and wave number σ. Notice that σ is low for long waves and high for short waves. Figure 6.7b shows that the celerity of long waves is $c = 5/3V$, as expected from the Kleitz-Seddon law. There are slight deviations as a function of Froude number. The celerity of very short waves ($\sigma \rightarrow \infty$) becomes $c = V(1 + 1/\text{Fr})$. Most notably at very low Froude number, the celerity can be much larger than the flow velocity. Obviously, a wave propagating in still water (Fr = 0) will propagate at a speed that is infinitely larger than the flow velocity. Figure 6.7c is most interesting; here, wave amplification ($\delta_L > 0$) occurs when Fr > 1.5 and the maximum amplification factor is $\delta_{L\,\max} = 0.53$, obtained when Fr = 2.66. For most rivers, Fr < 1, short waves will attenuate very rapidly when $\sigma > 1$. Essential knowledge is thus gained from this observation that waves shorter than $\lambda < 2\pi h/S_0$ are expected to attenuate quickly and only long waves will propagate over long distances in rivers.

6.6 Flow Pulses in Rivers

Flow pulses can affect the dynamics of river systems. Let us first consider alternating high flows Q_H and low flows Q_L with a discharge ratio $r_Q = Q_H/Q_L$. The total duration of the pulse is t times the duration of low flows. For the example sketched in Figure 6.8a, $t = 3$ and $r_Q = 4$, and the average discharge $\overline{Q} = 1$. The volume of water from the pulse and the average condition are obviously identical. From the sediment-rating curve given as a power function of discharge, $Q_s = Q^b$, the volume of sediment for the flow pulse in Figure 6.8b, $\tilde{\forall}_S = 2^2 + 2(0.5)^2 = 4.5$, is different from the volume of sediment carried from steady uniform flow, $\forall_S = 3$. The ratio of the sediment volume, $K_S = \tilde{\forall}_S/\forall_S = 1.5$, can be significant in some cases.

This concept has been generalized by Shin and Julien (2011) to give K_S as a function of r_Q and t:

$$K_S = \frac{(r_Q^b + t - 1)t^{b-1}}{(r_Q + t - 1)^b}. \tag{6.17}$$

Typical values of K_S are plotted in Figure 6.8c and d. For a fixed volume of water, $K_S > 1$ when $b > 1$, and unsteady flow carries a larger sediment load than steady flow.

Ephemeral streams are a particular case where $Q_L = 0$ and r_Q becomes extremely large. Equation (6.17) reduces to $K_S \simeq t^{b-1}$ for ephemeral streams. As an example, let us consider an ephemeral stream flowing for only 5 days

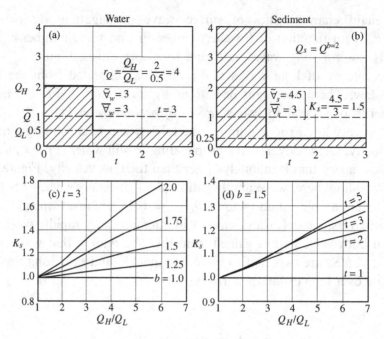

Figure 6.8. Ratio of sediment volume K_S for flow pulses and steady flow: (a) flow pulse diagrams for water, (b) sediment, (c) with sediment transport ratio as a function of the sediment rating curve exponent, and (d) time ratio

per year, which gives for $b \simeq 2$, $K_S \simeq (365/5)^{b-1} \simeq 72$. The volume of sediment transported in very flashy streams can be orders of magnitude larger than the sediment transported from the steady flow of the same amount of water.

In summary, Figure 6.9 sketches the effects of diffusive and dynamic wave propagation on flood hydrographs and sediment transport. The diffusive-wave approximation applies to most rivers and the peak water and sediment discharges decrease in the downstream direction. While the volume of water is the same, the amount of sediment transport decreases as a result of the flood-wave attenuation and this results in gradual riverbed aggradation. Conversely, for dynamic waves typically seen in mountain streams, the peak water and sediment discharges increase in the downstream direction. Mountain-stream floods tend to be very flashy and highly erosive. Mountain rivers tend to scour severely during large floods because of flood wave amplification.

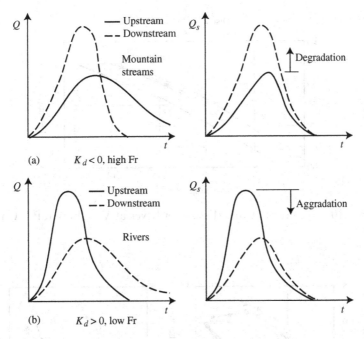

Figure 6.9. (a) Dynamic and (b) diffusive wave propagation characteristics

6.7 Loop-Rating Curves

We learned that the friction slope depends on the gradual changes in flow depth. The effect of flood-wave propagation on the river discharge Q can be written as

$$Q = \frac{A}{n} R_h^{2/3} \left(S_0 - K_d \frac{\partial h}{\partial x} \right)^{1/2}. \tag{6.18}$$

In most rivers, the flow is subcritical and Fr is sufficiently small to use $K_d \cong 1$. At a given stage, the rising limb of a flood wave conveys a higher discharge than the falling limb does. This implies that the stage–discharge relationship will not be unique, and many rivers will show a counterclockwise loop, as shown in Figure 6.10 for the Mississippi River. Notice that the stage difference at a given discharge can be as high as 6 ft (~2 m) in this case. Notice that the maximum discharge will be reached before the maximum stage.

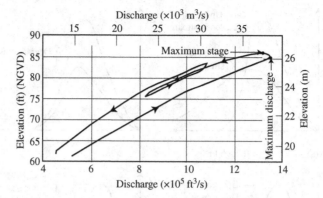

Figure 6.10. Hysteresis of the Mississippi River at Vicksburg (after Combs, 1994)

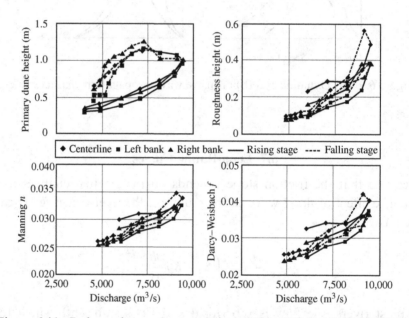

Figure 6.11. Bed roughness characteristics of the 1998 flood of the Rhine River, with dashed lines for falling stages (after Julien et al., 2002)

Other hydraulic characteristics are also known to change during floods. For instance, the large 1998 flood of the Rhine River in Figure 6.11 shows a counterclockwise hysteresis loop for the dune height while the roughness height, Manning n, and Darcy–Weisbach f varied with discharge without any hysteresis effect.

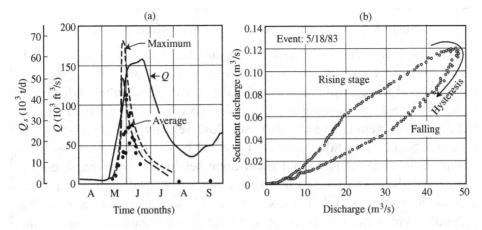

Figure 6.12. Hysteresis of the (a) Bell River and (b) Goodwin Creek

The loop-rating effects on shear stress and sediment transport can also be analyzed in a similar manner. Bed shear stress τ_0 based on the friction slope is given by

$$\tau_0 = \gamma R_h S_f = \gamma R_h \left(S_0 - K_d \frac{\partial h}{\partial x} \right). \tag{6.19}$$

Shear stress at a given flow depth will be larger on the rising limb than on the falling limb. Bedload sediment transport also increases with bed shear stress, resulting in larger sediment transport during the rising limb at a given flow depth than for the falling limb. Sediment transport is described by the sediment-rating curve (sediment discharge Q_s versus water discharge Q). Although both Q_s and Q increase during the rising stages, Q_s increases faster than Q. The sediment-rating curve is characterized by a clockwise loop-rating effect, as shown in Figure 6.12. This effect is predominant for streams transporting mostly bedload. Streams carrying predominantly washload may not respond to changes in bed shear stress, although floodwaters from upland areas may carry large quantities of fine sediments during the rising stages.

Finally, the case of reverse hysteresis is possible when considering dam break events, when large volumes of water and sediment surge into a channel. In such cases, the flood wave may travel faster than the silt-laden water and leave the sediment at the trailing end of the flood wave. This would result in a reverse hysteresis of the sediment-rating curve with higher sediment loads during the falling limb of the hydrograph. Case Study 6.1 presents fairly detailed data on the propagation of flood waves in the Rhine River in 1998.

Case Study 6.1 Fluvial Data Set for the Rhine River, The Netherlands

This case study presents detailed data on the 1998 Rhine River flood. The river stage measurements in Table CS6.1.1. show the propagation of the flood wave along the Rhine-Waal River in The Netherlands during the first week of November. Hydraulic flow properties including discharge, flow depth, stage, and mean flow velocity are summarized in Table CS6.1.2. This table shows data on the hysteresis effects of bedform changes in terms of large and small dune height and wavelength during the flood. Several particle-size distributions of the bed material are presented in Table CS6.1.3. Three sets of particle-size distributions along the channel center line can be compared with the distributions closer to the right bank and the left bank. Additionally, velocity and sediment-concentration profiles are shown in Table CS6.1.4. The navigable width of the Rhine River is maintained at 260 m. The collaboration with G. Klaassen at Delft Hydraulics and W. ten Brinke at the Rijkswaterstaat is most gratefully acknowledged.

Table CS6.1.1. *Flood stages along the Rhine-Waal River during the 1998 flood*

Date	Time	Lobith 862 km (cm)	Pannerdense Kop 867 km (cm)	Nijmegen 885 km (cm)	Tiel 913 km (cm)	Zaltbommel 935 km (cm)	Vuren 952 km (cm)
Oct. 31	09:00	1,372	1,318	1,118	776	467	224
Nov. 1	12:00	1,457	1,395	1,188	840	536	277
2	12:00	1,514	1,450	1,231	890	585	315
3	12:00	1,554	1,486	1,266	939	646	377
4	11:00	1,573	1,502	1,283	955	670	409
4	16:00	**1,574**	1,502	1,284	958	673	409
4	17:00	1,573	1,502	1,284	958	673	412
4	18:00	1,573	1,502	1,284	958	674	414
4	19:00	1,572	1,502	1,284	958	674	415
4	20:00	1,572		1,283	**959**	675	416
5	07:00	1,563	1,494	1,277	958	675	418
5	08:00	1,562	1,494	1,278	957	675	419
5	09:00	1,561	1,493		957	674	418
5	12:00	1,558	1,489	1,274	956	673	414
6	12:00	1,515	1,455	1,248	933	657	402

Table CS6.1.2. *Flood of the Rhine River near the Pannerdens Canal*

Date in 1998	Discharge (m^3/s)	Stage (m)	Depth (m)	Velocity (m/s)	Slope (cm/km)	Large length (m)	Dune height (m)	Small length (m)	Dune height (m)
29 Oct.	4,077	11.71	8.51			7.43	0.34		0
30	4,783	12.33	9.13			8.31	0.41		0
31	6,180	13.42	10.22		10.42	11.12	0.48		0
1 Nov.		13.96	10.76		11.77				
2	8,119	14.48	11.28		12.35	16.24	0.74		0
3	9,045	14.88	11.52	1.82	13.12	19.78	0.87		0
4	**9,464**	**15.01**	11.49	1.75	13.70	22.57	0.97		0
5	9,149	14.91		1.71	13.32	23.90	1.08		0
6	8,267		11.17	1.71	11.58	26.02	1.13		0
7	7,273	14.14	10.94		10.61	28.84	1.15		0
8		13.76	10.56		9.84				
9		13.46	10.05	1.63	9.07				
10	5,640	13.11			8.49	32.42	0.93		0
11		12.85	9.82	1.53	7.91				
12	5,122	12.69	9.49			34.78	0.79	6.94	0.27
13	4,850	12.46	9.25			36.63	0.74	6.72	0.25
16	4,522	12.18	9.1			40.01	0.66	6.61	0.28
19	4,527	12.17	9.1			42.28	0.56	7.95	0.26

Table CS6.1.3. *Particle-size distribution of the bed material*

Position (m from axis)	d_{90} (mm)	d_{84} (mm)	d_{75} (mm)	d_{65} (mm)	d_{50} (mm)	d_{35} (mm)	d_{25} (mm)	d_{16} (mm)	d_{10} (mm)
−33 percent left	7.81	5.52	3.00	1.86	0.95	0.61	0.45	0.37	0.32
−33 percent left	11.48	8.77	5.30	2.54	0.98	0.65	0.47	0.38	0.32
−33 percent left	7.89	5.77	3.04	1.47	0.81	0.56	0.45	0.38	0.33
Center line	11.51	8.81	5.77	3.35	1.90	1.31	0.95	0.76	0.64
Center line	12.19	9.91	6.40	3.04	1.18	0.71	0.50	0.40	0.34
Center line	12.00	9.60	4.00	1.85	0.94	0.70	0.54	0.41	0.34
33 percent right	11.79	9.26	6.80	4.80	2.69	1.11	0.68	0.46	0.40
33 percent right	11.77	9.23	6.11	3.23	0.90	0.47	0.40	0.33	0.30
33 percent right	5.64	3.72	2.62	1.93	1.44	0.97	0.78	0.60	0.49
Average (mm)	10.23	7.84	4.78	2.68	1.31	0.79	0.58	0.46	0.38

Table CS6.1.4. *Velocity v and concentration C profiles during the 1998 Rhine River flood*

z (m)	C (mg/l)	v (m/s)	z (m)	C (mg/l)	v (m/s)	z (m)	C (mg/l)	v (m/s)	z (m)	C (mg/l)	v (m/s)
October 31, $Q = 6{,}180$ m³/s, $H = 8.9$ m			November 3, $Q = 9{,}045$ m³/s, $H = 9.9$ m			November 5, $Q = 9{,}149$ m³/s, $H = 10.5$ m			November 7, $Q = 7{,}273$ m³/s, $H = 9.4$ m		
0.2	233	0.97	0.3	494	0.74	0.2	216	1.27	0.2	108	0.84
0.2	246	1.02	0.3	488	0.81	0.2	212	1.33	0.2	140	0.89
0.2	159	1.00	0.3	498	0.72	0.3	250	1.30	0.2	99	0.85
0.3	83	1.28	0.4	432	0.47	0.5	227	1.47	0.6	96	1.11
0.3	75	1.23	0.5	398	0.84	0.5	218	1.36	0.7	53	1.02
0.4	97	1.25	0.8	293	1.22	0.5	230	1.47	1.1	31	1.35
1.1	39	1.50	0.9	185	1.34	0.8	157	1.41	1.3	24	1.46
1.2	9	1.45	1.3	134	1.47	0.9	160	1.41	2.1	20	1.45
2.5	4	1.73	2.2	83	1.63	1.9	71	1.66	2.5	23	1.46
3.0	5	1.71	3.5	49	1.92	3.1	40	1.81	3.3	21	1.43
4.4	4	1.72	4.0	43	1.85	3.7	41	1.82	4.7	11	1.76
6.5	3	1.90	4.1	43	1.86	5.1	29	2.06	5.1	8	1.77
7.3	2	1.87	5.3	35	1.99	5.5	23	2.00	5.9	8	1.79
7.3	3	1.85	6.0	33	1.98	5.9	28	1.94	6.8	7	1.79
8.2	2	1.96	7.3	26	2.08	6.8	20	1.99	7.9	5	1.77
8.3	2	1.98	8.0	25	2.04	7.9	18	2.03	9.0	4	1.84
			9.0	23	1.90	9.0	8	2.07			

Exercise 6.1

Consider that the at-a-station hydraulic geometry relationship between mean flow velocity and discharge can be written as $V = aQ^b$. Demonstrate that flood-wave celerity becomes equal to $c = [1/(1-b)]V$. [*Hint*: Use the Kleitz-Seddon law.]

◆Exercise 6.2

Determine the relationship between celerity and flow discharge for the at-a-station hydraulic geometry of the Matamek River described in Section 5.1.1.

[*Answer*: $c \cong 2.17V = 0.048Q^{0.54}$]

♦♦*Exercise 6.3*

Can you explain why the temporal changes (e.g. $\partial h/\partial t$) are always opposite to the spatial changes (e.g. $\partial h/\partial x$)? Also, from Example 6.1, why is term (5) of the Saint-Venant Equation (6.5) always $-\beta$ times larger than term (4)?

♦♦*Exercise 6.4*

Demonstrate Equation (6.11b) from Equation (6.11a) by using the derivatives of the resistance relationship $V = \alpha h^{\beta-1}$ in the continuity equation $\partial h/\partial t = -\partial q/\partial x$.

$$\left[Hint: \frac{1}{g}\frac{\partial V}{\partial t} = (\beta - 1)\frac{V}{gh}\frac{\partial h}{\partial t}, \quad \frac{\partial h}{\partial t} = -\frac{\partial q}{\partial x} = -\left(\frac{V\partial h}{\partial x} + \frac{h\partial V}{\partial x}\right). \right]$$

Exercise 6.5

Demonstrate both sides of Equation (6.12) for Manning's equation in a wide rectangular channel in SI after substituting $\alpha = S^{1/2}/n$ and $\beta = 5/3$.

♦♦♦*Problem 6.1*

Consider a 100-ft wide main channel with a bankfull depth of 5 ft. The floodplain width is 500 ft, as sketched in Figure P6.1. Develop a spreadsheet to calculate the cross-section area and wetted perimeter as a function of flow depth every 4″ until a stage of 10 ft. Consider Manning $n = 0.02$ in the main channel and a slope $S_0 = 0.005$. Repeat the calculations for two values of Manning n on the floodplain $n = 0.02$ and $n = 0.1$. Calculate the flow velocity, discharge, and flood-wave celerity $c = \Delta Q/\Delta A$ for these two floodplain conditions.

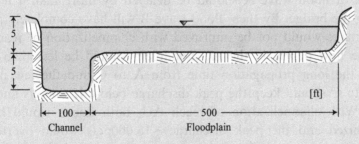

Figure P6.1. Geometry of a floodplain section

[*Answer*: The celerity $c \sim 23$ ft/s slightly below bankfull depth and suddenly drops to 5 ft/s slightly above bankfull depth. The celerity is still only around 10 ft/s when the stage is 10 ft on vegetated floodplains.]

◆*Problem 6.2*

Examine the flood-wave propagation characteristics upstream of town C, sketched in Figure P6.2. The town bridge at C has a maximum capacity of 12,000 cfs. An analysis of the 50-year flood hydrographs from the mountain areas at points A and B are sketched below, with peaks of 15,000 and 7,500 cfs, respectively. The two reaches AC and BC are 15 mi and 2 mi long, respectively. If both river reaches with floodplains are identical and as described in Problem 6.1 with $n = 0.02$, determine the celerity of the flood wave once it exceeds bankfull discharge. How long does it take the flood wave A to reach the bridge? Would re-channelization with levees for both reaches AC and BC have an impact on the bridge?

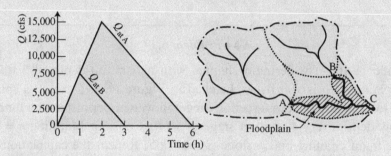

Figure P6.2. Example of flood-wave propagation with floodplains

[*Answer*: From Problem 6.1, the overbank flow celerity is $c \sim 5$ ft/s and the peak of flood wave A should be delayed by more than 4 h before reaching the bridge. By then, flood wave B will have completely passed, and drainage would not be improved with channelization of reach BC. Also, the flow depth on the AC floodplain should be less than ~ 1 ft. Finally, the long propagation time from A to C and the few minutes from B to C should keep the peak discharge below 12,000 cfs under the bridge. With channelization of reach AC, both floods would become synchronized and the peak discharge $\sim 18,000$ cfs would overtop the bridge.]

◆Problem 6.3

Several mountain streams in Colorado have a local slope $S_0 \simeq 0.04$ and Manning $n \simeq 0.03$. If the peak flow depth is 6 ft, estimate the velocity and Froude number. Calculate the wave number of a 1,000-ft-long flood wave. Determine whether this wave would attenuate or amplify. If the wave is initially 1 ft high, can you estimate its height 1,000 ft downstream? Why did several mountain rivers cause severe roadway damage during the 2013 Colorado floods?

[*Answer*: $V \simeq 33$ ft/s, Fr $\simeq 2.36$, $\sigma \simeq 1$, and $c \simeq 1.46V$, $\sigma \simeq 1$, and $\delta_1 \simeq 0.34$ and the wave will amplify from 1 ft to $y_d \simeq e^{0.34} = 1.4$ ft within a distance of 1,000 ft.]

◆Problem 6.4

In a mountain debris flow sketched in Figure P6.4, determine the friction slope and shear stress at points A and B at the same flow depth, $h = 5$ ft. Use the Shields diagram to estimate the particle size at incipient motion. Find the Froude number and Darcy–Weisbach f, and say if the wave celerity exceeds the fluid velocity?

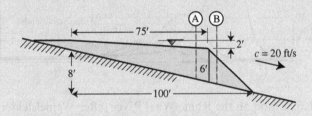

Figure P6.4. Longitudinal profile of a debris flow event

[*Answer*: $S_f \simeq 0.32$, $\tau \simeq 4.8$ kPa and the wave front at B can mobilize boulders, with $f \simeq 1$ and Fr $\simeq 1.5$, this wave should neither amplify nor attenuate.]

◆◆Problem 6.5

Consider the Missouri River flow in Figure P5.2, controlled by clear water releases from a reservoir. Assume a constant Manning n and a constant cross section without lateral inflow along the reach and determine the following: (a) the celerity c versus Q from the relation between the cross-section area and discharge; (b) the celerity of the wave generated from a sudden increase in discharge by 10,000 ft³/s; (c) estimate the relative magnitude of the terms of the Saint-Venant equation when the discharge increases from 25,000 to 35,000 ft³/s in 1 day; (d) what approximation of the Saint-Venant equation is best suited for flood-wave routing?; and (e) what is the flow discharge at a cross section located 10 miles upstream when the discharge is $Q = 31,600$ ft³/s and the water level rises at a rate of 4 ft per day?

[*Answers*: (b) $c \simeq 8$ ft/s, (d) diffusive-wave.]

◆*Problem 6.6*

From the Rhine River data in Case Study 6.1, estimate the flood-wave celerity of the 1998 flood. A flood wave in the same reach is shown in Figure P6.6 from a dam break on the Möhne River in Germany. Comment on whether the wave is amplified or attenuated, and roughly estimate the flood-wave celerity.

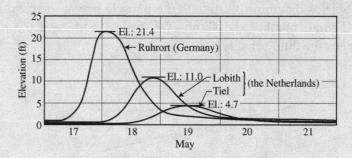

Figure P6.6. Flood on the Rhine/Waal River (after Wemelsfelder, 1947)

[*Hint*: Locate the Ruhr area on a map and use Table CS6.1.2 for Lobith and Tiel.]

◆*Problem 6.7*

Examine the velocity and concentration profiles in Table CS6.1.4, and determine whether there is hysteresis between discharge and sediment transport.

7

Mathematical River Models

Numerous river-engineering problems can be solved with mathematical models. A river model provides a numerical solution in space and time to differential equations describing conservation of mass and momentum, resistance to flow, and sediment transport. This chapter focuses on the finite-difference method. This chapter describes finite-difference approximations for flood-wave propagation in Section 7.1, followed by solutions to the advection–dispersion of contaminants in Section 7.2, aggradation–degradation in rivers in Section 7.3, and numerical river models in Section 7.4. The algorithms to be used depend on the type of differential equation to be solved. Table 7.1 provides a classification of typical river-engineering problems.

7.1 River Flood-Wave Propagation

This section looks at using the finite difference method for the simulation of flood-wave propagation in a one-dimensional river channel. Finite-difference

Table 7.1. *Differential equation types in river engineering*

Equation type	Equation	River-engineering problem
Hyperbolic	$\dfrac{\partial \phi}{\partial t} + v\dfrac{\partial \phi}{\partial x} = 0$	Advection (v constant)
	$\dfrac{\partial^2 \phi}{\partial t^2} = c^2\dfrac{\partial^2 \phi}{\partial x^2}$	Flood-wave propagation (c constant)
Parabolic	$\dfrac{\partial \phi}{\partial t} = K_d\dfrac{\partial^2 \phi}{\partial x^2}$	Diffusion–dispersion (K_d constant)
Elliptic	$\dfrac{\partial^2 \phi}{\partial x^2} + \dfrac{\partial^2 \phi}{\partial y^2} = 0$	Flow net

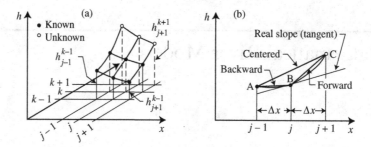

Figure 7.1. Finite-difference grid and approximation: (a) space and time differential discretization and (b) backward and forward differences

approximations are followed by presentations on finite-difference schemes (Section 7.1.1), consistency and convergence (Section 7.1.2), a linear stability analysis (Section 7.1.3), and higher-order approximations (Section 7.1.4).

7.1.1 Finite Differences

Let us consider a function $h(x, t)$ defined in space x and time t, where for instance h could describe the flow depth. We may divide the $x-t$ plane into a grid, as shown in Figure 7.1a with a grid spacing Δx and a time increment Δt. The value of the variable h at a given location and time is described as h_j^k, where the subscript j refers to the spatial location and the superscript k describes the time.

Boundary conditions define nodal values of $h_{j=0,n}^k$ at the upstream and/or downstream end/s of the spatial domain. The initial condition defines the $h_j^{k=0}$ values at the beginning of the simulation. The objective is to develop algorithms to compute values at the unknown time level h_j^{k+1} from known values at time h_j^k. A marching procedure then successively progresses from known to unknown time levels.

The finite-difference method is based on a Taylor-series expansion of the variable h_{j+1}^k written as a function of h_j^k as

$$h_{j+1}^k = h_j^k + \Delta x \left(\frac{\partial h}{\partial x}\right)_j^k + \frac{\Delta x^2}{2!} \left(\frac{\partial^2 h}{\partial x^2}\right)_j^k + \frac{\Delta x^3}{3!} \left(\frac{\partial^3 h}{\partial x^3}\right)_j^k + 0(\Delta x^4), \qquad (7.1)$$

where the derivative $(\partial h/\partial x)_j^k$ is evaluated at grid point j and time level k and $0(\Delta x^m)$ designates m-order truncation terms. For instance, if the term $\Delta x = 0.01$, a term of order $0(\Delta x^2)$ is of the order of 0.0001. As Δx becomes small, higher-order terms become negligible. Similarly, Taylor series can also be expanded to define h_{j-1}^k from h_j^k as

$$h_{j-1}^k = h_j^k - \Delta x \left(\frac{\partial h}{\partial x}\right)_j^k + \frac{\Delta x^2}{2!}\left(\frac{\partial^2 h}{\partial x^2}\right)_j^k - \frac{\Delta x^3}{3!}\left(\frac{\partial^3 h}{\partial x^3}\right)_j^k + 0(\Delta x^4). \qquad (7.2)$$

Rearranging Equations (7.1) and (7.2) and dividing by Δx gives, respectively,

$$\left(\frac{\partial h}{\partial x}\right)_j^k = \underbrace{\frac{h_{j+1}^k - h_j^k}{\Delta x}}_{\substack{\text{forward} \\ \text{difference}}} \underbrace{- \frac{\Delta x}{2!}\left(\frac{\partial^2 h}{\partial x^2}\right)_j^k - \frac{\Delta x^2}{3!}\left(\frac{\partial^3 h}{\partial x^3}\right)_j^k + 0(\Delta x^3)}_{\text{truncation error}}$$

$$\qquad (7.3)$$

$$\simeq \underbrace{\frac{h_{j+1}^k - h_j^k}{\Delta x} + 0(\Delta x)}_{\text{downwind}}$$

and

$$\left(\frac{\partial h}{\partial x}\right)_j^k = \underbrace{\frac{h_j^k - h_{j-1}^k}{\Delta x}}_{\substack{\text{backward} \\ \text{difference}}} \underbrace{+ \frac{\Delta x}{2!}\left(\frac{\partial^2 h}{\partial x^2}\right)_j^k - \frac{\Delta x^2}{3!}\left(\frac{\partial^3 h}{\partial x^3}\right)_j^k + 0(\Delta x^3)}_{\text{truncation error}}$$

$$\qquad (7.4)$$

$$\simeq \underbrace{\frac{h_j^k - h_{j-1}^k}{\Delta x} + 0(\Delta x).}_{\text{upwind}}$$

The partial derivative in Equation (7.3) is written as a forward (or downwind) difference with a first-order truncation error $0(\Delta x)$. Similarly, Equation (7.4) defines a backward (or upwind) difference with a truncation error $0(\Delta x)$. In both cases, the truncation error $0(\Delta x)$ approaches zero as Δx becomes very small. Therefore, both forward and backward finite differences are first-order approximations. A central finite-difference approximation can then be obtained by taking half of the sum of Equations (7.3) and (7.4) to cancel out the $0(\Delta x)$ term and result in

$$\left(\frac{\partial h}{\partial x}\right)_j^k = \frac{h_{j+1}^k - h_{j-1}^k}{2\Delta x} + 0(\Delta x^2). \qquad (7.5)$$

This is called a second-order approximation, because the truncation error is $0(\Delta x^2)$. Figure 7.1b sketches the forward, backward, and central finite

Table 7.2. *Explicit and implicit finite differences*

Finite difference	Explicit	Implicit
Backward (or upwind)	$\dfrac{\partial h}{\partial x} \simeq \dfrac{h_j^k - h_{j-1}^k}{\Delta x}$	$\dfrac{\partial h}{\partial x} \simeq \dfrac{h_j^{k+1} - h_{j-1}^{k+1}}{\Delta x}$
Forward (or downwind)	$\dfrac{\partial h}{\partial x} \simeq \dfrac{h_{j+1}^k - h_j^k}{\Delta x}$	$\dfrac{\partial h}{\partial x} \simeq \dfrac{h_{j+1}^{k+1} - h_j^{k+1}}{\Delta x}$
Central	$\dfrac{\partial h}{\partial x} \simeq \dfrac{h_{j+1}^k - h_{j-1}^k}{2\Delta x}$	$\dfrac{\partial h}{\partial x} \simeq \dfrac{h_{j+1}^{k+1} - h_{j-1}^{k+1}}{2\Delta x}$

differences with the real slope being the tangent of the function at B. The forward difference is the slope of BC, the backward difference is the slope of AB, and the central difference is the slope of the secant AC. Although all three approximations become exact as Δx goes to zero, the second-order central difference is closer than both first-order approximations.

Explicit formulations refer to partial derivatives at the known level k, while implicit formulations refer to the unknown level $k + 1$. Table 7.2 lists a few explicit and implicit approximations for $\partial h/\partial x$ at grid point (j, k).

7.1.2 Consistency and Convergence

Four properties, consistency, stability, convergence, and accuracy, are important in numerical analysis. A numerical scheme is said to be convergent when the difference between the solutions of the differential equations and the finite difference formulation tends to zero as the time step goes to zero. It is generally sufficient to check consistency and stability to ensure convergence.

The following formulation solves the flood-wave propagation problem where h is the flow depth and c is the flood wave celerity discussed in the previous chapter. Note that this algorithm also solves the simple advection problem when c is replaced with the flow velocity v. Hence, let us solve the flood-wave propagation equation from Chapter 6,

$$\frac{\partial h}{\partial t} + c \frac{\partial h}{\partial x} = 0, \tag{7.6}$$

with a forward difference in time and backward difference in space (FTBS)

$$\frac{h_j^{k+1} - h_j^k}{\Delta t} + c \frac{h_j^k - h_{j-1}^k}{\Delta x} = 0. \tag{7.7}$$

Rearranging to find the flow depth at the unknown level $k + 1$ as a function of the flow depth at the known level k, we obtain the following explicit formulation

$$h_j^{k+1} = C_c h_{j-1}^k + (1 - C_c) h_j^k, \tag{7.8}$$

where $C_c = c\Delta t/\Delta x$ is the Courant number. For us to get started, the initial condition must specify flow depth h for all j values at $k = 0$. A marching procedure is then obtained with one upstream boundary condition for all k values at $j - 1 = 0$.

Consistency is the property of a finite-difference scheme to reduce to the partial differential equation as the truncation error disappears. For example, the values of h_j^{k+1} and h_{j-1}^k in Equation (7.8) are replaced with Equations (7.1) and (7.2),

$$h_j^k + \Delta t \left(\frac{\partial h}{\partial t}\right)_j^k + \frac{\Delta t^2}{2!} \left(\frac{\partial^2 h}{\partial t^2}\right)_j^k + 0(\Delta t^3)$$
$$= (1 - C_c) h_j^k + C_c \left[h_j^k - \Delta x \left(\frac{\partial h}{\partial x}\right)_j^k + \frac{\Delta x^2}{2!} \left(\frac{\partial^2 h}{\partial x^2}\right)_j^k + 0(\Delta x^3) \right]. \tag{7.9}$$

Rearranging the equation and dividing by Δt results in

$$\underbrace{\left(\frac{\partial h}{\partial t}\right)_j^k + c\left(\frac{\partial h}{\partial t}\right)_j^k}_{\text{original equation}} + \underbrace{\left[\frac{\Delta t}{2!} \left(\frac{\partial^2 h}{\partial t^2}\right)_j^k - \frac{c\Delta x}{2!} \left(\frac{\partial^2 h}{\partial x^2}\right)_j^k + 0(\Delta t^2) + 0(\Delta x^2) \right]}_{\text{truncation error}} = 0,$$
$$\tag{7.10}$$

where the first part is the original equation. This numerical scheme is therefore unconditionally consistent with the partial differential equation, because the truncation error vanishes as Δt and Δx approach zero.

7.1.3 Linear Stability Analysis

Once a numerical scheme has been tested for consistency, a linear stability analysis is conducted to ensure convergence. The stability of a finite difference scheme examines the propagation of errors introduced in the numerical calculations (Abbott and Basco, 1989). Numerical schemes are unstable when perturbations grow as the calculations progress in time, even when $\Delta x \to 0$ and $\Delta t \to 0$. If the errors do not grow as they propagate in the numerical domain, the finite-difference scheme is stable.

The linear-stability analysis, also called the von Neumann procedure, examines the response of the finite-difference scheme to input perturbations described by a Fourier series in complex form as

$$h_j^k = \sum_{\tilde{n}=1}^{N/2} \zeta_{\tilde{n}}^k e^{\frac{i\tilde{n}2\pi j\Delta x}{L}}, \tag{7.11}$$

where N is the number of points per wavelength $L = N\Delta x$, $i = \sqrt{-1}$, and $\tilde{n}$ is the wave-number index. The complex function $e^{i\tilde{\alpha}j\tilde{n}}$ can be separated into a real and an imaginary part according to Euler's relation $e^{i\tilde{\alpha}j\tilde{n}} = \cos(\tilde{\alpha}j\tilde{n}) + i\sin(\tilde{\alpha}j\tilde{n})$. When viewed in the complex plane, the Fourier coefficients $\zeta_{\tilde{n}}^k$ exhibit an amplitude and a phase angle $\tilde{\alpha} = 2\pi/N$. The linear-stability analysis examines how each Fourier coefficient changes in time for any wave-number index $\tilde{n}$. It is usually sufficient to look only at $\tilde{n} = 1$, or

$$h_j^k = \zeta^k e^{i\tilde{\alpha}j}. \tag{7.12}$$

Accordingly, we can define $h_j^{k+1} = \zeta^{k+1} e^{i\tilde{\alpha}j}$ and $h_{j-1}^{k+1} = \zeta^k e^{i\tilde{\alpha}(j-1)}$ from Equation (7.12). The stability of Equation (7.8) is examined by substituting these terms back into Equation (7.8) to give

$$\zeta^{k+1} e^{i\tilde{\alpha}j} = C_c \zeta^k e^{i\tilde{\alpha}(j-1)} + (1 - C_c)\zeta^k e^{i\tilde{\alpha}j}. \tag{7.13}$$

After canceling the common term $e^{i\tilde{\alpha}j}$, we obtain

$$\zeta^{k+1} = [(1 - C_c) + C_c e^{-i\tilde{\alpha}}]\zeta^k = A_\alpha \zeta^k. \tag{7.14}$$

The term in the brackets of Equation (7.14) is the error amplification factor A_α, which is a complex number. The modulus $|A_\alpha|$ determines whether the Fourier coefficients grow (when $|A_\alpha| > 1$), or decay (when $|A_\alpha| < 1$). A finite-difference scheme is stable as long as $|A_\alpha| \leq 1$. Our example is sketched in the complex plane in Figure 7.2a, where it is given that $A_\alpha = (1 - C_c + C_c \cos\tilde{\alpha}) + iC_c \sin\tilde{\alpha}$ and $|A_\alpha| = \sqrt{[1 - C_c(1 - \cos\tilde{\alpha})]^2 + C_c^2 \sin^2\tilde{\alpha}}$. This simply describes a circle of magnitude C_c centered at $1 - C_c$. The effects of the number of points per wavelength N is also examined given $\tilde{\alpha} = 2\pi/N$ in Figure 7.2b.

Numerical stability $|A_\alpha| \leq 1$ corresponds to $C_c = c\Delta t/\Delta x \leq 1$, which is referred to as the Courant–Friedrichs–Lewy condition, in short the Courant condition. It is interesting to note that numerical stability requires the celerity c of the analytical solution to be less than the celerity $(\Delta x/\Delta t)$ of the numerical solution. Simply stated, once the grid size Δx is known, the time increment Δt for stable numerical simulations is specified as

$$\Delta t < \frac{\Delta x}{c}. \tag{7.15}◆◆$$

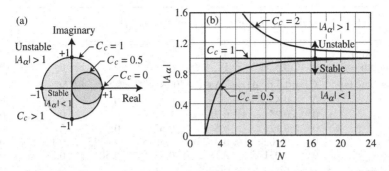

Figure 7.2. Stability diagram: (a) numerical stability in the complex plane and (b) amplification factor as a function of the number of nodes (after Abbott and Basco, 1989)

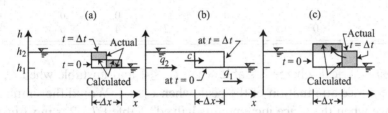

Figure 7.3. Effect of C_c on numerical calculations: (a) stable $C_c < 1$, (b) $C_c = \frac{c\Delta t}{\Delta x} = 1$, and (c) unstable $C_c > 1$

The physical interpretation of this stability condition is that the numerical time step Δt may not exceed the characteristic time step $\Delta x/c$. Otherwise, all the physical information does not have sufficient time to propagate to the next time step, and this will manifest itself as an instability. The effects of C_c on numerical simulations of flood wave propagation are detailed in Example 7.1 and the results are sketched in Figure 7.3.

Example 7.1 Flood-wave propagation with the FTBS scheme

In this example, let's apply the FTBS scheme from Equation (7.8) to solve for flood-wave routing at a constant celerity $c = 2$ m/s. Consider $\Delta x = 2{,}000$ m and examine the numerical propagation of a triangular wave with $C_c = 0.5$, 1.0, and 2.0 in Table E7.1.1. Comparing the results at $t = 2{,}000$ s shows a wave correctly propagated when $C_c = 1.0$ while the peak is numerically attenuated when $C_c = 0.5$ and amplified when $C_c = 2.0$.

Table E7.1.1. *Flood-wave propagation at $\Delta x = 2,000$ m*

Distance (km)	0	2	4	6	8	10	12
Initial depth (m)	0	0.5	1	0	0	0	0

Routing time (s): $C_c = 0.5$ or $\Delta t = C_c\Delta x/c = 500$ s; $h_j^{k+1} = 0.5h_{j-1}^k + 0.5h_j^k$

	0	2	4	6	8	10	12
0	0	0.5	1	0	0	0	0
500	0	0.25	0.75	0.5	0	0	0
1,000	0	0.125	0.5	0.625	0.25	0	0
1,500	0	0.0625	0.3125	0.5625	0.4375	0.125	0
2,000	0	0.03125	0.1875	0.4375	0.50	0.28125	0.0625

Routing time (s): $C_c = 1.0$ or $\Delta t = 1,000$ s; $h_j^{k+1} = h_{j-1}^k$

	0	2	4	6	8	10	12
0	0	0.5	1	0	0	0	0
1,000	0	0	0.5	1	0	0	0
2,000	0	0	0	0.5	1	0	0

Routing time (s): $C_c = 2.0$ or $\Delta t = 2,000$ s; $h_j^{k+1} = 2h_{j-1}^k - h_j^k$

	0	2	4	6	8	10	12
0	0	0.5	1	0	0	0	0
2,000	0	−0.5	0	2	0	0	0
4,000	0	0.5	−1	−2	4	0	0

The numerical scheme is stable when $C_c \leq 1$ and unstable when $C_c > 1$, but we notice attenuation of the peak when $C_c < 1$. Would the simulations improve when the space increment is halved? Table E7.1.2 is most instructive: a shorter Δx results in a better approximation of the peak value when $C_c = 0.5$. Specifically, at $t = 2,000$ s, the peak value $h_{max} = 0.61$ m with $\Delta x = 1$ km is closer to the correct value $h_{max} = 1$ m than $h_{max} = 0.5$ at $\Delta x = 2$ km. Stable numerical simulations improve as $\Delta x \to 0$ and better accuracy is gained through more extensive calculations.

Table E7.1.2. *Flood-wave propagation at $\Delta x = 1,000$ m*

Distance (km)	0	1	2	3	4	5	6	7	8	9	10
Initial depth (m)	0	0.25	0.5	0.75	1	0.5	0	0	0	0	0

Routing times (s): $C_c = 0.5$ or $\Delta t = 250$ s, $h_j^{k+1} = 0.5h_{j-1}^k + 0.5h_j^k$

	0	1	2	3	4	5	6	7	8	9	10
0	0	0.25	0.5	0.75	1	0.5	0	0	0	0	0
250	0	0.125	0.375	0.625	0.875	0.75	0.25	0	0	0	0
500	0	0.0625	0.25	0.5	0.75	0.81	0.5	0.125	0	0	0
750	0	0.031	0.156	0.375	0.625	0.78	0.656	0.31	0.06	0	0
1,000	0	0.016	0.094	0.266	0.5	0.7	0.72	0.484	0.19	0.03	0
1,250	0	0.008	0.055	0.18	0.383	0.6	0.71	0.6	0.336	0.11	0.01
1,500	0	0.004	0.031	0.127	0.28	0.49	0.655	0.655	0.468	0.22	0.06
1,750	0	0.002	0.016	0.07	0.20	0.39	0.57	0.655	0.56	0.344	0.14
2,000	0	0.001	0.009	0.04	0.145	0.29	0.48	0.61	0.61	0.45	0.24

Routing times (s): $C_c = 2.0$ or $\Delta t = 1000$ s, $h_j^{k+1} = 2h_{j-1}^k - h_j^k$

	0	1	2	3	4	5	6	7	8	9	10
0	0	0.25	0.5	0.75	1	0.5	0	0	0	0	0
1,000	0	−0.25	0	0.25	0.5	1.5	1	0	0	0	0
2,000	0	+0.25	−0.5	−0.25	0	−0.5	2	2	0	0	0
3,000	0	−0.25	1	−0.75	−0.5	0.5	−3	2	4	0	0
4,000	0	+0.25	−1.5	2.75	−1	−1.5	4	−8	0	8	0

In comparison, the case with $C_c = 2$ shows that decreasing Δx by one half does not increase the accuracy of the calculations. Indeed, at $t = 4,000$ s, $h_{max} = 8$ m for $\Delta x = 1,000$ m, compared with $h_{max} = 4$ m for $\Delta x = 2,000$ m. Therefore, decreasing Δx and Δt does not necessarily improve convergence, even if the numerical scheme is consistent. A consistent numerical scheme can be convergent only when it is stable.

7.1.4 Higher-Order Approximations

Finite differences with higher-order approximations are obtained by introducing additional nodes to cancel out the higher-order terms in the truncation error. For instance, a Taylor-series expansion between nodes j and $j + 2$ gives

$$h_{j+2}^k = h_j^k + 2\Delta x \left(\frac{\partial h}{\partial x}\right)_j^k + \frac{4\Delta x^2}{2!} \left(\frac{\partial^2 h}{\partial x^2}\right)_j^k + \frac{8\Delta x^3}{3!} \left(\frac{\partial^3 h}{\partial x^3}\right)_j^k + 0(\Delta x^4). \quad (7.16)$$

A second derivative is obtained after Equation (7.1) is doubled and subtracted from Equation (7.16):

$$\left(\frac{\partial^2 h}{\partial x^2}\right)_j^k = \frac{j_{j+2}^k - 2h_{j+1}^k + h_j^k}{\Delta x^2} - \Delta x \left(\frac{\partial^3 h}{\partial x^3}\right)_j^k + 0(\Delta x^2). \quad (7.17)$$

The truncation error of this approximation is only accurate to the first order, i.e. function of Δx. However, substituting Equation (7.17) back into the right-hand side of Equation (7.3) yields

$$\left(\frac{\partial h}{\partial x}\right)_j^k = \frac{-h_{j+2}^k + 4h_{j+1}^k - 3h_j^k}{2\Delta x} + \left(\frac{\Delta x^2}{2!} - \frac{\Delta x^2}{3!}\right)\left(\frac{\partial^3 h}{\partial x^3}\right)^k. \quad (7.18)$$

As a result, the first derivative is now turned into a second-order approximation, i.e. a function of Δx^2, after an additional grid point is added to the numerical scheme. Higher-order finite-difference approximations can be obtained in a similar fashion, and several useful finite-difference schemes of the first and the second derivatives are listed in Table 7.3. For instance, Equation (7.18) corresponds to the second-order forward difference of the first derivative.

Table 7.3. *First- and second-order finite-difference schemes*

	ϕ_{j-3}	ϕ_{j-2}	ϕ_{j-1}	ϕ_j		ϕ_j	ϕ_{j+1}	ϕ_{j+2}	ϕ_{j+3}

(a) First-order backward differences $0(\Delta x)$

$$\Delta x \frac{\partial \phi}{\partial x} = \qquad\qquad -1 \quad\; 1$$
(coefficients: $\phi_{j-1} = -1$, $\phi_j = 1$)

$$\Delta x^2 \frac{\partial^2 \phi}{\partial x^2} = \qquad 1 \quad -2 \quad\; 1$$
(coefficients: $\phi_{j-2} = 1$, $\phi_{j-1} = -2$, $\phi_j = 1$)

(b) First-order forward differences $0(\Delta x)$

$$\Delta x \frac{\partial \phi}{\partial x} = \qquad -1 \quad\; 1$$
(coefficients: $\phi_j = -1$, $\phi_{j+1} = 1$)

$$\Delta x^2 \frac{\partial^2 \phi}{\partial x^2} = \qquad 1 \quad -2 \quad\; 1$$
(coefficients: $\phi_j = 1$, $\phi_{j+1} = -2$, $\phi_{j+2} = 1$)

(c) Second-order backward differences $0(\Delta x^2)$

$$2\Delta x \frac{\partial \phi}{\partial x} = \qquad 1 \quad -4 \quad\; 3$$
(coefficients: $\phi_{j-2} = 1$, $\phi_{j-1} = -4$, $\phi_j = 3$)

$$\Delta x^2 \frac{\partial^2 \phi}{\partial x^2} = \quad -1 \quad\; 4 \quad -5 \quad\; 2$$
(coefficients: $\phi_{j-3} = -1$, $\phi_{j-2} = 4$, $\phi_{j-1} = -5$, $\phi_j = 2$)

(d) Second-order forward differences $0(\Delta x^2)$

$$2\Delta x \frac{\partial \phi}{\partial x} = \quad -3 \quad\; 4 \quad -1$$
(coefficients: $\phi_j = -3$, $\phi_{j+1} = 4$, $\phi_{j+2} = -1$)

$$\Delta x^2 \frac{\partial^2 \phi}{\partial x^2} = \quad\; 2 \quad -5 \quad\; 4 \quad -1$$
(coefficients: $\phi_j = 2$, $\phi_{j+1} = -5$, $\phi_{j+2} = 4$, $\phi_{j+3} = -1$)

(e) Second-order central differences $0(\Delta x^2)$

$$2\Delta x \frac{\partial \phi}{\partial x} = \quad -1\phi_{j-1} + 0\phi_j + 1\phi_{j+1}$$

$$\Delta x^2 \frac{\partial^2 \phi}{\partial x^2} = \quad 1\phi_{j-1} - 2\phi_j + 1\phi_{j+1}$$

7.2 Advection–Dispersion of River Contaminants

This section examines the one-dimensional propagation of contaminants in rivers. The presentation focuses on the following: (1) numerical diffusion, in Section 7.2.1; (2) higher order schemes to eliminate numerical diffusion, e.g. the Leonard scheme, in Section 7.2.2; (3) boundary conditions, in Section 7.2.3.

Consider the combination of advection and dispersion as a mechanism for the transport and spreading of contaminants at a concentration ϕ in a river. The contaminant could be very fine sediment particles in a river, i.e. washload, but it could also be any other conservative pollutant. A substance is said to be conservative when the total mass remains constant and therefore without sedimentation, adsorption, or chemical reaction, that would cause the decay/growth of the substance. The governing one-dimensional advection–dispersion equation is written as

$$\frac{\partial \phi}{\partial t} + v \frac{\partial \phi}{\partial x} = K_d \frac{\partial^2 \phi}{\partial x^2}, \qquad (7.19)$$

where v is the mean flow velocity in the downstream x direction and $K_d \simeq 250 h u_*$ is the dispersion coefficient. Note that a similar approach could be used for turbulent diffusion; however, it is assumed here that the substance is well mixed in the vertical and lateral directions and the concentration is uniform at a given cross section, i.e. a low Rouse number (Julien, 2010). It can also be noted that when K_d is very small, Equation (7.19) reduces to the flood-wave propagation analysis of Section 7.1, after v is replaced with c. The mathematical interest in Equation (7.19) arises from the fact that the equation is a hybrid between a hyperbolic equation, when $K_d = 0$, and a parabolic equation, when $v = 0$. In rivers, the flow velocity is usually important and the FTBS scheme developed in Section 7.1 can serve as a basis for further analysis.

7.2.1 Numerical Diffusion

Let's adopt the FTBS scheme for the advection term; a second-order central-difference approximation from Table 7.3e is used for the dispersion term. The resulting finite-difference scheme is written as

$$\frac{\phi_j^{k+1} - \phi_j^k}{\Delta t} + v\frac{\phi_j^k - \phi_{j-1}^k}{\Delta x} = K_d\frac{\phi_{j+1}^k - 2\phi_j^k + \phi_{j-1}^k}{\Delta x^2}, \qquad (7.20)$$

from which an explicit scheme is obtained after defining $C_u = v\Delta t/\Delta x$ and $C_k = K_d\Delta t/\Delta x^2$,

$$\phi_j^{k+1} = (C_u + C_k)\phi_{j-1}^k + (1 - C_u - 2C_k)\phi_j^k + C_k\phi_{j+1}^k. \qquad (7.21)$$

Although the dispersion term is a second-order approximation, the advection term is approximate to the first order only. The consistency analysis in Section 7.1.2, is applied and Equation (7.10) can be rewritten, after ϕ is substituted for h and v for c, as

$$\left(\frac{\partial\phi}{\partial t}\right)_j^k + v\left(\frac{\partial\phi}{\partial x}\right)_j^k + \frac{\Delta t}{2!}\left(\frac{\partial^2\phi}{\partial t^2}\right)_j^k - v\frac{\Delta x}{2!}\left(\frac{\partial^2\phi}{\partial x^2}\right)_j^k + 0(\Delta t^2) + 0(\Delta x^2) = 0. \quad (7.22)$$

The advection scheme can be written as

$$\frac{\partial\phi}{\partial t} = -v\frac{\partial\phi}{\partial x} + 0(\Delta t, \Delta x). \qquad (7.23)$$

After taking space and time derivatives of Equation (7.23) with constant v, we obtain

$$\frac{\partial^2\phi}{\partial t\partial x} = -v\frac{\partial^2\phi}{\partial x^2} + 0(\Delta t, \Delta x) \qquad (7.24a)$$

and

$$\frac{\partial^2\phi}{\partial t^2} = -v\frac{\partial^2\phi}{\partial x\partial t} + 0(\Delta t, \Delta x), \qquad (7.24b)$$

which can be recombined to obtain the wave equation

$$\frac{\partial^2 \phi}{\partial t^2} = v^2 \frac{\partial^2 \phi}{\partial x^2} + 0(\Delta t, \Delta x). \tag{7.25}$$

Substituting Equation (7.25) back into Equation (7.22) gives

$$\underbrace{\left(\frac{\partial \phi}{\partial t}\right)_j^k + v\left(\frac{\partial \phi}{\partial x}\right)_j^k}_{\text{advection term}} = \underbrace{\left(\frac{-v^2 \Delta t}{2} + \frac{v \Delta x}{2}\right)\left(\frac{\partial^2 \phi}{\partial x^2}\right)_j^k}_{\text{truncation error}} + 0(\Delta t^2, \Delta x^2). \tag{7.26}$$

This leads to the interesting finding that the truncation error of the first-order advection scheme includes a numerical diffusion term K_{num}, which can be quantified as

$$K_{\text{num}} = \frac{-v^2 \Delta t}{2} + \frac{v \Delta x}{2} = \frac{v \Delta x}{2}(1 - C_u). \tag{7.27}$$

It is now becoming clear that the proposed scheme in Equation (7.20) really solves the following equation:

$$\frac{\partial \phi}{\partial t} + \frac{v \partial \phi}{\partial x} = (K_d + K_{\text{num}})\frac{\partial^2 \phi}{\partial x^2}. \tag{7.28}$$

The numerical diffusion term in Equation (7.27) vanishes as C_u approaches unity, hence the interest in running computer models with values of C_u close to unity. This explains why the simulations in Example 7.1 were flawless when $C_c = 1$, while numerical diffusion viewed as an attenuation of the peak was observed for $C_c = 0.5$.

In practice, modelers must guard against simulations in which $K_{\text{num}} > K_d$. For instance, consider a model with $v = 1$ m/s, $K_d = 100$ m^2/s, $\Delta x = 2,000$ m, and $\Delta t = 1,000$ s. The numerical scheme described in Equation (7.20) is stable, $C_u = v\Delta t/\Delta x = 0.5 < 1$, but the numerical diffusion $K_{\text{num}} = (v\Delta x/2)$ $(1 - C_u) = 500$ m^2/s is five times larger than the physical dispersion K_d, and such simulations would be numerically stable and convergent, but physically meaningless. An attempt to control the magnitude of the numerical diffusion may be possible. For instance, the numerical diffusion could be set to be small compared to the physical dispersion $K_{\text{num}} = \alpha K_d$, where $\alpha < 1$. One could then determine the grid spacing Δx and time increment Δt as functions of C_u and K_d at a desirable level α from:

$$\Delta x = \frac{2\alpha K_d}{v(1 - C_u)}, \tag{7.29a}$$

$$\Delta t = \frac{2\alpha K_d C_u}{v^2(1 - C_u)}. \tag{7.29b}$$

For instance, setting up a model for $C_u = 0.9$ when $v = 1$ m/s and $K_d = 100$ m^2/s, while limiting K_{num} to 50 m^2/s, i.e. $\alpha = 0.5$, requires $\Delta x = 1{,}000$ m and $\Delta t = 900$ s, which is quite reasonable. It is clear that further reduction of K_{num} would require a finer grid spacing Δx. Also, both Δx and Δt may become excessively small when K_d decreases. Finally, explicit models are most sensitive to instabilities near the peak discharge of a flood hydrograph because the flow velocity is the largest and models are most likely to become unstable near the peak discharge.

7.2.2 Leonard Scheme

Leonard (1979) developed a third-order numerical scheme to solve the 1-D advection–dispersion Equation (7.19). Canceling higher-order space derivatives provides a higher accuracy for the advection term. The algorithm also completely eliminates the numerical diffusion. The explicit algorithm for ϕ_j^{k-1} is a function of four nodal values ϕ_{j-2}^k, ϕ_{j-1}^k, ϕ_j^k, and ϕ_{j+1}^k, given $C_u = v\Delta t/\Delta x$ and $C_k = K_d\Delta t/\Delta x^2$:

$$\phi_j^{k+1} = \left[C_k C_u + \frac{C_u}{6}(C_u^2 - 1) \right] \phi_{j-2}^k + \left[C_k(1 - 3C_u) - \frac{C_u}{2}(C_u^2 - C_u - 2) \right] \phi_{j-1}^k$$

$$+ \left\{ 1 - \left[C_k(2 - 3C_u) - \frac{C_u}{2}(C_u^2 - 2C_u - 1) \right] \right\} \phi_j^k$$

$$+ \left[C_k(1 - C_u) - \frac{C_u}{6}(C_u^2 - 3C_u + 2) \right] \phi_{j+1}^k. \tag{7.30}$$

The grid Peclet number $P_\Delta = C_u/C_k = v\Delta x/K_d$ is a measure of the ratio between advection $C_u = v\Delta t/\Delta x$ and dispersion $C_k = K_d\Delta t/\Delta x^2$. The linear stability analysis shows a wide stability range, as shown in Figure 7.4.

We can determine Δx and Δt as a function of K_d and v, C_u, C_k and P_Δ from

$$\Delta x = \frac{K_d C_u}{v C_k} = P_\Delta \frac{K_d}{v}, \tag{7.31a}$$

$$\Delta t = \frac{K_d C_u^2}{v^2 C_k} = \frac{K_d P_\Delta C_u}{v^2}, \tag{7.31b}$$

and numerical models with high values of C_u and P_Δ will maximize Δx and Δt.

As dispersion gradually tends toward a normal distribution, the dispersion standard deviation $\sigma_{\Delta t}$ increases with time Δt (e.g. Julien, 2010). Approximately

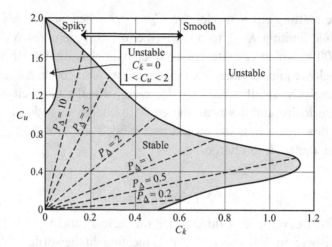

Figure 7.4. Stability diagram of the Leonard scheme (after Leonard, 1979)

95 percent of the material dispersed during one time step is contained within a plume of length $\pm 2\sigma_{\Delta t}$. From $K_d \Delta t = C_k \Delta x^2$, one can simply obtain a relationship between $\sigma_{\Delta t}$ and Δx as

$$\sigma_{\Delta t} = \sqrt{2 K_d \Delta t} = \sqrt{2 C_k} \Delta x. \tag{7.32}$$

It is therefore the value of C_k that determines the number of points that defines the bell-shaped distribution of the dispersion process. Numerical simulations will look "spiky" when $C_k < 1/8$, because $\Delta x > 2\sigma_{\Delta t}$. On the other hand, simulations with $C_k > 0.5$ will look "smooth," because $\Delta x < \sigma_{\Delta t}$.

Let us consider two numerical schemes to illustrate the concept of spiky and smooth simulations. The first algorithm uses $C_u = 1$, $C_k = 0.1$, and $P_\Delta = 10$. The grid size and time step are obtained from Equation (7.31) as $\Delta x = 10 K_d / v$ and $\Delta t = 10 K_d / v^2$. The numerical scheme in Equation (7.30) simply reduces to

$$\phi_j^{k+1} = 0.1 \phi_{j-2}^k + 0.8 \phi_{j-1}^k + 0.1 \phi_j^k. \tag{7.33a}$$

Notice that the term ϕ_{j+1}^k is eliminated because $C_u = 1$. This algorithm should be convenient as long as the spatial changes in ϕ are gradual. This simulation will also display spiky results with large variability between successive nodes because C_k is small. A second algorithm could be considered for numerically smooth simulations. For instance, a stable model could use $C_u = 0.8$ and $C_k = 0.4$ with a corresponding grid spacing $\Delta x = 2 K_d / v$ and time step $\Delta t = 1.6 K_d / v^2$. The coefficients of Equation (7.30) reduce to

$$\phi_j^{k+1} = 0.272 \phi_{j-2}^k + 0.304 \phi_{j-1}^k + 0.376 \phi_j^k + 0.048 \phi_{j+1}^k. \tag{7.33b}$$

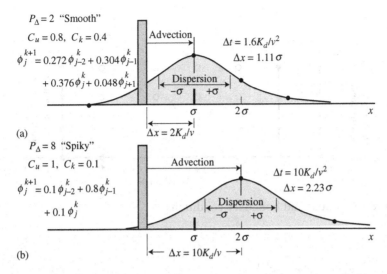

Figure 7.5. Advection-dispersion examples for (a) smooth and (b) spiky simulations

These two algorithms are compared in Figure 7.5; note the number of nodes describing the bell-shaped dispersion calculations for smooth and spiky simulations. Also, the reader will notice that the second algorithm for smooth simulations requires a finer grid and shorter time steps.

Notice that flood-wave propagation could be simulated in a similar manner. This would be a particular case of pure advection ($K_d = C_k = 0$) with the flow velocity v replaced with celerity c. This algorithm is stable only when $C_u = 1$ (based on celerity c) and reduces to the FTBS scheme with $C_c = 1$ discussed in Section 7.1.

7.2.3 Boundary Conditions

Higher-order numerical schemes require additional boundary conditions. For instance, consider the general form of the Leonard scheme in Equation (7.30)

$$\phi_j^{k+1} = a_{j-2}\phi_{j-2}^k + a_{j-1}\phi_{j-1}^k + a_j\phi_j^k + a_{j+1}\phi_{j+1}^k. \tag{7.34}$$

The presence of any negative coefficient $a_i < 0$ is an indicator that the scheme is unstable, or perhaps the coefficient has not been properly calculated. Also notice that the sum of all coefficients $\sum a_i = 1$. As sketched in Figure 7.6, this algorithm requires an initial condition at all nodes h^0 at time $t = 0$. This algorithm also requires a double upstream boundary condition at h_{-2} and h_{-1} and the single downstream boundary condition is because of the term h_{+1}.

Simple numerical schemes typically require a single boundary condition. For instance, the scheme from Equation (7.8), i.e. $\phi_j^{k+1} = 0.5\phi_{j-1}^k + 0.5\phi_j^k$,

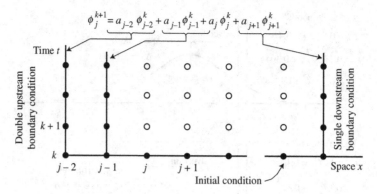

Figure 7.6. Initial and boundary conditions

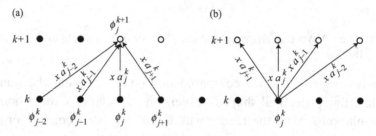

Figure 7.7. Origin and destination of numerical information: finite difference schemes in terms of (a) point of origin and (b) destination

requires: (a) a single upstream boundary condition $\phi - 1$ at every time step and (b) initial condition ϕ^0 over the entire domain. A downstream boundary condition is not required because $a_{j+1} = 0$.

It is interesting to examine the origin and destination of information in numerical schemes as sketched in Figure 7.7. The zones of influence of upstream boundary conditions are shown to propagate very quickly in the downstream direction when $a_{j-2} \neq 0$. Likewise, the downstream boundary condition would not have any influence on the calculations when $a_{j+1} \neq 0$. Higher-order approximations thus provide improved calculation results at the expense of requiring detailed boundary conditions.

The problem of interference between boundary and initial conditions occurs when the information propagates in the upstream direction. To illustrate this point, consider the simple propagation of contaminant at a value $\phi^0 = 1,000$ for a single time step. Without any other source of contamination,

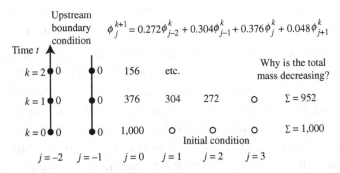

Figure 7.8. Interference between initial and boundary conditions

both the upstream and downstream boundary conditions should be zero. For smooth simulations from Equation (7.33b), we obtained $\phi_j^{k+1} = 0.272\phi_{j-2}^k + 0.304\phi_{j-1}^k + 0.376\phi_j^k + 0.048\phi_{j+1}^k$ and the calculations are illustrated in Figure 7.8. Can the reader explain why the total mass at $k = 1$ is not preserved? Well, this is because of the interference between the initial and upstream boundary conditions. Indeed, from the destination chart in Figure 7.7b, we observed that a quantity $a_{j+1}\phi_1^0 = 0.048 \times 1,000 = 48$ should have been allocated to ϕ_{-1}^1 and this quantity is the missing part to conservation of mass. The missing information is because the upstream boundary condition is not calculated. This information loss at the upstream direction can be referred to as "upstream leakage" and is due to the term $a_{j+1} \neq 0$. This interference possibility can cause issues in terms of conservation of mass that can be very difficult to detect in complex numerical problems.

Similar interference problems can also be encountered near the downstream end, because a downstream boundary condition is required when $a_{j+1} \neq 0$. This example illustrates quite well the trade-offs that exist between "smooth" and "spiky" higher-order numerical schemes. The additional boundary and initial conditions need to be properly worked out. For instance in this case, it would be possible to add sufficient nodes upstream and downstream of the simulation domain to keep the physical domain of interest farther away from the zones of influence of the boundary conditions. Numerical schemes where $a_{j+1} = 0$, e.g. simulations from Equation (7.33a), can often avoid the intricate boundary conditions required for "smooth" simulations from Equation (7.33b). Example 7.2 illustrates the calculations of advection and dispersion of contaminants in a river.

Example 7.2 Advection and dispersion of contaminants in a river

Consider advection and dispersion in a rough mountain channel with a slope $S = 4 \times 10^{-3}$. The flow depth h is 3 m, the mean flow velocity is $V = 2$ m/s, and the dispersion coefficient $K_d \simeq 250 h u_* \simeq 257$ m^2/s. At the upstream end of a fairly uniform 30-km reach, contaminant (or very fine sediment) is released in two pulses: (a) a first 15-min pulse at a concentration of 100,000 mg/l and (b) a second 3-min pulse at 200,000 mg/l starting 1 h after the initial release. Determine the maximum concentration 10 km downstream of the release. How long will the concentration exceed 10,000 mg/l at a point located 20 km downstream of the release?

The FTBS scheme in Section 7.2.1 can be simply defined with $\Delta t = 180$ s $= 0.05$ h and $\Delta x = 500$ m, the Courant number $C_u = V \Delta t / \Delta x = 0.72$ ensures stability. The dispersion number $C_k = K_d \Delta t / \Delta x^2 = 0.185$, and the grid Peclet number is $P_\Delta = 3.9$. A single upstream boundary condition must be specified; however, some numerical diffusion, equal to about $K_{num} \simeq 140$ m^2/s may be affecting the results.

The Leonard scheme eliminates numerical diffusion at the expense of more complex boundary conditions. Simulations should be rather spiky (low C_k), and we check that all coefficients $a_{j-2} = 0.075607$, $a_{j-1} = 0.577656$, $a_j = 0.337868$, and $a_{j+1} = 0.008869$ are positive and sum up to unity. The initial sediment concentration is set at 0, and the upstream boundary condition is 1×10^5 mg/l for the first five time steps and a single value of 2×10^5 mg/l at the 20th time step – all other values are zero. The Leonard scheme requires a second upstream boundary condition which can satisfy that no sediment leaks upstream from the point of release. At the downstream end, advection is dominant and the concentration could be set identical to the value calculated one grid space upstream at the previous time step. Additional grid spaces upstream and downstream could also be used to reduce interference.

Both algorithms yield similar results. As shown in Figure E7.2.1, the maximum concentration 10 km downstream is $C_{max} = 60,000$ mg/l at 1 h 30 min. At a location 20 km downstream, a concentration exceeding 10,000 mg/l will last about 90 min from 2 h 30 min until 4 h after the first release. The analysis of two pulses is quite instructive, because the very high concentration of the second pulse drops below the levels of the first pulse within a very short distance. The two pulses merge within a distance of 20 km.

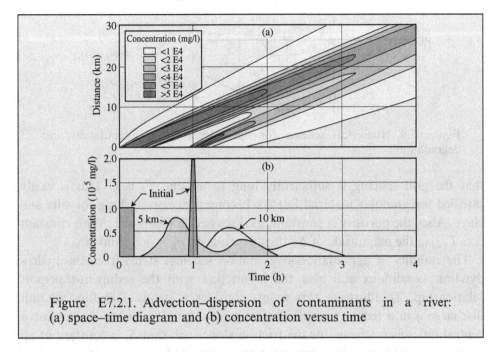

Figure E7.2.1. Advection–dispersion of contaminants in a river: (a) space–time diagram and (b) concentration versus time

7.3 Aggradation–Degradation in Rivers

This section examines the one-dimensional riverbed elevation changes as a result of the processes of aggradation and degradation. Let's focus on the following: (1) a numerical procedure for aggradation and degradation in rivers, in Section 7.3.1 and (2) a discussion of the needs for calibration and validation of river models, in Section 7.3.2.

7.3.1 Aggradation–Degradation Scheme

The analysis of riverbed elevation changes in deformable channels involves the combined effects of water-surface calculations and bed elevation changes. The water-surface calculations were examined with backwater curves of Section 5.2.3. The changes in bed elevation Δz are obtained from the sediment continuity relationship [Equation (2.53)], or

$$\Delta z = -\frac{T_E}{(1 - p_0)} \frac{(q_{sj+1} - q_{sj})\Delta t_s}{\Delta x},\qquad (7.35)$$

where $T_E = 1 - \exp[(-\Delta x \omega)/q]$ is the trap efficiency, p_0 the porosity of the bed material, ω the settling velocity, q the unit flow discharge, Δx the grid spacing, and Δt_s the time increment for sediment. Modelers usually check

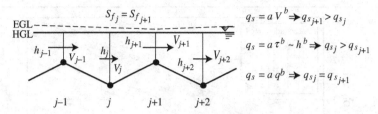

Figure 7.9. Numerical scheme for the calculation of aggradation and degradation

that the grid spacing is sufficiently long to assume $T_E \simeq 1$. This is easily satisfied for granular material but Δx becomes excessively long for silts and clays. Also, the porosity is constant, as discussed in Chapter 1. With constant Δx, T_E, p_0, the magnitude of Δz then depends on $q_{sj+1} - q_{sj}$ and Δt_s.

The stability of aggradation–degradation schemes stems from the hydrodynamic conditions and also the interaction with the sediment-transport relationship. To illustrate this point, consider a steady 1-D flow of unit discharge q in a rectangular channel at a low Froude number. Let us use a central difference scheme for the friction slope such that S_f is constant at all nodes. Assume alternating perturbations of the bed elevation at point j, as sketched in Figure 7.9.

The unit sediment discharge q_{sv} by volume is the subject of discussion here because $h_j > h_{j+1}$ and $V_j < V_{j+1}$, while $S_{fj} = S_{fj+1}$. Therefore, if the sediment-transport relationship is proportional to the flow velocity, i.e. $q_s \sim V^b$, then $q_{sj} < q_{sj+1}$, while if sediment transport depends on shear stress, i.e. $q_s \sim a\tau^b \sim a(hS)^b$, then $q_{sj} > q_{sj+1}$. Well, the choice of the sediment-transport formula can yield opposite results in terms of the numerical stability of the aggradation–degradation scheme.

To complicate things further, the backward difference Δz_{j+1} could be replaced with a forward difference Δz_j and the numerical stability could be reversed. For the above reasons, it is recommended to calculate Δz from Equation (7.35) and split Δz as $\Delta z_j = \alpha \Delta z$ and $\Delta z_{j+1}(1 - \alpha)\Delta z$. The distribution factor $0 < \alpha < 1$ distributes the bed elevation change between forward and backward differences.

It is also clear from Equation (7.35) that the magnitude of the aggradation–degradation increment Δz is linearly proportional to Δt_s. When the change in bed elevation Δz during Δt_s is small compared with the flow depth, the assumption of a rigid boundary calculation is justified and the hydraulic and sediment equations are uncoupled. This means that the aggradation–degradation calculations do not affect the backwater equations during that time interval. It is also advisable to keep Δt_s sufficiently short, because

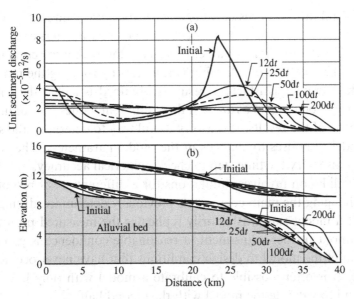

Figure 7.10. Example of (a) sediment transport and (b) aggradation and degradation in a deformable channel

several backwater algorithms do not perform well under the adverse slopes generated when Δz is too large.

An example of aggradation–degradation calculations is shown in Figure 7.10. The formulation is uncoupled, and the unit discharge is constant throughout the reach. The upstream sediment discharge depends on the local upstream slope, and the upstream bed elevation is fixed. A sand-transport equation is used for the calculations. Over time, aggradation takes place as sediment transport decreases in the downstream direction. The reach becomes increasingly uniform with a new bed slope several times milder than the initial reach slope. The sediment-transport capacity also gradually becomes fairly uniform and slightly decreasing in the downstream direction. This numerical example illustrates the important ability of alluvial rivers to develop gradual longitudinal profiles with very gradual changes in depth, velocity, slope, and sediment transport.

7.3.2 Model Calibration and Validation

Once a river-engineering problem has been defined and a mathematical model chosen, field data are needed to describe initial and boundary conditions, geometrical similitude, material properties, and design conditions. Additional data (e.g. discharge hydrographs and water-surface profiles) are required for calibration and verification.

Model calibration is usually necessary because empirical parameters describe resistance to flow and sediment transport. Parameters can be adjusted to ensure that the numerical model represents the prototype conditions very accurately. The method of adjusting parameters by running the model at different values until a satisfactory result is obtained is called the calibration. For instance, models can be run with three different values of Manning's *n* for a comparison of the results. It is a very useful way to determine the sensitivity of the model responses to changes in the model parameters. The calibration phase may also vary the time step to check numerical accuracy.

Model validation involves simulations for a different set of prototype data without changing the main coefficients previously determined during the calibration. If a model run satisfactorily replicates the measured prototype conditions without further adjustment, a reasonable confidence is gained in the application of the model to design conditions that have never occurred in the prototype. It is often possible to calibrate a model with only half of a field data set and to validate the model with the second half.

7.4 Numerical River Models

The fast development of computers allows efficient numerical solutions to river-engineering problems. A complete description of multidimensional river models is beyond the scope of this chapter, but our brief discussion uses the following classification: (1) one-dimensional models, in Section 7.4.1; (2) two-dimensional models, in Section 7.4.2; (3) three-dimensional models, in Section 7.4.3.

7.4.1 One-Dimensional Models

Typical one-dimensional models contain components for: (1) steady flow water surface profile computations; (2) unsteady flow simulation; (3) movable-boundary sediment-transport computations; and (4) water-quality analysis. A key element is that all four components use a simplified geometric representation and hydraulic computation routines. Commonly used 1-D models include HEC-RAS, GSTARS, FLUVIAL12, and other commercial codes like Mike11.

7.4.2 Two-Dimensional Models

Two-dimensional river models are typically depth-integrated. They are useful when the vertical variations in flow velocity, turbulence, and concentration

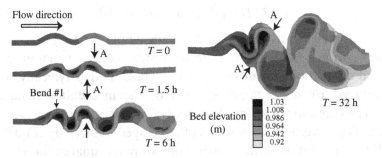

Figure 7.11. Lateral migration of a meandering channel (after Duan and Julien, 2010)

profiles are of lesser interest. The following assumptions for depth-integrated 2-D models are usually considered: (1) the radius of curvature is much larger than the channel width and secondary flows can be ignored; (2) the width–depth ratio of meandering or braiding streams is large enough to neglect sidewall effects; (3) the vertical velocity component can be disregarded; (4) the pressure distribution is hydrostatic; and (5) the grid size is much longer than bed deformations, and bedforms are considered as roughness elements.

Two-dimensional river models solve the continuity and the momentum equations after an empirical relationship is assumed for resistance to flow (e.g. Manning or Darcy–Weisbach/Chézy). The calculations can be performed in a raster-based or vector-based format, depending on the GIS database available for the calculations. Finite-difference schemes are well suited to raster-based data sets. The finite-element method is more appropriate for vector-based data sets and complex geometries. The calculations are repeated at successive time steps for unsteady-flow calculations. Sediment-transport calculations can effectively yield aggradation and degradation results with uncoupled formulations and relatively short time steps. A 2-D simulation example for the lateral migration of a meandering channel is shown in Figure 7.11.

Quasi-3D flow models simulate depth-averaged mathematical flow models in combination with the depth-integrated velocity profile. The continuity and the momentum equations are solved assuming empirical logarithmic velocity profiles. Models can solve 2-D sediment continuity and calculate aggradation and degradation from changes in bed load and suspended sediment transport using empirical formulas. Because the model calculations are in two dimensions, secondary currents in river bends cannot be properly simulated with quasi-3D models.

7.4.3 Three-Dimensional Models

Three-dimensional steady-state models are generally used for turbulent-flow simulations. In κ–ε models, the state of turbulence is characterized by the energy and dissipation parameters κ and ε. These models typically solve the depth-averaged Reynolds approximation of the momentum equation for velocity. The depth-averaged mass conservation determines the water-surface elevation. The deviation from the depth-averaged velocity is computed for each cell by the solution of the conservation of mass equation in conjunction with a κ–ε closure for vertical-momentum diffusion. Sedimentation computations are based on 2-D solid mass conservation for the channel bed and the exchange of sediment between bed load and suspended load.

Data required for running multidimensional models include: (1) channel geometry with cross sections; (2) upstream/downstream boundary conditions in terms of discharge and stage as functions of time for unsteady-flow models and flow-velocity profiles for 3-D models; (3) particle-size distribution of the bed material; (4) upstream/downstream sediment load (some models require both bed load and suspended load); and (5) suspended sediment concentration profiles for 3-D models. Models may not necessarily handle the data in the most appropriate manner. For instance, some κ–ε should account for the turbulence generated behind bedforms. Lumped values of κ and ε will be assumed in the model even if longitudinal profile data showing bedforms are available. Steady 3-D models are applied to estimate the initial rate of sedimentation and erosion in a given situation. The reason for this is the vast amount of computer time required for stabilizing the models under steady-state conditions. Case Study 7.1 illustrates how 3-D steady models can be used to solve complex river-engineering problems.

**Case Study 7.1 Lower Mississippi River Sediment Study,
Louisiana, United States**

The diversion of water and sediment from the Mississippi River into the Atchafalaya River has been closely monitored with physical and mathematical models (USACE, 1999). River models near the Old River Control Complex required extensive field and laboratory measurements. Besides the bathymetry, daily water and sediment discharge records over a period of 50 years served to calibrate long-term one-dimensional models (e.g. HEC-RAS) of the Mississippi and Atchafalaya Rivers.

Three-dimensional models, e.g. CH3D-SED, provide more details on velocity and sediment-concentration profiles. In general, 3-D steady-flow models require accurate field measurements of velocity profiles with Acoustic Doppler Current Profilers (ADCP), and sediment-concentration profiles by size fractions, measured with a P-63 sampler.

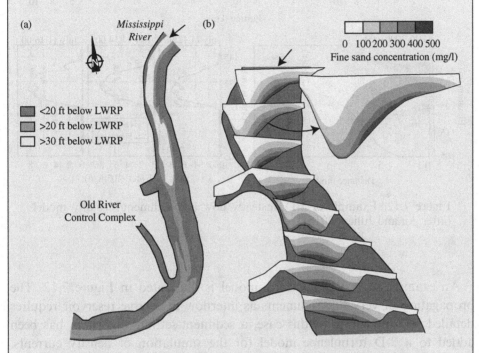

Figure CS7.1.1. (a) Bathymetry and (b) 3-D calculation of sediment concentration profiles in the Mississippi near the Old River Control Complex, Louisiana (after USACE 1999)

Figure CS7.1.1a shows the results of bathymetric changes from the physical model at a discharge of $\sim 1,000,000$ ft^3/s. Notice the sediment accumulation and the nonuniform thalweg depths near the downstream sharp bend in comparison to the gradual and deep profiles generated in the gently curved upstream bend. The suspended sediment-concentration profiles for fine sand are shown in Figure CS7.1.1b. Notice the high concentrations of fine sands on the point bars as compared to the thalweg areas.

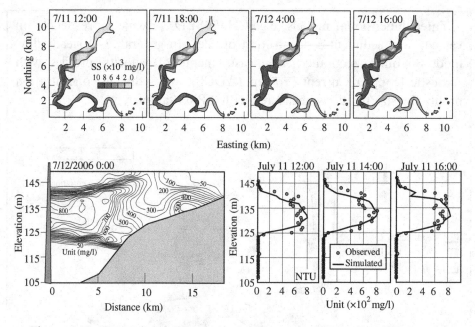

Figure 7.12. Example of 3-D unsteady flow and sediment/turbidity model (after An and Julien, 2014)

An example of a 3-D unsteady model is illustrated in Figure 7.12. The propagation of very fine sediments as interflow in a large reservoir requires detailed 3-D modeling. In this case, a sediment-settling algorithm has been added to a 3-D turbulence model for the simulation of density currents. This type of model must include the effects of density differences from water temperature as well as sediment concentration for proper simulation of the propagation of turbidity in density currents. More details on the comparison of this model with laboratory experiments can be found in An et al. (2012) and on field applications to a prototype reservoir in An and Julien (2014).

Numerous river models have been developed and some are readily available in the public domain. Table 7.4 provides a list of river-engineering models subdivided into: freeware, where executables can be downloaded at no charge, and commercial codes. The list is far from exhaustive but is representative of the current state of the art. Web users can readily access a wealth of information regarding the details of each model.

Table 7.4. *List of river models*

Freeware	Type	Features	Source/Reference
HEC-RAS	1-D	River analysis	USACE – Hydrologic Eng. Center
TREX	2-D	Watershed runoff, sediment, metals	CSU – Velleux et al. (2006, 2008, 2012), England et al. (2007), Ji et al. (2014)
GSTARS	2-D	Sediment transport in rivers and reservoirs	USBR-CSU – developed by Molinas, and Yang and Simoes (2000)
SRH	2-D	Rivers and sediment	USBR – Lai and Yang (2000), Lai (2008)
BRI-STARS	2-D	Bridges and rivers	FHWA – Molinas (2000)
iRIC	2-D, 3-D	River flow and riverbed variation	USGS-Hokkaido, Nelson et al. (2006), Shimizu et al. (2000)
Commercial	type	Features	Source/Reference
FLUVIAL12	1-D	Rivers, sediment	Chang (2006), e.g. Julien et al. (2010)
WMS, TABS	2-D	Watershed, rivers, coastal	USACE, ERDC-CHL
HydroSed	2-D	Rivers, sediment	Duan (2001), Duan and Julien (2005, 2010)
FLO-2D	2-D	Mudflows, floodplain	FLO-2D software, O'Brien et al. (1993)
MIKE11, 21	2-D	Rivers, sediment	DHI
CCHE	2-D	Rivers, sediment	CCHE, developed by Jia et al. (2001)
CH3D-SED	3-D	Large rivers	USACE-CHL, Gessler et al. (1999)
FLOW-3D	3-D	Fluid mechanics, reservoir density currents	FLOW Science, e.g. An et al. (2012), An and Julien (2014)

◆*Exercise 7.1*

Define Δx and Δt for $K_{num} = 50$ m^2/s from Equation (7.29a and b) when $C_u = 0.9$, $v = 1$ m/s and $K_d = 100$ m^2/s. Compare with Δx and Δt for $K_{num} = 5$ m^2/s, and discuss the results.

Exercise 7.2

Plot the modules of the amplification factor $|A|$ of Equation (7.14) on Figure 7.2(b) as a function of $2 < N < 20$ for values of $C_c = 0.25$, 0.5, 0.75, 1.0, 1.25, 1.5, and 2.0. Show that $|A|$ approaches unity as $N \to \infty$ for all values of C_c.

◆◆Exercise 7.3

With reference to Example 7.1, repeat the calculations for $\Delta x = 500$ m and $\Delta x = 250$ m for values of $C_c = 0.5$ and 2.0, respectively. Compare the results with those of Tables E7.1.1 and E7.1.2. Plot the results at $t = 2{,}000$ s for $C_r = 0.5$ and $\Delta x = 2{,}000$, 1,000, 500, and 250 m and show that the results converge. Also, plot the corresponding results for $C_c = 2.0$ and show that the results diverge as Δx decreases even if the scheme is consistent.

◆Exercise 7.4

Determine the coefficients for a stable Leonard scheme from Equation (7.30) with $C_u = 0.5$ and a grid Peclet number $P_\Delta = 0.5$. Discuss whether the simulations will be smooth or spiky. Also discuss the required boundary conditions.

[*Answer*: $\phi_j^{k+1} = 0.4375\phi_{j-2}^k + 0.0625\phi_{j-1}^k + 0.0625\phi_j^k + 0.4375\phi_{j+1}^k$.   The simulations will be very smooth. However the boundary conditions require caution because of potential leakage upstream and interference downstream.]

◆Computer Problem 7.1

Develop a stable numerical algorithm to solve the flood-wave propagation problem in Example 7.1. For instance, demonstrate that simulations with the Leonard scheme eliminate numerical diffusion.

◆◆Computer Problem 7.2

From Example 7.2, replace the first pulse (15 min pulse at 100,000 mg/l) with a 3-min pulse at 500,000 mg/l and compare the results.

◆◆◆Computer Problem 7.3

With reference to Computer Problem 5.1, calculate sediment transport using Equation (2.44) given $d_s = 0.5$ mm over the 25-km reach. Use the aggradation–degradation algorithm to calculate changes in bed elevation. Adjust the water-surface elevation after bed-elevation changes. Provide graphical output of the sediment-transport capacity, bed elevation, and hydraulic grade line at three different times.

[*Answer:* See similar results in Figure 7.10.]

8

Hillslope and Revetment Stability

This chapter discusses the stability of hillslopes and revetments. Section 8.1 covers hillslope stability, followed with revetment stability in Section 8.2.

8.1 Hillslope Stability

This section looks at geotechnical properties regarding hillslope stability and only essential knowledge is introduced here. The concept of Mohr circles and shear strength of soils is covered in Section 8.1.1, followed with a brief overview of infinite slope stability (Section 8.1.2), finite slope stability (Section 8.1.3), and rotational failure (Section 8.1.4).

8.1.1 Geotechnical Properties

Figure 8.1a illustrates two types of applied force per unit area, called stresses. The normal stress σ is applied perpendicular to the surface (+ in compression) and tangential stress τ along the cut surface. The principal stresses are the normal stresses on planes with zero shear stress. Normal and shear stresses are graphically represented by the Mohr circle.

As sketched in Figure 8.1b, the normal and tangential stresses on a plane at an angle θ from the principal stresses are obtained either graphically from the Mohr circle or from the following relationships:

$$\sigma_\theta = \frac{\sigma_1 + \sigma_3}{2} + \frac{\sigma_1 - \sigma_3}{2}\cos 2\theta \tag{8.1a}$$

$$\tau_\theta = \frac{\sigma_1 - \sigma_3}{2}\sin 2\theta \tag{8.1b}$$

note that τ_θ is positive counterclockwise (+CCW) when θ is +CCW from the direction σ_3. Accordingly, the maximum shear stress is equal to the Mohr circle radius $(\sigma_1 - \sigma_3)/2$.

205

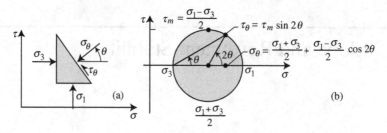

Figure 8.1. Mohr circle describing normal and shear stresses of soils: (a) shear-pressure diagram and (b) Mohr–Coulomb diagram

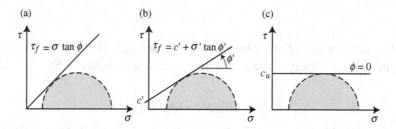

Figure 8.2. Mohr–Coulomb failure criteria for (a) noncohesive granular soils, (b) cohesive drained soils, and (c) cohesive undrained soils

The strength of soils can be described by the Mohr–Coulomb failure law:

$$\tau_f = c + \sigma \tan \phi \tag{8.2}$$

where τ_f is the failure shear strength of the soil, c is the cohesion, σ is the normal stress, and ϕ is the friction angle. As sketched in Figure 8.2, three type of condition are of interest here: (a) properties of granular materials; (b) effective stresses for drained cohesive soils; and (c) undrained conditions for clay soils.

The strength of noncohesive granular soils has been studied since Coulomb in 1776 and the failure of granular soils, e.g. sands and gravels, is

$$\tau_f = \sigma \tan \phi, \tag{8.3}$$

where ϕ is the friction angle. Figure 8.3a shows the Coulomb failure criterion for granular materials with the Mohr circle and critical angle θ_c of the failure plane. The friction angle of sands and gravels decreases with porosity, shown in Figure 8.3b.

Some interesting properties of granular soils can be defined from Figure 8.3. For instance, in conjunction with Figure 8.1 we obtain

$$\sin \phi = \frac{(\sigma_1 - \sigma_3)/2}{(\sigma_1 + \sigma_3)/2} = \frac{(\sigma_1/\sigma_3) - 1}{(\sigma_1/\sigma_3) + 1} \tag{8.4a}$$

and

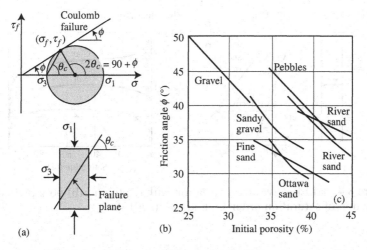

Figure 8.3. (a) Mohr–Coulomb failure criteria for granular soils and (b) friction angle versus porosity for sands and gravels (after Lambe and Whitman, 1969)

$$\frac{\sigma_1}{\sigma_3} = K_p = \frac{1 + \sin\phi}{1 - \sin\phi} = \tan^2(45° + \phi/2) = \tan^2\theta_c. \tag{8.4b}$$

The critical angle of the failure plane is given by $\theta_c = 45° + \phi/2$ and the coefficient of passive stresses is defined as $K_p = \sigma_1/\sigma_3$. Soil compaction will reduce the porosity and hence increase the friction angle. The particle angularity also increases the friction angle, such that compacted coarse angular materials have greater friction angles than loose fine and rounded particles.

The strength of cohesive soils depends on several factors including pore pressure, and for clays, the degree of consolidation and drainage condition. In general, the effective stress needs to be considered in the soil-strength analysis. As defined by Terzaghi, and sketched in Figure 8.4a, the effective stress σ' is obtained by subtracting the pore pressure u from the total normal stress σ, or $\sigma' = \sigma - u$. The Mohr–Coulomb criterion for effective stress then becomes

$$\tau_f = c' + \sigma' \tan\phi'. \tag{8.5}$$

The order of magnitude for the negative pore pressure of partially saturated soils is -2 kPa for sands, -10 kPa for silts, and -100 kPa for clays, as shown in Figure 8.4b. This means that partially saturated soils will have a higher strength than fully saturated soils. However, under hydrostatic pressure conditions, the effective stresses of saturated soils can be significantly lower than the total stresses.

Under drained conditions and large rates of deformation, the ultimate strength of clays is quite comparable to the behavior of sands, i.e. $c' = 0$.

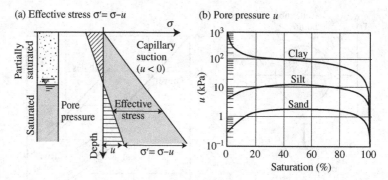

Figure 8.4. (a) Effective stress variation with depth and (b) pore pressure as a function of saturation (after Lu and Godt, 2013)

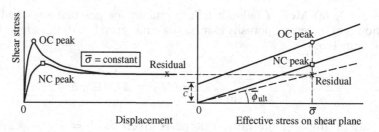

Figure 8.5. Residual and peak stresses (after Lambe and Whitman, 1969)

As sketched in Figure 8.5, the shear strength under large displacements is called the residual strength, which is less than the peak shear stress. The residual strength is less than the peak shear stress for normally consolidated (NC) or over-consolidated (OC) clays.

When considering the effective shear stress and drained conditions, it should be noted that the friction angle of normally consolidated clays, as in Figure 8.6a, varies with the plasticity index, i.e. the difference between the liquid limit and the plastic limit of a soil. The friction angle is typically greatest ($\phi' > 35°$) when the plasticity index is less than 20 percent and lowest ($\phi' < 25°$) when the plasticity index exceeds 50 percent. The ultimate friction angle of sand-clay mixtures depends on the percentage of clays finer than 2 μm, as shown in Figure 8.6b. Skempton (1964) considered the effective stress and drained conditions, and found that the ultimate friction angle is greatest for pure sands ($\phi' > 30°$) and lowest for pure clays ($\phi' < 10°$).

Undrained soil conditions represent cases when the loadings are applied faster than in the time required for the soil to consolidate. This type of situation is less relevant for the analysis of hillslope stability.

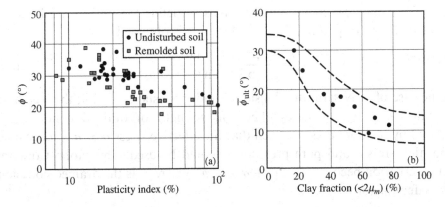

Figure 8.6. Friction angle of normally consolidated clays and ultimate friction angle: (a) angle of repose as a function of the plasticity index and (b) clay fraction (after Skempton, 1964)

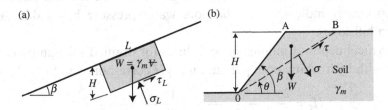

Figure 8.7. (a) Infinite and (b) finite slopes

8.1.2 Infinite Slope Stability Analysis

Following the analysis of Lu and Godt (2013), the long hillslope in Figure 8.7a shows the weight W of soil per unit width $W = \gamma_m LH \cos \beta$, where H is the vertical depth and γ_m is the specific weight of the soil. The force is made of two components: a normal component $W \cos \beta$ and a tangential component $W \sin \beta$.

Consequently, the applied stresses (force per unit area) are the normal stress σ and the tangential stress τ defined as

$$\tau = \gamma_m H \sin \beta \cos \beta, \qquad (8.6a)$$

and

$$\sigma = \gamma_m H \cos^2 \beta. \qquad (8.6b)$$

The safety factor SF of an infinite hillslope is therefore obtained from the ratio of the shear strength from Equation (8.2) to the applied shear stress in Equation (8.6), or

$$SF = \frac{\tau_f}{\tau} = \frac{c + \gamma_m H \cos^2 \beta \tan \phi}{\gamma_m H \sin \beta \cos \beta} = \frac{\tan \phi}{\tan \beta} + \frac{2c}{\gamma_m H \sin 2\beta}. \qquad (8.7)$$

One of the main practical applications is to define the critical hillslope soil thickness H_c corresponding to SF = 1, which is obtained from:

$$H_c = \frac{2c}{\gamma_m \sin 2\beta} \frac{\tan \beta}{(\tan \beta - \tan \phi)}. \tag{8.8}$$

In the case of fully saturated hillslopes under constant rainfall, the weight of saturated soil, γ_m, increases. The soil strength of drained conditions is reduced by the pore pressure such that the effective stress $\sigma' = \sigma - u$ obtained from the stress σ and pore pressure u should be used. The Mohr–Coulomb failure criterion becomes $\tau_f = c' + \sigma' \tan \phi'$, where c' is the drained cohesion. The safety factor is:

$$SF = \frac{c' + [(\gamma_m H \cos^2 \beta) - u]\tan \phi'}{\gamma_m H \sin \beta \cos \beta} = \frac{\tan \phi'}{\tan \beta} + \frac{2(c' - u\tan \phi')}{\gamma_m H \sin 2\beta}. \tag{8.9}$$

Besides the increased soil weight, the negative sign in the last term of this equation clearly indicates that the pore water pressure has a destabilizing effect on hillslopes.

The critical depth of infinite-slope failures for drained soils under constant rainfall with hydrostatic pore pressure $u = \gamma H$ corresponds to SF = 1, or

$$H_c = \frac{2c'}{\gamma_m} \left[\frac{1}{\tan \phi' + 2\cos^2 \beta (\tan \beta - \tan \phi')} \right]. \tag{8.10}$$

For instance, a drained soil with $c' = 2$ kPa and friction angle $\phi' = 30°$ and unit weight $\gamma_m = 21$ kN/m^3 on a slope $\beta = 30°$ becomes unstable when $H >$ 0.33 m, and thus such soils are prone to shallow landslides during intense rainstorms. For instance, if the soil porosity is 0.3, the infiltration of 100 mm of rain would yield shallow landslides and mudflows.

8.1.3 Finite Slope Stability Analysis

The finite slope analysis describes the stability of riverbanks. Consider the riverbank sketch in Figure 8.7b with a slope angle β and vertical height H with a soil of specific weight γ_m and friction angle ϕ. The weight per unit width W of a soil wedge OAB resting on a potential failure plane OB is given by

$$W = \frac{\gamma_m (AB) H}{2} = \frac{\gamma_m H^2}{2} (\cot \theta - \cot \beta). \tag{8.11}$$

The two stresses (normal and tangential) to the OB plane are therefore

$$\sigma = \frac{\gamma_m H}{2} (\cot \theta - \cot \beta)\cos \theta \sin \theta \quad \text{and} \tag{8.12a}$$

and

$$\tau = \frac{\gamma_m H}{2} (\cot \theta - \cot \beta)\sin^2 \theta. \tag{8.12b}$$

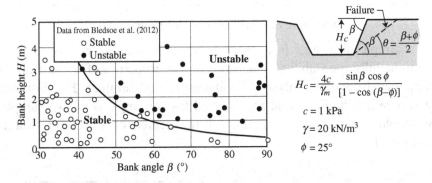

Figure 8.8. Stream-bank stability example

Substitution into the Mohr–Coulomb failure criterion Equation (8.2) gives

$$\tau = \frac{\gamma_m H}{2}(\cot\theta - \cot\beta)\sin^2\theta = c + \frac{\gamma_m H}{4}(\cot\theta - \cot\beta)\sin(2\theta)\tan\phi \quad (8.13a)$$

or

$$c = \frac{\gamma_m H}{4}(\cot\theta - \cot\beta)\left[2\sin^2\theta - \sin(2\theta)\tan\phi\right]. \quad (8.13b)$$

The critical-failure plane angle θ_c is obtained from $\partial c/\partial\theta = 0$, which yields the critical hillslope height H_c and θ_c as

$$H_c = \frac{4c}{\gamma_m}\frac{\sin\beta\cos\phi}{[1 - \cos(\beta - \phi)]}, \quad (8.14a)$$

$$\theta_c = \frac{\beta + \phi}{2}. \quad (8.14b)$$

For instance, a vertical cut in a soil of cohesion $c = 10$ kPa, friction angle $\phi = 30°$, and unit weight $\gamma_m = 20$ kN/m^3 gives $H_c = 3.46$ m and $\theta_c = 60°$. One example with field measurements from Bledsoe et al. (2012) is shown in Figure 8.8, where the geotechnical values are $c = 1$ kPa, $\phi = 25°$, and unit weight $\gamma_m = 20$ kN/m^3.

Two cases are particularly interesting: (1) granular soils ($c = 0$) give $\theta_c = \beta = \phi$ and (2) poorly drained soils, such as clays with $\phi = 0$ and undrained cohesion c_u, give a maximum vertical cut of $H_c = 4c_u/\gamma_m$. For instance, the case of a stiff clay with $c_u = 70$ kPa and $\gamma_m = 18.6$ kN/m^3 yields $H_c = 18$ m. Clearly, the case of undrained clay soils could result in deep landslides. Finally, Example 8.1 illustrates how the values of cohesion and friction angle for partially saturated soils can be used to define the hillslope stability.

Example 8.1 Hillslope stability analysis

This example illustrates how the infinite-slope analysis can be used to determine hillslope stability. The geotechnical properties of the soil can be determined from the analysis of undisturbed soil samples. For instance, the apparent cohesion c and the internal friction angle ϕ of two samples are shown to vary as a function of the degree of soil saturation in Figure E8.1.1.

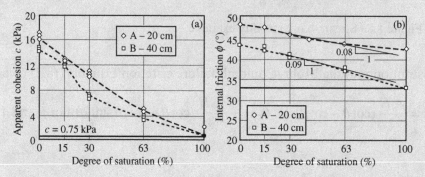

Figure E8.1.1. Soil (a) cohesion, and (b) friction angle versus saturation (after Kim et al., 2011)

The cohesion and friction angle both decrease considerably with the level of saturation. Fully saturated soils yield $c = 0.75$ kPa and $\phi = 33°$. The factor of safety of soils then depends on infiltration depth H and hillslope angle β. Kim (2012) used the following relationship from Equation (8.9) for the safety factor

$$\mathrm{SF} = \frac{c + (\gamma_m - \gamma)H\cos^2\beta\tan\phi}{\gamma_m H \sin\beta\cos\beta}. \qquad (E8.1.1)$$

This relationship assumes drained soil conditions where the pore pressure is given by $u = \gamma H \cos^2 \beta$. It also considers that the infiltration depth is significant during rainstorms and the soil is saturated from the failure plane to the free surface. The safety factor can then be plotted as a function of hillslope angle for different infiltration depths. When $c = 0.75$ kPa and $\phi = 33°$, the results as shown in Figure E8.1.2a indicate that hillslopes are unstable at slopes as low as 20°.

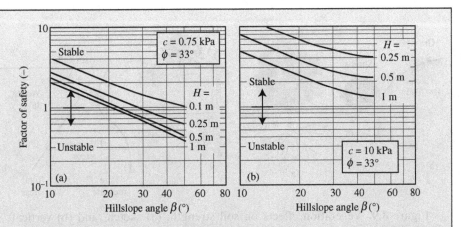

Figure E8.1.2. Factor of safety vs hillslope angle for (a) low and (b) high apparent cohesion (after Kim, 2012)

However, when the apparent cohesion is increased to 10 kPa, hillslopes become very stable even at very steep angles, as shown in Figure E8.1.2b. The increase in apparent cohesion can be found in unsaturated soils, well drained hillslopes, highly cohesive soils, over-consolidation, and/or the presence of roots.

8.1.4 Effects of Roots and Vegetation on Hillslope Stability

The effect of vegetation on hillslopes is primarily to strengthen the soils via the root network. The tensile strength of roots per se can be up to 40 MPa. However, in terms of soil strength, living plants can contribute to as much as tens of kPa in the root zone near the hillslope surface. The soil strength can increase by about 10 kPa up to a depth of 10–20 cm for grasses, up to 1 m for shrubs, and a couple of meters for trees. As sketched in Figure 8.9, vegetation thus has a very positive effect on strengthening soils near the surface.

It should also be considered that large trees will also add weight and increase the active force. The applied stress below the root zone increases in forested areas. The weakest soil layer will often be found just below the root zone. If the saturated zone can extend below the root zone during severe storms, shallow landslides are likely to occur.

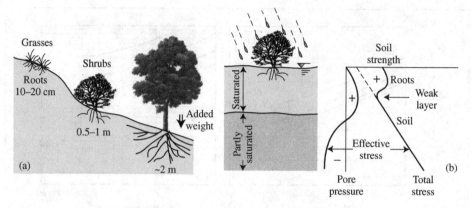

Figure 8.9. Vegetation effects on soil strength: (a) sketch, and (b) vertical profile

8.2 Revetment Stability

This section looks at the stability of a layer of stones or broken rock/concrete, called riprap, normally placed to protect river banks or structures from scour. A review of the theoretical stability of single stones (Section 8.2.1) is followed with detailed riprap size calculations from shear stress (Section 8.2.2) and flow velocity (Section 8.2.3) in straight channels, and for flow conditions in curved channels (Section 8.2.4). More practical considerations are provided on the gradation of riprap (Section 8.2.5), riprap filters (Section 8.2.6), and methods to prevent riprap failure (Section 8.2.7).

8.2.1 Stability of Single Stones

A stream is gently sloping downstream at a bed-slope angle θ_0. A stone rests on the stream embankment, inclined at a side-slope angle θ_1. Figure 8.10 illustrates the forces acting on the particle including the: (1) lift force F_L; (2) drag force F_D; (3) buoyancy force F_B; and (4) particle weight F_W. The submerged weight $F_S = F_W - F_B$ is obtained by subtracting the buoyancy force from the stone weight. The lift force is normal to the embankment plane and the drag force acts along the plane in the same direction as the velocity field surrounding the particle. In a bend, the streamlines are deflected from the downstream direction by an angle λ defined positive downward, i.e. toward the thalweg.

The stone stability analysis is slightly simplified here for river-engineering applications. For small (less than $\sim 20°$) angles θ_0 and θ_1, the projection of the submerged weight vector into the embankment plane is $a_\Theta \cong \sqrt{\cos^2 \theta_1 - \sin^2 \theta_0}$, and the second geometric parameter combines the two components of F_S along the embankment plane to define $\tan \Theta \cong \sin \theta_0 / \sin \theta_1$.

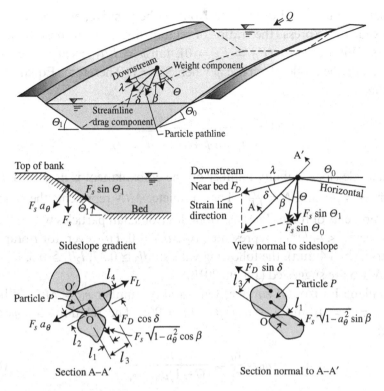

Figure 8.10. Particle stability diagram

There are two conditions for the stone equilibrium from the sum of moments:
(1) in the direction of particle motion along line A–A′ and (2) in the direction
normal to A–A′. From the first condition, the stability against rotation occurs
when the moments about the point of rotation 0 are equal. With the moment
arms l_1, l_2, l_3, and l_4 shown in Figure 8.10, the sum of moments gives

$$l_2 F_S a_\Theta = l_1 F_S \sqrt{1 - a_\Theta^2} \cos \beta + l_3 F_D \cos \delta + l_4 F_L. \qquad (8.15)$$

The left-hand side of Equation (8.15) is the stabilizing moment from the
particle weight. The last term on the right-hand side denotes the lift moment,
which always destabilizes the stone. The other terms on the right-hand side
may become negative, depending on the geometry. The stability factor SF is
defined as the ratio of the resisting moments to the moment generating
motion. The stability factor SF is the ratio of the sum of counterclockwise
moments to the sum of clockwise moments:

$$\text{SF} = \frac{l_2 F_S a_\Theta}{l_1 F_S \sqrt{1 - a_\Theta^2} \cos \beta + l_3 F_D \cos \delta + l_4 F_L}. \qquad (8.16) \blacklozenge\blacklozenge$$

Note that each term in Equation (8.16) must be positive, else the formulation is changed to express the ratio of stabilizing (+) to destabilizing (–) moments. Without flow ($F_D = F_L = 0$), $\tan\phi = l_2/l_1$ because the angle θ_0 (or θ_1) equals the angle of repose ϕ when SF = 1. Therefore, Equation (8.16) transforms to

$$SF = \frac{a_\Theta \tan\phi}{\eta_1 \tan\phi + \sqrt{1 - a_\Theta^2}\cos\beta} \qquad (8.17)\blacklozenge$$

where the side-slope stability number on the embankment is $\eta_1 = M + N\cos\delta$, in which $M = \frac{l_4 F_L}{l_2 F_S}$ and $N = \frac{l_3 F_D}{l_2 F_S}$. The parameter M/N represents the ratio of lift to drag moments of force. The case of no lift (for small particles within the laminar sublayer $d_s < 4\nu/u^*$) is represented by $M/N = 0$. However, for riprap (particles coarser than 4 mm), the following values: $l_1/l_3 \cong 0.37$, $l_2/l_3 \cong 0.3$, $l_4/l_3 \cong 2.6$, and $M/N \cong 5$ are suggested (Julien, 2010).

On a plane horizontal surface, the stability number $\eta_0 = M + N$ because ($\theta_0 = \theta_1 = \delta = 0$) and $\lambda + \delta + \beta + \Theta = 90°$. The relations for η_0 and η_1 become:

$$\eta_0 = \frac{\tau_0}{\tau_c} = \frac{\tau_0}{(G-1)\rho g d_s \tau_{*c}}, \qquad (8.18)\blacklozenge$$

$$\eta_1 = \eta_0 \left[\frac{(M/N) + \sin(\lambda + \beta + \Theta)}{1 + (M/N)}\right]. \qquad (8.19)\blacklozenge$$

When the flow is fully turbulent over a hydraulically rough horizontal surface, incipient motion corresponds to $\tau_{*c} \cong 0.047$ with SF = η_0 = 1.

The second equilibrium condition along the section normal to A–A' is

$$l_3 F_D \sin\delta = l_1 F_S \sqrt{1 - a_\Theta^2} \sin\beta. \qquad (8.20)$$

This second condition therefore defines the orientation angle β as

$$\beta = \tan^{-1}\left[\frac{\cos(\lambda + \Theta)}{\dfrac{(M+N)\sqrt{1 - a_\Theta^2}}{N\eta_0 \tan\phi} + \sin(\lambda + \Theta)}\right]. \qquad (8.21)\blacklozenge$$

This method thus defines the direction β at which unstable particles would move. Stones move at an angle β from the direction of steepest descent. A detailed calculation example is presented in Example 8.2.

Example 8.2 Particle stability analysis and moving particle direction

This example shows detailed calculations of particle stability in a river bend by solving successively Equations (8.18), (8.21), (8.19), and finally (8.17). It is assumed that $M/N = 5$ for coarse material. A 16-mm gravel particle would be near incipient motion on a bed slope of 6.8 ft per river mile under a bed shear stress of $\tau_0 = 12$ Pa. If the particle is placed on an embankment at a side-slope angle $\theta_1 = 15°$, calculate the orientation angle β when the streamlines are deflected at an upward angle $\lambda = -10°$ (e.g. toward the free surface). The calculation procedure is:

Step 1: The particle size is $d_s = 16$ mm.
Step 2: Let's assume an angle of repose $\phi = 37°$, and specific gravity $G = 2.65$.
Step 3: The side-slope angle is $\theta_1 = 15°$.
Step 4: The downstream slope angle is $\tan^{-1}(6.8/5280)$, or $\theta_0 = 0.074°$.
Step 5: The angle $\Theta = \tan^{-1}(\sin\theta_0/\sin\theta_1) = 0.28°$ is negligible in this case.
Step 6: The geometric factor $a_\Theta = \sqrt{\cos^2\theta_1 - \sin^2\theta_0} = 0.965$ is very close to 1.
Step 7: The applied bed shear stress $\tau_0 = 12$ Pa corresponds to a flow depth ~ 1 m.
Step 8: The streamline deviation angle $\lambda = -10°$ gives a downstream shear-stress component $\tau_0\cos\lambda = 11.8$ Pa and a transverse component $\tau_0\sin\lambda = -2.08$ Pa toward the free surface (negative) and this should help stabilize the particle.
Step 9: From Equation (8.18) and $\tau_{*c} \cong 0.047$, the plane-bed-stability number is

$$\eta_0 = \frac{12\,\text{N}\,\text{m}^3\,\text{s}^2}{\text{m}^2(1.65)1{,}000\,\text{kg}\times 9.81\,\text{m}\times 0.016\,\text{m}\times 0.047} = 0.985.$$

Step 10: From Equation (8.21), assuming $M/N = 5$, the 16-mm direction angle is

$$\beta = \tan^{-1}\left[\frac{\cos(-10° + 0.28°)}{\dfrac{6\sqrt{1-(0.965)^2}}{0.985\tan 37°} + \sin(-10° + 0.28°)}\right] = 26.8°.$$

Step 11: Despite favorable upward streamlines, this heavy 16-mm gravel moves toward the thalweg because $\beta + \Theta < 90°$.
Step 12: The stability factor is calculated from Equation (8.17).

$$SF = \frac{0.965\tan 37°}{0.985\left[\dfrac{5 + \sin(-10° + 26.8 + 0.28°)}{6}\right]\tan 37° + \sqrt{1 - 0.965^2}\cos 26.8°} = 0.82.$$

This 16-mm gravel size is unstable (SF < 1) and is expected to move. The reader should repeat the calculations to show that a 32-mm particle would be stable.

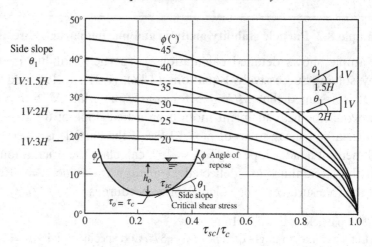

Figure 8.11. Ratio of critical shear stress on a side slope to critical shear stress on a horizontal surface

8.2.2 *Riprap Size Calculations from Shear Stress*

Lane's (1953) shear-stress analysis in a straight channel yields a simple first approximation when the lift–drag ratio becomes very large. From recent CSU experiments, Cox et al. (2014) analyzed the stability of various commercial block types for riverbank protection. Based on these experiments, the range of length ratios can be approximated by $l_1/l_3 \cong 0.5$–0.65, $l_2/l_3 \cong 1.3$–2.5, $l_4/l_3 \cong 1.3$–2.5, and the ratio M/N ranges from 1.2 to 17. For large M/N values, Lane's relationship yields an interesting first approximation for flow in straight channels ($\lambda = 0$):

$$\frac{\tau_{sc}}{\tau_c} = \eta = \cos\theta_1 \sqrt{1 - \left(\frac{\tan^2\theta_1}{\tan^2\phi}\right)} = \sqrt{1 - \left(\frac{\sin^2\theta_1}{\sin^2\phi}\right)}, \qquad (8.22)\blacklozenge$$

where τ_{sc} is the critical shear stress on the side slope and τ_c is the critical shear stress on a horizontal surface. This critical shear-stress ratio is plotted in Figure 8.11.

The effective rock size d_m stabilizing a riverbank under the applied shear stress τ_0 on a side-slope angle θ_1 can be simply estimated from

$$d_m \cong \frac{\tau_0}{\tau_{*c}\gamma(G-1)\left(\sqrt{1 - \dfrac{\sin^2\theta_1}{\sin^2\phi}}\right)}, \qquad (8.23)\blacklozenge$$

Example 8.3 Riprap size from shear stress in a straight channel

Determine the rock riprap size to stabilize the banks of a straight river given the width $W = 100$ m, flow depth $h = 3$ m, and channel slope $S = 800$ cm/km. The bank slope is $\theta_1 = 26°$, the rock density $G = 2.65$, and the angle of repose $\phi = 40°$.

Step 1: The applied shear stress is

$$\tau_0 = \gamma h S = 9,810 \text{ N/m}^3 \times 3 \text{ m} \times 800 \times 10^{-5} = 235 \text{ Pa}.$$

Step 2: The term in brackets of Equations (8.23) is calculated as

$$\eta = \sqrt{1 - \left(\frac{\sin^2 \theta_1}{\sin^2 \phi}\right)} = \sqrt{1 - \left(\frac{\sin^2 26°}{\sin^2 40°}\right)} = 0.77.$$

Step 3: An approximate riprap size is obtained from Equation (8.23) with $\tau_{*c} = 0.047$,

$$d_m \cong \frac{\tau_0}{\eta \tau_{*c}(\gamma_s - \gamma)} = \frac{235 \text{ Pa}}{0.77 \times 0.047} \frac{\text{m}^3}{(2.65 - 1)9,810 \text{ N}} = 0.4 \text{ m, or 16 in.}$$

where τ_0 is the applied shear stress, γ is the specific weight of water, θ_1 is the side-slope angle, ϕ is the angle of repose of the rock riprap, and τ_{*c} is the critical value of the Shields number (e.g. $\tau_{*c} \cong 0.047$). The term in brackets of Equation (8.23) becomes very small when the side-slope angle is approximately equal to the angle of repose. This means that extremely large riprap material will be required to stabilize steep riverbanks. Example 8.3 provides a riprap sizing example based on shear stress in a straight channel.

8.2.3 Riprap Size Calculations from Velocity

The stone size needed to protect a straight riverbank can also be estimated from the flow velocity. For stone riprap, the velocity at the top of the stone is called the velocity against the stone v_s and relates to the shear velocity u_* as $v_s \cong 5.75u_*$. Incipient motion on a plane surface $\tau_{*_c} = u_{*_c}^2/(G-1)gd_s = 0.06 \tan \phi$ can be rewritten in terms of a critical velocity against the stone v_{sc} as

$$v_{sc} \cong 5.75u_{*_c} \cong \sqrt{2(G-1)gd_s \tan \phi} \,. \tag{8.24}$$

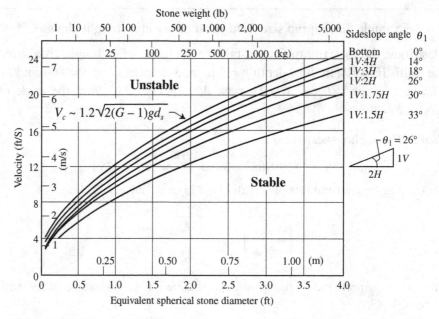

Figure 8.12. Mean flow velocity required to mobilize riprap stones at various side-slope angles

Alternatively, incipient motion can be written in terms of the mean flow velocity. For logarithmic velocity profiles, the velocity against the stone v_s is a function of the mean flow velocity V, flow depth h, and stone diameter d_s as $V \cong v_s \log(2h/d_s)$. After considering $K_c = \log(2h/d_s)\sqrt{\tan\phi}$, the critical mean flow velocity V_c on a horizontal surface becomes

$$V_c = K_c\sqrt{2(G-1)gd_s}. \tag{8.25}$$

On a side slope without secondary flows, the critical mean flow velocity V_c can be approximated according to Lane's approach by

$$V_c \cong K_c\sqrt{2(G-1)gd_s}\left(1 - \frac{\sin^2\theta_1}{\sin^2\phi}\right)^{1/4}. \tag{8.26}\blacklozenge$$

Figure 8.12 plots the critical mean flow velocity as a function of riprap size for representative riprap design conditions ($\phi \cong 40°$, $h \cong 10d_s$, or $K_c = 1.2$). At a given flow velocity, note how quickly the rock size increases as $\theta_1 > 20°$. Example 8.4 estimates the riprap size from the flow velocity in a straight channel.

Example 8.4 Riprap size from flow velocity in a straight channel

The mean flow velocity in a straight river reaches 18 ft/s (5.5 m/s) during floods at a Manning coefficient $n = 0.025$, the river is 900 ft wide (275 m) and 30 ft deep (9.1 m). Estimate the stable rock size at a side slope of $1V{:}2.25H$, or $\theta_1 = 24°$. The angle of repose of the angular rocks is $\phi = 40°$ and the specific gravity $G = 2.65$. Based on flow velocity, solving Equation (8.26) for d_s gives a riprap size estimate as

$$d_s \cong \frac{V_c^2}{K_c^2}\frac{\sin\phi}{2(G-1)g(\sin^2\phi-\sin^2\theta_1)^{1/2}}$$

$$= \frac{18^2\,\text{ft}^2\text{s}^2\sin 40°}{s^2 1.2^2\,2\times 1.65\times 32.2\,\text{ft}\sqrt{\sin^2 40° - \sin^2 24°}} = 2.7\,\text{ft} = 0.83\,\text{m}.$$

This value can also be graphically obtained from Figure 8.12.

8.2.4 Riprap Design in Sharp Bends

In sharp river bends, the streamlines near the outer bank are deflected downward at a positive angle λ given from $\tan\lambda \cong 11h/R$, where h is the mean flow depth and R is the bend radius of curvature. Recent research by C. Thornton and S. Abt at CSU defines the ratio K_b between the maximum shear stress in a bend compared to the shear stress in a straight channel. The results of Sin (2010) indicate that $K_b \cong 1 + 2W/R$, where W is the channel width and R is the radius of curvature. As an approximation to the detailed method in Section 8.2.1, Example 8.5 illustrates this riprap design procedure for a sharp river bend. Once the stone size is determined, the riprap gradation (Section 8.2.4) and filters (Section 8.2.5) must be considered. More information is found regarding traditional methods in Maynord (1988, 1992, 1995), with recent developments in Cox et al. (2014) for different block types.

8.2.5 Riprap Gradation

The calculation of a representative rock size for riprap design from previous sections is insufficient to ensure stability of riprap blankets. A riprap cover with stones of the same size scours to a greater depth (and at a lower flow velocity) than a mixture of stone sizes. Well-graded riprap mixtures develop armor layers which increase stone stability and reduce scour. The representative rock

Example 8.5 Riprap stability in a sharp river bend

The mean flow velocity in a sharp river bend reaches 16 ft/s ($V = 5$ m/s) during floods at a Manning coefficient of $n = 0.036$, the river is 325 ft wide ($W = 100$ m) and 10 ft deep ($h = 3$ m), the bed slope is $S = 0.008$, and the radius of curvature of 975 ft ($R = 300$ m). Estimate the stable rock size and the safety factor on a side slope $1V:2H$. The angle of repose of the material is $\phi = 40°$ and $G = 2.65$.

First, the hydraulic radius is $R_h = A/P = Wh/(W + 2h) = 2.83$ m, $V = 4$ m/s from Manning's equation, the shear stress is $\tau_0 = \gamma R_h S = 9,810 \times 2.83 \times 0.008$ Pa $= 222$ Pa, and $K_b \cong (R + 2W)/R = (300 + 200)/300 = 1.66$. The riprap size estimate is obtained after replacing the shear stress with $K_b\tau_0$ in the method for a straight channel with Equation (8.23) as in Example 8.3 as

$$d_m \cong \frac{K_b\tau_0}{\sqrt{1 - \frac{\sin^2\theta_1}{\sin^2\phi}}0.047(\gamma_s - \gamma)}$$

$$= \frac{1.66 \times 222 \text{ N m}^3}{\text{m}^2 \sqrt{1 - \frac{\sin^2 26°}{\sin^2 40°}}0.047 \times 1.65 \times 9,810 \text{ N}} \cong 0.63 \text{ m}.$$

The safety factor is obtained as follows from Example 8.2:

Step 1: The slope is 0.008, or 0.45°.

Step 2: The shear stress is $\tau_0 = 222$ Pa.

Step 3: The angle $\theta = \tan^{-1}(\sin 0.86°/\sin 26°) = 2°$.

Step 4: The factor $a_\theta = \sqrt{\cos^2\theta_1 - \sin^2\theta_0} = \sqrt{\cos^2 26° - \sin^2 0.45°} = 0.90$.

Step 5: The streamline angle is $\lambda \cong \tan^{-1}(11h/R) = \tan^{-1}[(11 \times 3)/200] = 9.4°$.

Step 6: $\eta_0 = \frac{21\tau_0}{(G-1)\gamma d_s} = \frac{21 \times 222 \text{ Nm}^3}{\text{m}^2 1.65 \times 9,810 \text{ N} \times 0.63 \text{ m}} = 0.46$.

Step 7: The angle $\beta = 8°$ is obtained assuming $M = 5$ N from

$$\tan\beta = \frac{\cos(\lambda + \theta)}{\left[\frac{6\sqrt{1 - a_\theta^2}}{\eta_0 \tan\phi} + \sin(\lambda + \theta)\right]} = \frac{\cos(9.4° + 2°)}{\left[\frac{6\sqrt{1 - 0.90^2}}{0.46 \tan 40°} + \sin(9.4° + 2°)\right]} = 0.14.$$

Step 8: $\eta_1 = \eta_0 \left[\dfrac{5 + \sin(\lambda + \beta + \theta)}{6} \right] = 0.46 \left[\dfrac{5 + \sin(9.4° + 8° + 2°)}{6} \right] = 0.41.$

Step 9: The stability factor is

$$SF = \frac{a_\theta \tan \phi}{\eta_1 \tan \phi + \sqrt{1 - a_\theta^2} \cos \beta} = \frac{0.9 \tan 40°}{0.41 \tan 40° + \sqrt{1 - 0.90^2} \cos 8°} = 0.97.$$

This is close to incipient motion and the riprap size should be $d_s \cong 0.7$ m = 27 in.

size d_m for riprap is larger than the median rock size d_{50}, and typically corresponds to d_{65}, such that $d_m \cong 1.25 d_{50}$ can be used. For instance, the recommended gradation from the FHWA HRE (2001) is shown in Figure 8.13. Riprap consisting of angular material is preferable to rounded stones.

A gradation curve specified by the US Army Corps of Engineers (1981) as shown in Table 8.1 is preferable to a single gradation curve. Any stone gradation within the limits is acceptable.

The riprap thickness should not be less than: (1) 12 in. (30 cm) for practical placement; (2) the upper limit of d_{100} stone; or (3) 1.5 times the upper limit of d_{50} stone, whichever is greatest. When riprap is placed under water, the thickness should be increased by 50 percent, and if it is subject to attack by large floating debris or wave action it should be increased by 6–12 in. (15–30 cm).

Riprap placement is usually accomplished by dumping directly from trucks. Draglines with orange-peel buckets, backhoes, and other power equipment can also be used advantageously to place riprap from floating barges. Rocks should never be placed by being rolled down a chute or pushed downhill with a bulldozer because these methods result in segregation of sizes.

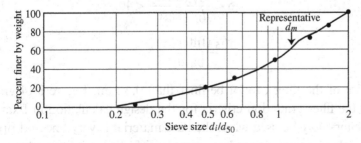

Figure 8.13. Recommended riprap gradation curve for sieved material

Table 8.1. *Typical riprap gradation bands from the US Army Corps of Engineers (1981)*

d_{100} max (in.)	Stone weight (lb) for percent lighter by weight[a]						d_{30} max (ft)	d_{90} max (ft)
	15		50		100			
	min	max	min	max	min	max		
12	5	13	17	26	35	86	0.48	0.70
15	11	25	34	50	67	169	0.61	0.88
18	18	43	58	86	117	292	0.73	1.06
21	29	69	93	137	185	463	0.85	1.23
24	43	102	138	205	276	691	0.97	1.40

[a] Assuming $G = 2.65$, or $\gamma_s = 165$ lb/ft^3 = 26 kN/m^3, 1 ft = 12 in. = 30.5 cm. The relationship between diameter and weight is based on a spherical shape.

8.2.6 Riprap Filters

Filters are used under the riprap revetment to allow water to drain easily from the bank without carrying out soil particles. Filters are required when the d_{15} of the riprap gradation exceeds five times the d_{85} of the bank material. Filter blankets must meet two basic requirements: stability and permeability. The filter material must be fine enough to prevent the base material from escaping through the filter, but it must be more permeable than the base material. Two types of filters are commonly used: gravel filters and synthetic filter cloths.

A gravel filter consists of a layer, or blanket, of well-graded material placed over the base material of the embankment, and underneath the riprap cover. Typical gravel filter sizes range from 3/16 to 3.5 in. (5–90 mm). The filter thickness should not be less than 6–9 in. (20 cm). The main specifications for filter gradation are as follows:

$$\frac{d_{50} \text{ (filter)}}{d_{50} \text{ (base)}} < 40, \tag{8.27a}\blacklozenge$$

$$5 < \frac{d_{15} \text{ (filter)}}{d_{15} \text{ (base)}} < 40, \tag{8.27b}\blacklozenge$$

$$\frac{d_{15} \text{ (filter)}}{d_{85} \text{ (base)}} < 5. \tag{8.27c}\blacklozenge$$

The d_{15} size of the filter cannot be finer than 0.4 mm. Figure 8.14 graphically illustrates the filter gradation criteria for the design calculation in Example 8.6. Multiple filters may be used when the base material is very fine and riprap very large. In such rare cases, each layer must satisfy the stability and permeability requirements relative to the underlying layer.

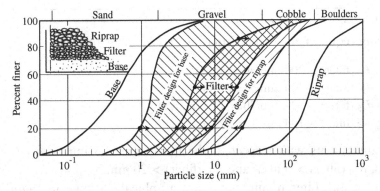

Figure 8.14. Example of riprap filter gradation curves

Example 8.6 Example of gravel filter design

The following filter design example involves the properties of the base material and the riprap given in Table E8.3.1.

Table E8.3.1. *Sizes of materials*

Base material Sand	Riprap Rock
$d_{85} = 1.5$ mm	$d_{85} = 400$ mm
$d_{50} = 0.5$ mm	$d_{50} = 200$ mm
$d_{15} = 0.17$ mm	$d_{15} = 100$ mm

From these gradation curves and Equation (8.27), the ratio d_{50} (riprap)/ d_{50} (base) > 40 and a filter is required. Also, the riprap does not contain sufficient fines, because d_{15} (riprap)/d_{85} (base) = 100/1.5 = 67, which is much greater than 5.

With reference to the base material, the specified gravel filter must satisfy the following requirements from Equation (8.27a–c), as follows:

Step 1: d_{50} (filter) < 40(0.5) mm = 20 mm,
Step 2: d_{15} (filter) > 5(0.17) mm = 0.85 mm,
Step 3: d_{15} (filter) < 40(0.17) mm = 6.8 mm, and
Step 4: d_{15} (filter) < 5(1.5) mm = 7.5 mm.

The requirements with respect to the base are: 0.85 mm < d_{15} (filter) < 6.8 mm, and d_{50} (filter) < 20 mm.

With reference to the riprap, the specified gravel filter must satisfy the following requirements from Equation (8.27a–c), as follows:

Step 5: d_{50} (filter) > 200 mm/40 = 5 mm,
Step 6: d_{15} (filter) < 100 mm/5 = 20 mm,
Step 7: d_{15} (filter) > 100 mm/40 = 2.5 mm, and
Step 8: d_{85} (filter) > 100 mm/5 = 20 mm.

The requirements with respect to the riprap are: 2.5 mm < d_{15} (filter) < 20 mm, d_{50} (filter) > 5 mm, and d_{85} (filter) > 20 mm.

These riprap filter requirements are graphically shown in Figure 8.14. Gravel filters having sizes within the double cross-hatched area are satisfactory. For instance, a good filter could have these sizes: d_{85} = 30 mm, d_{50} = 15 mm, d_{15} = 4 mm.

Synthetic filter cloths (plastic cloth and woven plastic materials) are also used as filters, replacing a component of a graded filter. Numerous plastic filter fabrics are commercially available. Synthetic filters must be permeable and minimize the possibility of clogging. It is often desirable to place a protective blanket of sand on the filter to prevent puncture when dumping the riprap. The sides and the toe of the filter fabric must be sealed or trenched so that the base material does not leach out around the filter cloth.

8.2.7 *Preventing Riprap Failure*

Typical riprap failure modes are illustrated in Figure 8.15 and identified as follows: (1) particle erosion; (2) translational slide; (3) modified slump; and (4) side slope.

Particle erosion by flowing water in Figure 8.15a is the most common erosion event. It can be initiated by the impingement of flowing water, flow in eddies, local flow acceleration, abrasion, freeze/thaw cycles, ice, and toe erosion. Probable causes of particle erosion include: (1) stone size that is not large enough or reduced by poor rock quality and/or abrasion; (2) very steep banks; (3) individual stones being removed by impact (e.g. logs, ice, and ships); and (4) riprap gradation that is too uniform. The solution requires coarser riprap material and appropriate riprap gradation and material angularity.

A translational slide in Figure 8.15b is a riprap failure caused by the downslope mass movement of stones. The initial phases of a translational

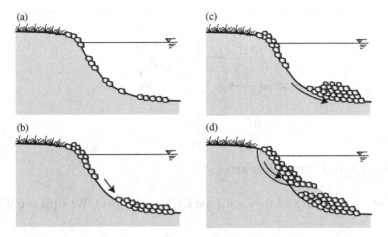

Figure 8.15. Examples of riprap failure modes: (a) particle erosion, (b) slide, (c) slump, and (d) sideslope failure

slide are indicated by cracks in the upper part of the riprap blanket that extend parallel to the channel. Slides are usually initiated by channel-bed degradation that undermines the toe of the riprap blanket. The causes of translational slides include: (1) very steep banks; (2) excessive pore pressure; and (3) lack of toe protection. The solution requires strengthening the toe of the riprap blanket.

A modified slump failure of riprap in Figure 8.15c is the mass movement of material only within the riprap blanket. Probable causes are: (1) a very steep bank and (2) lack of toe support. The solution requires coarse material to be added at the toe of the embankment and the side-slope angle of the embankment to be reduced.

A side-slope failure in Figure 8.15d is a rotational bank failure within the riverbank. The cause of slump failures is related to shear failure of the underlying base material that supports the riprap. Most likely causes of slump failures are: (1) excessive pore pressure in the base material and (2) side slopes that are too steep. The solution requires reduction of the embankment slope or possibly draining the base material. Riprap should not be used at slopes steeper than $1V{:}1.5H$.

♦*Exercise 8.1*

For the principal stresses sketched in Figure EX8.1, determine the normal and shear stress along plane B–B.

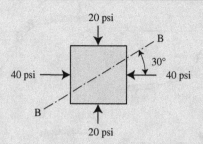

Figure EX8.1. Stress diagram

[*Answer*: $\sigma_\theta = 25$ psi and $\tau_\theta = -8.7$ psi CCW $= 8.7$ psi CW; 1 psi $= 6.9$ kPa.]

◆◆*Exercise 8.2*

With reference to Example 8.1, use the finite-slope analysis method to calculate and plot the maximum bank height as a function of the side-slope angle for three soil conditions: (a) dry soils; (b) 30 percent saturation; and (c) fully saturated soils. Can you explain why the conditions are so different in these three cases?

Exercise 8.3

With reference to Example 8.2, repeat the calculations for the 32-mm particle and show that the particle is stable.

◆*Problem 8.1*

Find the direction angle of a 10-mm particle under a shear stress $\tau_0 = 10$ Pa when the streamlines are deflected upward at $\lambda = -10°$. The downstream bed-slope angle is $\theta_0 = 0.05°$ and the side-slope angle is $\theta_1 = 10°$. Consider and compare the result with a 1-mm particle under identical flow conditions.

◆◆*Problem 8.2*

With reference to Example 8.3, determine the rock size that is stable and determine an appropriate gradation curve for the riprap blanket.

◆◆*Problem 8.3*

Use the channel sketched in Figure P8.3 and the following properties:

(a) specific weight of stone = 165 lb/ft^3 (26 kN/m^3);
(b) local depth at toe of outer bank = 25 ft (7.6 m);
(c) local depth at 20 percent upslope from toe = 20 ft (6.1 m);
(d) channel side slope = $1V{:}2H$;
(e) downstream channel slope = 2 ft/mile (38 cm/km);
(f) minimum centerline bend radius = 1,700 ft (520 m);
(g) average flow velocity = 7.2 ft/s (2.2 m/s); and
(h) surface channel width = 500 ft (150 m). Determine the size, thickness, and gradation of the riprap blanket required for the stabilization of the outer bank of this channel. If the riprap is placed over uniform 0.5-mm sand, design the gravel filter required to prevent leaching.

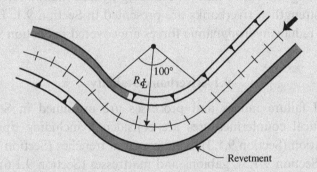

Figure P8.3. Flow in a curved channel with riprap in shaded areas

◆*Problem 8.4*

The gradation of a riprap sample is $W_{100} = 200$ lb, $W_{50} = 70$ lb, and $W_{15} = 30$ lb. Is this a satisfactory gradation if the representative rock size is 10 in.?

9

Riverbank Protection

Riverbank stabilization structures are designed to protect riverbanks and prevent the lateral migration of rivers. There are two different approaches: (1) strengthening the banks and (2) reducing hydrodynamic forces. Practical methods to strengthen riverbanks are presented in Section 9.1. Flow-control structures to reduce hydrodynamic forces are covered in Section 9.2.

9.1 Riverbank Stability

First, typical failure modes and processes are examined in Section 9.1.1. Several practical countermeasures are considered, including riprap (Section 9.1.2), vegetation (Section 9.1.3), windrows and trenches (Section 9.1.4), sacks and blocks (Section 9.1.5), gabions and mattresses (Section 9.1.6), articulated concrete mattresses (ACMs) (Section 9.1.7), soil cement (Section 9.1.8), retaining walls (Section 9.1.9), and miscellaneous considerations (Section 9.1.10).

9.1.1 Preventing Riverbank Failure

In general, the most erosive banks are sandy and/or silty, and the least erosive are clayey and/or gravelly. Riverbank failure is affected by active factors such as discharge magnitude and duration, flow velocity, and applied shear-stress magnitude and orientation. Additional active factors include seepage, piping, surface waves, and ice. Anthropogenic activities such as urbanization, drainage, floodplain farming, boating and commercial navigation, and water-level fluctuations from hydropower generation can have detrimental effects on bank stability. Passive components relate to bank-material size, gradation, weight, and cohesion. Biological factors such as vegetation and root strength play a significant stabilizing role.

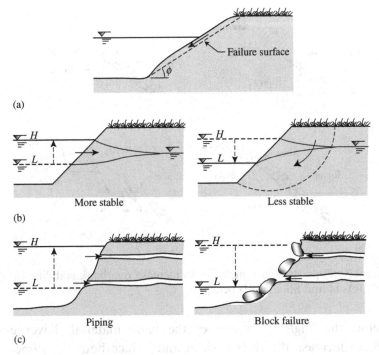

Figure 9.1. Typical riverbank failure modes: (a) noncohesive bank, (b) cohesive bank, and (c) stratified bank

Three typical modes of riverbank failure are sketched in Figure 9.1. With noncohesive granular material, grain removal at the toe of the outerbank induces sliding of the granular material as soon as the bank angle exceeds the angle of repose of the material (Figure 9.1a).

In the case of cohesive bank material, rotational failure is typical and the presence of tension cracks may accelerate the bank erosion process (Figure 9.1b). Bank stability depends on flood stages. During rising stages of flood hydrographs, the outflow of water from the river into the adjacent banks stabilizes the riverbanks. The converse is true during falling stages of flood hydrographs where riverbanks become less stable. In the case of strati-fied banks, lenses of noncohesive material can be mobilized through piping, which leaves the overlying cohesive material unsupported and subject to ten-sion cracks and cantilever failure (Figure 9.1c).

Successful countermeasures to increase riverbank stability include slope reduction and effective drainage. Figure 9.2 illustrates various types of slope-reduction methods where the objective is to reduce the embankment

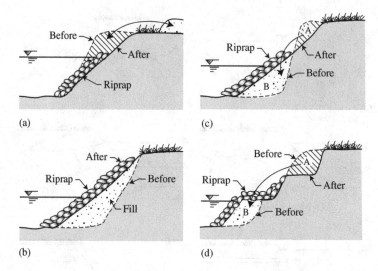

Figure 9.2. Slope reduction methods to increase riverbank stability: (a) cut, (b) fill, (c) cut and fill, and (d) benching

slope below the angle of repose of the bank material. River cuts (see Figure 9.2a) increase the river's width and reduce both the slope and the floodplain area. This method is usually preferred when adequate space is available on the floodplain. Where there is insufficient space, the slope may be flattened by use of fill material, which causes a reduction in channel width. When used extensively in urban areas, this river encroachment can have an adverse impact on the river's ability to carry floodwaters downstream and can cause major urban flooding problems. The cut-and-fill operations sketched in Figure 9.2c combine both approaches and balance out the amount of fill material. Benching is also an indirect method of slope reduction (Figure 9.2d). The benching method produces a series of stepped sections.

The control of groundwater within a slope can help stabilize the slope. Groundwater control may be achieved by two methods: (1) preventing infiltration of surface water into soils and (2) providing subsurface drainage to remove the water from the soil mass. The first method is generally accomplished by providing adequate surface drainage. The second method uses various subsurface drainage techniques which are effective when only a small quantity of water needs to be drained. A horizontal drain, as sketched in Figure 9.3a, consists of a slotted pipe to remove excessive pore water from cut slopes experiencing stability problems. Vertical drains, as shown in Figure 9.3b, can be utilized to intercept groundwater levels above the maximum river elevations.

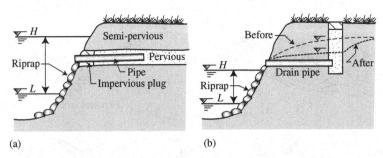

Figure 9.3. Slope-reduction methods to increase riverbank stability: (a) horizontal drain and (b) drain well

9.1.2 Riprap Revetment

Rock riprap is usually the most economical and widely used material for riverbank protection. A rock riprap blanket is flexible and is not weakened by settlement and differential settling. Local damage is easily repaired by the placement of more rock. Construction is not complicated, and special equipment is not necessary. Riprap is permeable and allows the free flow of water to and from the riverbanks. It is usually durable and recoverable and may be stockpiled for future use. The appearance of rock riprap is natural, and vegetation will grow between the rocks over time.

The important factors to be considered in designing rock riprap blanket protection are: (1) the velocity (both its magnitude and its direction) of the flow or shear stress in the vicinity of the rock; (2) the side slope of the bankline being protected; (3) the density of the rock; (4) the angle of repose for the rock, which depends on stone shape and angularity; (5) the durability of the rock; (6) the riprap blanket thickness; (7) the filter needed between the bank and the blanket to allow seepage but to prevent erosion of bank soil through the blankets; (8) the blanket stabilization at the toe of the bank; and (9) that the blanket must be tied into the bank at its upstream and downstream ends. Some of these factors were discussed at length in Section 8.2.

The upstream and the downstream ends of the riprap revetment should be tied into the bank to prevent stream currents from seeking a pathway behind the revetment and unraveling the riprap blanket. The most common method of tying into the bank is to dig a trench at both ends of the blanket, as shown in Figure 9.4a. The depth of a trench should be twice the blanket thickness, and the bottom width of the trench should be three times its thickness.

The most effective method for increasing the stability at the toe of the bank and preventing undermining is use of the launching apron sketched in Figure 9.4b. A flexible launching apron is laid horizontally on the bed at the

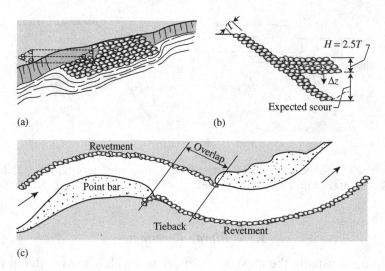

Figure 9.4. Tieback and launching apron for riprap revetments: (a) tieback, (b) launching apron, and (c) revetments

foot of the revetment, so that when scour occurs the materials will settle and cover the side of the scour hole on a natural slope. This method is recommended for noncohesive channel beds in which deep scour is expected. In cohesive channel beds, the revetment should be extended as far down as the expected worst scour level. Alternatives to use of the launching apron include driving a cut-off wall of sheet piles or continuing the revetment below the expected scour level or possibly down to bedrock.

Two examples of revetment profiles are shown in Figure 9.5: (a) for the case of a typical revetment on a cut/fill slope and (b) for the case of a trenchfill revetment. Trenchfill revetments are constructed on dry land after excavating a trench and placing the riprap blanket. Such structures are often encountered in irrigation canals, navigable waterways, and engineered meander cutoffs.

9.1.3 Vegetation

Vegetation is the most natural material for increasing stream stability and improving environmental conditions for wildlife. The presence of vegetation below the water surface can effectively protect a bank in two ways. First, the root system holds the soil together and increases the overall bank stability. Secondly, the exposed stalks, stems, branches, and foliage provide additional resistance to the streamflow, thus reducing flow velocities and inducing deposition.

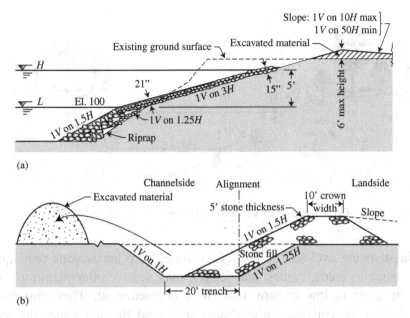

Figure 9.5. Examples of (a) riprap revetment and (b) trenchfill revetment

Vegetation is generally divided into three broad categories: grasses, shrubs, and trees. The grasses require a shorter period of time to become established. Woody plants offer greater protection against erosion because of their extensive root systems, which take time to develop. Under some conditions, the weight of the plant on steep slopes will offset the advantage of the root system. Vegetation is generally better suited to stabilizing stream banks than riverbanks. Also, bank failure in large rivers may yield large amounts of floating debris during floods.

Native plants should normally be used because they have become adapted to the climate, soils, and other ecological characteristics of the area. Nonnative plants can also be met with local opposition because invasive species may cause problems to other riparian users. Nitrogen-fixing plants may also be required in poor soils and cold regions. Other limitations regarding vegetation include: (1) failure to grow, depending on soil and changing climatic conditions; (2) undercutting before the roots fully develop; (3) extensive wetting and drying periods; (4) damage from anchor ice and freeze–thaw cycles; and (5) damage from livestock and wildlife.

Typical riparian vegetation zones are sketched in Figure 9.6. The splash zone is located between normal high water and normal low water in the zone of highest stress. The splash zone is exposed frequently to wave wash, erosive river currents, ice and debris movement, wet–dry cycles, and

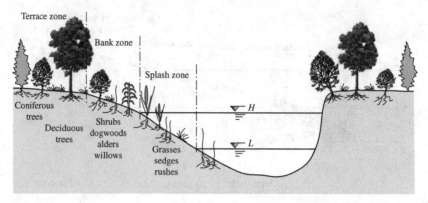

Figure 9.6. Typical riparian vegetation zones

freezing-thawing cycles. For the splash zone, only herbaceous semiaquatic plants, such as reeds, rushes, and sedges, may survive after planting. Reeds typically grow in low current areas with fine sediment. They also protect with their roots, rhizomes, and shoots, and bind the soil under the water. During floods, reeds tend to bend and reduce resistance to flow while protecting the underlying soils.

The bank zone lies above the normal high-water level and is typically inundated for at least a 60-day duration once every 2–3 years. This zone can be exposed periodically to wave wash, erosive river currents, ice and debris movement, and traffic by animals or humans. Both herbaceous and woody plants are used in the bank zone. These should still be flood-tolerant and able to withstand partial to complete submergence for up to several weeks. The water table in the bank zone is often quite close to the soil surface. Shrub, willow, dogwood, and alder transplants or 1-year-old rooted cuttings are effectively used in the bank zone. At high flows, shrubs reduce the speed of the current and increase friction and thereby decrease the erosive force of the water.

The terrace zone, inland from the bank zone, is not subjected to erosive action of the river except during extreme floods. The terrace zone contains species that are less flood-tolerant than those in the bank zone. The tree species also become taller and more massive. The combination of trees, shrubs, and grasses in this zone improves wildlife habitat diversity.

Grasses can be planted by hand seeding, transplanting clumps of grass or herbaceous plants, planting stems or rhizomes, or by mechanically spreading organic mulches containing seeds and fertilizers. Hydroseeding can be a useful and effective means of direct seeding, particularly on steep slopes. Often barges with hydroseeders can be floated directly to the remediation sites.

Seeds in a water slurry should be sprayed first and then covered with mulch to reduce soil moisture loss. Commercial manufacturers also market erosion-control matting that will hold the seed and soil in place until new vegetation is fully established. Reed rolls combine sections of sod, rhizomes, and shoots, which are enclosed within a net, and placed in a trench. Some of these methods can be labor-intensive. Case Study 9.1 illustrates the effectiveness of vegetation protection on the neck of Thompson Bend on the Mississippi River.

Case Study 9.1 Thompson Bend on the Mississippi River, Missouri, United States

Thompson Bend is located on the right descending bank of the Mississippi River between river miles 30 and 45, above the confluence of the Mississippi and Ohio Rivers; see Figure CS9.1.1.

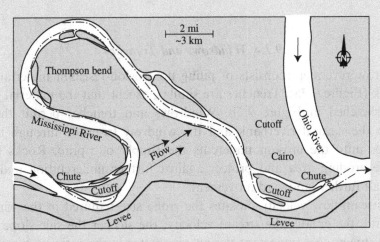

Figure CS9.1.1. Thompson Bend of the Mississippi River

The river flows in a gooseneck encompassing approximately 10,000 acres (40 km^2) of valuable agricultural land. At the throat of the bend, the linear overland distance is approximately 2 miles (3.2 km). The river distance along the thalweg is approximately 14 miles (22.5 km). During large floods the river naturally tries to flow overbank straight across Thompson Bend. The water-surface drop along the river is 7 ft (2.1 m) for a slope 0.5 ft/mile (9.5 cm/km) along the river and 3.5 ft/mile (66 cm/km) across the neck.

In the early 1980s, severe erosion of the upper bankline began along the right descending bank in the upper reaches of the bendway. In addition, localized surface erosion reached an estimated rate of 40,000 tons/acre

---

during this event (1,600,000 tons/km^2). Continued erosion could have allowed for development of a chute cutoff across the bend.

The revegetation process began in 1985. The results were immediately evident and very little erosion was observed during the fall flood of 1986. The area was tested again in the flood of 1990, when very little erosion occurred. The Thompson Bend also suffered very little visible damage during the Great Flood of 1993, when record high stages occurred, and the duration of the overland flows reached an unprecedented 130 days. However, the flood took its toll on the vegetation. Numerous trees that were inundated for most of the 130 days died. The flood of 1994 prevented any significant revegetation from occurring. However, the erosion was very minor compared with the massive amounts of scour that occurred in the early 1980s. Thompson Bend clearly illustrates the vital importance that vegetation exerts in controlling overbank scour.

9.1.4 Windrows and Trenches

A windrow revetment consists of piling up erosion-resistant material along the bank (Figure 9.7a). Trenches are similar, except that the material is buried as sketched in Figure 9.7b. Windrows and trenches permit the area between the natural riverbank and the windrow to erode through natural processes until the erosion undercuts the supply of riprap. Rocks launch onto the eroding area and protect against further undercutting and eventually halt further landward movement.

The stream-flow velocity dictates the stone size required in the windrow. In general, the greater the stream velocity, the steeper the side slope of the final revetment. An important design parameter is the ratio of the relative thickness of the final revetment to the stone diameter. Large stones will require more material than smaller stones to produce the same relative thickness. A well-graded stone material is important to prevent leaching of the bank material.

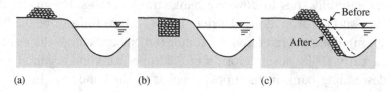

(a)   (b)   (c)

Figure 9.7. Windrows and trenches: (a) windrow, (b) trench, and (c) launching

9.1.5 Sacks and Blocks

Burlap sacks filled with soil or sand–cement mixtures are often used for emergency protection against levee breaches and local urban flooding. Most types of sacks are easily damaged and can deteriorate rapidly enough to be unsuitable for long-term protection.

Sacks and blocks do have certain advantages over stone riprap, as follows: (1) they allow possible placement on steep slopes; (2) they use locally available materials; (3) they result in a smooth boundary, if channel conveyance is a major consideration; and (4) they may be considered to be more aesthetic. The preferred placement technique is sketched in Figure 9.8. A rule of thumb is to consider flat placement only if the bank slope is less than $1V:2.5H$. On slopes around $1V:2H$, the bags should be overlapped by being placed with the long dimension pointing toward the bank. On slopes steeper than $1V:2H$, the bags should be overlapped with the short dimension pointing toward the bank. The maximum slope should be $1V:1H$.

Precast cellular sacks can be manufactured with locally available sand, cement, and aggregates or can be obtained from commercial sources. Blocks are durable under the effects of freeze/thaw cycles and are less likely to be lifted off the bank by ice. Blocks conform to minor changes in bank shape. Channel boundary roughness is less than that of riprap, and blocks may become very slippery when covered with algae near the water surface. Weepholes must be included in the revetment to allow drainage of groundwater and prevent pressure buildup that could cause revetment failure. A permeable fabric or a gravel blanket can be used as a filter under the blocks to prevent seepage. Blocks should not be grouted.

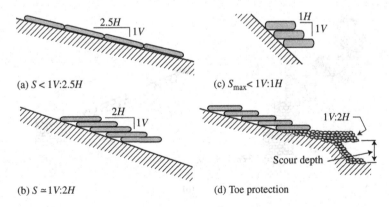

(a) $S < 1V:2.5H$

(b) $S \simeq 1V:2H$

(c) $S_{max} < 1V:1H$

(d) Toe protection

Figure 9.8. Sacks and blocks

Sack revetments also have disadvantages compared with stone riprap: (1) they are highly labor-intensive, and thus are generally more costly than stone; (2) they tend to act monolithically and are less flexible than riprap; (3) they are more sensitive to excess hydrostatic pressure; (4) uniformly sized sacks are not as effective against erosion and leaching as well-graded stones; and (5) synthetic bags may be vulnerable to hazards such as fire, ice, vandalism, and livestock traffic.

9.1.6 Gabions and Mattresses

Gabions are patented rectangular wire baskets filled with relatively small stones typically less than 8 in. (20 cm) in diameter. Where flow velocities are such that small stones would not be stable if used in a riprap blanket, the wire boxes provide an effective design option. The recommended maximum velocity for gabions ranges from 8 to 15 ft/s (2–5 m/s). The baskets are commercially available in a range of standard sizes and are made of galvanized steel wire with a polyvinyl chloride (PVC) coating. The baskets are typically 0.5 m × 1 m × 2 m and are folded flat to be manually assembled on site. A horizontal gravel filter under the basket prevents leaching of the base material. Examples of gabions and mattresses are shown in Figure 9.9.

Box gabions can be stacked on relatively steep slopes to form a massive structure capable of resisting the forces of both river flows and also unstable bankline materials. The flexibility of their mesh and filler stone allows them to maintain their structural integrity even after significant displacement, undercutting, or settlement. Box gabion structures are aligned along the

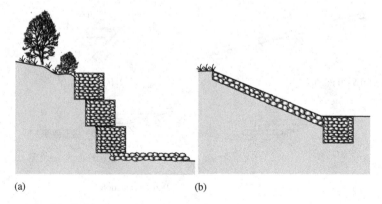

(a)   (b)

Figure 9.9. (a) Gabions and (b) mattresses

riverbank toe to form a retaining wall. They can also form dikes for diverting flows away from the bank. Mattresses are shaped into shallow, broad baskets and are tied together side by side to form a continuous protective blanket. They are normally placed on a smoothly graded riverbank slope.

Gabions and mattresses are labor-intensive and among the expensive methods of riverbank protection. Gabion structures need to be periodically inspected, and proper maintenance ensures reliable performance. The wire mesh is subject to damage from floating debris, water pollutants, corrosion, wear from high-velocity sediments, and vandalism. Any cracks or breaks in the PVC coating will expose the wire to corrosive elements. Many structures do not perform well in cold regions, particularly when entire box gabions filled with water freeze completely.

9.1.7 *Articulated Concrete Mattresses*

In large rivers, precast concrete blocks held together by steel rods or cables have been used to form flexible articulated mats, as sketched in Figure 9.10. The use of ACMs has been limited primarily to the Mississippi River. This is due to the very high cost of the plant required for the placement of the mattress beneath the water surface. The basic unit of this mattress is 4 ft (1.3 m) wide × 25 ft (7.6 m) long × 3 in. (7.5 cm) thick. ACMs are flexible, strong, and durable. Approximately 8 percent of an ACM's surface area is open, which permits fines to pass through. The open areas are undesirable, but necessary to facilitate placement of the mattress in swift deep water, to relieve hydrostatic pressure, and ensure flexibility. Yearly maintenance runs approximately 2 percent of in-place work. This type of revetment has an excellent service record and is considered the standard for the Lower Mississippi River.

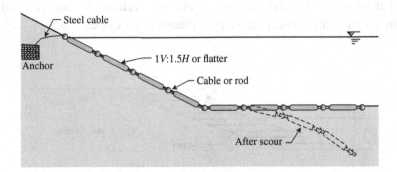

Figure 9.10. Articulated concrete mattresses

9.1.8 Soil–Cement

In arid areas with scarce riprap, the combination of on-site soil with cement provides a practical alternative. A soil–cement blanket with 8–15 percent cement may be an economical and effective riverbank protection method for use in well-drained sandy soils without vegetation. The key to success is proper recognition of the rigid nature of the material in a flexible boundary channel.

Soil–cement can be placed and compacted on slopes as steep as $1V:2H$. The best results are achieved on slopes no steeper than $1V:3H$. Figure 9.11 sketches a typical soil–cement construction for bank protection. For use in soil–cement, soils should be easily pulverized and contain at least 5 percent, but not more than 35 percent, silt and clay (material passing through the No. 200 sieve). Finer soils usually are difficult to pulverize and require more cement. When velocities exceed 6–8 ft/s and the flow carries sufficient bedload to be abrasive, the aggregates should contain at least 30 percent gravel particles retained on a No. 4 (4.75 mm) sieve.

A stair-step construction is recommended on steep channel embankments with the placement of 6–9 ft-wide (2–3 m) × 6 in.-thick (15 cm) horizontal layers of soil–cement. Special care should be taken to prevent raw soil seams between successive layers. A sheep's-foot roller should be used on the last layer of the day to interlock with the next layer. The completed soil–cement installation must be protected from drying for a seven-day hydration period. After completion, the material has sufficient strength to serve as a roadway along the embankment.

Soil–cement is most effective in arid areas. However, soil–cement has three major disadvantages: impermeability, low strength, and susceptibility to cold temperature variations. Failure may occur when the bank behind the blanket becomes saturated and cannot drain. Also, because a soil–cement blanket is relatively brittle, very little traffic (vehicular, pedestrian, or livestock) can be sustained without cracking the thin protection veneer. In northern climates the blanket can easily break up during freeze–thaw cycles.

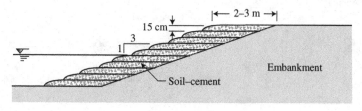

Figure 9.11. Soil cement

9.1.9 Retaining Walls

Retaining walls are near-vertical structures designed to prevent riverbank erosion or failure. Vertical retaining walls are valued for protecting waterfront property and improving access to water. Retaining walls have been classified into three distinct types: (1) gravity walls; (2) cantilever walls; and (3) sheet-piling walls. Examples of gravity walls are shown in Figure 9.12.

Crib walls, shown in Figure 9.12a, are constructed with interlocking structural members to form "boxes" filled with stones to increase the mass necessary for stability. The toe of the crib should always be protected with riprap. The most common cause of failure is scour around the pilings, and excess

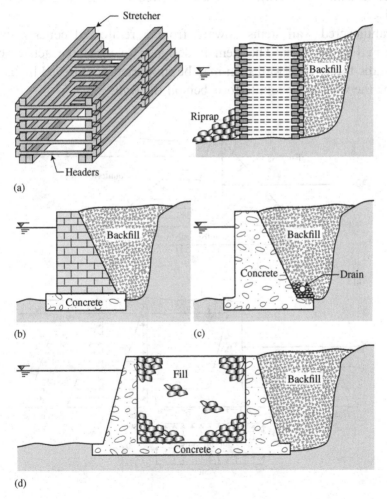

Figure 9.12. Gravity walls: (a) crib wall, (b) masonry wall, (c) concrete wall, and (d) caisson

pressure on the bank side. The fill material should drain freely. A filter fabric or gravel filter can prevent fine soils from leaching through. The bulkhead should be tied into the bank at the upstream and the downstream ends of the structure to prevent flanking.

Masonry walls (as in Figure 9.12b) typically consist of blocks, bricks, or concrete cinder blocks are similar to, but weaker in strength than, concrete walls, as shown in Figure 9.12c. Tension cracks may develop when there is no reinforcing steel in their construction. This type of wall has been successfully adapted to the construction of very large walls. Caissons, or large "boxes" of reinforced concrete, as shown in Figure 9.12d, are constructed on land and can be floated to the site. The caisson is then filled with concrete or compacted sand to provide the mass required to sink the continuous wall into place.

A cantilevered wall stems upward from a reinforced concrete base, as shown in Figure 9.13a. The stem is designed to resist the active soil and hydrostatic forces. The wall stem may be supported or stiffened by buttresses fronting the wall or by counterforts behind the wall.

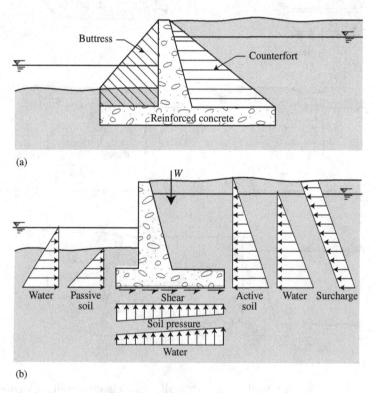

Figure 9.13. Cantilevered walls (a) with and (b) without buttress

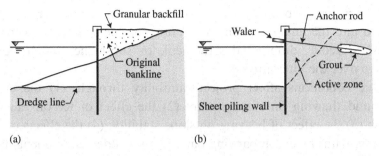

Figure 9.14. (a) Sheet-piling and (b) anchored walls

The external forces that act on a retaining wall are sketched in Figure 9.13b. The lateral soil force behind the structure is called active, compared to the passive force in front of the wall. Hydrostatic pressures result from groundwater behind the wall and channel- flow depth. The active and the passive earth pressures are usually computed based on methods originally proposed by Rankine and modified by others. Lateral earth pressures resulting from granular soils are usually more accurately predictable than those resulting from cohesive soils. For this reason and for drainage purpose, granular material is usually selected to backfill retaining walls.

The rise and fall of river stages causes an imbalance in horizontal hydrostatic forces on the wall. The vertical hydrostatic forces at the base of the wall also reduce the effective weight of the structure and lessen its ability to retain the soil. For groundwater control, the backfill should be a permeable granular material. Drains or weepholes through the wall improve drainage and lower the groundwater table behind the wall during falling river stages.

Sheet-piling walls are sometimes referred to as flexible walls or bulkheads. As shown in Figure 9.14, the walls are normally constructed by driving steel sheet piles. Because of the limited stiffness of the sheet piles, the height of these walls is somewhat restricted. Anchors increase the allowable height of cantilevered walls by providing horizontal restraining force to the wall. Walls of this sort are often referred to as "anchored" or "tied-back" walls, as shown in Figure 9.14b.

9.1.10 Other Considerations for Riverbank Protection

The essence of a successful bank-stabilization project is that it should be both effective and environmentally sound before the economics are considered. The engineering challenge is to determine the most suitable technique to solve a specific problem. Environmental and economic factors can then be integrated into the final selection. A stabilization method that has the ability to adjust to scour, differential settlement, or subsidence has a significant advantage over the alternatives.

Climatic conditions affect project durability through: (1) the effects of freezing and thawing on stone riprap; (2) the effect of ice on the selected structure; (3) the effect of heaving on slope stability; (4) the effects of wetting and drying, with the accompanying damage to wooden components by algae and bacterial growth and insects; (5) the effect of sunlight on synthetic materials; and (6) the effect of corrosion on wire meshes for gabions and mattresses.

Debris carried by the flow, usually in the form of uprooted trees, can cause such extensive damage on some streams as to rule out certain techniques and may prohibitively raise the cost of structures designed to withstand debris loads.

Animals can cause problems in a stabilization method. Beavers have a remarkable talent for girdling, felling, or eating vegetation of all sizes and species. Cows, deer, and rabbits may also find tender young vegetation on new stabilization plantings to their liking. Insect damage can be a problem for wooden components or vegetation. Preservative treatment for wooden components is common practice, and chemical treatment of vegetation at vulnerable stages may be feasible. Water-quality and environmental considerations may prohibit these options.

Vandalism and theft can be reduced by selection of a technique that minimizes temptation. Some materials that are obvious targets for vandals and thieves are posts, boards, concrete blocks and stones of attractive size and shape, small cables and wire, and easily removable fasteners. It is worth considering an increase in the size and the weight of components to make their removal or destruction more tedious.

Riparian users often dictate the type of structure to be adopted, whether it is in industrial areas, urban areas, boardwalks, fish and wildlife needs, etc. Restricted rights of way on riverbanks may allow or prohibit the passage or use of machinery needed for the construction of the selected structure. Restrictions on disposing of the excavated material may also be an issue.

The project life of most river protection works typically exceeds 100 years, but it can also be very short for emergency stabilization works during an extreme flood. Operation and maintenance costs can sometimes be traded against initial construction costs. If the project sponsor can monitor the condition of the work and maintain it as required, a less durable technique may be preferable to a "disaster-proof" method requiring a higher initial investment.

9.2 River Flow-Control Structures

Flow-control structures reduce hydrodynamic forces against riverbanks by controlling the direction, velocity, or depth of flowing water. While impermeable structures may entirely deflect a current, permeable structures may only reduce the flow velocity. Typical flow-control structures include hardpoints (Section 9.2.1), spurs (Section 9.2.2), guidebanks (Section 9.2.3), retards (Section 9.2.4), dikes (Section 9.2.5), fences (Section 9.2.6), jetties (Section 9.2.7), vanes (Section 9.2.8), bendway weirs (Section 9.2.9), and grade-control structures (Section 9.2.10).

9.2.1 Hardpoints

Hardpoints consist of stone fills spaced along an eroding bank line (Figure 9.15).

Most of the structure cannot be seen as the lower part consists of rock placed underwater, and the upper part is covered with topsoil and seeded with native vegetation. Hardpoints protrude only short distances into the river channel and are supplemented with a root section or tie-back extending landward into the bank to preclude flanking. Hardpoints are most effective in long and fairly straight river reaches not subject to direct attack.

9.2.2 Spurs or Groynes

A spur, also called a groyne, is a structure or embankment projected a fair distance from the bank into a river to deflect flowing water away from the bank. By deflecting the current away from the bank and causing sediment deposits, spurs prevent erosion of the bank and establish a more desirable width and channel alignment, as sketched in Figure 9.16.

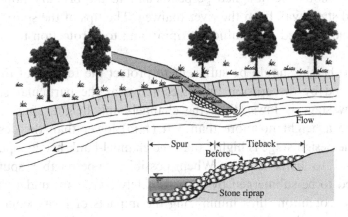

Figure 9.15. Hardpoint

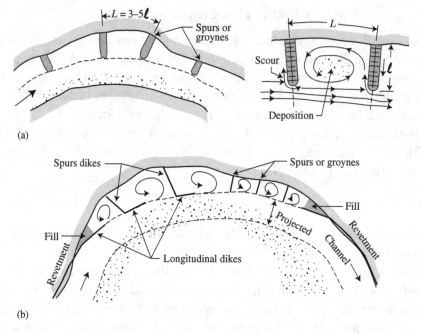

Figure 9.16. (a) Spur dikes or groynes and (b) groynes and longitudinal dikes

As the spur length increases, the severity of flow deflection increases, and the scour depth at the spur tip increases. The projected length should be less than 15 percent of the channel width at bankfull stage for impermeable spurs, and less than 25 percent of the width for permeable spurs. Spurs angled downstream require additional material but produce a less severe flow constriction than those normal to the flow. Spurs angled upstream will typically raise the water level near the outside bank of the river channel. Most spurs should be designed perpendicular to the primary flow direction and should stay away from the river thalweg. The tips of the spurs should be rounded and covered with sufficient riprap and toe protection to prevent toe scour.

The spur height should be sufficient to protect the regions of the channel bank affected by the erosion processes active at the particular site. If the design flow stage is lower than the channel-bank height, spurs should be designed to a height no more than 3 ft (1 m) lower than the design flood stage. If the design stage is higher than the channel-bank height, spurs should be designed to bank height. When possible, impermeable spurs should be designed to be submerged by approximately 3 ft (1 m) under their worst design flow condition, thus minimizing the impacts of local scour and flow concentration at the spur tip. Impermeable spurs should be designed with a slight fall toward the spur head to allow different amounts of flow constriction with stage (particularly important in narrow channels). Permeable spurs should be designed to permit passage of heavy debris over the spur crest without structural damage.

A series of spurs may protect the riverbank more effectively. Spurs on streams with suspended sediment discharge induce sedimentation to establish and maintain the new alignment. The spacing of spurs in a bank-protection scheme should be three to five times the projected length normal to the flow direction. Reducing the spacing between individual spurs results in a reduction of the local scour at the spur tip and causes the flow thalweg to stabilize farther away from the concave bank toward the center of the channel. Very short spurs are similar to hardpoints (Section 9.2.1). Very long spurs in shallow areas are spur dikes, discussed under dikes and jetties (Sections 9.2.5 and 9.2.7).

9.2.3 Guide Banks

Guidebanks are placed near the ends of approach embankments of bridge crossings, as shown in Figure 9.17. They guide the river through the bridge opening and align the streamlines to reduce head losses. The recommended shape of a guide bank is a quarter ellipse with a major to minor axis ratio of 2.5. The major axis is aligned with the main flow direction.

Design flows should not overtop the guide banks and the crest elevation should be 1 ft (30 cm) higher than the design flood stage. Guide banks also

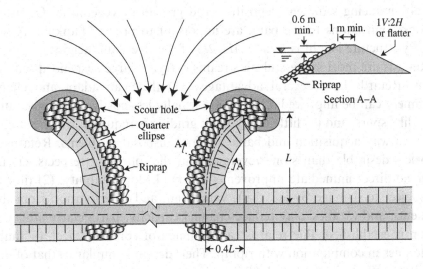

Figure 9.17. Guide banks

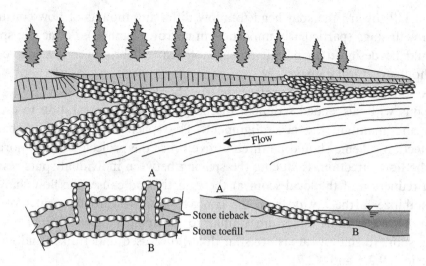

Figure 9.18. Retards

protect the highway embankment and reduce embankment scour. They also move the scour hole away from the abutments.

9.2.4 Retards

A retard is a low permeable structure aligned with the flow and located near the toe of the bank, as shown in Figure 9.18. A retard decreases the flow velocity behind the structure and eliminates erosive secondary currents, thereby inducing sediment deposition and growth of vegetation. Occasional tieback connections to the bank are important to prevent flanking. A satisfactory structure height is usually 1/3–2/3 of the riverbank height.

Retards are most successful on streams carrying large suspended sediment loads. Retards have several advantages: (1) they can adapt and channel alignment can be improved; (2) they are usually less costly than higher structures like spurs; and (3) little if any bank grading is required, which simplifies rights-of-way acquisition and bank material disposal problems. Retards are also less desirable than stone revetments in the following respects: (1) they offer no direct immediate improvement in bank-slope stability; (2) they can be subject to damage by ice, drift, vandalism, and deterioration from these elements; and (3) they may reduce channel capacity, particularly after vegetation is established. Pile retards can be made of concrete, steel, or timber, sometimes in combination with riprap. Their design is similar to that of dikes and jetties (Sections 9.2.5 and 9.2.7).

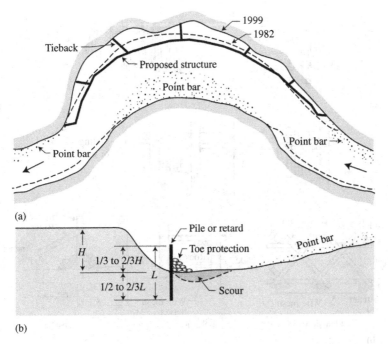

Figure 9.19. Bank stabilization sketch for dikes: (a) plan view and (b) cross section

9.2.5 Dikes

"Dike" is sometimes spelled "dyke," and "dijk" in the Netherlands. There are two principal types of dike: (1) stone-fill dikes and (2) timber-pile dikes. Stone-fill dikes are very long spur dikes. Except for their length, their characteristics are described in Section 9.2.2.

The design approach for dikes typically requires a comparison of recent and old aerial photographs to quantify bank-erosion rates. The general alignment is made with a preliminary location of upstream and downstream termination points. The upstream terminating point should correspond to the downstream end of the upstream point-bar deposition zone. The downstream termination should be near the point bar of the downstream bend. A constant radius of curvature is not required, but the alignment should be smooth and gradual, as sketched in Figure 9.19.

Timber-pile dikes may consist of closely spaced single, double, or multiple rows of timber piles, as shown in Figure 9.20. Double rows of timber piles can form timber cribs to be filled with rock riprap. Timber-pile dikes are vulnerable to failure through scour, which can be overcome if the piles can be driven to a greater depth. The base of the piles should be protected from scour with dumped rock. Timber-pile dikes are most effective in rivers with

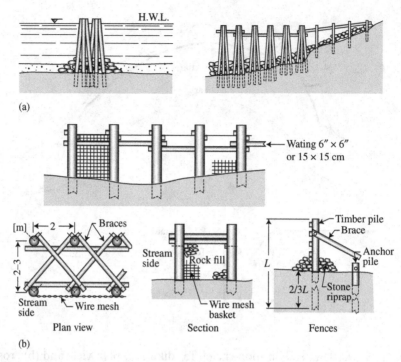

Figure 9.20. (a) Pile clusters and (b) timber pile fence

moderate flow velocities, shallow sand/silt-bed rivers with high suspended-sediment concentrations. Sediment deposition behind the dike field is a consequence of reduced velocities. Timber-pile dikes are not effective: (1) in deep rivers; (2) at high flow velocities; and (3) in rivers with clear water.

9.2.6 Fences

Fences can be used as a low-cost bank-protection technique on small- to medium-sized streams that are wide and shallow. Both longitudinal (parallel to stream) fence retards and lateral (perpendicular to stream) fences are successful when: (1) the channel gradient is stable with a low Froude number; (2) toe scour protection is provided; (3) tiebacks prevent flanking and promote suspended sediment deposition behind the fence; (4) vegetation can become established; and (5) metal, concrete, or composite materials, rather than wooden fences (which are liable to rot), reduce ice damage. Fences may not be very effective in areas subject to ice jams and large floating debris.

9.2.7 Jetties

Jetty fields add roughness mostly on a channel floodplain and along the riverbanks, as shown in Figure 9.21a. The added roughness along the bank reduces velocity and induces sedimentation to protect the riverbanks. Rows of jacks are usually angled approximately 45°–70° downstream from the bank. The spacing varies between 50 and 250 ft (15 and 80 m) depending on the debris and suspended sediment in the stream. Jetty fields should be used on both sides of the river, because the river may otherwise develop a chute channel across the point bar during floods.

Jetty fields are usually made up of steel jacks forming tetrahedral units as shown in Figure 9.21b. Kellner jacks are similar units, except that three L-shaped steel bars are welded or tied together with cables, as shown in Figure 9.21c. The aesthetic value of steel jacks is minimal and they should be avoided in all recreational rivers used for boating, fishing, rafting, and kayaking.

Jetty fields proved effective in controlling braided channels with wide floodplains and high sediment loads, e.g. the Middle Rio Grande in New Mexico.

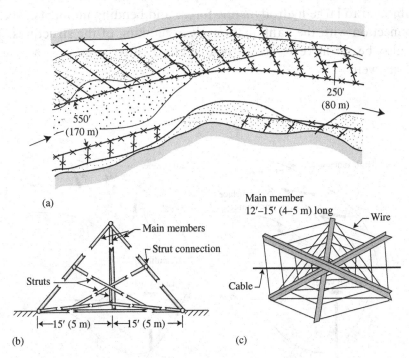

Figure 9.21. (a) Jetty field, (b) tetrahedron, and (c) Kellner jack

In a levees system where the river would locally impinge against the levees, the jetty fields kept the main channel within the central portion between the levees. Jetty fields can trap large quantities of sediment and organic debris that stimulate vegetation growth on the floodplain. The increased floodplain roughness attenuates flood waves and helps maintain a stable nonvegetated channel width. Jetty fields are not effective below dams where the sediment load has been greatly reduced. Like gabions and mattresses, ineffective jacks can leave a mess of tangled wires and steel.

9.2.8 Vanes

Vane dikes deflect the flow away from an eroding bank line (see Figure 9.22). Vanes do not connect with the riverbanks. There are two types of vane: (1) high vanes and (2) low vanes. High vanes guide the near-surface streamlines toward the centerline of the channel. The top elevation of high vanes is constructed below the design water-surface elevation. The ends of the vanes are subjected to local scour, which can be prevented by using riprap. High vanes can be effective when the rivers are not too deep. However, they often are built near the river thalweg and can be subjected to large hydrodynamic forces and bending moments. The lack of connection with the bank may result in flanking of the structures. They may also be sensitive to ice in cold regions. Low vanes are also called bendway weirs.

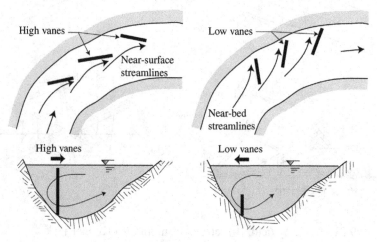

Figure 9.22. Vanes

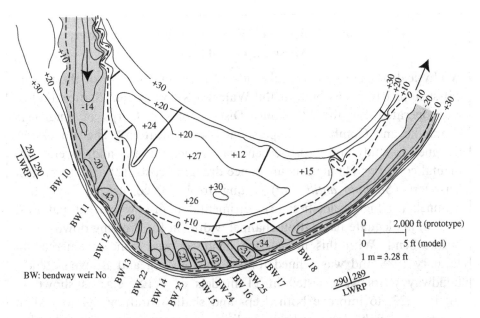

Figure 9.23. Bendway weir example on the Mississippi River (after Pokrefke, personal communication)

9.2.9 Bendway Weirs

Low vanes are often called bendway weirs, and act as low sills in channel bends. In navigable channels, the crests of bendway weirs are low enough to allow normal river traffic (Figure 9.23). Bendway weirs are usually angled 20° to 30° upstream and typically built in sets (4–14 weirs per bend). Longer weirs perform better and their spacing should be two to three times their length. Bendway weirs do not protect riverbanks and cannot replace revetments. However, secondary flows are reduced at high flows, and the streamlines are redirected perpendicular to the weir at low flows. Bendway weirs in large navigable channels result in: (1) improved navigation; (2) sediment deposition at the toe of the revetment; (3) more uniform surface water velocities; (4) flow patterns generally parallel with the banks; and (5) the thalweg being moved away from the outerbank. Case Study 9.2 discusses some features of the bendway weirs of the Mississippi River.

9.2.10 Grade-Control Structures

Grade-control structures stabilize the banks and bed of a channel by reducing stream slope and flow velocity. Drop structures generate abrupt drops

Case Study 9.2 Bendway Weirs of the Mississippi River, Missouri, United States

A physical movable-bed model study of a 20-mile section of the middle Mississippi River was built at the Waterways Experiment Station (WES) and operated for the St. Louis District of the Corps of Engineers (Pokrefke and Combs, 1998, personal communication). The model was designed to study and improve bendway navigation and address environmental concerns: (1) high-maintenance dredging costs; (2) need to protect least tern (*Sternula antillarum*) nesting areas; (3) constricted navigation channel; (4) high velocities; (5) detrimental high-flow current patterns; and (6) inadequate navigation channel widths in the crossing downstream of the bend. With this model, the bendway weir concept emerged in January 1988 and was refined for Price's Landing and Brown's Bends. Bendway weirs were tested in 11 models at WES, e.g. as shown in Figure 9.23, to improve both deep- and shallow-draft navigation, align currents through highway bridges, divert sediment, and protect docking facilities. From 1989 to 1995, over 120 weirs have been built in 13 bends of the Mississippi River. Analysis of the five oldest weir installations shows that from 1990 to 1995 dredging was reduced by 80 percent, saving US$3 million. In addition, tow delay times and incidence of towboat accidents were reduced, sediment and ice management was improved, least tern nesting areas were undisturbed, the aquatic habitat area was increased, and fish size and density in the weir fields increased (fivefold in some areas). Additional information on bendway weirs is available in the ASCE Manual of Practice No. 124 (2013). Numerical simulations were also conducted by Jia et al. (2001).

in bed elevation and must locally dissipate the energy while protecting both the bed and the banks upstream and downstream of the structure. The efficiency of grade-control structures decreases with increasing stream size.

Log and timber drop structures are appropriate in ephemeral streams and field gullies (Figure 9.24a). A series of small drops can control the grade in a relatively steep channel. Drop structures may take the form of a series of weirs, small check dams, gabions, cribs, sheet piles, etc. Behind check dams, riser pipes can be installed to drain the ponded water after the floods.

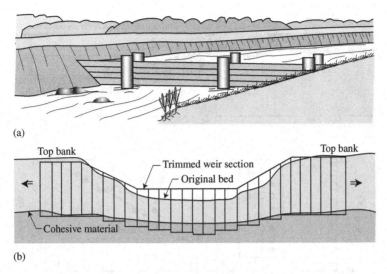

Figure 9.24. (a) Log crib drop structure and (b) sheet pile weir

Weirs are widely used grade-control structures. Vertical concrete and sheet pile weirs, gabion and mattress drop structures can be designed after considering the stability of the structure and the depth of the scour hole at the toe of the structure. High-drop structures (Figure 9.24b) typically require sheet piling or cutoff walls, because the groundwater table will induce seepage and piping under and around the structure, that can result in undermining and failure. The sheet-pile weir should be extended into both banks. Sheet piles are driven to two to three times the anticipated depth of maximum scour, or to refusal, and are trimmed at the top. Riprap protection is suggested both upstream and downstream of this type of weir. The USDA–ARS developed several low-drop and high-drop structures that have been field tested.

Broad-crested weirs, sloping sills or rock-fill weirs, and gradient restoration facilities are all structures where the drop in elevation takes place over longer distances. These structures, sketched in Figure 9.25, are typically made of riprap. The stone size for these structures can be estimated from the formula of Abt and Johnson (1991) $d_s = 5.23 \times S^{0.43} q_{ft^2/s}^{0.56}$, which was tested for near-prototype flow conditions. For instance, a flow rate of 100 cfs over a width of 10 ft at a 20 percent slope would require $d_s = 5.23 \times 0.2^{0.43} 10_{ft^2/s}^{0.56} = 10$ in. riprap size.

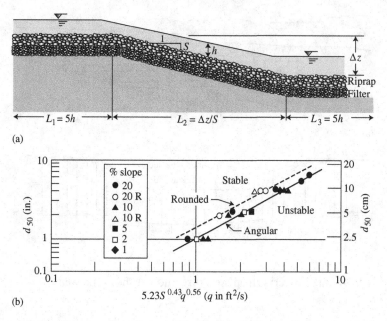

(a)

(b)

Figure 9.25. Overtopping flows: (a) longitudinal profile, and (b) design riprap size

♦*Problem 9.1*

Access a report available on the web on bank-protection measures. Examine five photographs of different structures discussed in this chapter and describe the pros and cons of each structure.

♦*Problem 9.2*

Examine the photographs of different bank-protection structures from one of the reports available on the web, or in your library. Identify one type of structure that has not been covered in detail in this chapter. Read some more material about this type of structure and define the pros and cons. For instance, can you find more about kicker dikes, chevrons, or bull noses, etc.?

♦♦*Problem 9.3*

Find a team partner in your class and prepare a 20-slide PowerPoint presentation on one of the following topics: (1) riprap design; (2) bank protection other than riprap; (3) flow-control structures. Your presentation should include some design elements, and pros and cons of selected structures. At least 10 slides should show photographs of structures that have been built to protect riverbanks. You should also include two slides showing structures which did not perform as expected, and explain why.

◆◆*Problem 9.4*

It may be time for a field trip in collaboration with your instructor, who may assign your team partner, find an appropriate field site, and conduct a trip to an area where riverbank protection structures have been constructed. Write a 12 p. (maximum) field report indicating your observations, illustrating different types of structures, and providing some information about their design characteristics, indicating if the structures worked properly or are in need of repair.

10

River Equilibrium

This chapter deals with the concept of equilibrium for deformable channels where particles on the wetted perimeter move and the channel reaches an equilibrium state. This is contrasting with the previously discussed concept of stability, where particles do not move. The concept of equilibrium can be compared to a beach where each wave carries sediment back and forth while the beach itself retains essentially unchanged features over long periods of time. Alluvial rivers can retain similar features over time while their banks and bed deform. Equilibrium simply requires that the amount of sediment entering and leaving a given reach remains the same. Section 10.1 covers the geometry of irrigation canals, followed by a presentation of downstream hydraulic geometry in Section 10.2, and river meandering in Section 10.3.

10.1 Irrigation Canal Geometry

First, the regime conditions for the design of irrigation canals are examined, in Section 10.1.1, followed with the analysis of ideal cross-section geometry (Section 10.1.2).

10.1.1 Regime Equations

The construction of irrigation canals in India and Pakistan fostered investigations to determine the optimum size (width, depth, and slope) of irrigation canals under poised (nonsilting and nonscouring) conditions. The terms "poised" and "graded" are synonyms with equilibrium conditions.

Empirical relationships called regime equations were proposed by Kennedy (1895), Lacey (1929), Inglis (1947, 1949), and Blench (1969), among others. In these equations, the Lacey silt factor f_l referred to "Kennedy's standard silt" from the Upper Bari Doab Canal. The geometry of canals with different bed-material sizes was then compared in terms of different values of f_l. The silt factor was eventually shown to increase with grain size as $f_l \simeq 1.59 d_{mm}^{1/2}$. The key relationships from Lacey determined the mean flow velocity V in ft/s, the hydraulic radius R_h in ft, the cross-section area A in ft^2, the wetted perimeter P in ft, and the dimensionless slope as a function of the design discharge Q in ft^3/s, and the Lacey silt factor f_l:

$$V = 0.794 Q^{1/6} f_l^{1/3}, \tag{10.1}$$

$$R_h = 0.47 Q^{1/3} f_l^{-1/3}, \tag{10.2}$$

$$A = 1.26 Q^{5/6} f_l^{-1/3}, \tag{10.3}$$

$$P = 2.66 Q^{1/2}, \text{ and} \tag{10.4}$$

$$S = 0.00053 f_l^{5/3} Q^{-1/6}. \tag{10.5}$$

In wide-shallow channels, the hydraulic radius is approximately equal to the flow depth, and the channel width can be approximated by the wetted perimeter. From the regime equations, the width–depth ratio thus increases slightly with discharge and grain size. An important limitation of the regime equation is that once the discharge and grain size are known, a unique value of bed slope is possible, from Equation (10.5). Consequently, the velocity in Equation (10.1) does not depend on slope, which is another limitation of this approach.

10.1.2 Ideal Canal Shape

Lane (1953) proposed a simplified but elegant relationship to define the ideal cross-section geometry of a straight canal. Figure 10.1 sketches the geometry of a canal with all particles of submerged weight F_s at incipient motion.

Lane assumed that $\tau_c \sim F_s \tan \phi$ at the free surface. He also assumed that the resultant on the side slope, $R/F_s \cos \theta_1 = \tan \phi$. This simplified formulation was discussed in Section 8.2.2.

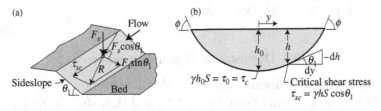

Figure 10.1. Ideal canal geometry: (a) force diagram, and (b) cross-section

$$\frac{\tau_{sc}}{\tau_c} = \sqrt{1 - \frac{\sin^2 \theta_1}{\sin^2 \phi}} = \cos \theta_1 \sqrt{1 - \frac{\tan^2 \theta_1}{\tan^2 \phi}}.$$ (8.22)◆

At the centerline of a straight canal, the applied shear stress $\gamma h_0 S$ equals the critical shear stress on a horizontal slope τ_c. On the side slope, the applied shear stress $\gamma h S \cos \theta_1$ corresponds to τ_{sc}. Thus, $\tau_c = \gamma h_0 S$ and $\tau_{sc} = \gamma h S \cos \theta_1$ result in the following identity:

$$\tau_{sc} h_0 = \tau_c h \cos \theta_1.$$ (10.6)

The differential equation for the ideal cross-section geometry is obtained from combining Equations (8.22) and (10.6) with $\tan \theta_1 = -dh/dy$

$$\left(\frac{dh}{dy}\right)^2 + \left(\frac{h}{h_0}\right)^2 \tan^2 \phi = \tan^2 \phi.$$ (10.7)

The ideal cross-section geometry in which all particles are at incipient motion is simply a cosine function of the lateral distance y from the centerline

$$\frac{h}{h_0} = \cos\left(\frac{y \tan \phi}{h_0}\right).$$ (10.8)◆

The centerline flow depth h_0 at $y = 0$ corresponds to incipient motion on a horizontal plane surface $h_0 \approx 0.047(G - 1)d_s/S$. The surface width $W = \pi h_0 \cot \phi$ is obtained from the values of y in Equation (10.8) where $h = 0$. The width–depth ratio W/h_0 is constant at $\pi \cot \phi$, the cross-section area is $A = 2h_0^2 \cot \phi$, the average flow depth becomes $\bar{h} = 2h_0/\pi$, and the hydraulic radius is $R_h = 8h_0 \cos \phi/\pi(4 - \sin^2 \phi)$. With the mean flow velocity V obtained from the Manning equation, the flow discharge is then AV (see Problem 10.1).

In practice, only coarse-bed channels may reach incipient motion at bankfull discharge. Sediment transport in all sand-bed channels cannot be ignored. The "ideal" geometry typically corresponds to very narrow channels with $W/h < 10$ when $\phi > 25°$. The width–depth ratio of almost all rivers far exceeds the ideal geometric conditions.

10.2 Downstream Hydraulic Geometry

This section includes discussions of bankfull conditions (Section 10.2.1), equilibrium in river bends (Section 10.2.2), and downstream hydraulic geometry equations (Section 10.2.3).

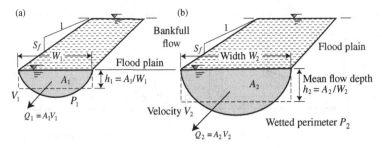

Figure 10.2. Downstream hydraulic geometry: (a) small stream, and (b) large river

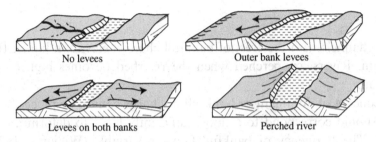

Figure 10.3. Natural levees and perched channel

10.2.1 Bankfull Conditions

Many rivers flow in their own deposits, called alluvia, to form alluvial channels. The equilibrium of alluvial channels implies a balance between incoming and outgoing water discharge and sediment load. As shown in Figure 10.2, we can compare two channels under bankfull conditions: a small river located in the upper part of a watershed, and a large river at the basin outlet. The downstream hydraulic geometry refers to the channel conditions in terms of width, mean flow depth, mean flow velocity, and slope at bankfull discharge. A channel under bankfull condition has a single value of channel width, depth, velocity, and slope. Therefore, it is important to recognize the difference between downstream and at-a-station hydraulic geometry, discussed in Section 5.1.1. Downstream hydraulic geometry describes unique channel properties under bankfull conditions at different locations, as opposed to at-a-station hydraulic geometry, describing channel properties at different discharges at a given cross section.

The bankfull discharge is the flow rate that a channel conveys when reaching the floodplain level. Natural levees result from the deposition of coarser fractions of the suspended load on the floodplain during overbank flows. As sketched in Figure 10.3, natural levees can form on one or both sides of

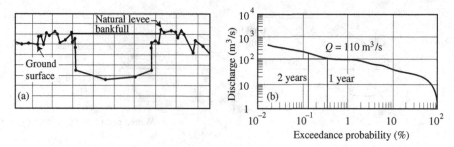

Figure 10.4. Examples of (a) bankfull and (b) dominant discharge (after Bender, 2011; Shin and Julien, 2010)

alluvial channels with the crest of bankfull channel elevation higher than the floodplain. Rivers are perched when the riverbed becomes higher than the floodplain.

The concept of dominant or bankfull discharge cannot be precisely quantified. It should correspond to a flood sufficiently large to deform an alluvial channel. The frequency of bankfull flows is variable (Williams, 1978) and remains an active research subject. In general, the bed should be mobilized frequently enough to prevent vegetation from becoming fully established. Periods of return of 1.5 and 2 years are usually appropriate, but mean annual floods, mean annual flows, and 5-year floods have also been reported in the literature. Examples for the determination of bankfull discharge and dominant discharge are illustrated in Figure 10.4.

The channel width, average flow depth, and slope vary with the dominant discharge. The data set of Lee and Julien (2006b) in Figure 10.5 includes 1,485 rivers worldwide. The reader will notice the variability in the measurements, particularly for the channel slope, when a single variable (flow discharge) is considered.

10.2.2 Equilibrium in River Bends

For flow in bends, the relative magnitude of radial acceleration terms in cylindrical coordinates, from Equation (2.15a), indicates that the centrifugal acceleration is counterbalanced primarily by pressure gradient and radial shear stress, as suggested by Rozovskii (1957):

$$-\frac{1}{\rho}\frac{\partial \tau_r}{\partial z} = \frac{V_\theta^2}{r} - gS_r, \tag{10.9}$$

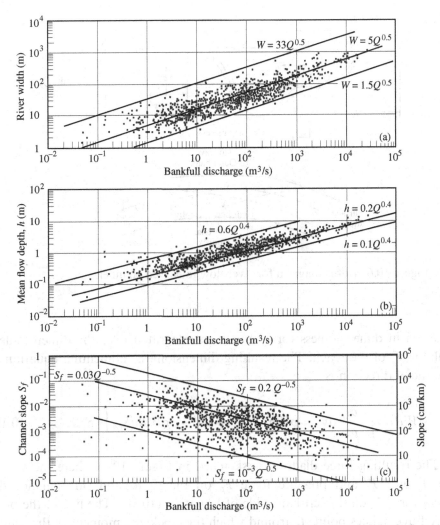

Figure 10.5. Downstream hydraulic geometry vs discharge: (a) width, (b) flow depth, and (c) channel slope (data from Lee and Julien, 2006b)

where the local downstream velocity V_θ, the radial shear stress τ_r, and the radial water-surface slope S_r vary with the radius of curvature r, as sketched in Figure 10.6.

Scaling factors are used to define dimensionless parameters for channel width $w^* = w/W$, flow depth $z^* = z/h$, radius of curvature $r^* = r/R$, velocity $v^* = v/\overline{V}$, radial shear stress $\tau_r^* = \tau_r/\tau_{rR}$, and the radial surface slope $S_r^* = S_r/S_{rA}$. The element of fluid volume $d\forall = dx\, dy\, dz$ for a reach of given length $dx = R\, d\theta$ is reduced to a dimensionless volume $d\forall^* = d\forall/WRh$. The radial equation of motion [Equation (10.9)] is multiplied by ρ and $d\forall$,

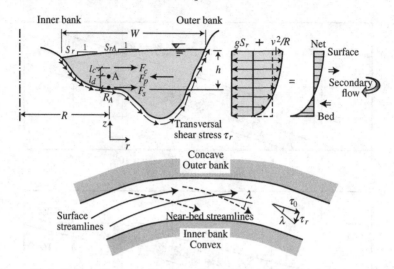

Figure 10.6. Force diagram for river flow in curved channels

reduced in dimensionless form, and then integrated over the dimensionless volume $\forall^*$ of the reach. The resulting dimensionless momentum equation in the radial direction is

$$-WR\tau_{rR}\int_{\forall^*}\frac{\partial\tau_r^*}{\partial z^*}\,d\forall^* = \rho Wh\overline{V}^2\int_{\forall^*}\frac{v^{*2}}{r^*}\,d\forall^* - \rho g WRhS_{rA}\int_{\forall^*}S_r^*\,d\forall^*. \quad (10.10)$$

The resulting force diagram is sketched in Figure 10.6, where the centrifugal force F_c and pressure force F_p on the right-hand side balance the shear force F_s on the left-hand side of Equation (10.10). The line of the pressure force defines point A, around which the clockwise moment of the centrifugal force with moment arm l_c balances the counterclockwise shear stress moment applied at the bed. The sum of moments about A gives

$$\Omega_R = \frac{\rho h\overline{V}^2}{R\tau_{rR}} = \frac{l_d\int_{\forall^*}\frac{\partial\tau_r^*}{\partial z^*}\,d\forall^*}{l_c\int_{\forall^*}\frac{v^{*2}}{r^*}\,d\forall^*}. \quad (10.11)$$

The dimensionless parameter Ω_R denotes the ratio of the centrifugal force generating secondary motion to the shear force abating the motion and dissipating energy. Equilibrium is obtained when Ω_R is constant and the corresponding magnitude of the radial shear stress is

$$\tau_{rR} \sim \frac{\rho\overline{V}^2}{\Omega_R}\left(\frac{h}{R}\right). \quad (10.12)$$

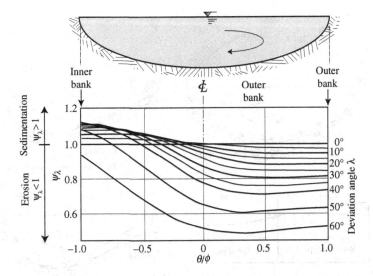

Figure 10.7. Safety factor of bed particles in river bends with and without secondary flows

The resulting ratio of radial shear stress τ_{rR} to the downstream bed shear stress $\tau_\theta = \rho u_*^2$ defines the deviation angle λ of the streamlines near the bed. Therefore, given $V/u_* = a(h/d_s)^m$ from Equation (5.8), we obtain

$$\tan \lambda = \frac{\tau_{rR}}{\tau_\theta} = \frac{1}{\Omega_R} \frac{V^2}{u_*^2} \frac{h}{R} = \left[\frac{a^2}{\Omega_R} \left(\frac{h}{d_s} \right)^{2m} \right] \frac{h}{R}. \qquad (10.13)$$

Secondary flows are most important when h/d_s is large, i.e. during floods. Rozovskii (1957) found that the term in brackets of Equation (10.13) is close to 11. Slightly different values have been proposed by Engelund (1974), de Vriend (1977), Odgaard (1981), and Hussein and Smith (1986). The deviation angle λ depends primarily on the ratio of flow depth to radius of curvature:

$$\tan \lambda \cong 11 \frac{h}{R}. \qquad (10.14) \blacklozenge\blacklozenge$$

The effect of secondary flows on particle stability in curved channels is examined through the influence of the deviation angle λ on the stability factor SF from Section 8.2. The downslope lateral shear stress destabilizes bed particles. Assuming that all particles on the wetted perimeter are initially at incipient motion, Figure 10.7 shows the relative particle stability ratio $\Psi_\lambda = \text{SF}_0 \, (\lambda \neq 0)/\text{SF}_0 \, (\lambda = 0)$ defined as the ratio of the stability factors with ($\lambda \neq 0$) and without secondary circulation ($\lambda = 0$).

Figure 10.7 shows that for small secondary flow angles, i.e. $\lambda < 15°$, secondary flows decrease particle stability near the outer bank and increase particle

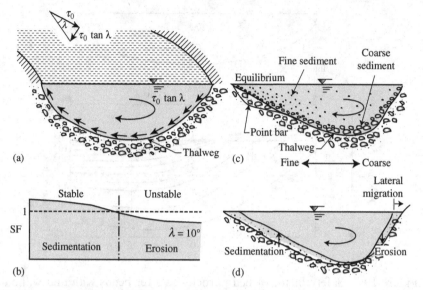

Figure 10.8. River bend: (a) secondary flow, (b) particle stability, (c) gradation, and (d) deformation

stability near the inner bank. When the strength of secondary circulation increases, and $15° < \lambda < 55°$, asymmetry in the particle stability curves develop and most of the wetted perimeter becomes unstable. Therefore, rivers in equilibrium will maintain low values of the secondary flow angle, typically $\lambda < 10°$.

Under significant secondary flows, the cross-sectional geometry becomes asymmetric, with the thalweg moving toward the outer bank, as sketched in Figure 10.8. Near the outer bank, only coarser particles can counterbalance the higher erosive forces. Conversely, finer particles will typically deposit to form a point bar near the inner bank.

For well-graded bed material where armoring is possible, coarser grains can be found near the outer bank and finer grains near the inner bank. In the case of uniform erodible bed material, secondary flows will scour the toe of the outer bank, leading to bank caving and lateral migration. Equilibrium is possible when bank caving balances inside deposition on the point bar.

It is also important to note that the change in cross-sectional geometry depends on the magnitude of the streamline deviation angle λ. From Equation (10.14) we learned that the angle λ decreases with the radius of curvature. The rate of lateral migration of a channel is therefore expected to be largest in sharp bends (low R). Finally, since the angle λ increases with flow depth h, strong secondary flows are thus expected during floods. At low flows, low values of the angle λ will result in fairly symmetrical cross-sectional geometry, as shown in Case Study 10.1.

Case Study 10.1 Sediment Transport of the Fall River, Colorado, United States

Fall River is a meandering stream flowing through Horseshoe Park in the Rocky Mountain National Park, Colorado (Figure CS10.1.1a). During the late spring snowmelt, the bankfull flow discharge is ~7 m³/s while the winter lows are less than 0.5 m³/s. The failure of the Lawn Lake Dam, located at the headwaters of Roaring River, caused a large flood on July 15, 1982. This formed a large alluvial fan that supplies coarse sediment for bedload transport to the Fall River. At a slope of 0.0013, it carries up to 0.6 kg/ms of sediment with sizes from 0.125 to 32 mm.

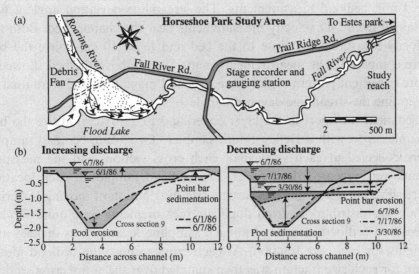

Figure CS10.1.1. Fall River (a) study area and (b) changes in cross-sectional geometry (after Julien and Anthony, 2002)

During the summer seasons of 1986 and 1987, detailed measurements were taken along two consecutive bends at 22 cross sections (Anthony, 1992). Measurements included bed and free-surface topography, velocity profiles, and sediment transport by size fraction. Anthony and Harvey (1991) examined the patterns of internal cross-sectional adjustments from high to low flows. Topographic measurements included bed and water-surface elevations at 1-m intervals across the channel for each cross section. Near the apex, the cross-sectional geometry features a deep thalweg along the outside of the bends and a pointbar along the inside of the bends. As shown in Figure CS10.1.1, the cross-sectional geometry of the Fall River bends depends on discharge. Increasing discharge causes erosion of the thalweg and sedimentation on the point bar. The larger flows strengthen the secondary flows, leading to more triangular cross sections. During falling discharges, secondary flows weaken and the cross section gradually becomes more rectangular through erosion of the point bar and sedimentation in the pools.

At each cross section, 2-D velocity profiles were measured with a Marsh–McBirney current meter at 1-m intervals across the channel. Both transverse and longitudinal velocities were measured with a 30-s sampling duration, starting from the bottom (with the current meter resting on the bottom) and then at approximately 10-cm intervals up to the water surface. Measurements were repeated at least three times for vertical velocity profiles (for both longitudinal and transverse flow) and six times for the 1-min bedload measurements. The streamline deviation angle λ was obtained from 2-D velocity measurements near the channel bed. Both the velocity measurements close to the bed and those 10 cm above the bed were compared. The measurements made 10 cm above the bed showed a more consistent pattern of flow direction and magnitude and were used to determine the streamline deviation angle λ.

Sediment transport in the layer covering 3 in. (7.67 cm) above the bed was measured with a Helley–Smith sampler. Two 1-min bedload samples were collected at each vertical of each cross section, and the measurements were repeated three times. The bedload samples at each location were sieved to determine the particle-size distribution of moving material. Grain-size distributions from duplicate measurements were quite similar. Measured bedload movement at each sampling location was divided into weights for each size fraction. At each cross section, the bedload movement for each size fraction was then summed over the entire cross section. For a given grain size, the relative percentage of the cross-section total material transported was then calculated for each sampling point. After repeating this process at each cross section, a bedload percentage map over the entire reach was obtained for that particular size fraction. Bedload percentage maps were produced for each size fraction, i.e. 16, 8, 4, 2, 1, 0.5, 0.25, and 0.125 mm. Two typical bedload percentage maps are shown in Figure CS10.1.2a and b to illustrate the motion of coarse grains ($d_{95} \cong 8$ mm) and fine grains ($d_{10} \cong 0.25$ mm) in sharp river bends.

The values of bedload percentages at each cross section were used to calculate the position of the center of mass of bedload transport for each size fraction. It is interesting to note that the location of the center of mass for fine grains is different from that for coarse grains. The lines linking the successive positions of the center of mass for different grain sizes are shown in Figure CS10.1.2c. Near the crossing, the bedload center of mass curves for each size fraction are fairly parallel and oriented in the downstream direction. Near the apex, the bedload center-of-mass curves are shifting across the channel.

The angle difference between the motions of coarse and of fine grains can be calculated from the angle β of the particle-stability analysis in Example 8.2. The particle-stability analysis determines the mean particle-direction angle β, given the particle grain size, side-slope angle, and downstream slope angle, and the shear stress. The results of calculations with the method in Example 8.2 with different size fractions for the Fall River bend are shown in Figure CS10.1.2d.

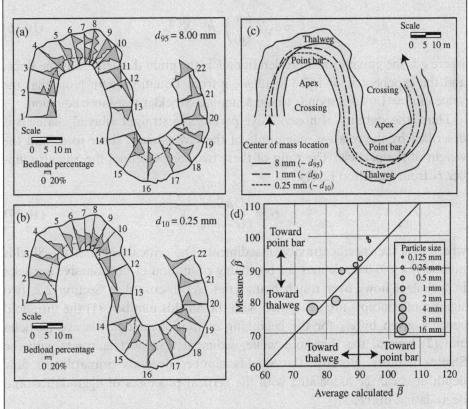

Figure CS10.1.2. Fall River (a) sediment transport for 8-mm particles, (b) sediment transport for 0.25-mm particles, (c) mean bedload particle trajectory, and (d) near apex mean particle direction angle (after Julien and Anthony, 2002)

10.2.3 Downstream Hydraulic Geometry of Rivers

The downstream hydraulic geometry of rivers can be examined by combining four relationships.

First, under steady-uniform bankfull flow conditions, the dominant discharge Q is

$$Q = WhV, \tag{10.15}$$

where the mean velocity V is normal to the cross-section area, h is the mean flow depth obtained from the cross-section area divided by the bankfull width W. The width–depth ratio of most rivers is sufficiently large that it may be assumed that the hydraulic radius R_h is approximately equal to h.

Second, the power form of the resistance equation from Equation (5.8) is

$$V = a\sqrt{g}\left(\frac{h}{d_s}\right)^m h^{1/2} S^{1/2}, \tag{10.16}$$

where g is the gravitational acceleration, d_s is the grain diameter, S is the slope, and the exponent $m = 1/[2.3 \log(2h/d_{50})]$ from Equation (5.9). Note that the value of $m = 1/6$ corresponds to the Manning–Strickler resistance equation.

Third, the stability of noncohesive particles in straight alluvial channels is described by the relative magnitude of the downstream shear force and the weight of the particle. The ratio of these two forces defines the Shields number τ_* from Equation (2.41):

$$\tau_* = \frac{hS}{(G-1)d_s}, \tag{10.17}$$

where G is the specific gravity of sediment. The critical value of the Shields number, $\tau_* \approx 0.047$, defines the beginning of motion of noncohesive particles in turbulent flows over rough boundaries. As discussed in Section 2.2, two significant concepts are associated with the Shields number: (1) the threshold concept given by τ_{*c} for the beginning of motion of noncohesive particles and (2) beyond the threshold value, sediment transport increases with the Shields number. Because the Shields number depends primarily on flow depth, it is can be associated with the vertical processes of aggradation and degradation in rivers.

Fourth, to describe flow in bends with a radius of curvature R proportional to the channel width W, Equation (10.13) shows that

$$\tan \lambda = b_r \left(\frac{h}{d_s}\right)^{2m} \frac{h}{W}, \tag{10.18}$$

where the value of $b_r = [(a^2 W)/(\Omega_R R)]$ is assumed to remain constant. Per Section 10.3.2, secondary flows can be associated with the lateral deformation of alluvial channels.

Equations (10.15)–(10.18) contain 13 variables, namely W, h, V, S, Q, d_s, τ_*, $\tan \lambda$, g, a, b_r, G, and m. The four equations allow the definition of the four dependent variables, W, h, V, and S, as functions of the others. Julien (1988) showed that Q, d_s, and τ_* were the primary independent variables, while the variability in the other parameters was comparatively small. Julien and Wargadalam (1995) determined empirical values for the remaining parameters from a large data set, including data from 835 rivers and canals.

The downstream hydraulic geometry of alluvial rivers is derived for flow depth h and surface width W in meters, average flow velocity V in meters per second, and friction slope S as

$$h \cong 0.133 Q^{\frac{1}{3m+2}} d_{50}^{\frac{6m-1}{6m+4}} \tau_*^{\frac{-1}{6m+4}}, \tag{10.19a}$$

$$W \cong 0.512 Q^{\frac{2m+1}{3m+2}} d_{50}^{\frac{-4m-1}{6m+4}} \tau_*^{\frac{-2m-1}{6m+4}}, \tag{10.19b}$$

$$V \cong 14.7 Q^{\frac{m}{3m+2}} d_{50}^{\frac{2-2m}{6m+4}} \tau_*^{\frac{2m+2}{6m+4}}, \tag{10.19c}$$

$$S \cong 12.4 Q^{\frac{-1}{3m+2}} d_{50}^{\frac{5}{6m+4}} \tau_*^{\frac{6m+5}{6m+4}}. \tag{10.19d}$$

From the equilibrium or dominant flow discharge Q in cubic meters per second, the median grain size $d_s = d_{50}$ in meters, and the Shields parameter $\tau_* = \gamma h S/(\gamma_s - \gamma)d_{50}$. The exponent m is calculated using $m = 1/[2.3\log(2h/d_{50})]$ from Equation (5.9). Figure 10.9 shows the predicted and measured values of river width, depth, velocity, and slope from Julien and Wargadalam (1995).

When the Manning–Strickler approximation is applicable, i.e. $m = 1/6$, a simplified form of Equation (10.19) is obtained in SI as

$$h \approx 0.133 Q^{0.4} \tau_*^{-0.2}, \tag{10.20a}\blacklozenge$$

$$W \approx 0.512 Q^{0.53} d_s^{-0.33} \tau_*^{-0.27}, \tag{10.20b}\blacklozenge$$

$$V \approx 14.7 Q^{0.07} d_s^{0.33} \tau_*^{0.47}, \tag{10.20c}\blacklozenge$$

$$S \approx 12.4 Q^{-0.4} d_s \tau_*^{1.2}. \tag{10.20d}\blacklozenge$$

The hydraulic geometry of stable channels is obtained from Equation (10.20a–d) when $\tau_* \cong 0.047$. Higher sediment transport rates require higher velocity and slope, and reduced width and depth. Example 10.1 calculates the downstream hydraulic geometry from Equations (10.19 and 10.20).

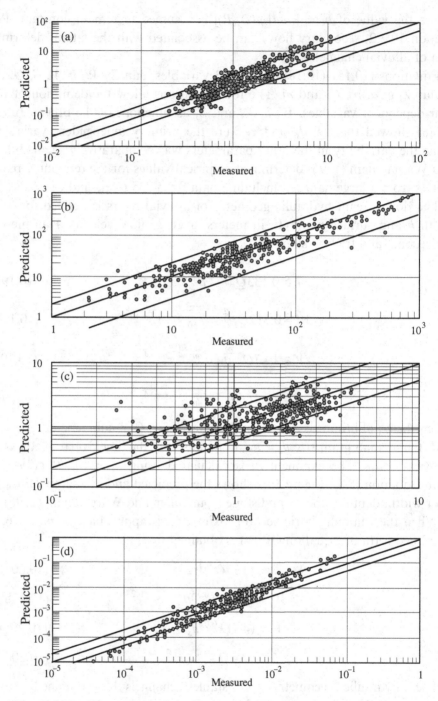

Figure 10.9. Comparison of downstream hydraulic geometry for (a) depth, (b) width, (c) velocity, and (d) slope (after Julien and Wargadalam, 1995)

> **Example 10.1 Application to stable channel geometry**
>
> Calculate the bankfull geometry of a gravel-bed stream at incipient motion, given $Q = 104$ m^3/s, $d_{50} = 56$ mm, and $\tau_\theta^* = 0.047$.
>
> *Step 1:* Using the Manning–Strickler approximation, the geometry from Equation (10.20a–d) gives
>
> $$h \approx 0.133(104)^{0.4}(0.047)^{-0.2} = 1.6\,\text{m},$$
>
> $$W \approx 0.512(104)^{0.53}(0.056)^{-0.33}(0.047)^{-0.27} = 35\,\text{m},$$
>
> $$V \approx 14.7(104)^{0.07}(0.056)^{0.33}(0.047)^{0.47} = 1.9\,\text{m/s},$$
>
> $$S \approx 12.4(104)^{-0.4}(0.056)(0.047)^{1.2} = 0.0028.$$
>
> *Step 2:* Check that the exponent m is close to 1/6 from
>
> $$m = \frac{1}{2.3 \log\left[\frac{2(1.6)}{0.056}\right]} = 0.25.$$
>
> *Step 3:* For improved accuracy, recalculate the flow depth from Equation (10.19a) with $m = 0.25$:
>
> $$h \cong aQ^b d_s^c \tau_\theta^{*d} = 0.133(104)^{0.363}(0.056)^{0.091}(0.047)^{-0.181} = 0.96\,\text{m}.$$
>
> *Step 4:* Repeat steps 2 and 3 with the calculated flow depth in step 3 until convergence:
>
> $$h = 0.96\,\text{m gives } m = 0.28, \text{ which is sufficiently close to 0.25.}$$
>
> *Step 5:* Calculate the channel width W, flow velocity V, and slope S by using $m = 0.25$ for the exponents of Q, d_s, and τ_θ^* in Equation (10.19a–d):
>
> $$W \cong 0.512(104)^{0.545}(0.056)^{-0.363}(0.047)^{-0.273} = 42\,\text{m},$$
>
> $$V \cong 14.7(104)^{0.091}(0.056)^{0.273}(0.047)^{0.455} = 2.5\,\text{m/s},$$
>
> $$S \cong 12.4(104)^{-0.364}(0.056)^{0.909}(0.047)^{1.182} = 0.0045.$$

Ideally, we may consider that rivers could form at any discharge on any granular material and on any surface slope. The flexibility of the system is demonstrated by solving Equation (10.19d) as $\tau_*(Q, d_s, \text{ and } S)$, and substituting back into Equation (10.19a–c). As a result, we can calculate flow depth h, width W, mean velocity V, and Shields number τ_* from Q, d_s, and S as

$$h \cong 0.2 Q^{\frac{2}{5+6m}} d_{50}^{\frac{6m}{5+6m}} S^{\frac{-1}{5+6m}}, \qquad (10.21\text{a})$$

$$W \cong 1.33 Q^{\frac{2+4m}{5+6m}} d_{50}^{\frac{-4m}{5+6m}} S^{\frac{-1-2m}{5+6m}}, \qquad (10.21\text{b})$$

$$V \cong 3.76 Q^{\frac{1+2m}{5+6m}} d_{50}^{\frac{-2m}{5+6m}} S^{\frac{2+2m}{5+6m}}, \qquad (10.21\text{c})$$

$$\tau_* \cong 0.121 Q^{\frac{2}{5+6m}} d_{50}^{\frac{-5}{5+6m}} S^{\frac{4+6m}{5+6m}}, \qquad (10.21\text{d})$$

where $m = 1/[2.3\log(2h/d_{50})]$ from Equation (5.9). The particular case in which $m = 1/6$ gives

$$h \approx 0.2 Q^{0.33} d_{50}^{0.17} S^{-0.17}, \qquad (10.22\text{a}) \blacklozenge$$

$$W \approx 1.33 Q^{0.44} d_{50}^{-0.11} S^{-0.22}, \qquad (10.22\text{b}) \blacklozenge$$

$$V \approx 3.76 Q^{0.22} d_{50}^{-0.05} S^{0.39}, \qquad (10.22\text{c}) \blacklozenge$$

$$\tau_* \approx 0.121 Q^{0.33} d_{50}^{-0.83} S^{0.83}, \qquad (10.22\text{d}) \blacklozenge$$

where the discharge Q is in cubic meters per second, the median grain size d_{50} of the bed material is in meters, and the channel slope is S. It is observed that the channel width and depth vary primarily with discharge, the grain size affects the Shields parameter and the slope primarily affects the Shields parameter and the flow velocity. The method used in Example 10.1 can also solve Equations (10.21 and 10.22).

10.3 River Meandering

This section presents a substantial analysis of river meandering. This includes an introduction to river morphology (Section 10.3.1), lateral river migration (Section 10.3.2), and an explanation why rivers meander (Section 10.3.3) with sine-generated curves (Section 10.3.4). This is followed with an analysis of downstream and lateral river mobility (Section 10.3.5) and a concept of timescale for meander evolution (Section 10.3.6).

10.3.1 Meandering River Morphology

Several geomorphic features of alluvial channels are illustrated in Figure 10.10. Bars refer to large bedforms that are often exposed during low flows. They are usually submerged at least once a year to prevent vegetation

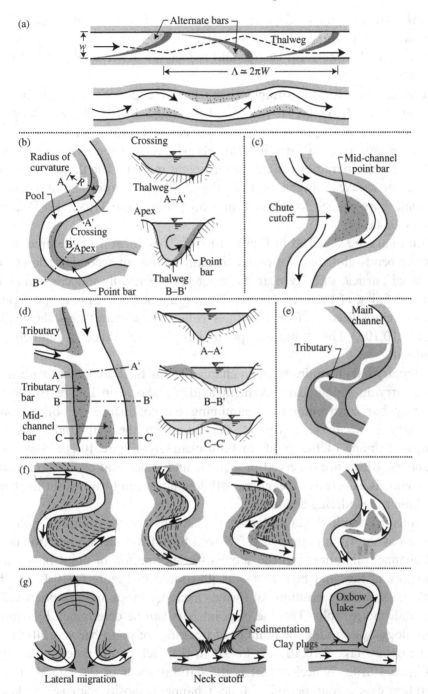

Figure 10.10. Geomorphic features of alluvial channels including (a) alternate bars, (b) point bars, (c) chute cutoff, (d) tributary bar, (e) submeanders, (f) scrolls, and (g) neck cutoff formation

growth. Bars are alluvial deposits that can be mobilized during floods. When bars do not get submerged approximately every year, vegetation grows and stabilizes the bars to form islands and reduce the active nonvegetated channel width of a river.

Alternate bars form in straight channels with deposits alternating between the right and left banks. As illustrated in Figure 10.10a, the wavelength of alternate bars Λ is proportional to the channel width W, i.e. $\Lambda \simeq 2\pi W$. Alternate bars tend to form in channels where the Froude number is high and the Shields parameter is close to incipient motion. The thalweg is said to wander or weave between both banks. Additional information regarding alternate bars can be found in Fujita and Muramoto (1982, 1985) and in Ikeda (1984a, b).

Point bars are sketched in Figure 10.10b. They form near the inner banks of river bends and their formation is akin to that of secondary flows. The increased particle stability near the inner bank generally induces sedimentation and fining of the bed deposits. During major floods, point-bar deposits can be remobilized to form channels called chute cutoffs, as sketched in Figure 10.10c. The truncated point bar is also called a mid-channel point bar.

Tributary bars form in the main channel below their confluence with tributaries carrying a significant sediment load. As shown in Figure 10.10d, the tributary bar contributes to streamlining the confluence of both streams. Tributary bars can be reworked, depending on the magnitude, sediment load, and timing of the floods in both channels. Figure 10.10e shows submeanders, which are observed in arroyos and ephemeral channels. In sand-bed channels, shortwave meanders with low flows can form within the banks of a larger meandering channel.

Scrolls are small ridges or terraces caused by the reworking of point-bar deposits as shown in Figure 10.10f. Scrolls form from the combination of lateral channel migration and the successive deformation of point bars during sequences of low and high flows, as illustrated in Figure CS10.1.1b. They mark the successive positions of former meander loops and are often visible on aerial photographs. The age of point bars can be determined by dendrochronology by counting the number of tree rings of point-bar vegetation.

The oxbow-lake formation is linked to the neck cutoff process illustrated in Figure 10.10g. A neck cutoff is the natural result of the lateral channel migration over a long period of time. Channel sinuosity increases as lateral migration of the outer bend progresses. When the channel sinuosity becomes very large, e.g. greater than three, the channel slope, flow velocity, and sediment transport capacity are greatly reduced. Consequently, the flooding risk

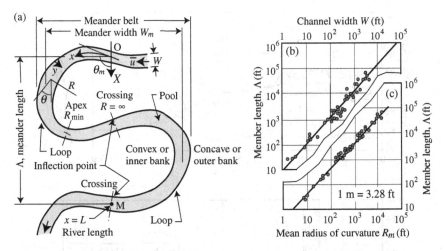

Figure 10.11. Definition sketch for (a) a meandering river and (b–c) empirical meander geometry (data from Leopold et al., 1960 and Ackers and Charlton, 1970)

is largely increased, and a high flood can cut through the neck of the long meander loops. Once the cutoff has occurred, silting of suspended sediment will plug both ends of the meander loop and result in the formation of an oxbow lake. Neck cutoffs rejuvenate the fluvial system by decreasing the sinuosity and increasing channel slope, velocity, and sediment transport. Over time, meander loops readjust their slope and sediment transport to reach a renewed state of equilibrium.

A meandering alluvial river reach is sketched in Figure 10.11a and two systems of coordinates are defined. The rectilinear axis X defines the downstream valley slope direction. In the curvilinear axis x follows the centerline of the meandering river path. The angle θ separates the directions X and x along the flow path. Measured at the river centerline, the radius of curvature R in the transverse direction y remains orthogonal to the downstream axis x. Both the magnitude and the direction of the radius of curvature R vary along the channel of width W and mean flow velocity V. The radius of curvature R_{min} is minimum at the apex and maximum $R = \infty$ at the crossing. Over a complete meander loop between O and M, the river length is L and the meander wavelength is Λ. The amplitude of the meander belt, or meander width, is W_m.

Leopold et al. (1960) empirically observed that the meander length Λ is ~10 times the channel width W, as shown in Figure 10.11b. The ratio of wavelength to the minimum radius of curvature Λ/R_m for meandering streams typically varies between 3 and 5. Field measurements from

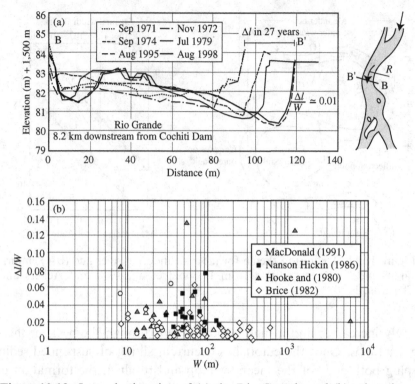

Figure 10.12. Lateral migration of (a) the Rio Grande and (b) other rivers (after Richard et al., 2005)

Leopold and Wolman (1960) indicate an average ratio of 4.7. Also, the mean radius of curvature $\overline{R}_m \simeq 2.3W$ is obtained from $\Lambda \simeq 10W \simeq 4.7\overline{R}_m$ in Figure 10.11b.

10.3.2 Lateral River Migration

The lateral migration of meandering rivers results from the erosion of the outer bank combined with sedimentation near the inner bank. This process is illustrated in Figure 10.12a, where the lateral migration Δl over a period of time Δt is shown to be a function of the radius of curvature R. In natural rivers; erosion rates also depend on bank-material strength, cohesion, armoring, and vegetation. Typical changes in cross-section hydraulic geometry are shown for the Rio Grande from 1971 to 1998. The lateral migration reached about 28 m in 27 years. The mean annual erosion rate divided by the channel width of 100 m is $\Delta l/W \simeq 0.01$. The relative annual lateral migration rates of 94 other rivers are shown in Figure 10.12b.

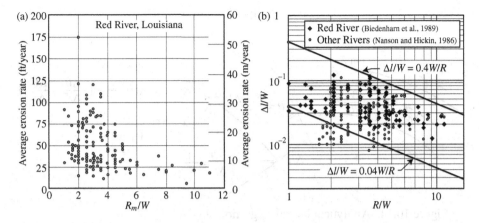

Figure 10.13. Lateral migration (a) annual rates and (b) relative migration (data from Biedenharn et al., 1989; Nanson and Hickin, 1986)

The ratio of radius of curvature to channel width can help predict erosion rates in rivers. Biedenharn et al. (1989) studied the effects of R_m/W and bank material on the erosion rates of 160 bends along the Red River in Louisiana and Arkansas. Per Figure 10.13a, the maximum erosion rates were observed in the range $2 < R_m/W < 4$.

The relative migration rate, defined as the annual migration rate divided by the channel width, can be developed as a function of R_m/W. Nanson and Hickin (1986) studied the relative migration rates of 18 rivers in Canada, and Figure 10.13b shows a comparison with the Red River. The relative migration rates are remarkably consistent and vary inversely with R/W.

$$\frac{\Delta l}{W} = p_{c1} \frac{W}{R},\qquad(10.23)$$

where p_{c1} is the mean annual percentage migration rate, which typically varies between 0.04 and 0.4. In other words, if the mean annual migration rate is $\Delta l = a/R$, then $a = p_{c1}W^2$.

10.3.3 Why Do Rivers Meander?

River meandering is characterized by a succession of alternating meander loops. A meander loop is defined as the channel reach between two inflection points. A meander consists of a pair of loops in opposite directions. Since the early explanation, based on de Baer theory, by Einstein (1926), different approaches were proposed by Rozovskii (1957), Yen (1970), de Vriend (1977), Odgaard (1981), Nelson and Smith (1989), and others. Several

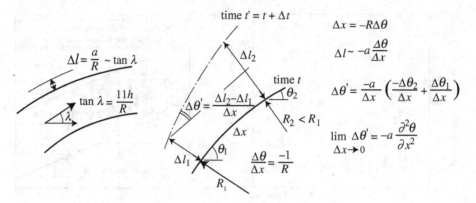

Figure 10.14. Analytical lateral migration sketch

attempts focused on the hydrodynamic stability of straight alluvial channels. A perturbation technique has been used to determine whether small oscillations amplify or decay. Examples can be found in Callander (1969), Anderson (1967), Engelund and Skovgaard (1973), Parker (1976), and Ikeda et al. (1981). Extremal hypotheses include the principle of minimum variance, first proposed by Langbein and Leopold (1966). The minimization involves the adjustment of the planform geometry and the hydraulic factors of depth, velocity, and local slope. Yang (1976) stated that the time rate of energy expenditure explains the formation of meandering streams. Other studies by Maddock (1970) and Chang (1980) use the principle of minimum stream power. Chang (1979a) argued that a meandering river uses less stream power per unit channel length. Julien (1985) treated meandering as a variational problem, and Yalin (1992) used Lagrange multipliers to explain the sine-generated curve. In recent decades, the equations of fluid motion in curved channels were primarily solved numerically (e.g. Duan and Julien, 2005 per Section 7.4.2).

To explain why rivers meander, a much simpler approach is proposed here. As sketched in Figure 10.14, the only relevant assumption is that the rate of lateral migration is proportional to the secondary flows, i.e. the mean annual river migration $\Delta l = a/R \sim \tan \lambda \simeq 11h/R$.

From $\Delta x = R(\theta_2 - \theta_1) = -R\Delta\theta$ and $\Delta l = -a\Delta\theta/\Delta x$, we can define the channel position at time $t' = t + \Delta t$. The rate of change in orientation angle $\dot{\theta} = \lim_{\Delta t \to 0} \Delta\theta'/\Delta t$ becomes

$$\dot{\theta} = \lim_{\Delta x \to 0} \left(\frac{-a}{\Delta x}\right)\left(\frac{\Delta\theta_2}{\Delta x} - \frac{\Delta\theta_1}{\Delta x}\right) = -a\frac{\partial^2\theta}{\partial x^2}. \tag{10.24}$$

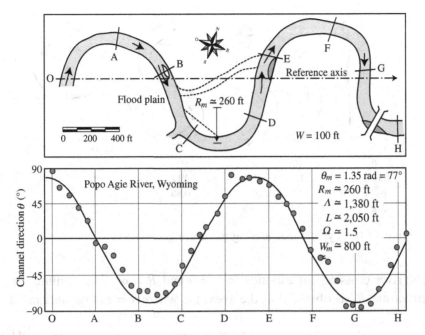

Figure 10.15. Meandering planform geometry (after Langbein and Leopold, 1966)

For any channel of the type $\theta = \sin kx$, the change in angle over time remains a sine function. The sine and cosine functions are thus self-generating functions describing river meandering.

10.3.4 Sine-generated Curves

When plotting the orientation angle θ of a river as a function of downstream distance x, Langbein and Leopold (1966) observed that θ varied as a function of the maximum angle θ_m set at the origin, the downstream distance x, and the sinuous river length L, as:

$$\theta = \theta_m \cos\frac{2\pi x}{L}. \tag{10.25}$$

This sine-generated curve is compared with an observed meandering pattern in Figure 10.15.

The radius of curvature R can be obtained from $1/R = -d\theta/dx$, which from Equation (10.25) gives

$$\frac{1}{R} = \frac{2\pi\theta_m}{L}\sin\left(\frac{2\pi x}{L}\right). \tag{10.26}$$

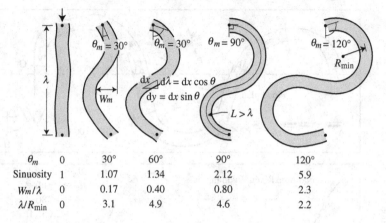

θ_m	0	30°	60°	90°	120°
Sinuosity	1	1.07	1.34	2.12	5.9
W_m/λ	0	0.17	0.40	0.80	2.3
$\lambda/R_{\min}$	0	3.1	4.9	4.6	2.2

Figure 10.16. Typical meandering planform geometry

The river crossing corresponds to $x = 0$ and $R = \infty$. The minimum radius of curvature R_m is obtained at the apex, i.e. where $x = L/4$ or $\sin(2\pi x/L) = 1$

$$R_m = \frac{L}{2\pi\theta_m}.$$ (10.27)

Several sine-generated curves are sketched in Figure 10.16. The meander length Λ is computed from the following relationship:

$$\Lambda = \int_0^L \cos\theta \, dx = \int_0^L \cos\left[\theta_m \cos\left(\frac{2\pi x}{L}\right)\right] dx = L J_0(\theta_m),$$ (10.28)

where $J_0(\theta_m)$ is the zero order Bessel function of the first kind. The function is currently available on Excel spreadsheets [as Besselj(x, 0)], and can be approximated as $J_0(\theta_m) \simeq 1 - 0.25\theta_m^2 + (1/64)\theta_m^4 - (1/2304)\theta_m^6 + \cdots$ with θ_m in radians (1 radian = 57.3°).

Similarly, the meander width W_m, defined in Figure 10.16, is evaluated by

$$W_m = 2\int_0^{L/4} \sin\theta \, dx = \int_0^L \sin\left[\theta_m \cos\left(\frac{2\pi x}{L}\right)\right] dx = \frac{L}{2} H_0(\theta_m),$$ (10.29)

where $H_0(\theta_m) = \frac{2}{\pi}\left(\theta_m - \frac{\theta_m^3}{3^2} + \frac{\theta_m^5}{15^2} - \frac{\theta_m^7}{105^2} + \cdots\right)$ is called the Struve function.

Three main dimensionless parameters describe meandering rivers, as shown in Figure 10.17:

The sinuosity

$$\Omega \equiv \frac{L}{\Lambda} = \frac{1}{J_0(\theta_m)},$$ (10.30a)

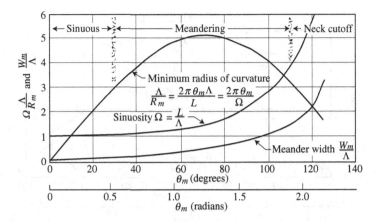

Figure 10.17. Three dimensionless parameters for meandering rivers

the relative meander width

$$\frac{W_m}{\Lambda} = \frac{H_0(\theta_m)}{2J_0(\theta_m)},$$ (10.30b)

and (3) the relative minimum radius of curvature

$$\frac{\Lambda}{R_m} = \frac{2\pi\theta_m\Lambda}{L} = \frac{2\pi\theta_m}{\Omega} = 2\pi\theta_m J_0(\theta_m).$$ (10.30c)

These properties of sine-generated curves are also supported by Leopold and Wolman (1960) who suggested $\Lambda/R_m = 4.7$ in Figure 10.11b.

10.3.5 Downstream and Lateral River Mobility

The effects of meandering on the downstream and lateral mobility of a river are examined through the effect of sinuosity on channel slope. The energy gradient of the valley S_0 is the energy loss ΔH over a meander wavelength Λ. The friction slope S of a meandering channel corresponds to the same loss ΔH over the length L, or $S = \frac{\Delta H}{L} = \frac{\Delta H}{\Lambda}\frac{\Lambda}{L} = \frac{S_0}{\Omega}$.

The influence of the sinuosity on the downstream hydraulic geometry of a meandering river can be examined through Equation (10.22a–d), which gives $h_R \sim \Omega^{1/6}$, $W_R \sim \Omega^{0.22}$, and $V_R \sim \Omega^{-0.39}$.

The downstream mobility of a channel is described by the relative Shields parameter τ_{*R}, defined as the Shields parameter for a sinuous channel divided by the Shields parameter of a straight channel. From

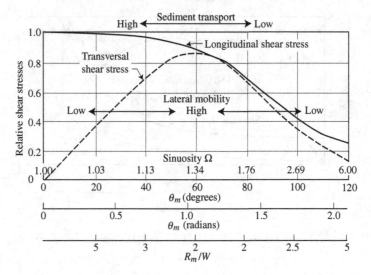

Figure 10.18. Relative downstream and lateral mobility of meandering rivers

Equation (10.22d), the ability to transport sediment in the downstream direction is given by:

$$\tau_{*R} = \Omega^{-5/6} = [J_0(\theta_m)]^{0.83}. \tag{10.31}$$

For comparison with a straight channel, a meandering channel with sinuosity $\Omega = 2$ would have a flow depth 12 percent larger and a channel width 16 percent larger. The flow velocity would decrease by 24 percent, and the shear stress would decrease by 44 percent. Thus, sinuosity increases cross-section area and decreases flow velocity and sediment transport. The downstream mobility of meandering channels is sketched in Figure 10.18. When $\theta_m > 1$ rad or 57.3°, meandering channels have a much reduced ability to transport sediment compared to straight channels.

The lateral mobility of a meandering river is described by the relative transversal shear stress $\tau_{*tR} = \tau_{*R} \tan \lambda = \tau_{*R} h_R / R_{mR}$. Considering a constant wavelength Λ, with $\tau_{*R} = \Omega^{-5/6}$ from Equation (10.31), $h_R \sim \Omega^{1/6}$, and $R_{mR} = \Omega/\theta_m$ from Equation (10.30c), we obtain

$$\tau_{*tR} = \frac{\theta_m \Omega^{1/6}}{\Omega \Omega^{5/6}} = \theta_m \Omega^{-1.67} = \theta_m [J_0(\theta_m)]^{5/3}. \tag{10.32}$$

The value of the lateral mobility of a meandering channel is represented by τ_{*tR} in Figure 10.18. High values of the transversal Shields parameter are found when $0.5 < \theta_m < 1.5$ rad (or $40° < \theta_m < 80°$). This corresponds to values of $\Lambda \sim 2.5\text{–}5R_m$ from Figure 10.17. Given $\Lambda \simeq 10W$ from Figure 10.11b, this leads us to the very important conclusion that the lateral mobility of meandering rivers should be high when $R_m \sim 2\text{–}4W$, which was observed in Figure 10.13a.

10.3.6 Timescale for Meander Evolution

Let's examine the gradual evolution of a river meander and define the rate of change in orientation angle for sine-generated curves by combining Equations (10.24, 10.25, and 10.30a),

$$\dot{\theta} = \frac{d\theta_m}{dt} = -a\frac{\partial^2\theta}{\partial x^2} = a\left(\frac{2\pi}{\Lambda}\right)^2 \theta_m J_0{}^2(\theta_m). \tag{10.33}$$

This can be solved for a constant Λ as a function of time (given $a = p_{c1}W^2$ and $\Lambda \simeq 2\pi W$):

$$t_{12} = \int_{t1}^{t2} dt = \frac{1}{a}\left(\frac{\Lambda}{2\pi}\right)^2 \int_{\theta_{m1}}^{\theta_{m2}} \frac{d\theta_m}{\theta_m J_0{}^2(\theta_m)} = \frac{1}{p_{c1}}\int_{\theta_{m1}}^{\theta_{m2}} \frac{d\theta_m}{\theta_m J_0{}^2(\theta_m)}$$

$$= \frac{1}{p_{c1}}\frac{\pi}{2}\left[\frac{N_0(\theta_{m2})}{J_0(\theta_{m2})} - \frac{N_0(\theta_{m1})}{J_0(\theta_{m1})}\right], \tag{10.34}$$

where $N_0(\theta_m)$ is the Neumann function, or the zero order Bessel Y function. It can be solved using Excel spreadsheets [as Bessely$(x, 0)$]. Some useful values of the Bessel, Struve, and Neumann functions are listed in Table 10.1.

As an example, assuming a continuous lateral migration process, we can estimate the time it takes for a meandering channel with $\theta_m = 40°$ to increase its sinuosity to $\theta_m = 80°$. If $p_{c1} = 0.1$ per year, from Figure 10.13b, Table 10.1 can be used to solve Equation (10.34) as $t_{40°-80°} = \frac{1}{p_{c1}}\frac{\pi}{2}\left[\frac{N_0(1.4)}{J_0(1.4)} - \frac{N_0(0.7)}{J_0(0.7)}\right] = \frac{1}{0.1}(0.94 - -0.34) \simeq 13$ years. Accordingly, sinuosity would increase from 1.13 to 1.76; and meander width would change from 0.24 to 0.63 times the meander wavelength. Notice that the time to change from $\theta_m = 80°$ to $\theta_m = 120°$ would be much longer, $t_{80°-120°} \simeq 10(4.9 - 0.9) = 40$ years. This supports the index in Figure 10.18, where the lateral mobility of a meandering river is high when $40° < \theta_m < 80°$ and low when $\theta_m > 80°$.

Table 10.1. *Useful values of the Bessel, Struve, and Neumann functions for meandering rivers*

θ_m radians	θ_m degrees	Bessel $J_0(\theta_m)$	Ω $\frac{1}{J_0(\theta_m)}$	Struve $H_0(\theta_m)$	$\frac{W_m}{\Lambda}$ $\frac{H_0(\theta_m)}{2J_0(\theta_m)}$	Neumann $N_0(\theta_m)$	Timescale $\frac{\pi}{2}\left[\frac{N_0(\theta_m)}{J_0(\theta_m)}\right]$
0.1	5.73	0.998	1.00	0.064	0.032	−1.534	−2.416
0.2	11.5	0.990	1.01	0.127	0.064	−1.081	−1.715
0.3	17.2	0.978	1.02	0.189	0.097	−0.807	−1.297
0.4	22.9	0.960	1.04	0.250	0.130	−0.606	−0.991
0.5	28.6	0.938	1.07	0.310	0.165	−0.445	−0.744
0.6	34.4	0.912	1.10	0.367	0.201	−0.309	−0.531
0.7	40.1	0.881	1.13	0.422	0.239	−0.191	−0.340
0.8	45.8	0.846	1.18	0.474	0.280	−0.087	−0.161
0.9	51.6	0.807	1.24	0.523	0.324	0.006	0.011
1.0	57.3	0.765	1.31	0.569	0.372	0.088	0.181
1.1	63.0	0.720	1.39	0.611	0.424	0.162	0.354
1.2	68.8	0.671	1.49	0.649	0.483	0.228	0.534
1.3	74.5	0.620	1.61	0.682	0.550	0.287	0.726
1.4	80.2	0.567	1.76	0.712	0.628	0.338	0.936
1.5	85.9	0.512	1.95	0.737	0.720	0.382	1.174
1.6	91.7	0.455	2.20	0.757	0.831	0.420	1.450
1.7	97.4	0.398	2.51	0.773	0.971	0.452	1.784
1.8	103.1	0.340	2.94	0.783	1.152	0.477	2.206
1.9	108.9	0.282	3.55	0.790	1.401	0.497	2.769
2.0	114.6	0.224	4.47	0.791	1.766	0.510	3.581
2.1	120.3	0.167	6.00	0.788	2.363	0.518	4.887

Problem 10.1

Define the ideal cross-section geometry for a 100-mm cobble-bed canal with all particles at beginning of motion. The slope of the channel is 0.01. Also estimate the flow discharge in this canal, using Manning–Strickler's approximation.

[*Answer:* $h = 0.78$ m, $W = 2.9$ m, $A = 1.45$ m^2, $R_h = 0.42$ m, $n = 0.044$, $V = 1.27$ m/s, $Q = 1.84$ m^3/s, $\tau_* = \frac{hS}{(G-1)d_s} = 0.047$ at incipient motion, and note that $W/h = 3.7$ is very low.]

◆◆Problem 10.2

Use the regime relationships to calculate the hydraulic geometry of an irrigation canal that conveys 5,000 ft^3/s in a very fine gravel-bed channel. Compare with the hydraulic geometry for stable channels.

[*Answer:* V = 4.6 ft/s, R = 5.7 ft, P = 188 ft, S = 6.9 × 10^{-4} from the regime equations; and V = 0.72 m/s, h = 1.78 m, W = 109 m, S = 1.3 × 10^{-4} from the Julien-Wargadalam equations (10.20) with d_s = 3 mm.]

◆Problem 10.3

The Cache la Poudre River near Rustic, Colorado, has a bankfull discharge of 17.6 m^3/s, a width of 12.8 m, a depth of 0.65 m, and a slope of 0.0048 with a grain size of 150 mm. Compare the actual geometry with the regime equations and downstream hydraulic-geometry relationships.

◆Problem 10.4

Examine the effects of channel sinuosity on downstream hydraulic geometry given the valley slope $S_0 = \Delta H / \Lambda$, and the slope of the meandering channel as $S = \Delta H / L = S_0 / \Omega$, where ΔH is the energy loss, Λ is the valley length, and L is the sinuous channel length.

[*Answer:* From the J–W equations with m = 1/6 for the analysis, one obtains $h \sim \Omega^{1/6}$, $W \sim \Omega^{0.22}$, $V \sim \Omega^{-0.39}$, and $\tau_* \sim \Omega^{-0.83}$. For instance, if $\Omega = 2$, the flow depth increases by 12 percent, the channel width increases by 16 percent, the flow velocity decreases by 24 percent, and the shear stress decreases by 44 percent. Sinuosity thus increases the cross-section area and decreases flow velocity and sediment transport.]

◆◆Problem 10.5

From the information on the Fall River CO, presented in Case Study 10.1, determine the following: downstream angle θ_0, side-slope angle θ_1 at high flow, bed shear stress τ_0, radius of curvature, and streamline deviation angle λ. Also use the downstream hydraulic geometry relationships for comparison with the field observations. Determine the: valley length Λ, river length L, sinuosity Ω, maximum deviation angle θ_m, and meander width W_m. Compare with the various plots of this chapter and determine whether lateral mobility is high or low.

[*Answer:* θ_0 = 0.0745°, $\theta_1 \simeq 15°$, $\tau_0 \cong \gamma h S \cong 12$ Pa, $R_m \simeq 15$ m, $\lambda \simeq \tan^{-1}(11h/R) = 36°$ at high flow and $\lambda \simeq 10°$ at low flow. From $Q = 7$ m^3/s, d_s = 8 mm and S = 0.0013. The J–W equations (10.22) give $h \simeq 0.52$ m, $W \simeq 22$ m, $V \simeq 0.55$ m/s, and $\tau_* \simeq 0.05$ which is close to incipient motion for bankfull flow. The actual cross-section geometry is slightly deeper and narrower than calculated. The measured planform geometry gives $\Lambda \simeq 50$ m, $L \simeq 103$ m, $\Omega \simeq 2$, $\theta_m \simeq 95° = 1.66$ rad, and $W_m \simeq 45$ m.

From the field measurements, we obtain $W_m \simeq 4.5W$, $\Lambda = 5W \simeq 3.5R_m$, $R_m \simeq 1.5W$, which is indicative of very sharp bends, as expected. Field measurements compare well with the figures and the lateral mobility is low.]

◆◆*Problem 10.6*

With reference to Case Study 9.1, compare the properties of the Mississippi River with the characteristics of meandering channels discussed in this chapter. Access Google maps and determine the following for Thompson Bend: (1) mean channel width; (2) minimum radius of curvature; (3) sinuosity; and (4) meander wavelength. Plot the orientation angle as a function of downstream distance for the two loops of the Thompson Bend. Determine the maximum angle θ_m and plot your results on as many figures as you can from this chapter.

◆*Problem 10.7*

Consider the aerial photograph of the Red River LA in Figure P10.7. Identify and locate the following geomorphic features: (1) river-bend apex; (2) river crossing(s); (3) point bars; (4) mid-channel bars; (5) scrolls; (6) oxbow lakes; (7) locate the areas where bank protection would be needed to limit lateral migration.

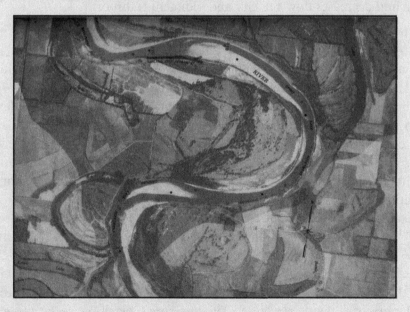

Figure P10.7. Reach of the Red River in Louisiana

♦*Problem 10.8*

Consider two photos (1972 on the left and 1992 on the right) of a reach of the Rio Grande in Figure P10.8. Examine the properties of the sine generated curves discussed in this chapter. Determine the relationships between the channel width and the wavelength, the radius of curvature and the channel width, the meander width and the wavelength, and the sinuosity. Estimate the maximum angle of the sine-generated curve θ_m for both years and find p_{c1} accordingly. Finally, from the radius of curvature and the channel width, estimate the mean annual lateral migration rate of this river as a percentage of the channel width for both years. Do you expect the rate of lateral migration to slow down or accelerate after 1992?

(a) (b)

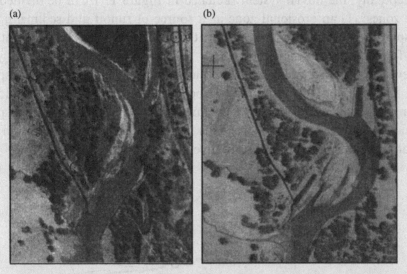

Figure P10.8. Reach of the Rio Grande in New Mexico in (a) 1972 and (b) 1992

♦♦*Problem 10.9*

Consider the aerial photo of the Red River in Figure P10.7. Locate the channel centerline in the downstream direction, plot θ as a function of the downstream distance x and check whether it follows a sine-generated curve. Define the: valley length Λ, river meander length L, sinuosity Ω, maximum deviation angle θ_m and meander width W_m. Compare with various plots of this chapter.

11

River Dynamics

Conceptually, the fluvial system sketched in Figure 11.1 can be divided into three zones: (1) an erosional zone as a source of runoff and sediment; (2) a transport zone for water and sediment conveyance; and (3) a depositional zone for runoff and sediment delivery.

Deviations from equilibrium conditions will trigger a dynamic response from alluvial rivers to restore the balance between the inflow and outflow of water and sediment. Section 11.1 examines the dynamics of a river response to changes in water and sediment loads. Sections 11.2 and 11.3 focus on river degradation and aggradation, respectively.

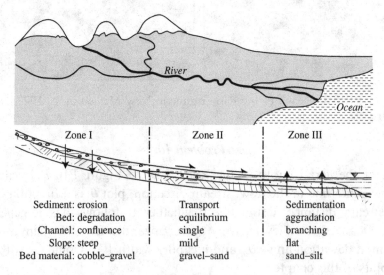

Zone I	Zone II	Zone III
Sediment: erosion	Transport	Sedimentation
Bed: degradation	equilibrium	aggradation
Channel: confluence	single	branching
Slope: steep	mild	flat
Bed material: cobble–gravel	gravel–sand	sand–silt

Figure 11.1. Erosion, transport, and sedimentation zones of a fluvial system

11.1 River Response

River databases for the definition of changes in hydraulic and sediment regimes are discussed in Section 11.1.1. This is followed with a geomorphic analysis of river response in Section 11.1.2, and a quantitative analysis in Section 11.1.3.

11.1.1 River Databases

Databases for the analysis of river dynamics include: (1) historical records; (2) maps and photos; (3) streamflow data; (4) sediment data; and (5) field surveys. Historical information regarding major floods affecting channel morphology and stability is desirable and upstream basin information should include flow diversions, hydraulic structures for flood control, irrigation, and navigation, and land-use and climate changes.

Topographic maps at various scales indicate the nature of the drainage area and the fluvial system. Geographic Information Systems (GIS) are useful in examining topography, soil types, and land-use data. The river planform geometry, longitudinal profile, and estimates of channel slope can be obtained from topo maps/GIS. Recent sediment deposits are usually clearly visible from aerial photographs. The analysis of lateral migration of rivers is usually possible from the comparison of several sets of cross sections, aerial photographs, satellite images, and Google maps. It is very often important to gather basic information on the cross-sectional geometry of a river including longitudinal and cross-section profiles, as well as non-vegetated planform geometry, bankfull conditions and floodplain elevation, and land use/vegetation of the river bank. Field surveys are most effective after a review of maps and photos. Field notes and sampling should include indications regarding: particle-size distribution of bed and bank material, relative stability/mobility of the riverbank and riverbed material, bank instability and mode of failure, stratigraphy and seepage, aggradation/degradation, sediment deposits, headcutting, bedrock control, and riparian vegetation.

The hydraulic information is very important to validate water-surface calculations from hydraulic models. River hydraulic considerations include main channel and floodplain roughness, flow velocities, high-water marks on bridge piers, structures, river choking, floating debris, ice cover, and ice jams. Streamflow data include discharge data on a daily basis for the entire period of record. The entire flow–discharge record can be used to determine the flow-duration curves and for the flood-frequency analysis. The bankfull discharge should normally fit within the range of 1–5 years in the flood-frequency analysis. Stream gauges are useful in determining the stage–discharge relationship. Loop-rating effects can be caused by aggradation and degradation, bedform changes, and floodwave propagation characteristics. At times, specific-gauge records can detect aggradation/degradation trends over long periods. Flights over the river are often informative regarding the overall planform stability of rivers during floods.

Sediment data include bed material and sediment transport; the particle-size distribution of the bed material should be determined as accurately as possible. The variability in bed material in alluvial rivers can be high. Several samples from different locations are often desirable. In degrading channels, careful attention should be paid to the coarse fractions of the surface material and underlying deposits. Sieve analyses are best suited to fine-grained streams, and gravel-bed and cobble-bed streams require the examination of large volumes of bed material to determine the median grain size. Laser diffraction methods are often employed for the fine-size fractions. Suspended sediment records should indicate the flux-average sediment concentration and the sediment load. Sediment load by size fractions are sometimes useful for the evaluation of the Rouse parameter. Sediment budgets by size fractions are sometimes most useful in determining the different patterns of sediment transport for washload and bed-material load. Good examples of river databases are found in Novak (2006), and Owen (2012), and for the Bernardo Reach of the Rio Grande in Case Study 11.1.

Case Study 11.1 Bernardo Reach of the Rio Grande, New Mexico, United States

The 10-mile-long Bernardo reach of the Middle Rio Grande in New Mexico is included in the habitat designation for two federally listed endangered species: the silvery minnow and the southwestern willow flycatcher. A complete database of flow and sediment transport in the Rio Grande below Cochiti Dam has been assembled at Colorado State University by G. Richard, C. Leon, and T. Bauer, in collaboration with D. Baird at the US Bureau of Reclamation. Besides the complete data reports (Leon et al., 1999, and Bauer et al., 2000), the geomorphic analysis of Richard et al. (2000) shows a 500-ft-wide river with a fine sand bed at $d_{50} \simeq 0.3$ mm. The channel width varies from 150 to 1,200 ft within a 10-mile-long reach. Discharge and sediment mass curves at Bernardo and San Acacia are shown in Problem 11.10. They respectively represent the cumulative runoff volume as a function of time and the cumulative sediment mass as a function of time. The influence of Cochiti Dam, built in 1975 is clearly identified.

Double mass curves represent the cumulative sediment load as a function of the cumulative runoff water volume. The Bernardo reach of the Rio Grande is illustrated in Figure CS11.1.1a. The slope of double mass curve defines average sediment concentrations in suspension. This curve highlights the reduced sediment concentration of the Rio Grande after the construction of Cochiti Dam in 1975. Mass difference curves show the difference between sediment inflow and outflow for a given reach. The mass-difference curves provide information on the net sediment balance of a given river reach and indicate whether a river is aggrading (+) or degrading (–) during that time period. Significant river degradation can be expected from the net sediment balance of the Bernardo reach of the Rio Grande, as shown in Figure CS11.1.1b. The degradation trend can be quantified as confirmed with the field measurements.

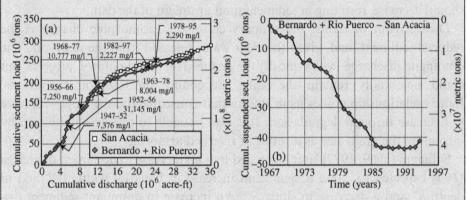

Figure CS11.1.1. Bernardo Reach of the Rio Grande (a) double mass curve and (b) mass difference curve (after Richard et al., 2000)

11.1.2 Geomorphic Analysis of River Response

Lane (1955b) proposed a relationship which empirically scaled the product of slope and discharge to the product of grain size and sediment discharge.

$$QS \sim Q_s d_s. \tag{11.1}◆◆$$

While Lane defined this empirical relationship, it is Borland who first sketched the widely used scale diagram (Pemberton and Strand, 2005).

The effect of dams on channel geometry can be examined from a geomorphic standpoint, based on Lane's relationship and Borland's scale. As sketched in Figure 11.2, the primary cause of the geomorphic change upstream of the dam

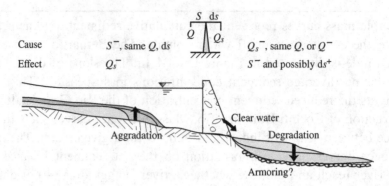

Figure 11.2. Geomorphic analysis of the impact of dams on river systems

is the reduced slope induced by the M-1 backwater profile. Under constant discharge and sediment size, Lane's relationship implies that the sediment load should decrease, resulting in sedimentation upstream of the dam.

Downstream of dams, the primary cause for geomorphic change is the reduced sediment load, given the clear water released from the reservoir. The dominant discharge will typically be reduced in flood-control projects but could potentially remain the same, depending on reservoir operations. Lane's relationship guides us in stating that consequently, the slope will have to decrease through degradation. It is also possible to restore equilibrium by increasing the grain size as a result of riverbed armoring.

Schumm (1969, 1977) later suggested that an increase in dominant discharge Q^+ is expected to cause a significant increase in bankfull width W^+ and in depth h^+ and a decrease in slope S^-. An increase in dominant sediment discharge Q_s^+ corresponds primarily to significant increases in slope S^+ and width/depth ratio. As sketched in Figure 11.3, changes in hydraulic geometry take place through reworking river alluvium. Flow depth can increase as the bed degrades, and flow depth decreases through aggradation. Channel widening will occur through bank erosion or overbank flows. Channel narrowing may result in the formation of mid-channel bars and islands, followed by incision of the main channel. In general, shoals are submerged sandbars at low flows in flat large rivers. Sandbars and gravel bars are submerged at high flow, and islands are covered with vegetation. Changes in flow velocity are quite naturally linked to bedforms, surface roughness, and sediment transport capacity.

11.1.3 *Quantitative Analysis of River Response*

In this section, we define the hydraulic geometry of alluvial channels as a function of flow discharge Q, bed-material grain size d_{50}, and the sediment

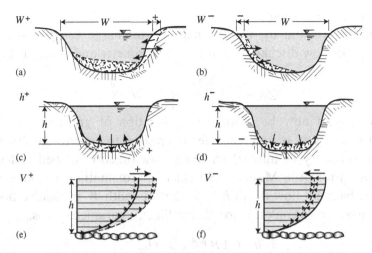

Figure 11.3. Increase and decrease in: (a), (d) river width, (b), (e) flow depth, and (c), (f) flow velocity

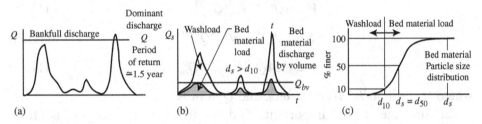

Figure 11.4. Time series in: (a) discharge, (b) sediment load, and (c) particle distribution

discharge Q_s. As sketched in Figure 11.4, the dominant discharge refers to bankfull discharge conditions or flood discharge with a period of return around 1.5 years. The sediment discharge includes two components: (1) washload for size fractions of bed material finer than d_{10} (Julien, 2010) and (2) bed-material load or discharge Q_{bv} for size fractions of the bed material coarser than d_{10}. The bed-material discharge is most important because it changes bed conditions and thus relates to hydraulic geometry.

Numerous quantitative relationships exist between bed-material sediment transport and hydraulic variables (Julien, 2010). The following empirical relationship, Equation (2.44), for bed-material transport is used as a first approximation:

$$Q_{bv} \simeq 18 W \sqrt{g} d_s^{3/2} \tau_*^2. \qquad (11.2a)$$

This relationship estimates the unit bed-material discharge by volume q_{bv} in m^2/s when $0.1 < \tau_* < 1$, where the grain diameter d_s is in meters.

Accordingly, the volumetric sediment discharge is $Q_{bv} = q_{bv}W$. From Equation (10.21b and d), the bed-material discharge can be written as a function of the flow discharge, bed-material diameter and channel slope as

$$Q_{bv} \simeq 1.1 Q^{\frac{6+4m}{5+6m}} d_{50}^{\frac{-2.5+5m}{5+6m}} S^{\frac{7+10m}{5+6m}}. \tag{11.2b}$$

The slope can now be written as a function of grain size and water/sediment discharges. It now becomes interesting to rewrite the downstream hydraulic geometry of alluvial channels as a function of bed-material discharge Q_{bv}. From the Manning–Strickler approximation, i.e. $m = 1/6$, we obtain the bankfull flow depth h and channel width W in meters, flow velocity V in m/s, dimensionless slope S, and Shields parameter τ_* as:

$$h \simeq 0.19 Q^{0.46} d_{50}^{0.13} Q_{bv}^{-0.12}, \tag{11.3a}$$

$$W \simeq 1.3 Q^{0.62} d_{50}^{-0.15} Q_{bv}^{-0.15}, \tag{11.3b}$$

$$V \simeq 4 Q^{-0.08} d_{50}^{0.02} Q_{bv}^{0.27}, \tag{11.3c}$$

$$S \simeq 1.2 Q^{-0.77} d_{50}^{0.19} Q_{bv}^{0.69}, \tag{11.3d}$$

$$\tau_* \simeq 0.14 Q^{-0.31} d_{50}^{-0.67} Q_{bv}^{0.57}, \tag{11.3e}$$

where Q_{bv} is in m³/s, d_{50} in meters, and Q in m³/s.

Perhaps the most important finding from this analysis is from Equation (11.3d), or

$$Q_{bv} d_{50}^{0.28} \simeq Q^{1.11} S^{1.44}. \tag{11.4}$$

This theoretical analysis supports Lane's relationship, Equation (11.1), which empirically scaled the product of slope and discharge to the product of grain size and sediment discharge.

As a useful alternative, we can examine the effects of changes in sediment concentration $C_{mg/l}$ instead of sediment discharge given $Q_{bv} = 3.8 \times 10^{-7} C_{mg/l} Q$. After substitution into Equation (11.3), we simply define the downstream hydraulic geometry relationships as a function of bankfull bed-material sediment concentration $C_{mg/l}$ in milligrams per liter, d_{50} in meters, and Q in cubic meters per second:

$$h \simeq 1.1 Q^{0.34} d_{50}^{0.13} C_{mg/l}^{-0.12}, \tag{11.5a}\blacklozenge$$

$$W \simeq 12 Q^{0.47} d_{50}^{-0.15} C_{mg/l}^{-0.15}, \tag{11.5b}\blacklozenge\blacklozenge$$

$$V \simeq 0.075 Q^{0.19} d_{50}^{0.02} C_{mg/l}^{0.27}, \tag{11.5c}\blacklozenge$$

$$S \simeq 4.4 \times 10^{-5} Q^{-0.08} d_{50}^{0.19} C_{\text{mg/l}}^{0.69}, \qquad (11.5d) \blacklozenge\blacklozenge$$

$$\tau_* \simeq 3 \times 10^{-5} Q^{0.26} d_{50}^{-0.67} C_{\text{mg/l}}^{0.57}. \qquad (11.5e) \blacklozenge$$

Note that $C_{\text{ppm}} \simeq C_{\text{mg/l}}$ when $C_{\text{ppm}} < 1{,}000$ ppm, and the difference between $C_{\text{mg/l}}$ and C_{ppm} is less than 10 percent at concentrations lower than 145,000 ppm (Julien, 2010).

The dynamic response of alluvial rivers to perturbations in dominant discharge, grain size, and sediment discharge can be summarized as follows:

$$Q^+ \rightarrow W^+ h^+ S^-, \qquad (11.6a)$$

$$Q_{bv}^+ \rightarrow S^+ \tau_*^+ V^+, \text{ and} \qquad (11.6b)$$

$$d_S^+ \rightarrow \tau_*^-. \qquad (11.6c)$$

We thus note that the river width and depth depend primarily on discharge. Increases in sediment discharges and sediment concentration are linked to channel slope and flow velocity. Grain-size effects on downstream hydraulic geometry are comparatively small.

As an example, an alluvial sand-bed channel has a bankfull width $W_1 = 300$ ft, a flow depth $h_1 = 10$ ft, slope $S_1 = 8$ ft/mile, and a flow velocity $V_1 = 10$ ft/s. We can now quantify the downstream hydraulic geometry when the dominant flow discharge is decreased by 50 percent and the sediment discharge is doubled. If the bed-material size remains the same, the changes in downstream hydraulic geometry with $Q_2/Q_1 = 0.5$ and $Q_{s2}/Q_{s1} = 2$ are estimated from Equation (11.3a–e): flow depth $h_2/h_1 = 0.5^{0.46} \times 2^{-0.12} = 0.67$; channel width $W_2/W_1 = 0.5^{0.62} \times 2^{-0.15} = 0.59$; velocity $V_2/V_1 = 0.5^{-0.08} \times 2^{0.27} = 1.27$; and slope $S_2/S_1 = 0.5^{-0.77} \times 2^{0.69} = 2.75$. To reach a new equilibrium, we expect: the flow depth to decrease to ~ 6.7 ft, the bankfull width should decrease to ~ 180 ft, the flow velocity should increase to ~ 12.5 ft/s, the bed slope should increase to ~ 22 ft/mile. While changes in width, depth, and slope are reasonable, can you explain how a river is going to drastically increase its bed slope?

We have described how meandering channels can handle a decrease in mobility and sediment transport in Section 10.3. The case of decreasing bed slope will be discussed further in Section 11.2. The case of increasing bed slope will be discussed in Section 11.3 for channel aggradation.

11.2 River Degradation

"River degradation" refers to the general lowering of the bed elevation owing to erosion. River degradation types are discussed in Section 11.2.1,

with focus on degradation below dams in Section 11.2.2. When the bed
material is coarse, an armor layer can form to limit degradation (Section
11.2.3). Headcuts and nickpoints are covered in Section 11.2.4.

11.2.1 River Degradation Types

River adjustments through degradation occur when the outgoing sediment
load exceeds the inflowing sediment load. Alluvial rivers scour bed material
and milder slopes are obtained through channel incision. Various types of
incision are sketched in Figure 11.5.

As sketched in Figure 11.5, incised channels are characterized by a shortage of sediment. They become narrower W^- and deeper h^+ than equilibrium
channels and their width/depth ratio $(W/h)^-$ decreases. In upland areas,
rills are small-scale channels, while gullies are larger-scale features.
Conventionally, rills can be crossed by farm machinery but gullies cannot.
In rivers, channel incision is found in arroyos and canyons. Arroyos are
ephemeral channels with flashy hydrographs carrying large sediment loads
during short periods of time. Many arroyos dry out in the downstream direction as a result of infiltration and evaporation. Canyons are often deeply

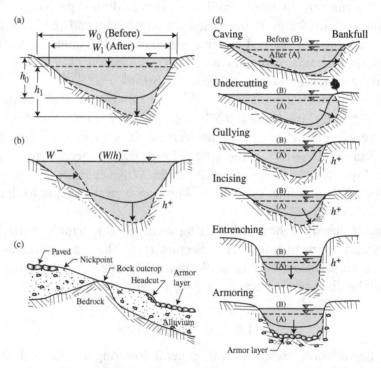

Figure 11.5. River bed degradation: (a) cross-section, (b) deformation,
(c) longitudinal profile, and (d) fluvial processes

entrenched in vertical bedrock walls. Channel degradation also causes the banks to become unstable and subject to failure. Gully-like incised channels become very unstable, and bank erosion may become a significant source of sediment to the channel.

11.2.2 Degradation below Dams

As sketched in Figure 11.6, clear water releases below dams alter the equilibrium between the flows of water and of sediment in alluvial channels. Reservoirs tend to decrease the magnitude of flood flows and increase low flows. There are three typical responses to the slope reduction via degradation below dams: (1) gradual degradation, discussed in this section; (2) rapid degradation leading to nickpoints and headcuts, discussed in Section 11.2.2; and (3) meandering, as discussed in Chapter 10.

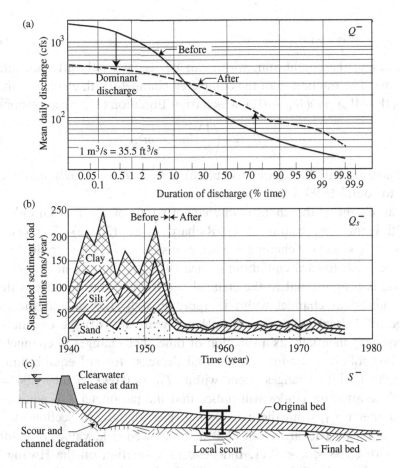

Figure 11.6. River degradation below a dam; (a) flod duration curves, (b) sediment discharge, and (c) longitudinal profile

The changes in river hydraulic geometry below dams can be studied analytically. Two main hypotheses are considered: (1) alluvial channels move toward equilibrium; and (2) the rate of change in hydraulic geometry is proportional to the deviation from equilibrium. For instance, the equilibrium width W_e can be defined from the equations of Julien and Wargadalam (Equations 10.20 and 10.22). The second hypothesis can be written as:

$$\frac{dW}{dt} = -k(W - W_e) \tag{11.7}$$

and from this hypothesis, the width changes can be determined as a function of time as

$$kt = \int_0^t k\,dt = -\int_{W_0}^W \frac{dW}{(W - W_e)} = -\ln\left(\frac{W - W_e}{W_0 - W_e}\right), \tag{11.8a}$$

or

$$W = W_e + (W_0 - W_e)e^{-kt}. \tag{11.8b}$$

Accordingly, the equilibrium width can be reached only after an infinitely long time. The fraction m of the completed channel-width change is obtained when $(W - W_e) = m(W_0 - W_e)$, which from Equation (11.8b) corresponds to

$$T_m = -\left(\frac{1}{k}\right)\ln(1 - m). \tag{11.9}$$

For example, half the change occurs at a time $T_{50} = 0.69/k$ and the time taken to complete 90 percent of the changes is $T_{90} = 2.3/k$.

As an example, the changes in channel width of the Rio Grande below Cochiti Dam were examined by Richard et al. (2005a). The results, in Figure 11.7, show: (a) channel degradation since 1971; (b) a gradual channel width decrease toward equilibrium since 1918; (c) the rate in channel-width decrease is proportional to the channel width and varies linearly with time. The equilibrium channel width is approximately 120 m and the rate of change in channel width is $k = 0.04$ per year; and (d) the channel width decreases exponentially as a function of time. The changes in channel width over time follow a gradual exponential decrease toward equilibrium. Half the channel width changes occur within $T_{50} = 0.69/k = 0.69/0.04 = 17.5$ years. The attentive reader will notice that the parameter k is equivalent to the parameter p_{c1}, from the meandering river analysis in Section 10.3. The investigation by Richard et al. (2005a, b) showed field values of k for four rivers with $0.038 < k < 0.11$. Independent validation on the Hwang River in South Korea by Shin and Julien (2010) showed $k = 0.029$. This type of

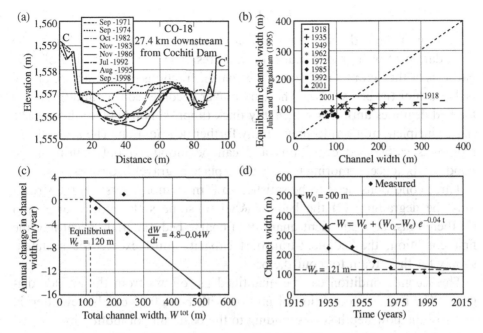

Figure 11.7. Morphological changes of the Rio Grande: (a) cross section, (b) channel width, (c) width changes, and (d) width decrease (after Richard et al., 2005a)

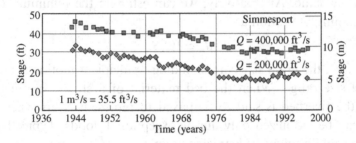

Figure 11.8. Specific-gauge records of the Atchafalaya River at Simmesport (after US Army Corps of Engineers, 1999)

analysis has also been successfully extended to changes in bed slope by Leon et al. (2009), with $k = 0.1$.

On large rivers, specific-gauge records are often used to determine whether a stream tends to aggrade or degrade over time. For instance, the specific-gauge record of the Atchafalaya River over a period of 50 years is shown in Figure 11.8. The water-surface elevation dropped approximately 15 ft between 1944 and 1980. Given that half the changes occurred in 24 years, we can estimate $k \simeq 0.69/T_{50} = 0.029$.

11.2.3 Armoring

Armoring of the bed layer refers to coarsening of the bed material from the degradation of well-graded sediment mixtures. The selective erosion of finer particles of the bed material leaves the coarser fractions of the mixture on the bed to armor the bed surface. The armor layer becomes coarser and thicker as the bed degrades until it is sufficiently thick (approximately twice the particle size at incipient motion) to prevent any further degradation. The armor layer represents stable bed conditions and can be mobilized only during large floods. In practice, armoring usually takes place in gravel-bed rivers.

Three conditions need to be satisfied to form armor layers: (1) the stream must be degrading; (2) the bed material must be sufficiently coarse; and (3) there must be a sufficient quantity of coarse bed material. Regarding the first condition, the sediment-transport capacity must exceed the sediment supply for the stream to scour the bed.

The second condition can be quantified as follows from the Shields diagram. The incipient condition of motion with $\tau_{*c} \simeq 0.05$ can be rewritten in terms of the flow depth corresponding to the beginning of motion by

$$h \simeq d_{sc}/10S, \tag{11.10a}$$

where d_{sc} is the grain diameter, h is the bankfull flow depth during floods, and S is the channel slope. It is noted that the units of grain size and flow depth are the same. Alternatively, we can estimate the minimum grain size at the beginning of motion as

$$d_{sc} \simeq 10hS. \tag{11.10b}$$

In most rivers, the grain size corresponding to incipient motion is typically in the gravel size range. When the bed material of a stream is much coarser than d_{sc}, the riverbed is said to be paved to represent rivers where the bed material can be mobilized only during exceptional floods. Typically, cobble-bed streams and boulder-bed streams are paved.

The third condition refers to the fraction of material Δp_c coarser than d_{sc} available in the bed material, as shown in Figure 11.9. When this percentage is large, the armor layer will form rapidly and the extent of degradation will be minimal. When this percentage is small, a large volume of bed material will be scoured before the armor layer can form. An armor layer of approximately twice the grain size d_{sc} will stabilize the bed.

From Figure 11.9, the scour depth Δz that will form an armor layer equal to $2d_{sc}$ can be estimated from

$$\Delta z = 2d_{sc}\left(\frac{1}{\Delta p_c} - 1\right). \tag{11.11}$$

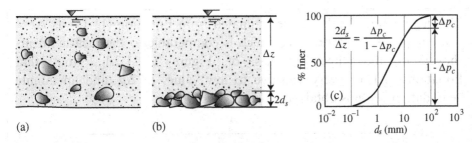

Figure 11.9. Armoring process: (a) initial, (b) final, and (c) particle distribution

The scour depth becomes large only when Δp_c is small, and it is therefore important to have a particle-size distribution that is truly representative of the deep layers. When several meters of degradation are expected, trenches and borings are useful in order to locate possible gravel and cobble layers that can limit the extent of degradation. The determination of bedrock elevation is most useful when severe degradation is expected.

Once an armor layer has formed, it plays a very important role in channel stability and morphology. Indeed, the riverbed is stable and the armor layer protects the bed against further degradation, except under large floods. Case Study 11.2 illustrates some of the main characteristics of river degradation and armoring.

Case Study 11.2 Degradation and Armoring of the Meuse River, The Netherlands

This example from Murrillo-Muñoz (1998) illustrates the gradual river changes associated with riverbed degradation over a 150-km reach of the Meuse River. The mean annual discharge is 230 m^3/s near Maastricht and exceeds 3,000 m^3/s during major floods. Near Maasbracht, the river shows a sharp transition from a gravel-bed toward a sand-bed river with d_{50} decreasing from ~ 15 to 3 mm and a corresponding bed slope decrease from 48 to 10 cm/km, as shown in Figure CS11.2.1. At a flow depth of 3 m, the critical grain size at incipient motion is calculated from the slope and Equation (11.10b) as $d_{sc} \simeq 10 \times 3$ m $\times 48 \times 10^{-5} = 15$ mm for the upper reach, and $d_{sc} \simeq 10 \times 3$ m $\times 10 \times 10^{-5} = 3$ mm for the lower reach. These size fractions correspond to d_{50} of the bed material. To examine whether armoring is possible during large floods, grain sizes exceeding 30 mm cannot be found in large quantities in the lower reach and degradation can be expected during floods. At river kilometer 110, only ~ 3 percent of the bed material is coarser than 30 mm, i.e. $\Delta p_c \simeq 0.03$.

The degradation depth can be estimated from Equation (11.11) as $\Delta z = 2 \times 0.03(-1 + 1/0.03) \simeq 2$ m, which is comparable to the measurements shown in the upper reaches of Figure CS11.2.1. Note that Equation (11.11) is very sensitive to low values of Δp_c and an infinite degradation depth is obtained when $\Delta p_c \to 0$.

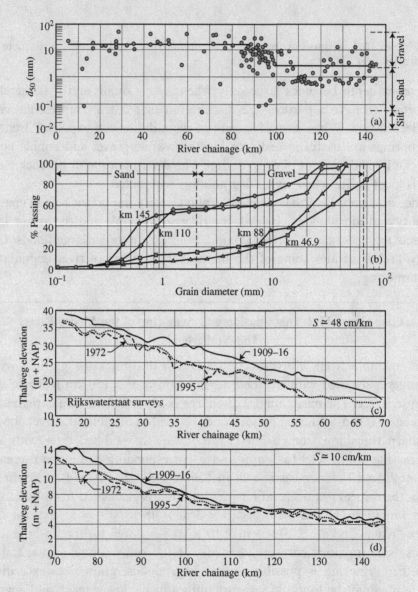

Figure CS11.2.1. Meuse River in the Netherlands: (a) median diameter, (b) particle distributions, (c), (d) longitudinal profile (after Murillo Munoz, 1998)

This case study also highlights another important property of fluvial armoring, i.e. the linear relationship between grain size and slope when h is constant. Indeed, from Equation (11.11a), the flow depth for the upper reach with $d_{50} \simeq 15$ mm and $S = 48$ cm/km is approximately $h \simeq 0.015/(10 \times 48 \times 10^{-5}) = 3.1$ m. The flow depth of the lower reach with $d_{50} \simeq 3$ mm and $S = 10$ cm/km is similar $h \simeq 0.003/(10 \times 10 \times 10^{-5}) = 3$ m. The empirical rule for downstream fining of armoring rivers is that S/d_{50} is constant.

11.2.4 Headcuts and Nickpoints

Headcuts and nickpoints can develop as the stream slope changes rapidly in the downstream direction (Figure 11.10). Headcuts are sudden drops in bed elevation, while nickpoints are sudden changes in bed slope. A headcut forms when the time taken to scour a vertical face from upstream exceeds the time taken to scour the same face from downstream. Stein and Julien (1993) differentiated headcuts and nickpoints based on the Froude number Fr. From their derivation, the critical ratio of drop height to flow depth D_{hc} is:

$$(D/h)_c \simeq 1.5 \left(\frac{0.4 + \mathrm{Fr}^2}{\mathrm{Fr}} \right)^2 \qquad (11.12)$$

As shown in Figure 11.10b, headcuts form above and nickpoints below the line.

Headcuts propagate in the upstream direction. They are typically observed in river tributaries after lowering the base elevation of the main river. As shown in Figure 11.11, this can cause structural damage to hydraulic structures and associated procedures along the tributaries such as bridges, drop structures, water intakes, and pumping levels. The upstream migration of headcuts and nickpoints is a characteristic feature of incised channels. The ensuing gullying in a tributary can result in significant bank instabilities and channel widening.

As sketched in Figure 11.12a, the migration rate of headcuts v_h is given by

$$v_h = \frac{1}{D} \left(\frac{q_{sd} - q_{su}}{1 - p_0} \right) \qquad (11.13a)$$

as a function of the volumetric unit sediment transport rates upstream q_{su} and downstream q_{sd} of the headcut of height D and porosity p_0.

In the case of a nickpoint, the relative degradation rate $\Delta z/D$ varies with the downstream distance x and time t, as shown in Figure 11.12b. The constant $k = q_{sv}/S = C_v q/S$ is a function of the volumetric unit sediment discharge q_{sv}, volumetric sediment concentration C_v, unit water discharge q,

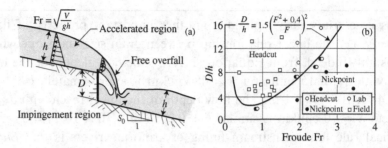

Figure 11.10. Hydraulic characteristics of headcuts and nickpoints: (a) longitudinal profile and (b) classification diagram (after Stein and Julien, 1993)

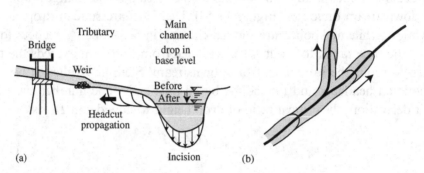

Figure 11.11. Headcut propagation: (a) from a downstream level change, and (b) migration up tributaries

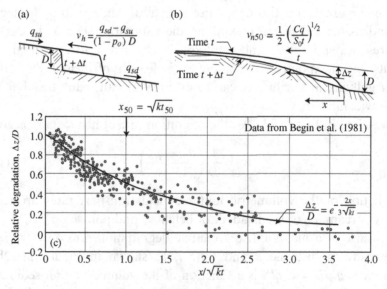

Figure 11.12. Propagation velocities of headcuts and nickpoints: (a) sketch for headcut, (b) nickpoint, and (c) propagation diagram

and the channel slope S. The measurements from Begin et al. (1981) demonstrated that the relative degradation rate varies with $x/\sqrt{kt}$ where $\Delta z/D \simeq e^{\frac{-2x}{3\sqrt{kt}}}$. Accordingly, the nickpoint propagation velocity at the point where $\Delta z/D = 0.5$ is located at $x_{50} = \sqrt{kt_{50}}$ and v_{n50} is

$$v_{n50} = \frac{1}{2}\left(\frac{k}{t_{50}}\right)^{1/2} = 0.5\left(\frac{C_v q}{S t_{50}}\right)^{1/2}. \tag{11.13b}$$

The propagation velocity of a nickpoint and the bed slope both decrease with time.

Gravel-mining operations remove the coarse armor layer from bed streams at low flows. This material is typically used for construction. Gravel mining is often viewed favorably as a means to reduce flood risks through channel degradation. However, unexpected damage can be seen after large floods have caused severe channel degradation. Headcut development and upstream migration can also result in the failure of upstream structures such as weirs and bridges. A list of potential problems associated with gravel mining is not limited to, but includes: (1) failure of bank protection structures requiring toe-protection measures such as aprons; (2) undermining of weirs and other river structures requiring drop structures and cut-off walls; (3) excessive pier scour causing structural instability problems at bridge crossings; (4) water levels too low for the intake of water supply and irrigation canals calling for pumping stations; (5) saltwater intrusion in river estuaries, requiring estuary barrages; (6) falling groundwater levels, which can affect agricultural practices; and (7) disconnect between the river and the floodplain for aquatic species and wildlife. A sound engineering approach to gravel mining could be to prohibit in-stream gravel mining and allow gravel mining in predesignated floodplain areas. Case Study 11.3 illustrates a more complex situation combining effects of dams, gravel mining, and headcuts in a river system.

Case Study 11.3 Gravel Mining Impact at Dry Creek, CA, United States

This case study illustrates the impact of degradation below dams which is compounded with gravel mining activities and headcutting. Dry Creek is a major tributary to the Russian River just south of Healdsburg, CA (Figure CS11.3.1). The Russian River Basin drains approximately 235 km^2 and Dry Creek drains 34 km^2. The Dry Creek valley is agricultural, with citrus fruits as the major crop. The upland areas are rugged, consisting primarily of hilly and mountainous terrain. The low-flow channel with a sinuosity of approximately 1.20 is incised within a wider-flow straight channel with a sinuosity close to 1.05. The low-flow channel is locally braided and anabranched. Tree cover is generally less than 50 percent of the bank line, and cut banks are evident. The bank material is mostly noncohesive silt, sand, gravel, and cobbles. Warm Springs Dam was only planned at the time of the analysis (Simons and Julien, 1983).

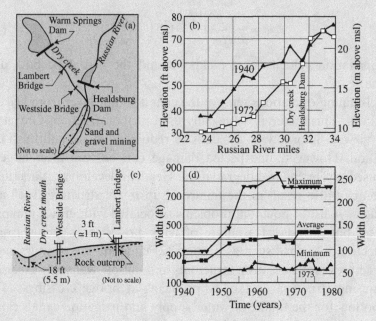

Figure CS11.3.1. Dry Creek: (a) location area, (b) longitudinal profile, (c) degradation impact, and (d) river width changes

The closing of Healdsburg Dam in 1952 resulted in a deepening of the Russian River. Gravel mining in the Russian River and other tributaries, such as Dry Creek, has been an important industry in Sonoma County since the early 1900s. Dry Creek degraded significantly in the 1950s and 1960s because of the drop in the base level of the Russian River. A total streambed elevation drop of 7 m has been recorded over a 0.6-km reach. This drop was the result of in-stream gravel mining downstream of Healdsburg Dam. The degradation of the Russian River induced lowering of the base of Dry Creek. The base-level drop of the Russian River initiated a headcut in Dry Creek that propagated 13 km upstream from the mouth in 23 years. With the deeper channel system and controlled flooding, higher banks became exposed to attack by the flowing water. Bank-erosion problems were detected more than 30 km upstream of the confluence with the Russian River. Abnormal flooding and fire sequences produced record runoff and sediment that has caused the deeply incised lower ends of Dry Creek to initiate channel widening. The analysis of aerial photographs indicates a significant increase in channel width.

11.3 River Aggradation

Channel aggradation refers to a gradual bed-elevation increase owing to sedimentation. Width–slope trade-offs are discussed in Section 11.3.1, river confluences and branches in Section 11.3.2, braided rivers in Section 11.3.3, and lateral shifts in Section 11.3.4.

11.3.1 Width–Slope Trade-Offs

The concept of a trade-off between channel width and slope in alluvial channels is introduced. Two conditions are examined: (1) equilibrium conditions for changes in width and slope for a given flow and sediment discharge: and (2) transient conditions describing a timescale for the changes in slope for a fixed channel width.

11.3.1.1 Equilibrium Width–Slope Trade-Offs

Figure 11.13 shows two long alluvial river reaches with channel widths W_1 and W_2.

Let's provide analytical support to this concept, considering a discharge $Q = \frac{1}{n} h^{5/3} S^{1/2} W$. The ratio of hydraulic parameters in successive reaches can be defined as $Q_r = Q_2/Q_1$, $h_r = h_2/h_1$, etc. To simultaneously satisfy the same discharge and Manning n in both reaches, $Q_r = n_r = 1$, we obtain

$$h_r = \left(\frac{1}{W_r^2 S_r} \right)^{3/10}. \tag{11.14a}$$

Likewise, for sediment transport with Equation (11.2a), i.e. $Q_{bv} \simeq 18 W \sqrt{g} d_s^{3/2} \tau_*^2$ with $\tau_* = hS/(G-1)d_s$, we can define the ratios that maintain the same water and sediment discharge in both reaches for a fixed sediment size and density, $Q_{sr} = d_{sr} = (G-1)_r = 1$,

$$h_r = \left(\frac{1}{W_r S_r^2} \right)^{1/2}. \tag{11.14b}$$

It becomes interesting to eliminate h_r from Equation (11.14a and b) to obtain the following important relationship between the width ratio W_r and the slope ratio S_r as

$$S_r = W_r^{1/7}. \tag{11.15a}$$

This defines the important width–slope trade-off, which implies that increasing the river width will require a steeper slope in order to pass the same sediment load through the wider reach. It is also possible to solve

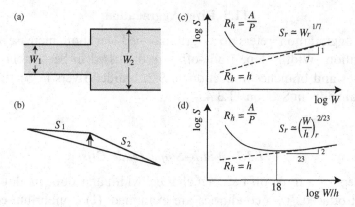

Figure 11.13. River equilibrium trade-offs: (a) width changes, (b) slope adjustments, (c) slope vs width, and (d) slope vs width/depth

Equation (11.14a and b) as $S_r = h_r^{-2/9}$ to rewrite the equilibrium slope ratio as a function of the width–depth ratio $\xi_r = W_r/h_r$ as

$$S_r = \left(\frac{W_r}{h_r}\right)^{2/23} = \xi_r^{2/23}. \tag{11.15b}$$

These two approximations sketched in Figure 11.13 are valid when the width–depth ratio is large enough that it can be assumed that the hydraulic radius approximately equals the flow depth. Leon et al. (2009) expanded this analysis to include the effects of hydraulic radius in rectangular channels. For instance, after the width–depth ratio is defined as $\xi_r = W_r/h_r$, the hydraulic radius becomes $R_h = h_r[\xi_r/(2 + \xi_r)]$, and we obtain

$$S_r = \frac{(2+\xi_r)^{20/23}}{\xi_r^{\,18/23}}, \tag{11.16}$$

which reduces to Equation (11.15b) when ξ_r is large.

11.3.1.2 Transient Width–Slope Trade-Offs

Since we found that wider channel reaches require steeper slopes to transport the same sediment load, we can expect different sediment transport rates if the slope remains unchanged. How long would it take for a river to adjust its slope? The analysis of sediment transport in long river reaches with different channel widths is considered. Solving the Manning equation as a function of the width–depth ratio $\xi = W/h$ gives the flow depth and channel width as follows:

$$h = \frac{n^{3/8} Q^{3/8} (2+\xi)^{1/4}}{S^{3/16} \xi^{5/8}}, \tag{11.17a}$$

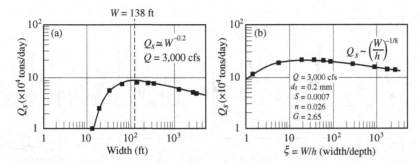

Figure 11.14. Sediment transport capacity as a function of (a) channel width and (b) width–depth ratio for the Rio Grande (after Park, 2013)

$$W = \frac{n^{3/8}Q^{3/8}\xi^{3/8}(2+\xi)^{1/4}}{S^{3/16}}. \tag{11.17b}$$

For sediment transport in rectangular channels, we obtain with $Q_{bv} \simeq 18W\sqrt{g}d_s^{3/2}\tau_*^2$ and $\tau_* = R_hS/(G-1)d_s$, given $R_h = h[\xi/(2+\xi)]$, and $\xi = W/h$:

$$Q_s = \frac{18\sqrt{g}S^2Wh^2\xi^2}{\sqrt{d_s}(G-1)^2(2+\xi)^2} = \left[\frac{18\sqrt{g}S^{25/16}n^{9/8}Q^{9/8}}{\sqrt{d_s}(G-1)^2}\right]\frac{\xi^{9/8}}{(2+\xi)^{5/4}}. \tag{11.18}$$

This approach is illustrated in Figure 11.14 with an example from the Rio Grande. The maximum sediment transport rate $Q_{s\,max}$ from Equation (11.18) is found when $\xi_{max} = 18$. Different values of this maximum can be found with different sediment-transport equations, but a minimum slope and maximum sediment transport rate are usually found for fairly narrow channels (e.g. Leon, 2003; Park, 2013). It is interesting to learn that when the term in brackets of Equation (11.18) is constant, the sediment transport capacity in wide channels, i.e. $\xi > 50$, decreases with the width–depth ratio to the power $-1/8$ or width to the power $-1/5$, i.e. $Q_s \sim \xi^{-1/8} \sim W^{-1/5}$.

As an example, a 35-km reach of the Rio Grande in the Bosque del Apache National Wildlife Refuge has been investigated over a long period of time. Wide river reaches with widths from 50 to 350 m have been maintained for a long period of time, as sketched in Figure 11.15a. The resulting channel slope in Figure 11.15b increased in the wider river reach. Leon et al. (2009) also examined the gradual changes in bed slope from a daily simulation over a 20-year period of time. In Figure 11.15d, it is found that the transient solution [$S = S_e + (S_0 - S_e)e^{-kt}$ from Equation 11.8b with slope instead of width] provides a very good approximation to the daily unsteady flow simulation (Leon, 2003). It is interesting to note that in this case, with daily

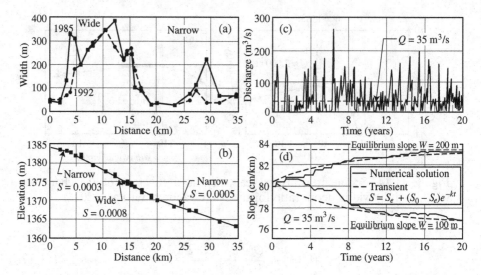

Figure 11.15. Width-slope trade-off of the Rio Grande: (a) river width, (b) longitudinal profile, (c) discharge vs time, and (d) channel slope vs time (after Leon et al., 2009)

changes in flow discharge from Figure 11.15c, half the slope change will take place in approximately 8–10 years.

11.3.2 River Confluences and Branches

The question of interest here is: what would happen when two channels, each carrying half the total discharge, merged into (or split from) a single channel? For channels with the same grain size and sediment concentration (i.e. constant d_s and $C_{mg/l}$), channel width, depth and slope can now be determined as a function of discharge from Equations (11.5a–d). With half the discharge, the width $W_{1/2} \simeq (0.5)^{0.47} W = 0.71 W$, the depth $h_{1/2} \simeq (0.5)^{0.34} h = 0.8 h$, and the slope $S_{1/2} \simeq (0.5)^{-0.08} S = 1.057 S$. Consequently, a single channel does efficiently transport the sediment load at a flatter bed slope. This process of river confluences and branching is sketched in Figure 11.16. River confluence points are stable and degradation is expected, while river branching points are less stable and induce channel aggradation. Case Study 11.4 provides a nice example of changes in sediment transport near river diversions.

River branches under equilibrium conditions convey water and sediment discharges in the downstream direction. From Equation (11.5a–e), with constant grain size and sediment concentration, the following changes are expected from river branching: (1) an increase in total river width; (2) a moderate decrease in flow depth and shear stress; (3) a significant increase in width–depth ratio; (4) a slight decrease in flow velocity; and (5) a slight increase in riverbed slope. The opposite applies to river confluences.

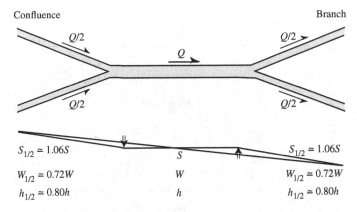

Figure 11.16. Sketch showing river confluences and branches

Case Study 11.4 Diversion of the Mississippi into the Atchafalaya River, Louisiana, United States

This example addresses the sediment diversion of the Mississippi River into the Atchafalaya River. Approximately 22 percent of the flow discharge of the Mississippi River is diverted into the Atchafalaya River. The sand concentration in the Mississippi River varies largely with discharge, as shown in Figure CS11.4.1a. Sand concentrations remain less than ~ 200 ppm at all discharges of less than 1,000,000 ft^3/s. Sand concentrations increase approximately linearly with discharge. The effect of a flow diversion should be to decrease the sand concentration downstream of the diversion.

An analysis of sediment transport by size fraction and a sediment budget are shown in Figure CS11.4.1b. The very fine sand and fine sand fractions are in reasonable equilibrium, and the inflowing sediment load equals the outflowing sediment load. For the medium sand and coarse sand fractions, however, the inflowing sand load of ~ 19 million yd^3/year (one cubic yard = 1 yd^3 = 0.76 m^3) far exceeds the outflowing of 2.25 million yd^3/year. We thus expect the sedimentation of ~ 17 million cubic yards of medium and coarse sand, 0.25 mm $< d_s <$ 1 mm, in the Mississippi River. In a river reach that is 262 miles long and assuming an average river width of $\sim 2,000$ ft, the accumulation of sediment represents an average accumulation of sand of 0.16 ft/year. At this rate, it would take 100 years to raise the bed by 16 ft. The tendency toward braiding is less likely to take place because the Mississippi River has a low width–depth ratio ($\sim 2,000$ ft wide and 50 ft deep).

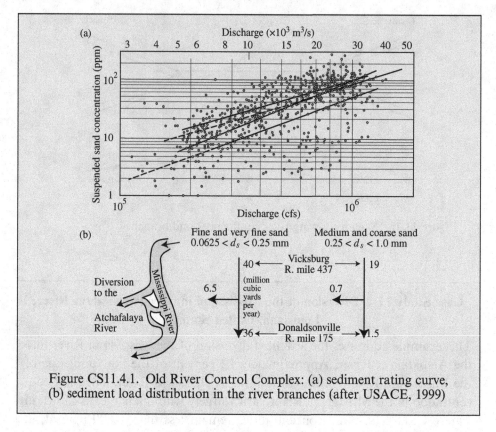

Figure CS11.4.1. Old River Control Complex: (a) sediment rating curve, (b) sediment load distribution in the river branches (after USACE, 1999)

Similarly, the case of alluvial islands can be examined by splitting the flow while considering the same grain size and sediment concentration. As sketched in Figure 11.17a, the converging point downstream of the island is a stable point, while the upstream diverging point becomes unstable. The example in Figure 11.17b and c nicely illustrates the stability of converging points and mobility of branching points of the Jamuna River.

It is interesting to note that the total width $2W_{1/2} \simeq 1.42W$ after splitting the flow is increased by 42 percent, while the slope increases by $\simeq 6$ percent. The trade-off between width and slope is defined as $S_r \simeq W_r^{1/7}$. This means that for a given grain size, narrower channels can carry the sediment load with flatter slopes. A large system of sand-bed river branches and confluences with the Ganges and Jamuna Rivers flowing into the Padma River is shown in Case Study 11.5.

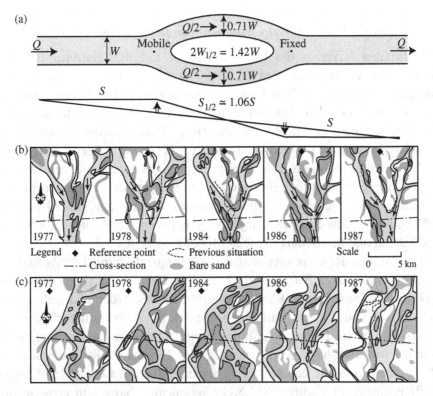

Figure 11.17. Examples of (a) flow around islands, (b) stable converging points, and (c) unstable branching point on the Jamuna River (after Klaassen et al., 1993)

11.3.3 Braided Rivers

Several geomorphic diagrams based on bankfull discharge and slope can be seen in Figure 11.18. A single channel with a dominant discharge was thought to meander on mild slopes and braid on steep slopes. This concept was developed by Lane (1957) who proposed a slope–discharge relation for sand-bed channels. Empirically, braided channels were observed when $SQ^{1/4} \simeq 0.01$ and channels were meandering when $SQ^{1/4} \simeq 0.0017$, given the slope S and the dominant discharge Q in cubic feet per second. Leopold and Wolman (1957) found a different relationship, and the variability in slope is significant. It remains difficult to separate braiding from meandering channels on the basis of slope. These $S - Q$ diagrams can only serve as a geomorphic index of river planform geometry.

Case Study 11.5 Alluvial Changes of the Jamuna River, Bangladesh

The Jamuna River is the lowest reach of the Brahmaputra River in Bangladesh. It drains an area of 550,000 km^2, and the mean annual discharge is 20,000 m^3/s. The main characteristics of the Ganges and the Padma Rivers are also summarized in Table CS11.5.1. It is interesting that the finer sediment size in the Padma River yields a milder channel slope. This large system of braided sand-bed river is shown in Figure CS11.5.1a, with a number of braids at low flows, typically varying between two and three. The total width of the braided channel pattern ranges from 5 to 17 km. At the confluence with the Ganges, the average annual flood is ~ 60,000 m^3/s, low-flow discharges vary between 4,000 and 12,000 m^3/s, and the maximum discharge reached 100,000 m^3/s in 1988.

The Jamuna River is very active, with frequent channel shifts and lateral migration rates E frequently exceeding 500 m/year (Klaassen et al., 1993). The annual lateral migration rate of first-order channels ranges between 75 and 150 m. Bank-erosion rates of second-order channels of 250 to 300 m are common. Annual lateral migration rates exceeding the channel width W have been measured and E/W generally decreases with R/W, where R is the channel radius of curvature. An example of lateral shifts is shown in Figure CS11.5.1b. Significant changes in cross-section geometry can take place within a few years.

The width, depth, flow velocity, and coarse sediment transport of the Jamuna River are shown in Figure CS11.5.1c. The exponents of at-a-station hydraulic geometry relationships are similar to the exponents describing downstream hydraulic geometry. This is because this braided river transports so much sediment that it can deform its channel quickly enough to maintain downstream hydraulic geometry relationships at both low and high flow regimes. The figures also display a significant variability in the measurements, which is typical of measurements in braided alluvial rivers. As much as there is debate regarding the exact value of dominant discharge in braided streams, the reader will notice a 200 percent variability in width and depth at a discharge of 50,000 m^3/s.

Table CS11.5.1. *Characteristics of the Jamuna, Ganges, and Padma Rivers*

River	Drainage area (km^2)	Mean annual discharge (m^3/s)	Bankfull discharge (m^3/s)	Slope (cm/ km)	d_s (mm)	Width (m)	Depth (m)	Velocity (m/s)
Jamuna	550,000	20,000	48,000	7.5	0.20	4,200	6.6	1.70
Ganges	1,000,000	11,000	43,000	5	0.14	3,700	6.5	1.78
Padma	1,550,000	28,000	75,000	4.5	0.10	5,200	7.5	1.93

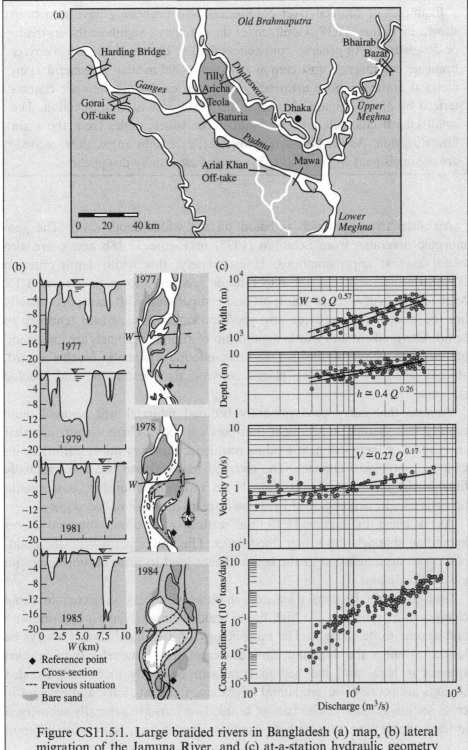

Figure CS11.5.1. Large braided rivers in Bangladesh (a) map, (b) lateral migration of the Jamuna River, and (c) at-a-station hydraulic geometry of the Jamuna River (after Klaassen et al., 1993)

Examples of braided river confluences and branching have also been shown in Figure 11.17. Confluences do not move significantly upstream or downstream. In general, confluences tend to be streamlined, while river branches can migrate upstream at a rate of ~900 m/year. In general, symmetrical branches move upstream. Asymmetrical bifurcations are characterized by one dominant channel in the main downstream direction. The small lateral channels tend to bifurcate at wider angles from the main flow direction. As the bifurcation approaches a right angle, these secondary channels tend to get smaller in size and eventually disappear.

An alternative approach is based on the width–depth ratio. The geomorphic diagrams from Schumm (1977) in Figure 11.18b and c are also useful as first approximations. Unfortunately, this width–depth criterion is not always valid, e.g. the Rio Grande reaches, shown in Figure 11.15. Measurements of bedload transport and comparisons with transport capacity are too uncertain to provide any good indication of a stream tendency to braid. The gradient and width–depth ratio of straight channels are intermediate between those of braided and meandering channels. In this regard, straight channels should be placed between the meandering and braided streams.

Streams that carry predominantly bedload material will respond quite rapidly to a change in sediment-transport capacity. There are streams that carry most of their sediment load as washload; this may not result in significant morphological changes. The riverbed material size becomes gradually finer in the downstream direction. From Lane's relationship, downstream fining is usually accompanied by a downstream decrease in bed slope.

As sketched in Figure 11.19, the settling of bedload material forces aggrading channels onto their floodplains. The flow spreads onto the floodplain with accumulation of the bed-sediment load to form natural levees on a wide floodplain.

Braided rivers become wide and shallow with bars subjected to rapid changes in morphology. At high flows, braided streams have a low sinuosity and appear to be straight. The morphology of a channel should always be looked at under the dominant flow discharge. The flow velocity of braided streams is high, and the bed material can be easily mobilized. Braided streams are prone to severe lateral migration, frequent shifts, and changes in cross-section geometry. The bars of braided streams are generally submerged at least once a year, thus without significant vegetation. Islands are different

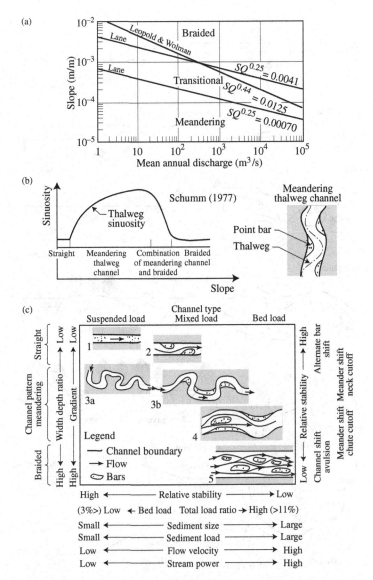

Figure 11.18. Geomorphic diagrams by (a) Lane (1957) and Leopold and Wolman (1957), (b) sinuosity diagram by Schumm (1977), and (c) relative stability diagram after Schumm (1977)

from bars in that they are stabilized by vegetation and rivers with multiple islands are anastomosed. Anastomosed rivers are usually more stable than braided channels because vegetation straightens the banks and stable islands control the flow between the branches. A discussion of anabranches and valley confinement for the Platte River is presented by Fotherby (2009). During

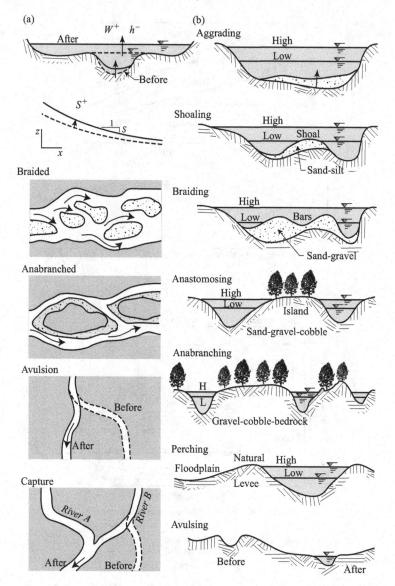

Figure 11.19. Geomorphic changes of aggrading channels: (a) main features, and (b) processes

floods, vegetated islands trap sediment and aggrade. Through aggradation and natural levee formation, a river raises its own bed elevation above the surrounding floodplain and can form a perched river. Perched rivers are stable as long as they cannot breach their levees. Perched rivers are prone to avulsion, in which rivers select a new flow path that can be located up to hundreds of kilometers away from their original location.

11.3.4 Lateral Shifts in Rivers

Lateral shifts in rivers can be associated with sediment plugs, alluvial fans, and tectonic activities. The formation of sediment plugs can be observed during the process of neck cutoffs and channel avulsions. Sediment plugs also occur when rivers transporting large sediment loads deposit large sediment volumes over short distances. Another important factor in the formation of sediment plugs is the Rouse number, which relates the settling velocity of transported material to the shear velocity. When $u*/\omega$ is less than 2.5, i.e. when the Rouse number is greater than 1, backwater can cause settling of large quantities of sediment over short distances. From the Exner equation, a timescale for the aggradation of the entire channel depth $\Delta z = h$ can be defined as

$$\frac{\Delta z}{\Delta t} = \frac{-1}{(1-p_0)} \frac{\Delta q_s}{\Delta x} => \frac{h}{t_{63}} \simeq \frac{0.63q_s}{(1-p_0)X_{63}}. \tag{11.19a}$$

For instance, the trap–efficiency relationship Equation (2.52), $T_E = 1 - e^{-\omega X/q}$, shows that in backwater areas, 63 percent of the transported material settles within a distance $X_{63} = q/\omega$. Accordingly, the timescale is obtained as

$$t_{63} \simeq \frac{(1-p_0)hq}{0.63\omega q_s} \approx \frac{h}{\omega C_v} \tag{11.19b}$$

For instance, in a river with fine sand, i.e. $\omega \simeq 0.01$ m/s, and a depth-averaged bed-material concentration $C_v \simeq 0.0005 \simeq 1,325$ mg/l, the backwater effects would fill a flow depth of 3 m within 1 week. Sediment plugs can form when Ro is high and clear water flows overbank while the sediment piles up on the channel bed.

Alluvial fans are found where steep mountain channels reach valley floors, as sketched in Figure 11.20. The sudden break in bed slope causes the bed material transported by the river to deposit. The accumulation of debris usually takes a conical shape. The volume of material in the alluvial fan is indicative of the sediment-transport capacity of the stream through geologic times. The aggradation takes place on the riverbed and on natural levees between the apex of the alluvial fan and the valley floor. Rivers on alluvial fans are typically braided. For example, the Kosi River in India, as shown in Figure 11.20b, carries about 190 million metric tons of sediment per year and has shifted laterally by 112 km since 1731. During floods, the river width can reach 16–20 km and the slope decreases from 95 cm/km near the apex to about 5 cm/km near the confluence with the Ganges River.

Tectonic activities often impact rivers via aggradation and degradation processes. As sketched in Figure 11.21, sudden uplift and subsidence may

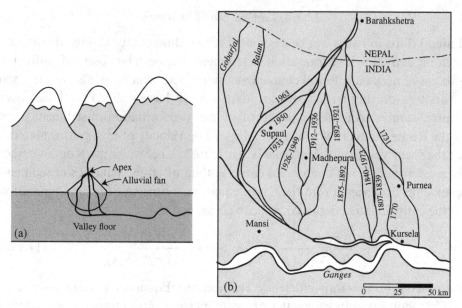

Figure 11.20. Alluvial fan (a) sketch and (b) example of the Kosi River (after Gole and Chitale, 1966)

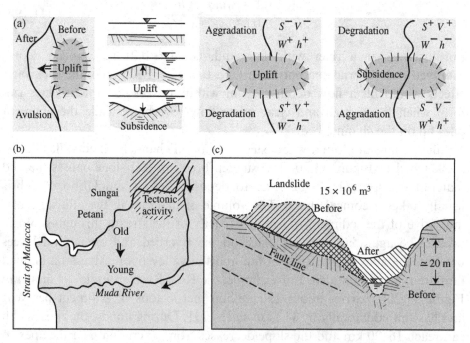

Figure 11.21. Tectonic effects on alluvial rivers (a) sketches; (b) lateral shifting of the Muda River in Malaysia; and (c) landslide along a fault line near Rio Grijalva in Mexico, 2007

result in avulsions of alluvial rivers. The uplift of an alluvial river should cause aggradation upstream and degradation downstream. Opposite effects are expected for subsidence. Subsidence from excessive groundwater pumping, and oil and gas extraction, can also affect river systems. Subsidence in very flat irrigation canals can reverse the flow direction and may call for the installation of pumping stations. On Figure 11.21b, Sungai Petani in Malaysia shows a tectonic lateral shift of the Muda River away from its natural estuary. Also, Figure 11.21c is a sketch of the complete blockage of Rio Grijalva in Mexico from a massive landslide along fault lines.

♦♦*Exercise 11.1*

Demonstrate that $SQ^{\frac{1}{2+3m}} \cong 12.4 \, d_{50}^{\frac{5}{4+6m}} \, \tau_*^{\frac{5+6m}{4+6m}}$ from Equation (10.19d). Derive Equation (11.2b) from Equation (10.21b and d) and Equation (11.2a). Show how Equation (11.4) resembles Lane's relationship.

Exercise 11.2

Derive hydraulic-geometry relations (11.5a–e) from Equation (11.3a–e) and the definition of concentration $Q_{bv} = 3.8 \times 10^{-7} C_{mg/l} Q$.

Exercise 11.3

Examine the combined effects of a 50 percent decrease in discharge Q^- and a 200 percent increase in sediment discharge Q_{bv}^+ on channel width, depth, velocity, slope, and Shields parameter.
[*Answers:* The new flow depth is 67 percent of the initial flow depth $h \simeq 0.67 h_0$, $W \simeq 0.58 W_0$, $V \simeq 1.27 V_0$, $S \simeq 2.75 S_0$, and $\tau_* \simeq 1.84 \tau_{*0}$.]

♦*Exercise 11.4*

A six-foot-high headcut is observed in a stream where the Froude number is near critical. At what flow depth would the headcut turn into a nickpoint?

♦*Exercise 11.5*

From the channel-slope data changes with time of the Rio Grande in Figure 11.15d, estimate the value of the coefficient k. Also estimate the time required for half the slope change and 75 percent of the slope change to take place.

◆*Exercise 11.6*

If the width–depth ratio of an alluvial river increases by two orders of magnitude, what is the expected change in channel slope for this river?

Exercise 11.7

With reference to the Jamuna River in Case Study 11.5, determine the planform geometry from the planform predictors based on discharge and slope.

◆*Exercise 11.8*

Consider a constant term in brackets in Equation (11.18), and determine the value of width–depth ratio that gives the maximum sediment-transport rate.

[*Answer: Q_s is maximum when $\xi_{max} = 18$.*]

Exercise 11.9

Estimate the average discharge and sediment concentration for the Bernardo Reach of the Rio Grande before and after 1975 from the data in Case Study 11.1 and Problem 11.10.

◆*Exercises 11.10*

Demonstrate Equation (11.17a and b) from Manning's equation in a rectangular channel.

◆◆*Problem 11.1*

Solve Equation (11.2a) for S and substitute into Equation (10.22a–d) to obtain Equation (11.3a–e).

◆*Problem 11.2*

Estimate the hydraulic geometry of an alluvial stream at a bankfull discharge of 4,500 ft^3/s with $d_{50} = 0.5$ mm and a bed-material concentration of 150 ppm.

[*Answer:* With $Q = 127$ m^3/s, $d_s = 0.0005$ m, and $C_{mg/l} = 150$, we obtain from Equation (11.5a–e), $h \simeq 1.2$ m, $W \simeq 170$ m, $V = 0.63$ m/s, $S \simeq 2.2 \times 10^{-4}$, and $\tau_* \simeq 0.3$.]

Problem 11.3

Consider the data in Figure 11.6 and assume that the sediment size remains unchanged. Estimate the relative changes in hydraulic geometry of the river below the dam.

◆◆Problem 11.4

The Bernardo reach of the Rio Grande in New Mexico features a channel width ranging from 150 to 1,200 ft with an average of ~ 500 ft. The average reach slope is 80 cm/km, with a median grain size of 0.3 mm, and a sand load up to 10,000 tons per day at discharges of $\sim$5,000 ft^3/s. Compare with the range of channel width and slope calculated at a sediment concentration of sand varying from 500 to 2,000 mg/l.
[*Answer*: From Equation (11.5) and $C = 500$ mg/l, $W = 163$ m and $S = 4.6 \times 10^{-4}$.]

◆Problem 11.5

With reference to the Jamuna, Ganges, and Padma Rivers' confluences in Case Study 11.5, apply the hydraulic-geometry relationships and compare with field measurements of W, h, V, and S. Compare the slopes upstream and downstream of the confluence. Would equilibrium require an increase or decrease in slope downstream of a confluence?

◆◆Problem 11.6

The Jamuna River is a large braided river with a median grain size of 0.2 mm. The river conveys $\sim 48,000$ m^3/s at bankfull flow and the corresponding average bed-material discharge is approximately 2.6 Mtons per day. Estimate the downstream hydraulic geometry of the river.
[*Answer:* Calculate by using Equation (11.3a–e) with $Q = 48,000$ m^3/s, $d_s = 0.0002$ m, and $Q_{bv} = 11.6$ m^3/s to give $h \simeq 6.7$ m, $W \simeq 2,500$ m, $V \simeq 2.8$ m/s, $S \simeq 3.2 \times 10^{-4}$, and $\tau_* \simeq 6$. Field measurements in Table CS11.5.1 indicate $h \simeq 6.6$ m, $W \simeq 4,200$ m, $V \simeq 1.7$ m/s, $S \simeq 7.5 \times 10^{-5}$, which gives $\tau_* \simeq 15$. The calculated equilibrium slope exceeds the measured slope, and the stream is suspected to be aggrading and braiding.]

◆*Problem 11.7*

Examine the possibility of fitting the exponential model for hydraulic geometry [Equation (11.9b)] to the specific gauge record in Figure 11.8. Determine the value of k and the period of time corresponding to 50 and 90 percent of the geomorphic adjustment.

Problem 11.8

From the data in Case Study 11.3, estimate the slope of the Russian River and estimate the grain size at incipient motion when the flow depth reaches 3 m.

◆*Problem 11.9*

From the information in Case Study 11.2, estimate the minimum grain size at beginning of motion during a flood ($Q = 100,000$ ft^3/s) on the Meuse River.

[*Answer:* The flow depth is ~ 25 ft during floods, or $h \simeq 7.6$ m, and the slope is ~ 4 ft/mile, or $S \simeq 7.6 \times 10^{-4}$. We obtain $d_{sc} \simeq 10hS = 0.058$ m or 60 mm. All sand and gravel sizes are in motion during floods. Only cobbles could armor the riverbed if available in sufficiently large quantity.]

◆◆*Problem 11.10*

Consider the mass curves for the flow and sediment discharge of the Rio Grande. Estimate the relative changes in discharge and sediment load at Bernardo before and after 1978. Infer the relative changes in downstream hydraulic geometry of the river. How much do you expect the width, depth, velocity, and slope to increase or decrease? See Figure P11.10.

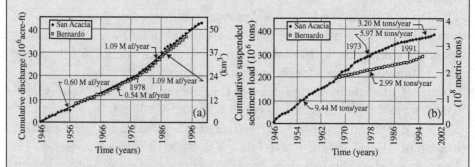

Figure P11.10. Rio Grande mass curves: (a) water, and (b) sediment (after Richard et al., 2000)

♦♦ *Problem 11.11*

Consider the Tanana River in Alaska with the data in the Figure P11.11 from Buska et al. (1984). Notice the sudden decrease in bed slope with downstream fining and braiding. Determine the approximate flow depth at which the bed material will be at incipient motion. Note that the bed slope is proportional to the grain size. Compare with the results of those of the degrading Meuse River in Case Study 11.2.

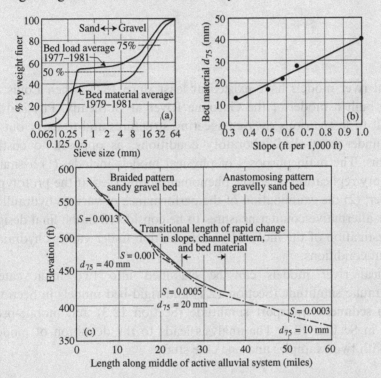

Figure P11.11. Morphology of the Tanana River: (a) bimodal particle distribution, (b) grain size vs slope, and (c) longitudinal profile (after Buska et al., 1984)

12

Physical River Models

Physical river models have existed at least since 1875, when Louis Jerome Fargue built a model of the Garonne River at Bordeaux. Physical models are built to test various river-engineering structures and to carry out experiments under controlled laboratory conditions, as opposed to costly field programs. The main purposes of physical models include: (1) a small-scale laboratory replication of a flow phenomenon observed at the prototype scale in a river; (2) the examination of the performance of various hydraulic structures or alternative countermeasures to be considered in the final design; and (3) investigation of the mobile-bed deformation under various hydraulic and sediment conditions.

Physical river models can be classified into two main categories: (1) hydraulic similitude (Section 12.1) and rigid-bed models in Section 12.2; and (2) sediment transport similitude (Section 12.3) and mobile-bed river models in Section 12.4. The analysis leads to the definition of model-scale ratios with two examples and one case study.

12.1 Hydraulic Similitude

As sketched in Figure 12.1, the prototype parameters with the subscript p describe field-scale conditions for which the hydraulic model with subscript m is built in the laboratory. Model scales with subscript r refer to the ratio of model to prototype parameters. For instance, the prototype flow depth h_p and the model flow depth h_m define the flow depth ratio $h_r = h_p/h_m$. The inverse ratio can be used as long as the definition is consistent.

For all scale models, the following considerations are relevant: (1) the scale ratios should be reasonable to ensure accuracy of measurements, e.g. a measurement error of 1 mm in a model at a scale of 1:100, or $z_r = 100$, represents an error of 10 cm in the prototype; (2) we must consider the

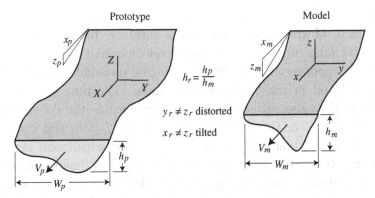

Figure 12.1. Physical river prototype and model

physical limitations in space, water discharge, and accuracy of instrumentation, e.g. the Mississippi River cannot be realistically modeled in a 100-m-long hydraulics laboratory; and (3) we must appropriately simulate the boundary conditions, e.g. the stage and discharge of the inflow and outflow.

Hydraulic models use water and the scales of mass density ρ_r and kinematic viscosity v_r are unity. Under constant gravitational acceleration, $g_r = 1$, the scales for specific weight γ_r and dynamic viscosity μ_r are also unity. Hydraulic models simply require $\rho_r = g_r = v_r = \gamma_r = \mu_r = 1$.

Geometric similitude describes the relative size of two Cartesian systems of coordinates (x, y, z). The vertical z_r-length scale is defined as the ratio of the prototype vertical length z_p to the model vertical length z_m such that $z_r = z_p/z_m$. For instance, a vertical length scale $z_r = 100$ indicates that a model flow depth of 20 cm corresponds to a prototype flow depth of 20 m. The horizontal length scales are defined in a corresponding manner in the downstream x and lateral y directions as x_r and y_r. It is often desirable to exaggerate the vertical scale of a model to avoid model flow depths that are excessively shallow and prevent accurate measurements. Model *distortion* implies that the vertical z_r and the lateral y_r scales are not identical. The distortion factor is obtained from the ratio of y_r/z_r and will affect the representation of cross-sectional areas. Model *tilting* results from different vertical z_r and downstream horizontal x_r scales. The downstream model slope $S_r = z_r/x_r$ is effectively tilted when the horizontal length scale is different from the vertical scale. The surface-area scales for planform and cross-sectional surfaces are, respectively, $A_r = x_r y_r$ and $A_{xr} = y_r z_r$. The volume scale for a distorted model corresponds to $Vol_r = x_r y_r z_r$. In the particular case where the horizontal and vertical length scales are identical, i.e. $L_r = x_r = y_r = z_r$, the corresponding area and volume scales are, respectively, $A_r = L_r^2$ and $Vol_r = L_r^3$.

Kinematic similitude refers to parameters involving time, e.g. velocity V, acceleration a, kinematic viscosity υ. For instance, the velocity scale V_r is defined as the ratio of prototype to model velocities as $V_r = V_p/V_m$. The time-scale $t_r = t_p/t_m$ appropriately describes kinematic similitude when fluid motion in the model and in the prototype are similar. The time scale is typically defined from the ratio of the downstream distance to the downstream flow velocity $t_r = x_r/V_r$. The two kinematic relationships $L = Vt$ and $V = at$ can be combined to cancel time and obtain $V_2 = aL$. Any experiment in which the gravitational acceleration is the same in the model and the prototype requires $a_r = g_r = 1$. With gravity acting in the vertical direction, this relationship yields the most important kinematic relationships for physical river models:

$$\mathrm{Fr}_r = \frac{V_r}{z_r^{0.5}} = 1. \tag{12.1} \blacklozenge\blacklozenge$$

This is known as the Froude similitude criterion. Since $V_r = g_r t_r$, the time and velocity scales for Froude similitude models are identical, $t_r = V_r = z_r^{0.5}$, when the model is not tilted, $z_r = x_r$. It is important to consider that distorted and tilted models could produce different timescales for flow velocities in the y and the z directions. Therefore, distorted and tilted models cannot appropriately account for 2-D and 3-D convective and turbulent accelerations. They should not be used to model vorticity, diffusion, and turbulent mixing.

Dynamic similitude implies a similarity in the dynamic behavior of fluids. Dynamic similitude refers to parameters involving mass, e.g. mass density ρ, specific weight γ, and dynamic viscosity μ. The basic concept of dynamic similitude is that individual forces acting on corresponding fluid elements must have the same force polygon in both systems. An individual force acting on a fluid element may be either a body force, such as weight in a gravitational field, or a surface force resulting from pressure gradients, viscous shear, or surface tension.

Gravitational and viscous effects are, respectively, described by the Froude number $\mathrm{Fr}_r = V_r/(g_r z_r)^{0.5}$ and the Reynolds number $\mathrm{Re}_r = V_r z_r/\upsilon_r$. In hydraulic models, the Froude and Reynolds similitude, $\mathrm{Fr}_r = \mathrm{Re}_r = 1$, can only be simultaneously satisfied when $V_r^2/z_r = V_r z_r = 1$, which is the trivial full scale $V_r = z_r = 1$. We thus conclude that exact similitude of all force ratios in hydraulic models is impossible except at the prototype scale. Hydraulic models can only simulate the dominant hydrodynamic forces in the system. In river models, gravitational effects are typically dominant and resistance to flow does not depend on viscosity as long as flows are hydraulically rough. In most river models, however, the balance of gravitational and inertial forces implies similitude according to the Froude criterion. The Froude similitude satisfies similarity in the ratio of inertial to weight forces, and it properly scales the ratio of velocity head to flow depth. Consequently the Froude similitude denotes similarity conditions in specific-energy diagrams for the model and the prototype. It also properly describes rapidly varied flow conditions. It is nevertheless required to check that the Reynolds number of a river model remains high enough for the model flow to be turbulent.

12.2 Rigid-Bed River Models

Rigid-bed models are built to simulate flow around river improvement works and hydraulic structures. A rigid boundary implies that the bed is fixed and without sediment transport. This is the case when the Shields parameter of the bed material is $\tau_* < 0.03$. Rigid-bed model scales can be determined in either one of two cases: (1) a Froude similitude model, in which resistance to flow can be neglected and (2) friction similitude, or similitude in resistance to flow, leading to distorted/tilted models. Froude similitude models (Section 12.2.1) are well suited to the analysis of 3-D flow around hydraulic structures without sediment transport. In long river reaches, resistance to flow cannot be neglected and both Froude and friction similitude can be simultaneously satisfied with tilted/distorted models (Section 12.2.2).

12.2.1 Froude Similitude $\mathrm{Fr}_r = 1$

The model scales for hydraulic models with exact geometric similitude can be determined from the Froude similitude criterion. The scale ratios for hydraulic models with $L_r = x_r = y_r = z_r$ reduce to $V_r = t_r = L_r^{0.5}$ and $M_r = L_r^3$. All other scale ratios can be derived from the length, time, and mass scales by use of the fundamental dimensions of any considered variable. Figure 12.2 shows photos of rigid boundary models at the CSU Hydraulics Laboratory. The models from C. Thornton illustrate the water intake to the penstock of a powerhouse, and river bends of the Middle Rio Grande.

The exact Froude similitude imposes constraints that can be difficult to work with. For instance, scaling the size of roughness elements according to the length scale will maintain the same resistance parameter as long as $\mathrm{Re}_* > 70$ for both the model and the prototype. In practice, this is possible only for very coarse bed channels, such as cobble- and boulder-bed streams. Applying the Froude similitude criterion without distortion and tilting changes the near-bed conditions, because the laminar sublayer thickness δ in hydraulic models is $\delta_r = u_{*_r}^{-1} = L_r^{-1/2}$. Accordingly, the strict geometrical similitude produces a disproportionally large laminar sublayer thickness in the model. For hydraulically smooth surfaces, resistance to flow varies inversely with the Reynolds number, and resistance to flow will be larger in the

Figure 12.2. Photos of rigid boundary physical models: (a) hydropower intake and (b) river bends of the Rio Grande (photos of the CSU Hydraulics Laboratory from C. Thornton)

model than in the prototype. The scale ratios for commonly used variables are listed in Column three of Table 12.1.

12.2.2 Friction Similitude $S_r = f_r \mathrm{Fr}_r^2$

The similitude in resistance to flow is considered based on the Darcy–Weisbach friction factor f. The governing equation to be preserved in gradually varied flow models with rigid boundaries is the resistance relationship $S_r = f_r \mathrm{Fr}_r^2$. Resistance to flow has been extensively discussed in Section 5.1.3 and can be defined as $\sqrt{8/f} = a(h/d_s)^m$ where $m = 1/2.3 \log(2h/d_{50})$ from Equations (5.8 and 5.9). In general, $S_r = z_r/x_r$ and $f_r = (d_{sr}/z_r)^{2m}$, and the governing equation of similitude for resistance to flow becomes

$$\mathrm{Fr}_r^2 = \frac{z_r}{x_r}\left(\frac{z_r}{d_{sr}}\right)^{2m}. \tag{12.2}$$

The Manning–Strickler similitude is particularly satisfied from Equation (12.2) when $m = 1/6$.

To simultaneously satisfy the Froude and Manning–Strickler similitude criteria, it is possible to solve Equation (12.2) for $\mathrm{Fr}_r = 1$ and obtain $d_{sr} = z_r^4/x_r^3$. For instance, using the same boundary roughness in the model and prototype, $d_{sr} = 1$, implies $z_r = x_r^{0.75}$ and requires tilting $z_r \neq x_r$ of the physical model.

Table 12.1. *Scale ratios for rigid-bed and mobile-bed river models*

		Rigid-bed (Froude)		Mobile-bed			
			Tilted	Complete $\mathrm{Fr}_r=1,\ S_r=f_r$ $d_{*r}=1,\ \tau_{*r}=1$		Incomplete similitude $\mathrm{Fr}_r\neq 1$ Same size $\leftrightarrow$ Different size	
	Scale	Froude $\mathrm{Fr}_r=1$ $x_r=z_r$	$\mathrm{Fr}_r=1$ $S_r=f_r$ $x_r\neq z_r$	m varies	$m=1/6$	$d_{sr}=1$	$d_{sr}\neq 1$
Geometric							
Length	x_r	L_r	x_r	x_r	x_r	x_r	x_r
Width	W_r	L_r	y_r	y_r	y_r	y_r	y_r
Depth	h_r	L_r	z_r	$x_r^{\left(\frac{1+m}{1+4m}\right)}$	$x_r^{0.7}$	$x_r^{0.5}$	$x_r^{0.5}d_{sr}^{-1}$
Particle diameter	d_{sr}	L_r	$z_r^4 x_r^{-3}$	$x_r^{\left(\frac{2m-1}{2+8m}\right)}$	$x_r^{-0.2}$	1	d_{sr}
X-section area	$W_r h_r$	L_r^2	$z_r y_r$	$y_r x_r^{\left(\frac{1+m}{1+4m}\right)}$	$y_r x_r^{0.7}$	$y_r x_r^{0.5}$	$y_r x_r^{0.5}d_{sr}^{-1}$
Volume	$x_r W_r h_r$	L_r^3	$x_r y_r z_r$	$y_r x_r^{\left(\frac{2+5m}{1+4m}\right)}$	$y_r x_r^{1.7}$	$y_r x_r^{1.5}$	$y_r x_r^{1.5}d_{sr}^{-1}$
Kinematic							
Time (flow)	t_r	$L_r^{1/2}$	$x_r z_r^{-1/2}$	$x_r^{\left(\frac{1+7m}{2+8m}\right)}$	$x_r^{0.65}$	$x_r^{1-0.5m}$	$x_r^{1-0.5m}d_{sr}^{1+2m}$
Time (bed)	t_{br}	–	–	$x_r^{\left(\frac{2+5m}{1+4m}\right)}$	$x_r^{1.7}$	$x_r^{1.5}$	$x_r^{1.5}d_{sr}^{-1}$
Velocity	V_r	$L_r^{1/2}$	$z_r^{1/2}$	$x_r^{\left(\frac{1+m}{2+8m}\right)}$	$x_r^{0.35}$	$x_r^{0.5m}$	$x_r^{0.5m}d_{sr}^{-1-2m}$
Shear velocity	u_{*r}	$L_r^{1/2}$	$z_r x_r^{-1/2}$	$x_r^{\left(\frac{1-2m}{2+8m}\right)}$	$x_r^{0.2}$	1	d_{sr}^{-1}
Settling velocity	ω_r	–	–	$x_r^{\left(\frac{1-2m}{2+8m}\right)}$	$x_r^{0.2}$	1	d_{sr}^{-1}
Discharge	Q_r	$L_r^{5/2}$	$y_r z_r^{3/2}$	$y_r x_r^{\left(\frac{3+3m}{2+8m}\right)}$	$y_r x_r^{1.05}$	$y_r x_r^{0.5+0.5m}$	$y_r x_r^{0.5+0.5m}d_{sr}^{-2-2m}$
Sediment discharge	q_{br}	–	–	1	1	1	1
Dynamic							
Mass	M_r	L_r^3	$x_r y_r z_r$	$y_r x_r^{\left(\frac{2+5m}{1+4m}\right)}$	$y_r x_r^{1.7}$	$y_r x_r^{1.5}$	$y_r x_r^{1.5}d_{sr}^{-1}$
Shear stress	τ_r	L_r	$z_r^2 x_r^{-1}$	$x_r^{\left(\frac{1-2m}{1+4m}\right)}$	$x_r^{0.4}$	1	d_{sr}^{-2}
Dimensionless							
Slope	S_r	1	$z_r x_r^{-1}$	$x_r^{\left(\frac{-3m}{1+4m}\right)}$	$x_r^{-0.3}$	$x_r^{-0.5}$	$x_r^{-0.5}d_{sr}^{-1}$
Darcy–Weisbach	f_r	—	$z_r x_r^{-1}$	$x_r^{\left(\frac{-3m}{1+4m}\right)}$	$x_r^{-0.3}$	x_r^{-m}	$x_r^{-m}d_{sr}^{4m}$
Froude	Fr_r	1	1	1	1	$x_r^{0.5m-0.25}$	$x_r^{0.5m-0.25}d_{sr}^{-0.5-2m}$
Reynolds	Re_r	$L_r^{3/2}$	$z_r^{3/2}$	$x_r^{\left(\frac{3+3m}{2+8m}\right)}$	$x_r^{1.05}$	$x_r^{0.5+0.5m}$	$x_r^{0.5m+0.5}d_{sr}^{-2-2m}$
Shields	τ_{*r}	–	–	1	1	1	1
Grain Reynolds	Re_{*r}	–	–	1	1	1	1
Sediment diameter	d_{*r}	–	–	1	1	1	1
Sediment density	$(G-1)_r$	–	–	$x_r^{\left(\frac{3-6m}{2+8m}\right)}$	$x_r^{0.6}$	1	d_{sr}^{-3}

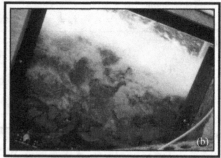

Figure 12.3. Very steep rigid boundary physical models: (a) CSU overtopping facility and (b) flow through riprap at CSU (models from J. Ruff and S. Abt, photos from J. Kuroiwa)

The user has two degrees of freedom and can arbitrarily select two of the three parameters d_{sr}, x_r, z_r. For instance, if the vertical scale z_r is 20 and the horizontal scale x_r is 100, the roughness scale d_{sr} should be 0.16, meaning that the boundary roughness of the model should be 6.25 larger than the prototype roughness. Figure 12.3 illustrates a very steep rigid-bed model for overtopping flow at the CSU Hydraulics Laboratory, from J. Ruff and S. Abt.

The modeler still has one degree of freedom in arbitrarily selecting y_r. For instance, when $x_r = y_r \neq z_r$, the plan view of the physical model will match the prototype but the cross-section geometry will be disproportionately deep. This tilted model will also be distorted. Model distortion is often encountered in practice. It is often beneficial to raise the model Reynolds number and maintain turbulent flow. The advantage of distorted models is also found in increasing flow depth and decreasing resistance to flow, as the modeler can then empirically increase the size of roughness elements until the model results compare with field measurements. Model roughness is typically increased with blocks, sticks, and/or blades until the stage–discharge relationship replicates the prototype conditions. The physical model is then said to be calibrated. Because the kinematic similarity is not exact, however, any attempt to determine kinematic properties such as streamlines and turbulent mixing cannot be properly scaled in distorted models. The scale ratios for a distorted rigid-bed model are presented in Column four of Table 12.1 and a calculation example is presented in Example 12.1.

Example 12.1 Froude and friction similitude: a tilted rigid-bed model

A 2,000-m-long gravel-bed river reach has a flow depth of 2 m, a width of 50 m, and a mean velocity of 0.3 m/s. If the prototype Manning coefficient $n = 0.025$ and $d_{50} = 5$ mm, find suitable scales for a physical model to fit within the 20-m space available in the laboratory. The prototype Froude number, slope, and Shields parameter are $\mathrm{Fr}_p = V_p/\sqrt{gh_p} = 0.3/\sqrt{9.81 \times 2} = 0.067$,

$$S_p \cong n^2 V^2/h^{4/3} = 0.025^2 \times 0.3^2/2^{4/3} = 2.23 \times 10^{-5},$$

$$\tau_{*p} = h_p S_p/(G-1)d_{s_p} = 2 \times 2.23 \times 10^{-5}/1.65 \times 0.005 = 0.0054 < 0.03.$$

A rigid-bed model is appropriate because the sediment is not moving.

An exact Froude similitude model would require a length scale $x_r = x_p/x_m = 2,000/20 = 100$ m and model depth of $h_m = h_p/h_r = 2/100 = 0.02$ m from $x_r = z_r = h_r$. The velocity ratio is $V_r = \sqrt{x_r} = 10$, giving a model velocity $V_m = V_p/V_r = 0.3/10 = 0.03$ m/s. The corresponding model Reynolds number is $\mathrm{Re}_m = V_m h_m/v = 0.03 \times 0.02/10^{-6} = 600$ with laminar flow in the model.

Model distortion is therefore necessary to increase the accuracy of flow-depth and velocity measurements and to guarantee turbulent flow in the model. The vertical scale is arbitrarily set as $h_r = z_r = 25$ to yield a model flow depth of $h_m = h_p/h_r = 2/25 = 0.08$ m. The model velocity is $V_m = V_p/V_r = V_p/\sqrt{z_r} = 0.3/\sqrt{25} = 0.06$ m/s. The model Reynolds number would then be $\mathrm{Re}_m = V_m h_m/v = 0.06 \times 0.08/10^{-6} = 4800$ and the model flow is turbulent.

Other model scales are calculated from Table 12.1, Column four for tilted rigid beds.

For example, planform geometry is similar when $y_r = x_r = 100$, and the timescale is $t_r = x_r z_r^{-1/2} = (100/\sqrt{25}) = 20$. The model discharge is $Q_m = Q_p/Q_r = V_p W_p h_p/y_r z_r^{3/2} = (0.3 \times 50 \times 2)/(100 \times 25^{3/2}) = 0.0024$ m^3/s. The tilted model slope is $S_m = S_p/S_r = S_p/z_r x_r^{-1} = 2.23 \times 10^{-5} \times 100/25 = 8.9 \times 10^{-5}$. The size of the roughness elements should be $d_{sm} = d_{sp}/d_{sr} = d_{sp} x_r^3/z_r^4 = 0.005 \times 100^3/25^4 = 13$ mm, which is coarser than the bed material of the prototype. In this type of model, roughness elements, pebbles, or small blocks could be placed on the smooth model surface until the stage–discharge relationship in the lab replicates the field measurements. This is a trial-and-error calibration procedure.

12.3 Sediment-Transport Similitude

This section first describes sediment transport similitude in terms of settling velocity (Section 12.3.1) and sediment transport (Section 12.3.2).

12.3.1 Settling Velocity Similitude $d_{*r} = 1$

The similitude criterion for sediment suspension is based on the settling velocity ω defined as $\omega = 8(v/d_s)[(1+0.0139d_*^3)^{0.5} - 1]$ from Equation (2.42). The term in brackets remains the same for the model and prototype when $d_{*r} = 1$, which yields $\omega_r d_{sr} = 1$.

The criterion for sediment suspension in a river requires identical values of the ratio of shear velocity to settling velocity, or $u_{*r}/\omega_r = 1$. By definition, the shear velocity ratio is $u_{*r} = (g_r z_r S_r)^{0.5} = z_r/x_r^{0.5}$ which yields the similitude relationship for suspended sediment transport as

$$\frac{u_{*r}}{\omega_r} = \frac{z_r d_{sr}}{x_r^{0.5}} = 1. \tag{12.3}$$

This leads directly to $d_{sr} = x_r^{1/2}/z_r$.

It is interesting to consider that similitude in dimensionless particle diameter $d_{*r} = 1$ does not necessarily impose the same sediment size in the model and prototype. In reality, the definition of $d_* = d_s[(G - 1)g/v^2]^{1/3}$ for hydraulic models imposes the following relationship between the particle diameter and density:

$$d_{sr}^3 = \frac{1}{(G-1)_r}. \tag{12.4}$$

Clearly, large sediment particles made of very light material can be used in river models. The properties of light material commonly used in practice are listed in Table 12.2.

12.3.2 Sediment Transport Similitude $\tau_{*r} = 1$

The similitude in sediment transport is considered based on the sediment-transport equation of Einstein–Brown, shown in Figure 2.16, which gives $q_{bv} = \omega d_s f(\tau_*)$. Therefore, the same value of the Shields parameter in the model and prototype will result in the same value of the dimensionless sediment-transport rate $q_{bv}/\omega d_s$. The similitude in the Shields parameter $\tau_{*r} = 1$ imposes

Table 12.2. *Lightweight sediment properties for mobile-bed models*

Material	Specific gravity G	Sediment size d_s (mm)	Model scale[a] x_r	Comment
Polystyrene	1.035–1.05	0.5–3	350–600	Durable but difficult to wet and tends to float
Gilsonite	1.04–1.1	–	100–500	Asphalt – avoid skin contact
Acrylonitrile butadiene styrene (ABS)	1.07–1.22	2–3	30–200	e.g. Lego blocks – add detergent against air-bubble adherence
PVC flexible	1.1–1.3	1.5–4	17–105	Hydrophobic
Nylon (polyamidic resins)	1.13–1.16	0.1–5	50–70	–
Lucite, perspex, plexiglass, acrylite	1.18	0.3–1	40	Dusty
Coal	1.2–1.6	0.3–4	5–33	Possible nonhomogeneity in specific gravity and sorting
Ground walnut shells	1.2–1.4	0.15–0.41	10–34	Deteriorate in 2–3 months, color water (dark brown)
PVC rigid	1.3–1.45	1.5–4	8–17	–
Bakelite	1.3–1.4	0.3–4.0	10–17	Porous, tends to rot, changes diameter, and floats
Pumice	1.4–1.7	–	4–11	Highly variable
Loire sand	1.5	0.63–2.25	7	Dusty
Lytag (fly-ash)	1.7	1–3	4	Porous
Quartz sand	2.65	0.1–1	1	Reference value

[a]Approximate scale based on $x_r = (G-1)_r^{-1.67}$ for complete similitude with $m = 1/6$.

the following relationship between the particle diameter and the slope similitude $S_r = z_r/x_r$:

$$\tau_{*r} = \frac{z_r^2}{x_r(G-1)_r d_{sr}} = 1, \qquad (12.5)$$

which can be solved as $d_{sr} = z_r^2/[x_r(G-1)_r]$.

It is important to notice the great compatibility between the two sediment similitude conditions, namely $d_{*r} = 1$ and $\tau_{*r} = 1$. Three main points need to be considered: (1) the grain shear Reynolds number $Re_{*r} = 1$ because $\tau_* d_*^3 = Re_*^2$ and therefore both the model and prototype conditions will be at

Figure 12.4. Mobile-bed physical models at CSU (a) riverbank erosion and
(b) upper-regime plane-bed for 0.2-mm white and 0.6-mm black sands

the same location on the Shields diagram; (2) combining Equation (12.5)
with Equation (12.4) results in $d_{sr} = x_r^{1/2}/z_r$ which also gives similitude in the
ratio of the suspended load to bed load from Equation (12.3); and (3) similar
bed-form configurations are expected because both the model and prototype
are at the same position on the Shields diagram (Julien, 2010). Illustrations
of mobile-bed physical models are shown in Figure 12.4. The models include
lateral deformation of a river and upper-regime plane bed with sediment
transport at the CSU Hydraulics Laboratory.

12.4 Mobile-Bed River Models

Mobile-bed models are useful when sediment transport is significant, e.g.
when $\tau_* > 0.06$. This section covers a definition of mobile-bed model scales.
Complete river model similitude is first discussed (Section 12.4.1), followed
with incomplete similitude for special river models where some similitude criteria can be simplified (Section 12.4.2).

12.4.1 Complete Mobile-Bed Similitude

The complete similitude for mobile-bed models simultaneously satisfies four
similitude criteria: (1) Froude similitude $Fr_r = 1$; (2) friction similitude
$S_r = f_r Fr_r^2$; (3) settling velocity similitude $d_{*r} = 1$; and (4) sediment transport
similitude $\tau_{*r} = 1$. As sketched in Figure 12.5, the four similitude
equations can be solved simultaneously. For hydraulic models, the four similitude equations involve five parameters: Fr_r, z_r, x_r, d_{sr}, $(G - 1)_r$. After eliminating Fr_r, d_{sr}, $(G - 1)_r$, this leaves one degree of freedom, i.e. either the
depth scale z_r or preferably the model length scale x_r. The river model that
satisfies complete similitude is tilted because $z_r \neq x_r$. The modeler may also
select the lateral scale y_r which is not specified by the equations. A distorted

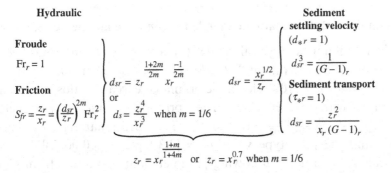

Figure 12.5. Four main equations for complete hydraulic and sediment similitude

model where $y_r = x_r \neq z_r$ is usually preferred in order to preserve the planform geometry.

In the practical case where the Manning–Strickler relationship is suitable ($m = 1/6$), the hydraulic similitude simplifies to $d_{sr} = z_r^4/x_r^3$, and after combining this with the sediment similitude $d_{sr} = x_r^{1/2}/z_r$, we simply obtain $z_r = x_r^{7/10}$. This condition further simplifies previous requirements as $d_{sr} = x_r^{-0.2}$ and $(G-1)_r = x_r^{0.6}$. This means that the sediment material in the laboratory model will be coarser and lighter than the prototype material.

Similitude in bed-material sediment transport is examined from the Einstein–Brown relationship with $q_{bvr} = \omega_r d_{sr} f(\tau_{*r})$, which results in $q_{bvr} = 1$ because $\omega_r d_{sr} = 1$ when $d_{*r} = 1$. This demonstrates why the two sediment-transport similitude criteria are so important in river models. A timescale for riverbed aggradation and degradation can be determined from the sediment continuity relationship Equation (2.54), $\partial z_b/\partial t_b = -(1 - p_0)^{-1}\partial q_{bv}/\partial x$, where q_{bv} is the unit bedload discharge by volume, p_0 is the porosity of the bed material, z_b is the bed elevation, and t_b refers to the timescale for the riverbed deformation. With the same porosity for the model and prototype, we obtain the timescale for the bed deformation as $t_{br} = z_r x_r/q_{bvr}$. This sedimentation timescale is useful in the analysis of local bed-elevation changes through local scour, bedforms, and changes in bedload transport. Notice that the timescale for bed-elevation changes is different from the hydraulic timescale obtained from the Froude similitude criterion.

In diffusion–dispersion studies (Julien, 2010, Chapter 10), the timescale for vertical mixing can be estimated from $t_{vr} = z_r/u_{*r}$, while the timescale for lateral mixing is given by $t_{tr} = y_r^2/z_r u_{*r}$. Of course these two scales are equivalent only for undistorted models, i.e. $y_r = z_r$. Also, the length scale for vertical mixing is $x_{vr} = z_r V_r/u_{*r}$, equivalent to the length scale for lateral mixing $x_{tr} = V_r y_r^2/z_r u_{*r}$ only for models without distortion. Example 12.2

Example 12.2 Calculation example for complete mobile-bed similitude

Consider building a physical model for a very wide fine sand-bed river with an 8-m flow depth and 2 m/s velocity. Given a 40,000 m^3/s flow discharge and 7×10^{-5} slope, determine the scale ratios for complete similitude with $x_r = 250$. The prototype Shields parameter is $\tau_{*p} = \frac{h_p S_p}{(G-1)_p d_{sp}} = \frac{8 \times 7 \times 10^{-5}}{1.65 \times 0.0002} = 1.7 > 0.03$. The bed material is definitely mobile and the prototype value $m = [2.3 \log(2 \times 8/0.0002)]^{-1} = 0.089$. The scale ratios for the following parameters are obtained from Column five of Table 12.1: (1) the depth ratio is $z_r = z_p/z_m = 250^{\left[\frac{1+0.089}{1+4 \times 0.089}\right]} \simeq 80$ and the model depth is $z_m = 8/80 = 0.1$ m; (2) the model particle diameter is $d_{sm} = 0.2$ mm/$250^{\left[\frac{-1+2 \times 0.089}{2+8 \times 0.089}\right]} \simeq 1.06$ mm; and (3) it is interesting to calculate $m = [2.3 \log(2 \times 0.1/0.0011)]^{-1} = 0.19$ for the model which is very close to the Manning-Strickler value $m = 1/6 = 0.167$.

Recalculating the scales now from Column six of Table 12.1 gives: (1) a model depth $z_m = 8/250^{0.7} = 0.167$ m; (2) a particle size $d_{sm} = 0.2$ mm/$250^{-0.2} \simeq 0.6$ mm; (3) the model sediment density $(G - 1)_m = 1.65/250^{0.6} \simeq 0.06$, or $G_m = 1.06$, for which gilsonite or polystyrene may be considered. The other parameters would be the hydraulic timescale $t_r = 250^{0.65} = 36$, and model velocity $V_r = 250^{0.35} = 6.9$ with a model velocity $V_m = 2/6.9 = 0.29$ m/s. The sediment timescale $t_{br} = 250^{1.7} = 11,900$ means that 1 h in the laboratory corresponds to 16 months of riverbed deformation. The model slope is $S_m = S_p/S_r = 7 \times 10^{-5}/250^{-0.3} = 37 \times 10^{-5}$ and selecting $y_r = x_r$ results in a reasonable model discharge $Q_m = Q_p/Q_r = 40,000/250^{2.05} = 0.48$ m^3/s.

---

illustrates how to calculate the scale ratios for mobile-bed models. In practice, complete mobile-bed similitude is somewhat restricted because large-scale models may require large volumes of light material.

12.4.2 Incomplete Mobile-Bed Similitude

When the conditions for complete similitude cannot practically be fulfilled, one constraint can sometimes be sacrificed in order to gain an additional degree of freedom. As the model further deviates from complete similitude, there is a greater risk that the model may yield incorrect results. Let's explore two types of models where the similitude in friction and sediment transport is preserved while the Froude similitude is sacrificed, i.e. $Fr_r \neq 1$: (1) same sediment-size model $d_{sr} = 1$ and (2) different sediment-size $d_{sr} \neq 1$.

12.4.2.1 Same Sediment-size Model, $d_{sr} = 1$ and $\mathrm{Fr}_r \neq 1$

It is very tempting to use the same sediment material for the model and the prototype, because the material can conveniently be found in large quantities. The similitude in dimensionless particle diameter $d_{*r} = 1$ implies $(G-1)_r = 1/d_{sr}^3$; thus the sediment of the same density must also have the same particle size.

Imposing the constraint of using the same sediment size implies that one of the four main similitude criteria has to be sacrificed in order to retain the necessary degree of freedom. With specific reference to Figure 12.5, it is obvious that the settling velocity similitude will be satisfied when $d_{sr} = 1$, but the similitude in sediment transport can be satisfied only when $z_r = x_r^{1/2}$. The sediment-transport similitude imposes a very high model distortion, which can be easily met in rigid-boundary models. However, in mobile-bed models, the deeply exaggerated vertical scale will be filled with sediment and the desirable model distortion may not be maintained for a long time.

An additional problem associated with using the same sediment is that the hydraulic similitude $z_r = x_r^{3/4}$ cannot be simultaneously satisfied. This implies that either (or often both) the criteria requiring similitude in resistance to flow and the similitude in Froude number will not be satisfied. The resistance Equation (12.2) with $z_r = x_r^{1/2}$ and $d_{sr} = 1$ can be rewritten in terms of the Froude number as $\mathrm{Fr}_r = x_r^{\frac{-1+2m}{4}}$. The Froude ratio is not unity but changes slightly with model scale. Other scale ratios for incomplete similitude models with same sediment are listed in Column seven of Table 12.1. Near-equilibrium mobile-bed river models can have different values of the Froude number as long as the flow remains subcritical. For instance, a large alluvial sandbed river with a low Froude number can be simulated with a higher subcritical Froude number.

12.4.2.2 Different Sediment Size Model, $d_{sr} \neq 1$ and $\mathrm{Fr}_r \neq 1$

It is possible to retain sediment-transport similitude with a different sediment size and density. The scales for such models are found in the last column of Table 12.1. This type of model offers similitude in bedload and suspended sediment transport. In comparison with the complete model similitude, the modeler gains one degree of freedom in selecting the particle size or density for the model material. This gain is achieved at the cost of not satisfying the Froude similitude. This type of model should appropriately simulate sediment transport and resistance to flow and may work for rivers with low Froude numbers. However, the force diagrams and the lateral/vertical accelerations are not appropriately simulated. Case Study 12.1 illustrates how the scales of a model that uses the same material for both systems can be determined.

Case Study 12.1 Mobile-Bed Model of the Jamuna River, Bangladesh

A physical model has been built for a proposed bridge crossing the large Jamuna River in Bangladesh (Klaassen, 1990, 1992). The Jamuna River is a fine sand-bed river and a mobile-bed model aimed at an examination of channel pattern deformation near the bridge training works. The limitations of this model included model dimensions at a maximum of 50×20 m^2 and the use of 0.2-mm sand because large quantities of lightweight material would be prohibitive.

The scaling procedure resulted in a considerably distorted model with the main model characteristics listed in Table CS12.1.1. A fair replication of the channel pattern was observed at a model slope of ~ 0.007. The objectives were satisfactorily met, but one of the main concerns about the model relates to local scour. Local scour can only be correctly reproduced in undistorted models that satisfy the Froude condition. Because both conditions were not fulfilled, local scour could not be scaled. Another concern is the local presence of supercritical flow. With a model Chézy coefficient C_m assumed to be ~ 25 m$^{1/2}$/s, the average channel velocity was estimated at ~ 0.4 m/s. This implies a model Froude number ~ 0.5, while the prototype value is Fr ~ 0.2. Locally, higher values of the Froude number could be found which induced scale effects.

Table CS12.1.1. *Characteristics of the Jamuna River model (after Klaassen, 1990)*

Parameter	Prototype	Scale factor	Model
Particle size	$d_{sp} = 0.2$ mm	$d_{sr} = 1$	$d_{sm} = 0.2$ mm
Density	$G_p = 2.65$	$(G-1)_r = 1$	$G_m = 2.65$
Slope	$S_p = 7 \times 10^{-5}$	$S_r = 0.01$	$S_m = 7 \times 10^{-3}$
Discharge	$Q_p = 10,000$ m^3/s	$Q_r = 10^6$	$Q_m = 0.01$ m^3/s
	$= 90,000$ m^3/s		$= 0.09$ m^3/s
Bankfull width	$W_p = 3,000$ m	$y_r = 1,000$	$W_m = 3.3$ m
Total width	15,000 m	$y_r = 1,000$	15 m
Flow depth	$h_p = 5.8$ m	$z_r = 200$	$h_m = 0.032$ m
Sediment transport	$q_{sp} = 1.4 \times 10^{-3}$ m^2/s	$q_{sr} = 80$	$q_{sm} = 1.34 \times 10^{-5}$ m^2/s
Flood duration	$T_p = 78$ days	$t_r = 2,500$	$T_m = 0.03$ day
Froude number	Fr$_p = 0.1$–0.2	Fr$_r = 0.25$	Fr$_m = 0.4$–0.8

◆*Exercise 12.1*

Derive the scale ratios for tilted rigid-bed models from similitude in Froude number and Manning–Strickler. Compare the results with those of Table 12.1 Column four. Also determine the results of the particular case in which $x_r = y_r = z_r = L_r$.

◆◆*Exercise 12.2*

Demonstrate that a hydraulic model with the same values in grain shear Reynolds number and Shields parameter also has the same dimensionless particle diameter.

Exercise 12.3

In Case Study 12.1, show that the grain shear Reynolds numbers of the model and of the prototype are quite similar despite the large model distortion.

◆◆*Exercise 12.4*

Derive the scale ratios for complete mobile-bed similitude from simultaneously solving for similitude in: (1) Froude number $Fr_r = 1$; (2) resistance equation; and (3) sediment transport $\tau_{*r} = d_{*r} = 1$. Compare the results with those of Table 12.1 Column five, and check the values for the Manning–Strickler relationship ($m = 1/6$).

◆*Exercise 12.5*

Derive the scale ratios for an incomplete mobile-bed model with similitude in: (1) the resistance equation and (2) sediment transport with $\tau_{*r} = d_{*r} = 1$. Compare the results with those of Table 12.1, Column (8), for $d_{sr} \neq 1$. Why is the Froude similitude not satisfied?

◆*Exercise 12.6*

In Case Study CS12.1, determine whether the scale factors in Table CS12.1.1 satisfy similitude for: (1) Froude; (2) Manning–Strickler; and (3) sediment transport.

Problem 12.1

The filling and emptying gates of a canal lock extend the full height of the lock chamber. When a vessel is lowered in the prototype lock, the waves and currents produced by the outflow cause the vessel to pull at its moorings. In a 1:25 Froude model without distortion and tilting, the maximum tension in the moorings is 1.6 lb (7.12 N) when the gates are opened at the proper rate. Determine the maximum mooring-line tension in the prototype.

[*Answer:* $F_p = F_m F_r = 1.6$ lb $L_r^3 = 25,000$ lb, or 112 kN.]

♦Problem 12.2

A hydraulic open-channel model is designed given the maximum laboratory discharge of 2 l/s with a maximum laboratory space of 60 m. If a 10-km river reach and a river flow rate of 300 m³/s are modeled, determine the suitable scaling length for the model. From this, determine the scaling ratios for time, discharge, and hydropower.

♦♦Problem 12.3

With reference to Case Study 12.1, demonstrate that the model scales in Table CS12.1.1 are fairly comparable with those obtained from Table 12.1, Column eight, with $x_r = y_r = 10,000$, $z_r = 100$, $d_{sr} = 1$, and show that $m = 0.09$.

[*Answer:* $t_r = 10,000^{1-0.5\times0.09} = 6,600$, $Q_r = 10,000^{1.5+0.5\times0.09} = 1.5 \times 10^6$, $S_r = x_r^{-0.5} = 0.01$, and $Fr_r = 10,000^{0.5\times0.09-0.25} = 0.15$.]

♦Problem 12.4

With reference to Case Study 12.1, calculate the ratio u_*/ω and the Shields parameter values for both the model and the prototype. Verify the similitude in sediment transport.

[*Answer:* $(u_*/\omega)_p = 2.52$, $(u_*/\omega)_m = 1.88$, $\tau_{*p} = 1.23$, and $\tau_{*m} = 0.67$. Sediment transport is not in exact similitude.]

♦♦Problem 12.5

On April 5, 1987, the center span of the 165 m-long bridge over Schoharie Creek, NY, collapsed during a near-record flood $Q_p = 1,750$ m³/s. Each rigid pier was supported on a spread footing bearing on a glacial till which the designers considered nonerodible. A physical model of the

Schoharie Creek Bridge was built at CSU (Santoro et al., 1991). The timing of the event and sediment-transport similitude are not required. You would like to replicate the secondary currents around the bridge pier footing and for this an exact Froude similitude model is desirable. If the scale ratio for this model is 1:50, determine the appropriate scales for velocity, time, and discharge.

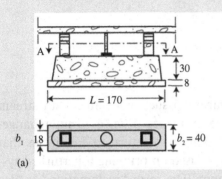
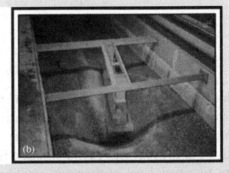

Figure P12.5. Schoharie Creek Bridge: (a) model scales, and (b) physical model (after Santoro et al., 1991)

[*Answer:* $t_r = V_r = 7.07$, and $Q_r = 50^{2.5} = 17,700$, with a model discharge $Q_m = 1,750/17,700 = 0.1 \text{ m}^3/\text{s}$.]

13

Stream Restoration

This chapter presents a discussion of various issues associated with stream restoration and rehabilitation. These two terms are often used interchangeably, but in general, restoration is the seeking of a return to conditions that existed in the past while rehabilitation is focused on bringing a perturbed system to a new level of dynamic equilibrium. The term "reclamation" is also encountered, signifying the adaptation of a natural resource for human purposes. Following the publication of *Kosmos* by Alexander von Humboldt (1769–1859) and *On the origin of Species* by Charles Darwin (1809–1882), it was Ernst Haeckel (1834–1919) who coined the term "ecology" to describe our shared "house" or environment. This chapter covers an environmental overview of the following types of restoration issues: (1) watershed environment, in Section 13.1; (2) channel rehabilitation, in Section 13.2; (3) aquatic environment, in Section 13.3; and (4) stream restoration, with guidelines, in Section 13.4.

13.1 Watershed Environment

This section discusses the watershed environment in terms of water quantity and quality. Mine reclamation is covered in Section 13.1.1, and urban drainage in Section 13.1.2.

13.1.1 Mine Reclamation

Watershed contamination is a major issue when dealing with acid mine drainage, highly contaminated sites, and radioactive waste in mine-closure and Superfund sites. For instance, acid mine drainage from exposed mine tailings can create a very toxic environment for aquatic species. The analysis of nonpoint source pollution is essential for locating the source areas of

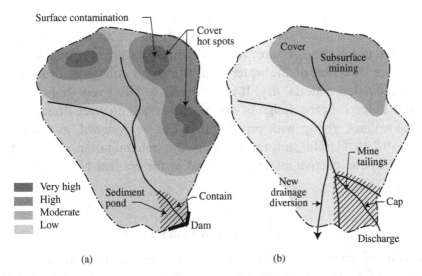

Figure 13.1. Land surface restoration from (a) surface contamination and (b) subsurface contamination

contamination. For instance, contaminated areas can be delineated by: (1) in-situ surveying methods, such as using gamma probes for Radium-226, and field portable X-ray fluorescence spectrometers for arsenic (Orechwa, 2015) or (2) remote sensing, such as can be achieved with the airborne visible/infrared imaging spectrometer, for heavy metals (Velleux et al., 2006). Eventually, detailed mapping of the highly contaminated surface areas can be performed with remote sensing techniques or drones, and the most contaminated areas can sometimes be covered. Detailed spatially distributed computer models like TREX can then be applied to examine the toxicity of metals and the impact on aquatic species. The toxicity of a mixture of metals is expressed by the cumulative criterion unit (CCU) index. It is the ratio of dissolved recoverable metals' concentration to a hardness-adjusted water-quality criterion, summed for all metals in the mixture. An example of distributed modeling at the watershed scale for an EPA Superfund site has been presented by Velleux et al. (2006, 2008).

In highly contaminated areas, two types of strategy can be developed, depending on whether we face ubiquitous surface contamination or localized subsurface mining, as sketched in Figure 13.1. In the case of widespread contamination, hot spots can sometimes be located and conveniently covered to reduce the contamination level of surface runoff.

Surface runoff is typically contained in sediment ponds to capture all contaminants. The design of these dams for highly contaminated sites with small drainage areas can go as far up as the Probable Maximum Flood (PMF).

In the case of subsurface mining, the mine itself can often be covered while the contaminated mine residues are typically stored in mine tailings. The typical design for mine closures requires capping of the mine tailings and an appropriate diversion of the water away from the tailings in order to keep the contaminated material dry. It is important to effectively drain the top of the mine tailings. In the design of drainage channels on steep drainage slopes of mine tailings, riprap with proper filters can be designed, as discussed in Section 9.2.10. Useful information on the rehabilitation of watersheds impacted by radioactive material is found in the manual NUREG-1623, by Johnson (2002).

13.1.2 Urban Drainage

Urban development typically increases surface runoff, because of the increase of impervious surface area in rooftops, roads, and paved parking areas. The most direct method for urban runoff control is the storage of surface flows with detention ponds, basins, and small reservoirs. As sketched in Figure 13.2, a detention basin or small reservoir fills up when the inflow exceeds the outflow. Detention basins induce groundwater recharge and evaporation, and also trap the incoming sediment load. Water storage decreases the outflowing peak discharges and increases low-level discharges. The peak discharge reduction will decrease flooding risks and will reduce the sediment load transported to lower watershed areas.

In urban areas and upper watersheds, a detention pond can be controlled by an outlet weir, or a small spillway. It is also possible to control the development of gullies by building a simple drop structure, as sketched in Figure 13.2b. In some cases, the flood volume is stored in the detention basin and gradually drained through riser pipes. On larger watersheds, we can gain control of the timing of the flood peak from each subwatershed. For the instance in Figure 13.2c, the flood at city A would be reduced by an early release from T1 and floodwater retention at T4. This approach can reduce flooding and sedimentation problems (Vischer and Hager, 1992).

Best management practices are developed to mitigate the increased runoff, primarily through detention storage. The main objective is typically to reduce the lower flow discharges (up to 80 percent on the flow duration curve) and improve water quality of low flows. The high flows (the upper 20 percent) should remain unchanged. Urban areas, nevertheless, increase the total volume of runoff. With proper control of the peak flows, there will be a significant increase in low flows and ephemeral streams may become perennial through clear-water releases and groundwater seepage flows. It is important

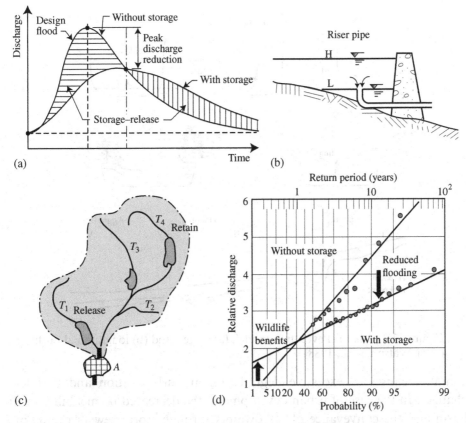

Figure 13.2. Flood storage from urban areas: (a) peak hydrograph reduction, (b) detention storage, (c) drainage basin, and (d) flood frequency curve

to consider that clear-water releases from detention ponds can induce gullying and channel instabilities downstream of detention basins. This can be controlled by the construction of check dams and grade-control structures.

13.2 Channel Rehabilitation

This section discusses channel rehabilitation methods and four main topics are discussed: channel evolution in Section 13.2.1, gravel mining in Section 13.2.2, overbank vegetation in Section 13.2.3, and bioengineering methods in Section 13.2.4.

13.2.1 Channel Evolution

The case of excessive channel degradation, as discussed in Section 11.2, often requires action toward rehabilitation to new natural conditions. Two of the

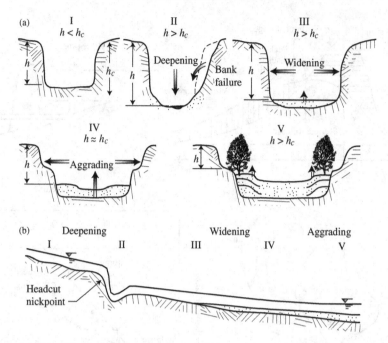

Figure 13.3. Channel evolution model: (a) stages, and (b) longitudinal profile (after Watson et al., 1988)

common causes of excessive degradation are urbanization and base-level changes. In the case of urban development, the decreased permeability of the basin and the conveyance of floodwaters through storm sewers results in a large increase in peak discharge, coupled with a reduced sediment load in the receiving channels (Annable et al., 2010). This will typically result in excessive degradation rates with the upstream propagation of headcuts. To alleviate this problem, as well as detention, storage can be incorporated in the urban development plan; and a stream rehabilitation program may also include drop structures. In the case of base-level changes, the channel evolution model should be considered in the development of an adequate remediation plan.

Channel degradation is the result of an imbalance in the sediment supply–capacity relationship. The channel-rehabilitation strategies will typically require a restoration of a new equilibrium. A channel evolution model has been proposed by Schumm et al. (1984), and reworked by Watson et al. (1988, 1999). It is based on the analysis of incised watersheds in Northern Mississippi under the Demonstration Erosion Control Project. As sketched in Figure 13.3, five stages of development are identified with incision (Stage I) until a critical height is exceeded (Stage II). This is followed with lateral extension with slab failure (Stage III) and rotational failure

(Stage IV). Finally, channel widening and slope reduction result in channel aggradation, where vegetated banks can develop. A valuable source of detailed information on channel-width adjustment processes is available in ASCE Task Committee on Hydraulics (1998 a, b). One of the solutions to channel degradation problems in the Demonstration Erosion Control area has been the development, testing, and construction of low- and high-drop grade-control structures (e.g. Abt et al., 1992).

For very high elevation drops, proper energy-dissipation measures below spillways and weirs must be considered (Reclamation, 1974, 1976, 1977). For moderate elevation drops, rigid structures can be designed, such as steel sheet-piles and concrete weirs. In raising the groundwater table over short distances around high-drop structures, care should be taken to prevent the water from seeking its way around the intended structure. Deep sheet piles can effectively prevent head-cutting and can also serve as cutoff walls to prevent groundwater flow around the structure and piping. It is preferable to consider several low-head structures than fewer large-drop structures. Vertical structures may also prevent fish passage and migration, such that gradual structures such as sloping sills, rock ramps, and gradient restoration facilities can be environmentally preferable. Gradual-drop structures can be made gabions, rock rip-rap, riffles and pools, or boulder clusters (Fischenich and Seal, 2000), as sketched in Figure 13.4.

13.2.2 Gravel Mining

In-stream gravel mining operations in rivers remove the coarse armor layer at low flows. The material can then be sieved and used for construction, and thus a source of revenues. In many places, gravel mining is viewed favorably as a means to reduce flood risks through channel degradation. However, unexpected damage can often be seen during subsequent floods in which large flood discharges cause severe channel degradation. Headcut development and upstream migration can also result in the failure of upstream structures such as weirs and bridges.

A list of potential problems associated with gravel mining is not limited to, but includes: (1) failure of bank protection structures and the need for toe-protection measures such as aprons; (2) undermining river structures requiring drop structures and cut-off walls; (3) excessive pier scour causing structural instability problems at bridge crossings needing additional protection against pier scour; (4) water levels too low for the intake of water supply and irrigation canals and the need for pumping stations; (5) headcut propagation and degradation up the tributaries calling for additional drop

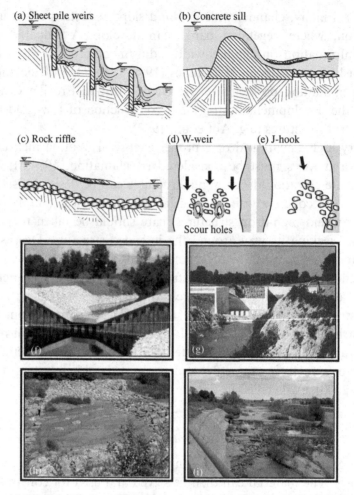

Figure 13.4. Grade control sketches (a–e), and illustrations: (f) low drop, (g) high drop, (h)riprap riffle, and (i) soil cement drop

structures; (6) saltwater intrusion in rivers estuaries requiring estuary barrages; (7) lowering groundwater levels which can affect agricultural practices; and (8) disconnect between the river and the floodplain for the aquatic species and wildlife.

A sound engineering practice to eliminate gravel mining problems is sketched in Figure 13.5. In-stream gravel mining is prohibited, however, off-site gravel mining in pre-designated floodplain areas is permitted. Mining operations are done without the use and contamination of the river waters. Once the material has been extracted, the excavation sites are turned into: (1) environmental ponds inside a recreational area or (2) valuable waterfront property. The Cache la Poudre River near Fort Collins is a good example of successful operations.

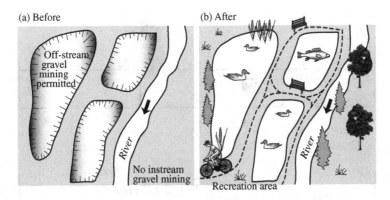

Figure 13.5. Off-stream gravel mining

13.2.3 Overbank Vegetation

The influence of vegetation on channel flow characteristics is discussed in this section. First, Figure 13.6a shows that not all types of vegetation will increase roughness. Tall grasses may stand tall and obstruct low flows, but they can bend and form a relatively smooth surface during floods. Here, a suggested approximation is $n \cong 0.03 + 0.35e^{-2h/h_g}$ where h is the flow depth and h_g is the grass height. Shrubs and conifers will tend to trap floating organic debris and will definitely increase roughness. Cylindrical hardwood trunks and stems will not cause a significant obstruction and may effectively decrease hydraulic roughness.

In the following analysis, we can gain basic understanding on the effect of roughness on channel flow with floodplains. Let us consider a straight channel with the geometric characteristics shown in Figure 13.6b. The channel and overbank area is wide and shallow, i.e. $W_c = W_o = 100$ m and main channel depth $H = 1$ m, slope $S = 70$ cm/km, and different roughness values for overbank and channel n_o and n_c, respectively. A simple approximation for the total flow discharge Q_t is

$$Q_t = Q_c + 2Q_o \cong \frac{WS^{1/2}}{n_c} \left[h^{5/3} + 2\frac{n_c}{n_o}(h-H)^{5/3} \right] \qquad (13.1)$$

As shown in Figure 13.6b for a floodplain without vegetation $n_o = n_c$ at $Q_t = 1{,}000$ m³/s, the solution is a flow depth of 2 m with the main channel carrying 62 percent of the total discharge. The flow velocity on the floodplain is close to 2 m/s. At the same discharge, the effects of increasing the floodplain roughness through vegetation in Figure 13.6c shows an increase in flow depth, a reduced velocity on the floodplain and an increased percentage of the total discharge (now 82 percent) in the main channel. It is most

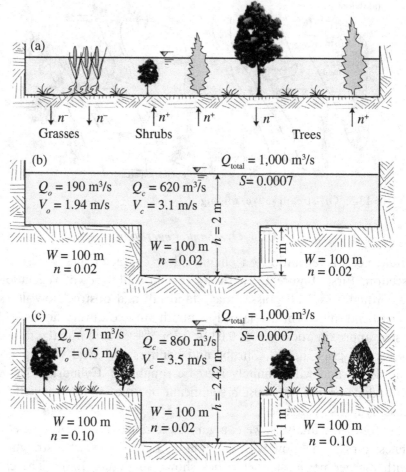

Figure 13.6. Effects of floodplain vegetation on hydraulic roughness: (a) floodplain vegetation, (b) channelization, and (c) channel with vegetated floodplain

interesting to learn that floodplain vegetation increases the flow velocity and shear stress in the main channel.

The concept of vegetation freeboard equivalence (VFE) has been proposed by Lee and Julien (2012b). The flood stage difference with and without vegetation (0.42 m in this case) is compared to the levee freeboard height.

13.2.4 Bioengineering

Stream restoration with vegetation requires knowledge of the plant tolerance to water for growth and resistance to hydraulic forces during floods.

Fabric Encapsulated Soils (FES) consist of fabric-wrapped soils with rooted plants easily transplanted near river banks. Bio-degradable fabrics provide temporary erosion resistance until the fabric decomposes. It is expected that by that time, the root zone will be sufficiently developed to sustain the high velocity and shear stresses during floods. The banks remain flexible and although vegetation slows down erosion, it will not provide a fixed river bank. Willow posts have been successfully used to stabilize river banks (Derrick, 1997).

Woody plants commonly found with very good ability to root from poles and live cuttings are: (1) the balsam poplar (*Populous balsamifera*); (2) the eastern cottonwood (*Populous deltoides*); and (3) the peachleaf willow (*Salix amygdaloides*). There is a very detailed list of vegetation performance for fascines, brush mattresses and brush layering in Reclamation (2015).

The permissible shear and velocity resistance of vegetated and bio-engineered surfaces has been compiled by Fischenich (2001). The properties against erosion of some natural, bio-engineering and Rolled Erosion Control Products (RECP) are listed in Table 13.1. In contrast to materials like steel and concrete, the resistance of vegetated surface decreases with time (Gray and Sotir, 1996). Within a period of 10 h, the allowable velocity can decrease by 30–50 percent, e.g. from 2 to 1 m/s for average grass and from 5 to 2.5 m/s for non-degradable soil reinforcement products (Fischenich, 2000). A lot more information can be found in Reclamation (2015).

Rootwads are placed into the banks of a channel with the root mass facing the upstream direction (Sylte and Fischenich, 2000). They work best in streams coarser than gravel and tend to erode in sand-bed streams. Cables should not be used. Log jams can also be used in steep and coarse mountain streams. They occur naturally in the Pacific Northwest and engineered log-jams can be designed with the roots facing upstream and away from the bankline. Figure 13.7 illustrates several types of woody debris structures for bankline protection. More design information can be found in D'Aoust and Millar (2000) and Reclamation (2015).

The durability of woody vegetation to cycles of wetting and drying is also examined and in general, cypress, catalpa, cedars, chestnut, junipers, honey-locust, redwood, and walnut are resistant to heartwood decay. In the Pacific Northwest, the western redcedar (*Thuja plicata*) has been considered to be the most desirable with survival up to 50–100 years (Johnson and Stypula, 1993). The Douglas fir (*Pseudostuga* sp.) and the Sitka spruce (*Picea sitchensis*) have been considered to be excellent. References regarding woody vegetation include Shields et al. (2008) and the recent Large Wood Manual (Reclamation and USACE, 2015).

Table 13.1. *Permissible velocity and shear stress for vegetated and bio-engineered surfaces (after Fischenich, 2001)*

Type	Velocity (ft/s)	Velocity (m/s)	Shear stress (lb/ft^2)	Shear stress (Pa)
Vegetation				
Reed plantings	–	–	0.1–0.6	4.8–29
Short native grass	3–4	0.9–1.2	0.7–0.95	33–46
Hardwood tree planting	–	–	0.41–2.5	20–120
Long native grass	4–6	1.2–1.8	1.2–1.7	58–82
Turf	3.5–8	1.1–2.4	1.0–3.7	48–178
Bio-engineering				
Wattles	3	0.9	0.2–1.0	10–48
Live brush mattress (initial)	4	1.2	0.41–4.1	20–200
Reed fascine	5	1.5	0.6–1.25	29–60
Live fascine	6–8	1.8–2.4	1.25–3.1	60–150
Live willow stakes	3–10	0.9–3.0	2.1–3.1	100–150
Coir roll	8	2.4	3–5	150–240
Vegetated coir mat	9.5	2.9	4–8	190–380
Live brush mattress (grown)	12	3.7	3.9–8.2	190–400
Brush layering	12	3.7	0.4–6.2	20–300
Temporary degradable Rolled Erosion Control Products (RECP)				
Jute net	1–2.5	0.3–0.8	0.45	22
Straw with net	1–3	0.3–0.9	1.5–1.6	72–80
Fiberglass roving	2.5–7	0.8–2.1	2.0	96
Coconut fiber with net	3–4	0.9–1.2	2.25	108
Non-degradable Rolled Erosion Control Products (RECP)				
Unvegetated	5–7	1.5–2.1	3.0	144
Partially established	7.5–15	2.3–4.6	4.0–6.0	190–290
Fully vegetated	8–21	2.4–6.4	8.0	380

13.3 Aquatic Environment

This section discusses the aquatic environment with specific focus on aquatic habitat in Section 13.3.1, ichthyomechanics in Section 13.3.2, dissolved oxygen in Section 13.3.3, advection dispersion in Section 13.3.4, contaminants and toxicity in Section 13.3.5, aquatic life cycle in Section 13.3.6, and endangered species in Section 13.3.7.

13.3.1 Aquatic Habitat

Stream restoration projects should consider the aquatic habitat of living species. Some of the very important factors to consider include: (1) minimum

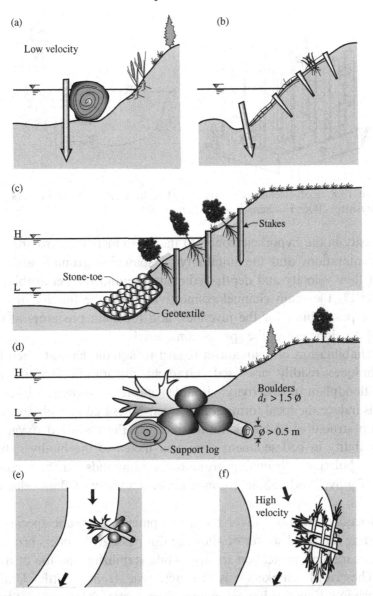

Figure 13.7. Stream rehabilitation with (a) coir roll, (b) brush mattress, (c) stone toe, (d and e) rootwads, and (f) log-jam

in-stream flow needs, appropriate water quality, temperature and dissolved oxygen; (2) possible contamination from acid mine drainage, heavy metals, chemicals, toxic waste and radionuclides; (3) excessive nutrients can also lead to algae bloom with an impact on dissolved oxygen levels; (4) infestations from invasive species, e.g. zebra mussels, and diseases can be devastating; (5) the food chain and food sources including macro-invertebrates and

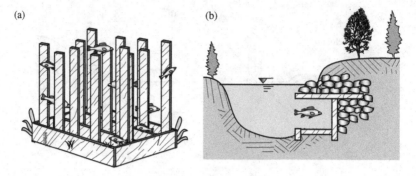

Figure 13.8. Fish-friendly structures: (a) stake beds and (b) lunkers (after Derbyshire, 2006; Fischenich and Allen, 2000)

bio-diversity in the hyporheic zone; (6) the reproductive environment including fish migrations and the suitability of spawning grounds and redds in terms of flow velocity and depth, sediment size and the permeability of the substrate; (7) the main channel connectivity with the floodplain can yield sheltering possibilities for the juveniles; and (8) main predators at different stages of development of the species considered.

The rehabilitation of a stream in regard to aquatic habitat often involves the techniques readily discussed. Gradient restoration facilities improve channel-floodplain connectivity. Boulder clusters, W-weirs, J-hooks and rootwads induce the local formation of scour holes and provide shelter areas. Low-drop structures and rock ramps increase the dissolved oxygen levels. The variability in bed sediment sizes also increases the biodiversity of the benthic populations. Spawning grounds for salmonids can be favored by a variety of gravel and cobble grounds under a variety of flow velocities and flow depths.

Fish-friendly structures favor the development of aquatic species by way of providing food and shelter or allowing fish passage. Lunkers provide overhanging shade and protection for fish while stabilizing the toe of a streambank. The word "lunkers" is an acronym from "Little Underwater Neighborhood Keepers Encompassing Rheotactic Salmonids." Stake-beds and lunkers sketched in Figure 13.8 can be constructed onshore and moved instream by a loader or backhoe at the installation site.

13.3.2 *Ichthyomechanics*

The mechanics of a fish swimming, or ichthyomechanics, is critical to the development of fish-protection technology. Katopodis (1992) compiled information on how long (endurance time) or how far (swimming distance) a

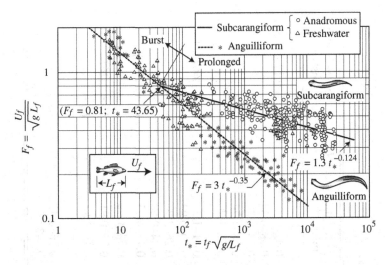

Figure 13.9. Fish endurance characteristics (after Katopodis, 1992)

particular fish can swim against a given water velocity. A fish of length L_f can swim at a velocity U_f over a duration t_f. Results for subcarangiform and anguilliform swimming modes are shown in Figure 13.9 and Table 13.2. Subcarangiform is an undulatory swimming mode with small side-to-side amplitude at the anterior and large amplitude in the posterior half of the body. In the anguilliform mode the entire body length participates in propulsion.

For fish swimming against the current at a flow velocity V, the dimensionless fish Froude number $F_{Vf} = V/\sqrt{gL_f}$ can be used to estimate the maximum endurance swimming distance $X_{max} \simeq 0.4 L_f F_{Vf}^{-7}$ (Katopodis, 1992). This approximation, valid for the subcarangiform mode, is only applicable when $F_{Vf} < 0.65$.

The maximum swimming distance against a current depends on flow velocity and fish size. Typical values are shown in Figure 13.10. Reducing flow velocity below 0.5 m/s will largely increase the mobility of a variety of fish sizes. Flow velocities exceeding 2 m/s will allow the passage of only large fish and only over a very short distance, of a few meters.

13.3.3 Dissolved Oxygen

The saturated dissolved oxygen concentration in rivers varies with temperature and altitude, as shown in Figure 13.11. The saturated dissolved oxygen level drops significantly with temperature and fish species seek cold waters during summers.

Table 13.2. *Swimming performance of some fish species (after Katopodis, 1992)*

Common name	Scientific name	Length L_f (m)	Endurance t_f (s)	Swimming speed U_f (m/s)	Velocity–length relationship
Subcarangiform swimming mode					
Atlantic salmon	*Salmo salar*	0.231	300	0.52	$U_f = 0.17 + 1.57L_f$
Chum salmon	*Oncorhynchus keta*	0.038–0.048	300	0.18–0.34	–
Coho salmon	*Oncorhynchus kisutch*	0.051–0.133	534–1,746	0.34–0.70	–
Pink salmon	*Oncorhynchus gorbuscha*	0.465–0.596	72–1,278	0.78–1.74	$\bar{U}_f = aL_f^{0.63}$ $a = 1.46 - 2.3$
Sockeye salmon	*Oncorhynchus nerka*	0.126–0.621	6–1,350	0.55–1.7	$U_f = 1.66L_f^{0.61}$
Arctic char	*Salvelinus alpinus*	0.080–0.420	6–1,089	0.41–1.3	–
Brook trout	*Salvelinus fontinalis*	0.041–0.172	10–1,800	0.20–0.93	–
Rainbow trout	*Oncorhynchus mykiss*	0.082–0.310	1–1,800	0.26–2.7	
Largemouth bass	*Micropterus salmoides*	0.081–0.224	300–1,800	0.34–0.59	
Flathead chub	*Platygobio gracilis*	0.170–0.300	600	0.43–0.63	$U_f = 1.45L_f^{0.67}$
Lake whitefish	*Coregonus clupeiformis*	0.060–0.510	72–1,278	0.34–1.02	$U_f = 0.91L_f^{0.35}$
Longnose sucker	*Catostomus catostomus*	0.040–0.530	600	0.23–0.91	$U_f = 1.26L_f^{0.53}$
Walleye	*Stizostedion vitreum*	0.080–0.380	600	0.38–0.84	$U_f = 1.37L_f^{0.51}$
Anguilliform swimming mode					
Burbot	*Lota lota*	0.120–0.620	600	0.36–0.41	$U_f = 0.44L_f^{0.07}$
Lamprey	*Petromyzon marinus*	0.145–0.508	0.8–1,635	0.3–3.96	

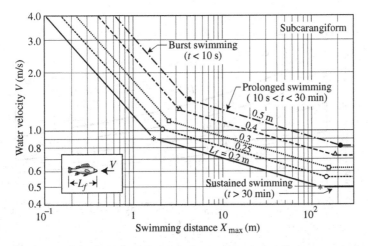

Figure 13.10. Typical swimming distance curve versus fish length (after Katopodis, 1992)

The oxygen deficit D indicates the shortage of dissolved oxygen in the water in comparison to its level in saturated condition. The impact of oxygen-depleting wastes is greatest during the summer, when the water temperature is high and river flows are low. When oxygen deficit develops in a river, the reaeration process replenishes dissolved oxygen from the free surface exchange at a rate proportional to the oxygen deficit. The reaeration rate decreases with flow depth and increases with flow velocity. Two approximations are useful: (1) $k_2 \cong 5 U_{m/s}/H_m^{1.67}$ from Churchill et al. (1962) for rivers and streams and (2) $k_2 \cong 4 U_{m/s}^{0.5}/H_m^{1.5}$ from O'Connor and Dobbins (1958) for lower velocities. The reaeration rate k_2 typically ranges from 0.1/day for relatively still and deep waters to about 5/day in turbulent shallow streams.

The Biochemical Oxygen Demand (BOD) is the amount of dissolved oxygen consumed by living organisms under aerobic conditions. Two types of BOD are differentiated, depending whether the source is carbonaceous (CBOD) from carbohydrates/organic matter or nitrogenous (NBOD) from the breakdown of organic-N (e.g. proteins) to ammonia, with subsequent conversion to nitrites from the *Nitrosomonas* bacteria, and subsequently to nitrates from the *Nitrobacter* bacteria. Domestic and industrial wastes often contain high levels of BOD, which can seriously deplete the dissolved oxygen content of surface waters and reduce the aquatic biodiversity.

The depletion of dissolved oxygen from the BOD is proportional to the load L and a reaction-rate coefficient k_1. For municipal sewage, the reaction coefficient for the CBOD is approximately $k_1 \cong 0.1$ day at 20°C and nearly

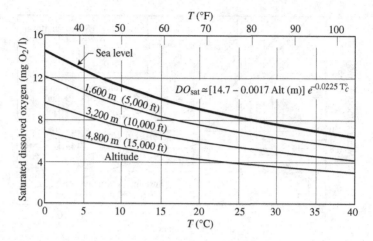

Figure 13.11. Saturated dissolved oxygen versus temperature and altitude

doubles for every 10°C in temperature, i.e. $k_1 \cong 0.2$ day at 30°C, according to the Arrhenius relationship $k_1(T_2) \cong k_1(T_1)\theta_A^{T_2-T_1}$, where the Arrhenius constant $\theta_A = 1.07$. When waste is added to a river, the rate of oxygen consumption may exceed the oxygen supply from reaeration.

The mass balance of the dissolved oxygen deficit can be written as

$$\frac{dD}{dt} = k_1 L - k_2 D. \tag{13.2}$$

This simple differential equation is solved to define the oxygen deficit D_t as a function of time t from the initial Load L_0 and oxygen deficit D_0. Recovery of the system happens when $k_2 > k_1$, which yields the oxygen deficit D_x as a function of the downstream distance x for a river flowing at a mean flow velocity V

$$D_x = \frac{k_1 L_0}{(k_2 - k_1)} \left(e^{-k_1 x/V} - e^{-k_2 x/V} \right) + D_0 e^{-k_2 x/V}. \tag{13.3}$$

From this relationship, also known as the Streeter–Phelps equation, we can calculate the downstream distance x_{max} with the maximum oxygen deficit from

$$x_{max} = \frac{V}{(k_2 - k_1)} \ln \left\{ \frac{k_2}{k_1} \left[1 - \frac{D_0(k_2 - k_1)}{k_1 L_0} \right] \right\}. \tag{13.4}$$

Of course, the dissolved oxygen content DO in the river corresponds to the initial dissolved oxygen level: $DO_x = DO_{sat} - D_x$.

The Federal water quality standard for early aquatic life stages is a daily minimum of 5 mg/l for warm water and 8.0 mg/l for cold water. The reproduction of fish and invertebrates can be greatly impaired when $DO < 4$–5 mg O_2/l. Under anaerobic conditions, many species are completely eliminated and the aquatic environment is plagued with odor and turbidity problems. Example 13.1 illustrates how to calculate the dissolved oxygen curve.

Example 13.1 Dissolved oxygen in Colorado streams

Examine the effects of water temperature on dissolved oxygen in mountain streams. Consider a constant CBOD loading of 20 mg O_2/l into the South Platte River near Denver, Colorado. Assume that the oxygen depletion rate $k_1 \cong 0.1$/day at 20°C doubles for every 10°C increase in temperature. The mean flow velocity in the river is held constant at 0.5 m/s or 43.2 km/day and the reaeration constant $k_2 = 0.6$/day is kept constant at any temperature. Use the Streeter–Phelps equations to calculate the dissolved oxygen concentration as a function of the downstream distance at temperatures from 10 to 30°C. Also determine at what distance the dissolved oxygen concentration will be at a minimum.

For instance, at 30°C, the dissolved oxygen $DO_{sat} \cong 6.05$ mg O_2/l at 1 mile high (1,600-m altitude) and the oxygen depletion rate is $k_1 \cong 0.2$/day. The distance at which the dissolved oxygen is at a minimum is calculated from Equation (13.4) as

$$x_{\max} = \frac{43.21 \text{ km}}{(0.6 - 0.2)} \ln \left\{ \frac{0.6}{0.2} \left[1 - \frac{0 \times (0.6 - 0.2)}{0.2 \times 20} \right] \right\} = 118 \text{ km.}$$

The maximum oxygen deficit is then calculated from Equation (13.3) as

$$D_x = \frac{0.2 \times 20}{(0.6 - 0.2)} (e^{-0.2 \times 118/43.2} - e^{-0.6 \times 118/43.2}) + 0 \times e^{-0.6 \times 118/43.2} = 3.85 \text{ mg } O_2/l.$$

The corresponding minimum dissolved oxygen concentration from Equation (13.3) is

$$DO_x = DO_{sat} - D_x = 6.05 - 3.85 = 2.2 \text{ mg } O_2/l.$$

Water temperature this high would most likely cause fish kill. The dissolved oxygen profiles at different water temperatures are shown in Figure E13.1.1.

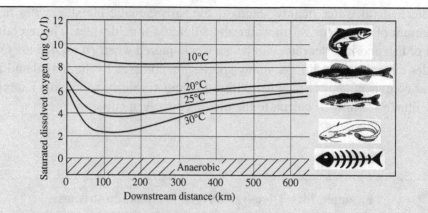

Figure E13.1.1. Spatial variability in dissolved oxygen as a function of temperature

It is clear that low water temperatures maintain high water quality throughout the entire reach. However, water quality can rapidly deteriorate during hot summers when DO < 4–5 mg O_2/l. The reader can also certainly foresee the potential impact of global warming on the quality of the aquatic environment in high mountain streams.

13.3.4 Advection and Dispersion in Rivers

The propagation of contaminants in natural rivers can be analyzed from solving the advection–dispersion equations as

$$\frac{\partial C}{\partial t} = -U\frac{\partial C}{\partial x} + K\frac{\partial^2 C}{\partial x^2} - kC, \tag{13.5}$$

where U in m/s is the mean river-flow velocity, K in m^2/s is the dispersion coefficient ($K \cong 12Uh$ where h is the flow depth), and k in s^{-1} is the settling rate of sediment, or the contaminant decay rate [$k = \omega/h$ where ω is the settling velocity from Equation (2.42)]. Two types of spill are considered: (1) an instantaneous spill and (2) a continuous spill.

For the instantaneous spill, the mass of contaminant M in grams is tracked at a distance x downstream from the spill as a function of time t in a river of width W and depth h and mean flow velocity U. Once the sediment is fully mixed over the entire cross-section area, i.e. $x > 10W^2/h$, the contaminant concentration is estimated from

$$C(x,t) = \frac{M}{2Wh\sqrt{\pi Kt}}e^{-\frac{(x-Ut)^2}{4Kt} - kt}. \tag{13.6}$$

The maximum concentration can be defined as a function of downstream distance x as

$$C_{\max} = \frac{M}{2Wh\sqrt{\pi Kt}}e^{-kt}. \tag{13.7}$$

As an example, six metric tons of contaminants are instantaneously spilled into a river 20 m wide, 1 m deep and flowing at 1.5 m/s. Dispersion starts after lateral mixing is complete, i.e. $x > 10 \times 20^2/1 = 4$ km, with $K \cong 12Uh = 12 \times 1.5 \times 1 = 18$ m^2/s. Assuming settling of silt-sized particles at $\omega = 1 \times 10^{-4}$ m/s and $k = \omega/h = 1 \times 10^{-4}$ s^{-1}, the maximum concentration 10 km from the spill is obtained at time $t = x/U = 10{,}000/1.5 = 6{,}667$ s from Equation (13.7), $C_{\max} = \frac{6 \times 10^6}{2 \times 20 \times 1\sqrt{\pi \times 18 \times 6{,}667}}e^{-6{,}667 \times 10^{-4}} = 125$ mg/l.

In the case of a continuous spill at a concentration C_{spill} and flow rate Q_{spill} in a river at a flow rate Q_{river}, the initial concentration is $C_0 = C_{spill}Q_{spill}/(Q_{spill} + Q_{river})$. A dimensionless settling parameter is defined as $\Gamma = \sqrt{1 + 4kK/U^2}$. The concentration for a constant contaminant release of duration T varies with distance x and time t as

$$C(x,t) = \frac{C_0}{2}\left\{ \begin{aligned} &e^{\frac{Ux(1-\Gamma)}{2K}}\left[\text{erfc}\left(\frac{x - Ut\Gamma}{2\sqrt{Kt}}\right) - \text{erfc}\left(\frac{x - U(t-T)\Gamma}{2\sqrt{K(t-T)}}\right)\right] \\ &+ e^{\frac{Ux(1+\Gamma)}{2K}}\left[\text{erfc}\left(\frac{x + Ut\Gamma}{2\sqrt{Kt}}\right) - \text{erfc}\left(\frac{x + U(t-T)\Gamma}{2\sqrt{K(t-T)}}\right)\right] \end{aligned} \right\}, \tag{13.8}$$

where erfc is the complementary error function $\text{erfc}(x) = 1 - \frac{2}{\sqrt{\pi}}\int_0^x e^{-a^2}\,da$, which is easily calculated with any mathematical package, or from the values given in Table 4.1. An example is shown in Figure 13.12. For complex cases, the numerical simulation using the Leonard scheme in Chapter 7 is also very helpful in practice.

13.3.5 Contaminants and Toxicity

The nitrogen cycle is important in a variety of river systems. Through bacterial fixation, plants can convert atmospheric nitrogen (N$_2$) into ammonia through a nitrogen-fixation process. Plants then incorporate ammonia and nitrates into nucleic acids and proteins that sustain the food chain. This

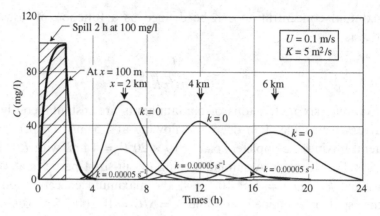

Figure 13.12. Example of advection–dispersion of contaminant

process is of environmental importance, and is related to the inverse break-down of organic-N into ammonia and then nitrites and nitrates. Ammonia is toxic to aquatic life in high concentrations. The transformation of ammonia to nitrates requires oxygen and the NBOD already discussed can lead to depletion of dissolved oxygen in rivers. The problem can also be exacerbated in agricultural areas by the excessive use of fertilizers, which can spur algae growth and cause water-quality problems.

Phosphorus is a nutrient-limiting plant growth in lakes and rivers. In general, phosphorus in parent minerals is not highly soluble, but potassium sulfates used in fertilizers are highly soluble. Also, urban runoff typically contains 0.6 mg/l (up to 1 mg/l) of total phosphorus, owing to the presence of phosphorus in cleaning agents and detergents. Untreated waters suffer from combined sewer overflows containing up to 15 mg P/l, which can easily cause major algae blooms, fish kill through oxygen depletion, and rotting algae mats on lake shores and river banks. This process of increasing the nutrients and organic matter while depleting dissolved oxygen in surface waters is called "eutrophication." Healthy phosphorus levels in surface waters are less than 0.01 mg P/l. There is a serious risk of eutrophication and fish kill when the phosphorus levels exceed 0.02 mg/l. For instance, consider a stream with a mean flow discharge of 2 m^3/s carrying 5 metric tons of phosphorus annually into a 10 km^2 lake. The annual water volume is $2 \times 365 \times 86{,}400$ m^3 and the loading is 5×10^6 g P/63×10^6 m^3 = 0.08 mg P/l, which is excessive and causing eutrophication. It is often desirable to identify the limiting nutrient, N or P, to prevent eutrophication. Given that the ratio N/P $\cong$ 7.2 in the biomass, it is considered that nitrogen is the limiting factor when the N/P < 7.2 in surface waters (Chapra, 1997).

Acid mine drainage is commonly the result of the exposure of pyrite (FeS_2) to the atmosphere. The oxidation process yields mine drainage that is acidic and rich in sulfates and ferric hydroxides. The main concern is the severely impaired water quality resulting from the coating of yellow ferric hydroxide on stream beds and the elimination of the macroinvertebrate habitat in the hyporheic zone. The low pH also dissolves metals from the parent rocks and minerals and increases both water hardness and dissolved solids.

Metals are found in dissolved, bound, or particulate form. It is useful to measure the ratio of the concentration in the sediment to the concentration in the water. The partition between phases depends on the chemical affinity of the surfaces and the presence of organic carbons. The soil–water partition coefficient defines the ratio of sorbed to dissolved from

$$K_d = \frac{[\ ]_{sorbed} \text{ in mg/kg}}{[\ ]_{dissolved} \text{ in mg/l}}. \tag{13.9}$$

as an example, K_d for zinc is given by the formula $K_d = [Zn_{sorbed} \text{ in mg/kg}]/[Zn_{dissolved} \text{ in mg/l}]$ such that K_d is measured in l/kg. Partition coefficients are large, such that log K_d is usually reported. Values of log K_d typically range from one to six, with low values indicating high solubility. Contaminants are primarily dissolved when K_d and suspended sediment concentrations are low. Conversely, contaminants adsorb to solids when K_d is high. For metals, the values of log K_d vary with pH and approximations such as log $K_d = A$ pH $+ B$ can be used for cadmium ($A_{Cd} = 0.49$ and $B_{Cd} = -0.6$), copper ($A_{Cu} = 0.27$ and $B_{Cu} = 1.49$), and zinc ($A_{Zn} = 0.62$ and $B_{Zn} = -0.97$) from Sauvé et al. (2000, 2003), Lu and Allen (2001), and Velleux (2005). As the pH increases downstream of mine waste piles, metals become less soluble and adsorb to solid surfaces or precipitate. Such analyses should not be over-simplified; the concentration of dissolved metals depends not only on the metal, but also on the mineral considered, the pH, and the content of organic material. Finally, it is equally important to determine the impact of metals concentration on living organisms.

The ability of a chemical to adsorb to organic matter can be measured by a simple experiment. A contaminant is mixed in a sealed container with equal amounts of octanol and water, with concentrations measured in both liquids after equilibration and separation. The ratio of the concentration in the octanol phase to the concentration in the water phase defines the octanol–water partition coefficient $K_{ow} = [\ ]_{octanol}/[\ ]_{water}$. The values of log K_{ow} for chemicals range from one to seven, with the greater values for hydrophobic contaminants. In the environment, the soil–water partition coefficient to organic carbons K_{oc} in l/kg is very closely correlated to K_{ow}, e.g. log $K_{oc} \cong$ 0.9 log K_{ow}. In soils containing more than 0.1 percent of organic carbons, the soil–water partition coefficient is obtained from $K = f_{oc}K_{oc}$, where f_{oc} is the fraction of organic carbon in the soil, and K_{oc} is the soil–water partition coefficient normalized to organic carbon in l/kg (or cm^3/g). The equilibrium contaminant concentration in the saturated pore waters is obtained from $C_{dissolved} = C_{adsorbed}/f_{oc}K_{oc}$. For instance, consider river sediment containing 5 percent of organic carbons and 12 mg/kg of contaminant with log $K_{oc} = 4$. The soil–water partition coefficient is calculated from $K = f_{oc}K_{oc} = 0.05 \times 10^4$ = 500 l/kg and the equilibrium contaminant concentration in the pore water is $C_{dissolved}$ = 12 mgkg/500 kgl = 0.024 mg/l. Wetlands can effectively remove contaminants, nitrates, and phosphates from surface waters. The very high fraction of organic carbons in wetlands increases the K value, which will increase adsorption and decrease the dissolved contaminant concentration. In comparison, sand-gravel filters (lower f_{oc}) would be less effective in removing hydrophobic contaminants.

The toxicity of contaminants in rivers is determined as the lethal concentration (LC_{50}) resulting in 50 percent mortality of a given aquatic organism. "Acute toxicity" refers to high concentrations over short (hours to days) periods and "chronic toxicity" refers to long durations (weeks to years). Acute concentration levels are much higher than chronic levels. For instance, the acute (2.4 mg/l) and chronic (0.012 mg/l) toxicity levels of mercury in a rainbow trout differ by two orders of magnitude. The duration of the exposure to a contaminant is also very important. For instance, the acute toxicity of copper to a rainbow trout decreases from $LC_{50} = 0.39$ mg/l for a 12-h exposure to 0.13 mg/l over 24 h, and to 0.08 mg/l over 96 h. In comparison, the 96-h toxicity of zinc and lead to a rainbow trout is 1 mg/l for Zn and 9.6 mg/l for Pb. For comparison, approximate median (and 90th-percentile) total metal concentrations from urban runoff are 0.05 (0.13) mg/l for Cu, 0.22 (0.7) mg/l for Zn, and 0.20 (0.50) mg/l for Pb. The handbook by Johnson and Finley (1980) provides a detailed list of acute toxicity levels to fish and invertebrates for a large number of chemicals.

"Bioaccumulation" refers to the accumulation of contaminants in the biomass of aquatic species from long-term exposure to contaminated waters. The bioconcentration factor (BCF) is the ratio of the equilibrium contaminant concentration in the biomass to the contaminant concentration in the ambient water. For the case of polychlorinated biphenyls (PCB), the BCF in l/kg is log BCF $\cong 4.4$. The BCF is equivalent to the partition coefficient and after a long period of exposure to contaminated waters, the PCB in a fish is

[PCB$_{fish}$ in mg/kg] = BCF[PCB$_{dissolved}$ in mg/l]. If the aqueous concentration of PCB is 12 ppt (1 part per trillion = 10^{-6} mg/l), the PCB concentration in a fish is [PCB$_{fish}$] = $10^{4.4}$[12 × 10^{-6} in mg/l] = 0.30 mg/kg. Eating 100 g of fish contains 0.03 mg of PCB, compared to 0.0012 mg in 100 liters of water.

Radionuclides only decompose according to the radioactive decay kinetics described by a decreasing exponential where $k = 0.693/t_{50}$ given the half-life t_{50}. The half-life of plutonium Pu^{239,240} (24,100 years) is extremely long compared to 432, 30, 28.8, and 22.3 years for Am^{241}, Cs^{137}, Sr^{90}, and Pb^{210}, respectively. The radioactivity unit is the curie (Ci), which corresponds to the decay rate of 1 g/s of radium. A related unit is the Becquerel (Bq), describing the decay of one atom per second, i.e. 1 Ci = 3.7 × 10^{10} Bq. The radioactivity level of natural rivers is very low and in describing it, Ci is commonly prefixed with terms like pico (pCi = 10^{-12} Ci), femto (fCi = 10^{-15} Ci), and atto (aCi = 10^{-18} Ci). The most soluble radionuclide is Sr^{90}, with a partition coefficient of 200 l/kg, while others are hydrophobic, with log K$_d$ = 5.7 l/kg for both Cs^{137} and Pu^{239,240}, and Pb^{210} at log K$_d$ = 7 l/kg.

In this section, the basic concepts presented can provide crude estimates while detailed analyses can be far more complicated. It is important to develop a feel for the issues involved and their complexities. To a large extent, each situation is fairly unique. Additional material can be found regarding modeling in Chapra (1997) and Di Toro (2001), regarding environmental engineering and water quality in Mihelcic (1999) and Novotny (2003), and regarding wetlands in Campbell and Ogden (1999) and WEF (2001).

13.3.6 Aquatic Life Cycle

The entire life cycles of aquatic species should be considered in many stream restoration projects. The life cycle of the Atlantic salmon is sketched as an example in Figure 13.13. The eggs hatch in early spring and feed off a relatively large yolk sac at the alevin stage. At the fry stage, they leave the breeding ground and look for food and shelter. They bear greenish colors at the parr stage and grow very rapidly as summer food abounds. After a couple years in the river, the smolts turn silvery in view of their seaward migration. They become vulnerable to ospreys during their period of adaptation to salt water in river estuaries. After a year at sea they may return as grilse to their native river. Adults will pursue the migration upriver to their spawning grounds and lay their eggs in gravelly spawning grounds called redds. The low river flows during winter and very cold winters can lead to freezing of the spawning grounds, as shown in Figure 13.13b. More information about

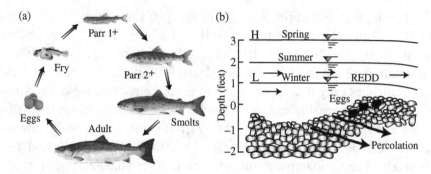

Figure 13.13. Life cycle of the Atlantic salmon and spawning grounds: (a) life cycle and (b) spawning ground

the Atlantic salmon is available in Bley and Moring (1988) and Hutchings and Jones (1998).

Habitat suitability curves can be established for different aquatic species. For instance, habitat suitability curves for the Atlantic salmon (*Salmo salar*) are shown in Figure 13.14 from Scruton and Gibson (1993) and Gibson and Cutting (1993). The survival rate in Figure 13.14d from Gibson (1993) indicates that the survival rate to spawning adult stage for the Atlantic salmon is less than 1 percent. The habitat suitability is also linked to the hydrologic cycle in terms of river discharge, as shown in Frenette et al. (1984). Critical periods for the salmon's survival can be identified as the period of deep freeze of the redds after spawning, low flows during the alevin and fry stages, food supply and predation during summer growth periods, predation during the seaward migration of smolts, and access to spawning grounds during the upstream migration.

13.3.7 *Endangered Species*

Endangered species is the second most threatened conservation status for wild populations, after critically endangered species. The introduction of nonindigenous species can disrupt ecosystems to the extent that native species become endangered. These invasive species compete with the native species and can in some cases upset the ecological balance leading to the decline of unexpected species.

In the United States, the Endangered Species Act is listing "endangered" or "threatened" species. The US Fish and Wildlife Service is responsible for classifying and protecting endangered species. The law is intended to protect

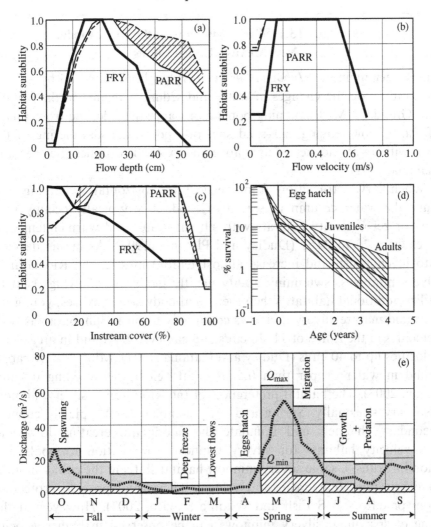

Figure 13.14. Atlantic salmon (a–c) habitat suitability curves, (d) survival rates, and (e) hydrologic cycle of the Matamek River (after Gibson, 1993; Gibson and Cutting, 1993; Frenette et al., 1984)

species that would otherwise be left alone for survival. There can be difficulties regarding the criteria for placing a species on (or removing it from) the endangered species list, and land restrictions and lobbying can be an obstacle. Several species have been delisted and the vast majority of listed species are recovering. The example of the silvery minnow (*Hybognathus amarus*) is briefly described in Case Study 13.1.

Case Study 13.1 Silvery Minnows of the Rio Grande, New Mexico, United States

The silvery minnow (*Hybognathus amarus*) has been impacted by natural and anthropogenic changes in water and sediment regime in the Middle Rio Grande, NM. Remaining populations appear to decline from a lack of warm, slow-moving, silt–sand substrate pools, and dewatering of the river with an abundance of nonnative and exotic fish species (Bestgen and Platania, 1991).

When larvae transition from vertical to horizontal swimming and leave the water column in search of food, they typically occupy very low- or zero-velocity habitats with silt substrates and warmer temperatures from 20 to 24°C (Dudley and Platania, 1997). As larvae develop into juvenile fish and increase almost tenfold in size, the Rio Grande silvery minnows' swimming ability greatly increases, allowing them to utilize additional habitat. When their total body size increases, young-of-year and mature silvery minnows commonly begin to inhabit areas with velocities in the range of 11–20 cm/s and can even be found in areas with velocities up to 40 cm/s (Dudley and Platania, 1997). Minnows are rarely found in water deeper than 0.5 m or at velocities exceeding 0.4 m/s (Leon, 2003). The habitat preferences of the Rio Grande silvery minnow also vary seasonally. Spring and early summer flows typically connect secondary channels and inundate the floodplain, creating low- and zero-velocity habitats for silvery minnows in addition to pools and shoreline runs. Usable silvery-minnow habitat shifts in the cooler months to debris piles that create low-velocity niches and provide protection from predators (US Fish and Wildlife Service, 2007). However, at the time of spawning, silvery minnows can be observed in higher-velocity areas commonly associated with main channel mesohabitats so that their semibuoyant, nonadhesive eggs are carried downstream (Platania and Altenbach, 1998).

In terms of substrate, though silvery minnows are almost always found over silt and sand substrates, individuals can be observed in mesohabitats such as pools and shoreline runs with sands and gravels, and, in extreme cases, cobbles (US Fish and Wildlife Service, 2007). As a small, heavy-bodied fish, up to 13 cm in length, the Rio Grande silvery minnow prefers habitats with low flow velocity, low to moderate flow depths, and silt or sand substrate (Cowley et al., 2006). Experiments showed that minnows can swim at velocities up to 0.4 m/s, with short sprints up to 1 m/s.

13.4 Stream Restoration Guidelines

In general, there is no "cookbook" approach to stream restoration projects. Environmental engineering practices are relatively new and rejuvenating perspectives are being developed and tested at this time. Solutions normally seek equilibrium conditions between the water/sediment regime and stream ecology. A "3E" rule is often applied to restoration projects: (1) Effective; (2) Environmentally sound; and (3) Economical. There is no point in using environmentally friendly measures if they are not effective. Costs can be considered only after environmentally acceptable and effective options are delineated. Stream restoration projects often have a relatively short timescale (~ 10 years). Valuable stream-restoration manuals include Reclamation (2015), NRCS (2007), and Brookes and Shields (1996). Ten guidelines are proposed to guide stream restoration practice:

(1) Objectives – the objective(s) of the project should be clearly delineated. Is the project focusing on restoration or rehabilitation?
(2) Past, present, and future – examine the past and present watershed and stream conditions. Did major events adversely impact the watershed (e.g. mining, forest fires, dams)? What are the prospective future conditions?
(3) Upper watershed – look at the geology, deforestation, land use changes, urbanization, climate, and extreme events. Examine water and sediment supply, flood frequency curves, sediment mass curves, water quality, etc.
(4) Downstream reach – look at possible changes in the downstream river reach in reservoirs, base level changes, headcutting, etc.
(5) Channel geometry – determine the equilibrium downstream hydraulic geometry in terms of width, depth, velocity, slope, and morphology.
(6) Aquatic habitat – define appropriate habitat conditions including low and high flows, pool-riffles, spawning grounds, shade, aeration, migration, predators, etc.
(7) Examine alternatives – identify several different stream rehabilitation schemes that would suit the engineering and environmental needs.
(8) Design selection – examine various alternatives, select the best and proceed with the design. Is the solution effective, environmentally sound, and economical?
(9) Construction – carefully plan the construction and consider the possible impact of possible extreme events during the construction period.
(10) Monitoring – things may not work as planned. Postconstruction analysis and monitoring should be pursued until the objectives have been met.

◆◆*Problem 13.1*

Consider the flow conditions described in Figure 13.6b and c in terms of flow velocity and shear stress on the floodplain. Can you explain why increasing floodplain vegetation at a given discharge level increases flow velocity and shear stress in the main channel? Do you also have an increase in velocity and shear stress on the floodplain?

Problem 13.2

For the channel conditions in Figure 13.6, calculate the flow depth, discharge, flow velocity, and shear stress in the main channel and the floodplain at a discharge of 900 m^3/s.
[*Answer:* For $n = 0.02$, $Q_0 = 164$ m^3/s, $v_c = 3$ m/s, and when $n = 0.1$, $Q_c = 778$ m^3/s, $v_0 = 0.46$ m/s.]

◆*Problem 13.3*

Consider the following data for the Rimac River in Lima, Peru from J. Kuroiwa (personal communication). The 5-year flow discharge is about 200 m^3/s and is passing through a 10-m-wide steep reach dropping 5 m of elevation over a distance of 130 m. Find the rock size needed for a rock ramp to stabilize the river bed. Also consider a local flow concentration up to 70 m^2/s. What rock size would be required to design a rock ramp to stabilize this reach?
[*Answer:* For the second case, $d_s = 52$ in.]

◆*Problem 13.4*

Consider the flow conditions described in Figure 13.6 in terms of flow velocity and shear stress on the floodplain. Which of the vegetation measures from Table 13.1 would be suitable to withstand the hydraulic forces?

◆◆*Problem 13.5*

During the migration of the Atlantic salmon (*Salmo salar*), the smolt length is approximately 15 cm and the returning adult length is approximately 70 cm. Estimate the swimming velocity and fish endurance from Figures 13.9 and 13.10 for these conditions.

◆*Problem 13.6*

When considering the floodplain flow conditions in Figure 13.6b, decide if the overbank flow conditions present suitable habitat for the Atlantic salmon given the characteristics shown in Figure 13.14a and b. What range of discharge would be at least 50 percent suitable for fry/parr?

◆◆*Problem 13.7*

Recalculate the dissolved oxygen from Example 13.1 when the Arrhenius constant θ_A is 1.047 instead of 1.07 and when the velocity is 1 m/s at a flow depth of 3 m.

◆◆*Problem 13.8*

Recalculate the example in Figure 13.12 when U and K are doubled. What would happen if k is also doubled? Could you also compare the calculations with the case where the entire mass of solids is injected instantaneously at a time of 1 h? What would be the difference in peak concentration downstream?

◆◆*Problem 13.9*

With the help of a classmate, find a topic of interest from this chapter. Carry out a review of five reports or papers from the literature and prepare a 20-slide presentation.

◆◆*Problem 13.10*

Marcos Palu (personal communication) reported that after the collapse of Fundao Dam in Brazil, the sediment concentration in the Doce River reached a value of 580 g/l for about 6 h. Consider the following river characteristics: river width 130 m, flow depth 3.5 m, flow velocity 1.1 m/s, bed slope 0.0005, and shear velocity 0.13 m/s. Use Equation 13.8 with the following dispersion coefficient, $K_d = 150$ m^2/s and the sediment settling rate $k = 0.0000036$ s^{-1} to estimate the sediment concentration as a function of time at Oculos station, located 94 km downstream. Compare the results with $k = 0$.

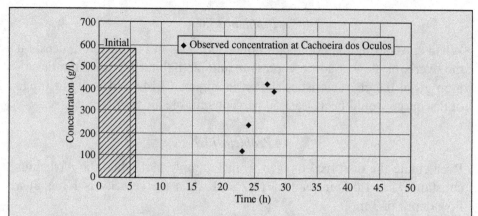

Figure P13.10. Dispersion of sediment in the Doce River

[*Answer:* The measured concentration was 118 g/l after 23 h, 239 g/l after 24 h, 419 g/l after 27 h and 387 g/l after 28 h.]

14

River Engineering

This chapter presents a discussion of various types of river-engineering structures. The reader explores a variety of solutions to a wide spectrum of river-engineering problems. This chapter covers an overview of the following river-engineering structures: (1) river flood control in Section 14.1; (2) river closure and jet scour in Section 14.2; (3) canal headworks in Section 14.3; (4) bridge crossings in Section 14.4; (5) navigation and waterways in Section 14.5; (6) dredging in Section 14.6; and (7) river ice in Section 14.7.

14.1 River Flood Control

This section discusses river-engineering solutions to flood control. The methods include floodways and channel conveyance in Section 14.1.1, and levees in Section 14.1.2.

14.1.1 Floodways and Channel Conveyance

Reservoirs provide direct flood control through surface runoff storage. A reservoir fills up when the inflow exceeds the outflow, thus reducing peak discharge and increasing low flows. An increase in flow storage is beneficial to hydropower, navigation, and irrigation. An example for the Mississippi River is shown in Figure 14.1.

Floodwave attenuation is also sometimes achievable by diverting floodwaters into a topographic depression near the river, e.g. into a detention basin, a swamp, or a bayou. The example of the Bonnet Carré Floodway near New Orleans is shown in Figure 14.2a. Floodway outlets consist of spillways or control gates, which are usually located on or near the floodplain to regulate the overbank flow discharge. Floodways are rarely used for long periods of time and it is important to operate the facilities periodically

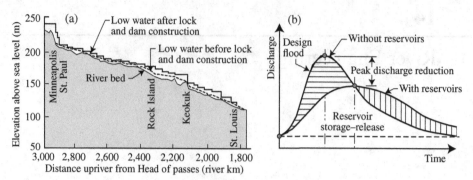

Figure 14.1. (a) Dams on the upper Mississippi River and (b) floodwave attenuation

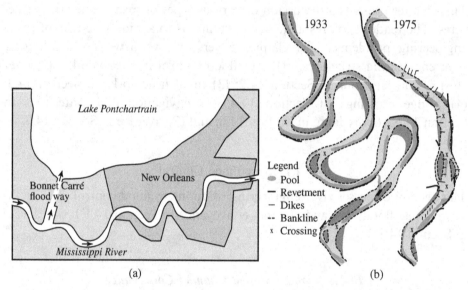

Figure 14.2. Flood control structures on the Mississippi River: (a) Bonnet Carré Floodway, Louisiana and (b) channel straightening near Greenville, South Carolina

to ensure proper usage in case of emergency. Floodwaters usually carry a significant suspended sediment load and sedimentation problems are possible near the structure. Local scour can also occur near the floodway control structures. Erosion problems may become very serious in the case of a possible capture of the river by the distributary.

Lowering the river flood levels is sometimes possible: (1) by reducing the riverbed roughness; (2) by enlarging the conveying cross section; and/or (3) by straightening the river channel and thus steepening the channel slope. Effective roughness reduction can include: (1) clearing the banks and floodplain from vegetation; (2) eliminating sandbars, islands, and rock out-crops; and (3) smoothing the banks with revetments. Removing bedforms in large rivers is not effective because dunes will shortly reappear.

Enlarging the conveying cross section can sometimes be done by deepening or widening the river channel. This approach can be successful when the sediment load is small and the bed is stable. However, widened cross sections typically fill up with sediment until the original bed level and slope have been restored. Dredging the riverbed is usually very costly and the channel improvement will likely be temporary.

Shortening of the river channel can be achieved by meander cutoffs, but such a channel rectification should be carried out with great care. If it is not fixed by embankments, the river might start meandering again. The example of the Mississippi River near Greenville, SC, is illustrated in Figure 14.2b. The locally steepened slope will increase the sediment load, causing erosion upstream and sedimentation downstream. In many cases, an increase in stream conveyance does not solve flood protection problems but merely passes them on downstream.

14.1.2 Levees

A levee is an earth embankment constructed along a stream to protect the land on the floodplain from being flooded (Figure 14.3). A floodwall is a concrete structure that serves the same purpose and is found in urban areas where insufficient space prohibits building a levee. For thousands of years, river levees have been built for the protection of people and their property against flooding. It still is the most expedient method for flood control.

Levees reduce the storage capacity of the floodplain and thus restrict the flow conveyance of the river. These reductions will induce higher water levels and limit floodwave attenuation. Because of the lateral migration of meandering streams, the levees should be placed at a fair distance from migrating channels, preferably outside the meander belt.

The elevation of levees is primarily determined by the stage of the design flood. River engineers perform a benefit–cost analysis to compare the flood-protection gain with the construction cost of raising the levee above the design level. In general, they measure the benefit of additional protection by considering the exceedance probability and the cost of flood damage at different water levels. Public, environmental, political, and military pressure may justify the large sums needed to protect against flooding. Other factors affecting the levee height include tides and wind waves in coastal areas and wide rivers. The design of the levee must prevent breaching as a result of

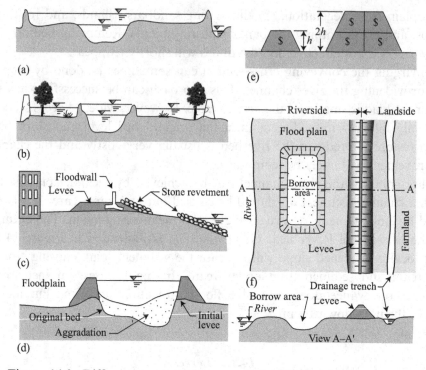

Figure 14.3. Different types of levees and floodwalls: (a) levee system, (b) double levees, (c) floodwall, (d) perched river, (e) levee material, and (f) borrowing pits

seepage, piping, sliding, slope failure, and revetment erosion. Levees should provide safety until their crests are overtopped.

The elevation of the levees may also depend on anticipated aggradation of the riverbed in coming years. River sedimentation within the levees causes the flow-rating curves to shift upward and levees must be raised beyond the initial level, as in e.g. the Yellow River and the Rio Grande. As sketched in Figure 14.3e, the volume of material required for levee protection increases with the square of the levee height, and the cost can become prohibitive in the case of perched rivers. The levee crown elevation is based on the design flood profile plus an allowance for settlement and freeboard.

Settlement of either the foundation or the levee embankment, or both, may result in a loss of freeboard, which provides additional levee height for factors that cannot be rationally accounted for in flood-profile computations. Compacted levees with sheep-foot or rubber-tired rollers require strong foundations of low compressibility and low water content of borrow material close to specified range. Borrow areas should be on the river side of the levee, and long, shallow borrow areas along the levee alignment are favored because there are fewer potential problems, as shown in Figure 14.3f.

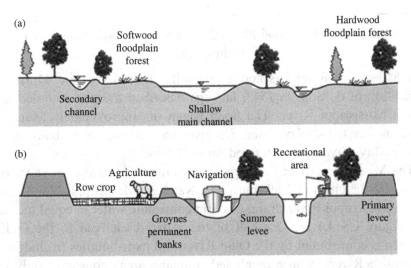

Figure 14.4. Double-leveed floodplain environment

Riverside borrow areas should be designed to fill slowly on rising stages and drain fully during the falling stages, and the bottom of the excavated area should keep distance from the levee.

For rivers where the flood hydrographs rise or fall quickly, the levee section should be analyzed for rotational instability following a rapid drawdown. Also, internal filters should be designed when less than desirable material is used for embankment construction. Underground seepage in pervious foundations beneath levees may result in: (1) the buildup of excessive hydrostatic pressure on the landside and (2) sand boils or in piping beneath the levee itself, unless seepage control measures are provided. Principal control measures include riverside impervious blankets, landside seepage berms, pervious toe trenches, pressure-relief wells, cutoff trenches, and sheet piling.

Agricultural use on the landside requires the control of the groundwater level. Local surface drainage of landside agricultural fields through the levees is usually possible with: (1) levees along the tributary channels; (2) culverts with flap gates; or (3) pumping plants. For major tributaries, levees are extended upstream to the limits of backwater influence. The drainage trench alongside the levee returns the landside waters into culverts equipped with flap gates that automatically open to permit outflow when the water level on the land side is higher than on the river side of the levee. Flap gates automatically close when the head differential is reversed. Ponding areas and pumping stations are required for land drainage when the main river stage remains high for extended periods of time. An example of environmentally friendly double leveed system is shown in Figure 14.4. Case Study 14.1 also illustrates the flood-control plan for the Mississippi River.

Case Study 14.1 Flood Protection Plan for the Mississippi River, United States

The Mississippi River has flooded its valley for a long time. During the expedition of de Soto in 1543, la Vega described the first recorded flood of the Mississippi River. The flood began on approximately March 10, 1543 and crested 40 days later. The river had returned to its banks by the end of May, having been in flood for ~80 days.

The Mississippi River discharges an average of 520 km^3 of water each year past the cities of Vicksburg and Natchez, Mississippi. Not all parts of the Mississippi River drainage basin contribute water in equal measure; see Figure CS14.1.1a. One-half of the water discharged to the Gulf of Mexico is contributed by the Ohio River and its tributaries (including the Tennessee River), whose combined drainage areas constitute only one-sixth of the area drained by the Mississippi River. By contrast, the Missouri River drains 43 percent of the area but contributes only 12 percent of the total water.

The Mississippi River now discharges an average of about 145 million metric tons of suspended sediment per year past Vicksburg and eventually to the Gulf of Mexico (Meade and Moody, 2010). The suspended sediment load carried by the Mississippi River to the Gulf of Mexico has decreased by one-half in the past 200 years. The decrease has happened mostly since the 1950s, as the largest natural sources of sediment in the drainage basin were cut off from the Mississippi River main stem by the construction of large reservoirs on the Missouri and Arkansas Rivers. This large decrease in sediment load from the western tributaries was counterbalanced somewhat by a five-to-tenfold increase in sediment load in the Ohio River as a result of deforestation and farming.

By the year 1879, the need for improvement of the Mississippi River had become widely recognized and Congress established the Mississippi River Commission (MRC) with the assignment and duty to protect the banks of the Mississippi River, improve navigation, prevent destructive floods, promote and facilitate commerce, trade, and postal service. Great floods in 1882 and subsequent years plagued the valley as levees were overtopped or crevassed. The rising flood heights between the levees caused many to question the total reliance on building levees to contain the river's floodwaters. Other approaches to improving flood protection were suggested but always rejected by the MRC in favor of a "levee-only" policy. The role of the MRC grew with each flood, finally culminating in the Flood Control Act of 1917, which authorized the MRC to construct an extensive program of flood protection with cost sharing by states and local interests. The program maintained the levee-only approach and included new levee construction and strengthening of existing levees to standards set 3 ft above the high water of 1912. By the end of 1926, the improved levee system had successfully passed several major high-water events. These successes convinced the MRC and the public that the flood-control problem was nearly solved.

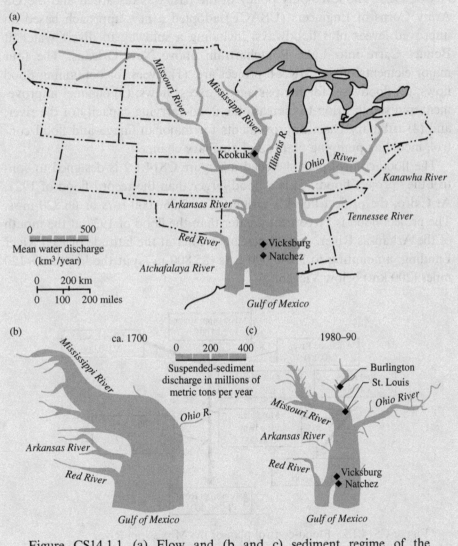

Figure CS14.1.1. (a) Flow and (b and c) sediment regime of the Mississippi River (after Meade and Moody, 2010)

The false sense of security in the Lower Mississippi Valley vanished in the flood of 1927, a natural disaster of great proportions. This tremendous flood extended over nearly 26,000 miles^2, killed more than 500 people, and drove more than 700,000 people from their homes. Thirteen crevasses in the main Mississippi River levees occurred, demonstrating that even the largest and strongest levees would not alone protect from flooding. To prevent a recurrence of the 1927 flood, Congress authorized the Mississippi River and Tributaries (MR&T) project in the Flood Control Act of 1928. The levee-only policy of the past was discarded and the US Army Corps of Engineers (USACE) adopted a new approach based on improved levees plus floodways, including a spillway to divert water at Bonnet Carré into Lake Pontchartrain above New Orleans. The four major elements of the MR&T project are: (1) levees for containing flood flows; (2) floodways for the passage of excess flows; (3) channel improvement and stabilization to increase the flood-carrying capacity of the river; and (4) tributary basin improvements for major drainage and flood control, including pumping plants and auxiliary channels.

The flood-control plan sketched in Figure CS14.1.2 is designed to control the "project flood." It is a flood larger than the record flood of 1927. At Cairo, the project flood is estimated at 2,360,000 ft^3/s or 66,820 m^3/s. The project flood is 11 percent greater than the flood of 1927 at the mouth of the Arkansas River, and 29 percent greater at the latitude of Red River Landing, amounting to 3,030,000 ft^3/s (85,800 m^3/s) at the location ~120 miles (200 km) below Vicksburg.

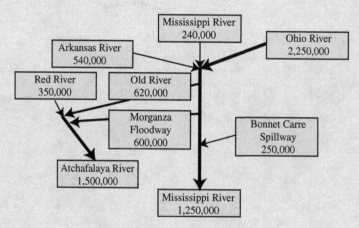

Figure CS14.1.2. Flood-control plan of the Mississippi River

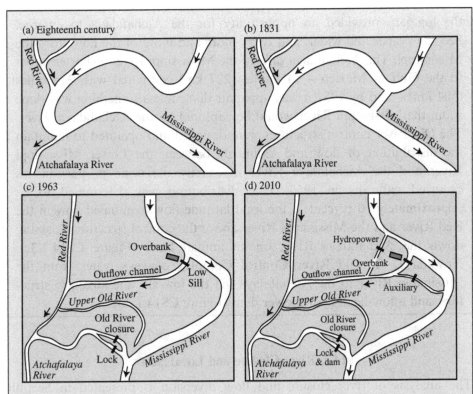

Figure CS14.1.3. Mississippi River near the Old River Control Complex, Louisiana

The Atchafalaya River. When the first European settlers arrived, they found the Red River emptying into the Mississippi at Turnbull's Bend and the Atchafalaya River a well-defined distributary flowing out of Turnbull's Bend a few miles to the South (Figure CS14.1.3a). In 1831, Captain Shreve dug a canal across the narrow neck of Turnbull's Bend. The river accepted the shortcut and abandoned its old channel, the upper part of which eventually silted up, leaving the lower section open, which became known as Old River (Figure CS14.1.3b). The Red River no longer flowed into the Mississippi River, but into the Atchafalaya River. Old River connected them to the Mississippi. The current usually flowed west from the Mississippi through Old River into the Atchafalaya. However, during floods on the Red River, the flow was sometimes reversed. For years the head of the Atchafalaya River was blocked by a massive "raft" – a 30-mile-long (50-km) log jam – that defied efforts of settlers to remove it. In 1839, the State of Louisiana began to dislodge the raft to open up the river as a free-flowing and navigable stream. The removal of the log jam provided an opportunity for the Atchafalaya to enlarge, become deeper and wider, and carry more and more of the flow from the Mississippi. The Atchafalaya offered the Mississippi River a shorter outlet to the Gulf of Mexico – 142 miles (227 km) compared with 315 miles (504 km) – and by 1951 it was apparent that, unless something was done soon, the Mississippi River would be captured by the Atchafalaya River. The Old River control structures were designed and operated to maintain the distribution of flow and sediments between the Lower Mississippi River and the Atchafalaya River in exactly the same proportions as occurred naturally in 1950. That distribution was determined to be approximately 30 percent of the total latitude flow (combined flow in the Red River and the Mississippi River above the control structures) passing down the Atchafalaya River on an annual basis (Figure CS14.1.3c). Nowadays, the Old River Control Complex conveys water from the Mississippi River to the Atchafalaya via the low sill, the auxiliary structure, and a low-head hydropower dam (Figure CS14.1.3d).

14.2 River Closure and Local Scour

The analysis of river closure and flow diversion is presented in Section 14.2.1. The partial river closure and use of cofferdams is covered in Section 14.2.2. Section 14.2.3 reviews methods to calculate local scour surrounding river closures.

14.2.1 *River Closure and Diversion*

The construction of dams requires closing off a river reach and diverting streamflow around a dry construction site. For the construction of earth dams and high concrete dams in deep, narrow canyons, the entire river channel is generally closed by building upstream and downstream earth and rock-fill cofferdams. For large dams on major rivers, the streamflow must be conveyed around the work site through diversion tunnels, as sketched in Figure 14.5. Such tunnels may serve a dual purpose: (1) flow diversion during construction and (2) regulated outlet works later. On small streams, a temporary flume or pipeline may be adequate to divert the streamflow around the construction site.

The size of a diversion tunnel and the cofferdam height depend primarily on the magnitude and timing of the design flood. The tunnel inlet must be low enough to allow flow through the tunnel as soon as the river is closed

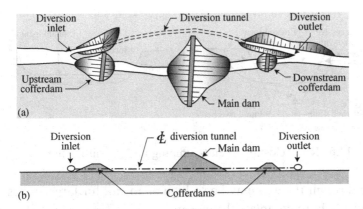

Figure 14.5. River diversion

off. A log boom or trash rack upstream of the inlet may be required for preventing partial blockage of a diversion tunnel by debris. Diversion tunnels are designed to be closed temporarily, usually by gates or stop logs, at the upstream portal to permit filling of the reservoir. If the tunnel is used solely for diversion, permanent tunnel closure is ensured with a concrete plug.

The construction sequence for channel diversions is generally as follows: (1) construct diversion tunnel; (2) construct temporary upstream cofferdam; (3) construct temporary downstream cofferdam; (4) dewater construction site within the cofferdams; (5) construct permanent upstream cofferdam; (6) construct main dam; (7) close the diversion tunnel; and (8) remove the downstream cofferdam. The initial closure is usually made with temporary cofferdams across the main river channel upstream and downstream of the construction site. The water in the work area between the temporary structures is pumped out for the construction of a permanent cofferdam in the dewatered area. The permanent upstream cofferdam can be constructed as an extension of the temporary upstream cofferdam or as a part of the main dam embankment section.

There are two effective procedures for river closures: (1) successively narrowing by end dumping from one or both riverbanks (Figure 14.6a) and/or (2) gradually raising a low sill across the channel by dumping uniformly across the opening from barges or a suspended cable (Figure 14.6b).

During a river closure by end-dumping, the upstream water level is raised from the backwater effect of choking the flow. The difference in water levels upstream and downstream of the closure causes a significant increase in flow velocity, causing material scour in the closure section. As velocity increases, all the fill material being dumped into the closure section may be transported downstream without closing the channel. Just before complete closure,

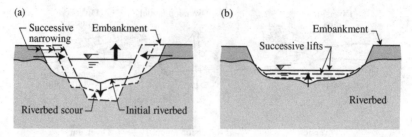

Figure 14.6. River closure (a) end-dumping and (b) sill-raising

velocities through the gap may be so high as to call for large stones and concrete blocks only to complete the closure.

In gradually raising the crown of a low sill across the opening, a mound is built up uniformly across the channel by dumping from a temporary bridge, a barge, or a suspension cable. As coarse material is added, the flow is choked and forces backwater effects upstream of the sill. Obstruction of the channel by the sill raises the upstream water surface and diverts part of the streamflow through the diversion tunnel. The reduced discharge over the sill eases the completion of the closure. A combination of sill-raising and end-dumping can be effective.

To determine the stability of material deposited in flowing water, Isbash's equation has been modified for river closures and stilling basins. The relationship between critical flow velocity V_c for beginning of motion and diameter of a stone d_s is

$$V_c = K_c [2g(G-1)]^{1/2} d_s^{1/2}, \qquad (14.1)$$

where K_c is a coefficient equal to 1.2 for river closure and 0.86 for stilling basins, g is the gravitational acceleration, and G is the specific gravity of the stone. The stone diameter d_s that can withstand a critical average flow velocity V_c is calculated from

$$d_s = \frac{1}{2g(G-1)} \left(\frac{V_c}{K_c}\right)^2. \qquad (14.2)$$

A simplified relationship in SI is that the stone diameter in m is approximately equal to $d_s \simeq V_c^2/40$. The weight F_W for a spherical stone of diameter d_s is calculated from

$$F_W = \gamma_s \frac{\pi d_s^3}{6} \qquad (14.3)$$

where γ_s is the specific weight of the stone. A stone size calculation for a river closure is presented in Example 14.1.

Example 14.1 River closure calculation

During the closure of a river, the water-surface elevation drops approximately $\Delta h \sim 6$ ft (1.82 m) within a short distance. Estimate the stable stone diameter d_s at a specific weight $\gamma_s = 165$ lb/ft³ or $G = 2.65$. Assuming conservation of energy in this short reach, the velocity gain in a 6-ft drop is

$$V_c = \sqrt{2g\Delta h} = \sqrt{2 \times 32.2 \times 6} = 19.6 \text{ ft/s (or 6 m/s)} \qquad \text{(E14.1.1)}\blacklozenge$$

In the case of a shallow opening, the velocity may correspond to the critical-flow depth, given the unit discharge q as $h_c = (q^2/g)^{1/3}$. The required stone size from Equation (14.2) is

$$d_s = \frac{1}{2 \times 32.2 \times 1.65}\left(\frac{19}{1.2}\right)^2 = 2.5 \text{ ft (or 0.77 m), and}$$

$$F_W = 165 \times \pi(2.5)^3/6 = 1{,}390 \text{ lb (or 6.2 kN).}$$

These results are comparable with those previously presented in Figure 8.12. We can substitute Equation (E14.1.1) into Equation (14.2) to find that the stone diameter increases linearly with Δh. The appropriate stone size is approximately two-fifths of the drop height.

14.2.2 Cofferdams

In large rivers, the entire flow cannot be diverted from the construction site and a partial river closure with flow contraction must be considered. Cellular cofferdams can enclose part of the streambed that can be dewatered by pumping to provide a dry construction area. The two-stage construction of a lock and dam is sketched in Figure 14.7.

Although cofferdams are temporary structures, each stage may be in place for several years, and they should be providing the optimum degree of protection. The cofferdam is generally one of the expensive items in river construction, particularly when flood stages are frequent or flashy in nature, and when the construction extends through several years. For preliminary studies, floods with the following recurrence intervals have been used for cofferdam design (Petersen, 1986): (1) a 3–5-year period of return for river construction not exceeding 2 years or (2) a 5–25-year flood when the construction period exceeds 2 years. Where long streamflow records are available, elevation of the second or the third largest historical flood can be used

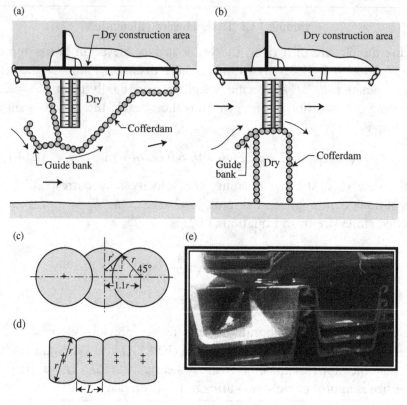

Figure 14.7. River closure with cofferdams: (a) stage I, (b) stage II, (c) circular cells, (d) diaphragm cells, and (e) sheet piles

to determine the cofferdam crest elevation including freeboard. For final design, the risk of potential damage from overtopping and the cost of a higher cofferdam can be assessed. If the risk of loss of life and property damage downstream is high, the standard project flood may be adopted for cofferdam design.

Cellular cofferdams are predominately designed with either circular or diaphragm cells (Figure 14.7c and d). As a rule of thumb, the cell width should approximately equal the cell height. The circular cells are connected by cells of circular arcs. Circular cells are stable so that each cell can be filled immediately upon completion, making it possible for equipment to work from one cell to the next. The diaphragm type consists of two walls of circular arcs connected by straight diaphragm sections, but individual cells are not stable. A template is used for guidance in driving the sheet piles, and when a cell is completed the template is removed and reused for the next cell. According to Petersen (1986), templates are somewhat easier to set for diaphragm cells than for circular cells. Several diaphragm cells are driven and then carefully filled by keeping the differences in fill elevation in adjacent cells within approximately 4–5 ft (1.2–1.5 m) to avoid distortion of the diaphragms. Circular cells usually require less steel for high structures. However, less steel is usually required for the diaphragm cells of low cofferdams. The failure of a circular cell is usually local, with damage limited to one cell, but diaphragm-cell failure may extend to adjacent cells.

Cofferdams are constructed by driving a wall of interlocking steel sheet piles through pervious material until the impervious clays or bedrock can be reached. Cofferdams are driven to bedrock whenever possible, and if the material is a soft shale, piles are driven 6 in.–1 ft (15–30 cm) into the shale. In pervious materials, the piles are driven as deep as possibly feasible to increase the seepage path and decrease the seepage flow into the work area and thus reduce the pumping costs. Following the completion of cofferdam construction, the work area inside the cofferdam is dewatered at a limited drawdown rate (usually not exceeding 5 ft in 24 h or 1.5 m per day). Sheet-pile cells are usually filled hydraulically with readily available local material such as sand, gravel, rock, or earth for stability. It is particularly important to fill the lower half of each cell with pervious material to facilitate drainage and avoid high hydrostatic forces on the cell walls. A concrete cap (~6 in. or 15 cm) is poured to protect the top of the cell fill against scour in the event the cofferdam is overtopped, to prevent infiltration of rainwater into the fill, and to provide a working surface for the contractor's equipment. The concrete cap also provides a base for sandbagging to provide additional protection height against overtopping beyond the cofferdam design flood elevation.

Considerable contraction and abutment scour can be expected around the tip of cofferdams during floods. It may be possible to change the cofferdam configuration to reduce local scour or to armor the bed where maximum scour is expected. In general, the scour model tests from Franco and McKellar (1968) indicated that: (1) the point of maximum scour is usually near the upstream corner of the cofferdam; (2) the depth, location, and area of maximum scour can be affected by the alignment of the upstream arm of the cofferdam; (3) the extent of scour along the cofferdam increases with the angle of the upstream arm; and (4) the area of maximum scour can be moved away from the cofferdam with a spur dike or guidebank at the upstream corner of the cofferdam. The scour-prevention guide banks are sketched in Figure 14.7a and b.

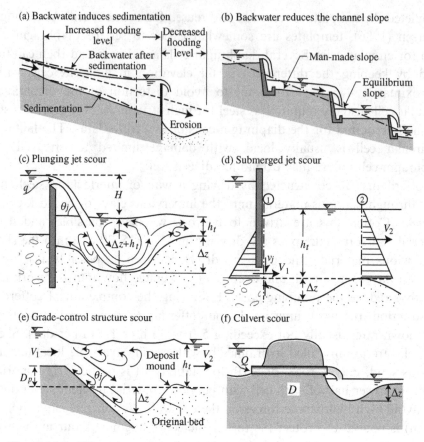

Figure 14.8. River slope changes and local jet scour

14.2.3 Local Scour

The construction of weirs and dams raises the upstream water level, thus artificially decreasing the channel slope and flow velocity. The artificial slope is therefore less than the equilibrium slope of the natural channel, as sketched in Figure 14.8a and b. There is also a need to dissipate energy and control scour. The bed scour caused by plunging jets or submerged jets needs to be considered in the stability analysis of hydraulic structures.

With reference to the four cases of jet scour illustrated in Figure 14.8, the scour depth Δz below plunging jets in Figure 14.8c can be estimated from the empirical equation of Fahlbusch (1994) as a function of unit discharge q, the jet velocity V_1 entering the tailwater depth h_t at an angle θ_j measured from the horizontal at the water surface, and the gravitational acceleration g:

$$\Delta z = K_p \sqrt{\frac{qV_1}{g} \sin \theta_j} - h_t. \tag{14.4}$$

The coefficient for plunging jet K_p depends on grain size with $K_p \simeq 20$ for silts, $5 < K_p < 20$ for sand, and $3 < K_p < 5$ for gravel.

Submerged jets flow entirely under the free surface, as sketched in Figure 14.8d. The discharge under a sluice gate downstream of a hydraulic structure has a considerable scour potential. Hoffmans and Verheij (1997) applied Newton's second law to a control volume and found the equilibrium scour depth Δz from

$$\Delta z = K_{sj} y_j \left(1 - \frac{V_2}{V_1} \right), \tag{14.5}$$

where V_2 is the outflow velocity, V_1 is the inflow velocity, y_j is the inflow jet thickness, and K_{sj} is a submerged jet-scour coefficient. The value of K_{sj} depends on the particle size and varies from $K_{sj} \simeq 50$ for silts to $20 < K_{sj} < 50$ for sand and $7 < K_{sj} < 20$ for gravel.

As sketched in Figure 14.8e, scour below grade-control structures and also sills and drop structures can be estimated from the method of Bormann and Julien (1991) as

$$\Delta z = \left\{ 1.8 \left[\frac{\sin \phi}{\sin(\theta_j + \phi)} \right]^{0.8} \frac{q^{0.6} V_1 \sin \theta_j}{[(G-1)g]^{0.8} d_s^{0.4}} \right\} - D_p, \tag{14.6}$$

where Δz is the scour depth, D_p is the drop height of the grade-control structure, q is the unit discharge, V_1 is the approach velocity, d_s is the particle diameter, g is the gravitational acceleration, G is the specific gravity, ϕ is the angle of repose of the bed material, and θ_j is the jet angle measured from the horizontal.

As sketched in Figure 14.8f, local scour below circular culvert outlets has been studied at CSU by Ruff et al. (1982). The scour depth Δz can be predicted as

$$\Delta z = 2.07 D \left(\frac{Q}{\sqrt{gD^5}} \right)^{0.45}, \tag{14.7}$$

where Q is the discharge, D the culvert diameter, and g the gravitational acceleration.

The local scour depths predicted from these empirical relationships give approximations subjected to improvements as more field and laboratory data become available. Hoffmans and Verheij (1997) discuss different methods, including the analysis of different data sets. Examples 14.2 and 14.3 illustrate how these empirical relationships can be used to estimate the scour depths below hydraulic structures.

Example 14.2 Scour depth below a sluice gate

A sluice gate sketched in Figure E14.2.1 is operated at an opening of 0.34 m. The upstream water level is 10 m and the downstream water level is 5 m. The unit discharge in this wide opening is 2 m²/s. The bed material is gravel with $d_{50} = 5$ mm and $d_{90} = 7$ mm. Estimate the scour depth:

Step 1: The flow velocity in the vena contracta is

$$V_1 = \frac{q}{y_j} = \frac{2}{0.34} = 5.9 \text{ m/s}.$$

Step 2: The outflow velocity is

$$V_2 = \frac{q}{h_t} = \frac{2}{5} = \frac{0.4 \text{ m}}{\text{s}}.$$

Step 3: The scour depth from Equation (14.5) with $K_{sj} = 15$ is

$$\Delta z = 15 \times 0.34 \left(1 - \frac{0.4}{5.9}\right) \simeq 4.8 \text{ m}.$$

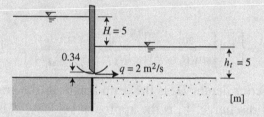

Figure E14.2.1. Scour below a sluice gate

Example 14.3 Scour depth below a grade-control structure

A broad-crested weir is built across a 50-m-wide river. The drop height is 2.25 m, and the face angle of the structure is 60°. The scour slope is approximately $1V{:}2H$ in non-cohesive material with $d_{50} = 2$ mm and $d_{90} = 2.5$ mm (see Figure E14.3.1). Estimate the scour depth when the river discharge is 160 m³/s:

Step 1: Determine $q = Q/W = (160/50) = 3.2$ m²/s, $D_p = 2.25$ m, $d_s = 0.002$ m, $\theta_j = \tan^{-1} 1/2 = 26°$, and $g = 9.81$ m/s².
Step 2: Estimate $\phi = 40°$ and $G = 2.65$.
Step 3: Assume critical flow at the sill crest with a flow depth $h_c = (q^2/g)^{1/3} = (3.2^2/9.81)^{1/3} = 1$ m.
Step 4: The approach flow velocity is $V_1 = q/h_c = 3.2/1 = 3.2$ m/s.

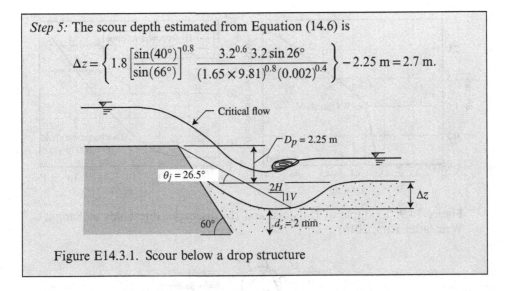

Step 5: The scour depth estimated from Equation (14.6) is

$$\Delta z = \left\{ 1.8 \left[\frac{\sin(40°)}{\sin(66°)} \right]^{0.8} \frac{3.2^{0.6}\, 3.2 \sin 26°}{(1.65 \times 9.81)^{0.8} (0.002)^{0.4}} \right\} - 2.25 \text{ m} = 2.7 \text{ m}.$$

Figure E14.3.1. Scour below a drop structure

14.3 Weirs and Canal Headworks

This section covers a brief presentation on the operation of weirs and canal headwork. The gate operation schemes for weirs are reviewed in Section 14.3.1. The canal headwork structures minimize sedimentation with sediment excluders, discussed in Section 14.3.2 and sediment ejectors, discussed in Section 14.3.3.

14.3.1 Weirs

Some river-engineering problems are inherent to the use of fixed weirs with variable discharge: (1) the water level upstream of the weir should be raised for water supply during periods of low river discharge; (2) during floods, high stages increase the risk of flooding and sedimentation; and (3) erosive forces near the weir at high discharge levels may also necessitate expensive bed and bank stabilization.

The operation of weirs on alluvial rivers can also be quite complex when sediment concentrations are high. For instance, it is usually desirable to maintain a high water level at a weir for irrigation, water supply, and run-of-the river hydropower production. However, high water levels increase the sediment-trap efficiency upstream of the weir. During floods, significant sedimentation volumes can require dredging costs far exceeding the benefits from hydropower generation. Kim (2016) defined the concept of hydraulic threshold for stage and discharge. Accordingly, the stage threshold is the stage above which hydropower revenues exceed sedimentation-dredging

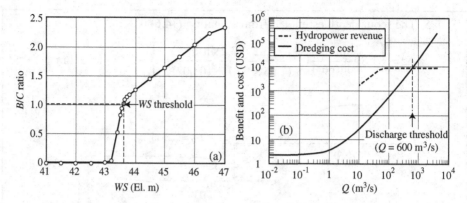

Figure 14.9. Example of (a) stage and (b) discharge thresholds at Sangju Weir (after Kim, 2016)

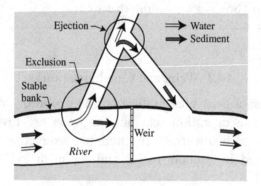

Figure 14.10. Sketch of sediment exclusion and ejection

costs. The discharge threshold is the flow rate below which dredging costs are less than hydropower revenues. In practice, the stage should be kept high at low flows, but the gates should be opened to keep the stage as low as possible when the discharge exceeds the discharge threshold. Figure 14.9 illustrates the concept of stage and discharge thresholds.

The regular supply of water to irrigation canals demands adequate control of the water level and sediment transport at the canal intake. Canal headworks control the water level with the construction of a weir. Sedimentation problems can be expected around canal headworks because the irrigation canals have milder slopes than natural channels. The decreased sediment-transport capacity upstream of the weir results in aggradation that will eventually reach the canal intake. As sketched in Figure 14.10, the two main strategies to control sedimentation problems near canal headworks are: (1) sediment exclusion to prevent the sediment from entering the canal and (2) sediment ejection to return sediment back to the natural river.

14.3.2 Sediment Exclusion

Sediment excluders prevent sediment from entering a canal by deflecting sediment away from the canal headworks, as sketched in Figure 14.10. Canal headworks may be located on the outside of a stable river bend. In river bends, the surface water (low sediment concentration) seeks the outside of the bend, while the sediment-laden streamlines near the bed are deflected away from the canal headworks. Secondary flows in river bends can be considered: (1) in stable rivers without lateral migration and (2) in rivers not carrying a significant load of floating debris, which would accumulate against the headworks.

In Figure 14.11a, natural flow curvature causes the near-bed sediment-laden streamlines to approach the convex bank of a river bend. In straight channels, guide walls can be designed to create favorable secondary flows (Figure 14.11b). Guide walls increase exclusion efficiency and are effective in continuous sluicing operations. Guide vanes can also be used to produce favorable secondary currents for sediment removal. Bottom and surface vanes, shown in Figure 14.11c and d, induce secondary currents that divert the bottom streamlines containing a heavy sediment load away from the canal headworks, and surface water containing a relatively light sediment load can be diverted through the canal headworks. Under steady flow, guide vanes (Figure 14.11e) can exclude practically the entire bedload, but they should be avoided when the flow is unsteady.

14.3.3 Sediment Ejection

Sediment ejectors remove sediment from the canal system. Ejection methods do not preclude other sediment-control methods, but provide a factor of safety should the exclusion devices fail to perform according to design. A sediment ejector should be located near the head regulator, where settling is predominant – otherwise the sediment deposited in the canal may not reach the ejector.

The tunnel ejector shown in Figure 14.12a can eject bedload from straight canals. A ramp is placed upstream of the tunnel ejector with a ramp height level with the roof of the ejector. The incoming bed material rolls down the ramp and enters the tunnel. It is then returned to the main river downstream of the headworks weir. For small discharges, the roof can be omitted and the guide vanes alone will deflect the bedload. The vortex tube in Figure 14.12b is efficient at removing coarse bedload material up to gravels in small canals. The main feature of the vortex tube is a pipe with a slit opening along the top side. As water flows over the tube, the shearing action sets up a vortex motion

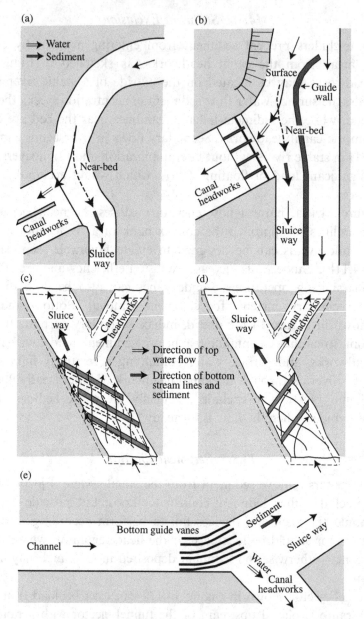

Figure 14.11. Sediment exclusion with (a) natural curvature, (b) guide walls, (c) bottom vanes, (d) surface vanes, and (e) bottom guide vanes

that catches the bedload as it passes over the lip of the opening. It is important to have sufficient hydraulic head difference for the vortex to be efficient.

Settling basins, sketched in Figure 14.12c, remain popular for the removal of both bedload and suspended load at canal headworks. The underlying

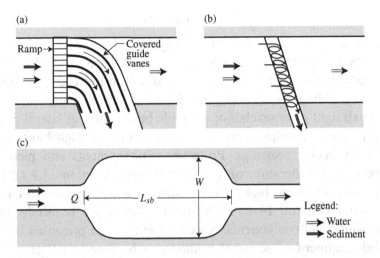

Figure 14.12. Sediment ejection with (a) tunnel ejector, (b) vortex tube, and (c) settling basin

principle is simply to provide a section wide and long enough to reduce flow velocity and allow sediment to settle out. The basic relation for the design of settling basins is the trap efficiency T_E, which represents the ratio of the weight of sediment settling in the basin to the weight of sediment entering the basin. The trap-efficiency relationship was derived in Equation (2.52):

$$T_E = 1 - e^{\frac{-\omega W L_{sb}}{Q}} = 1 - e^{\frac{-\omega A_{sb}}{Q}}, \tag{14.8}$$

where W is the settling basin width, ω is the settling velocity of sediment, L_{sb} is the length of settling basin, and Q is the flow discharge. At a given discharge and particle size, the efficiency depends on the surface area of the settling basin A_{sb}. A calculation example is detailed in Example 14.4.

Example 14.4 Calculation of a settling-basin length

Determine the length of a 30-m-wide settling basin designed to trap 90 percent of $d_s = 0.05$ mm silt entering a canal at a discharge of $Q = 3$ m³/s. The unit discharge $q = Q/W = 3/30 = 0.1$ m²/s. The settling velocity of the silt particles is approximately $\omega = 2$ mm/s.

The settling-basin length is calculated from Equation (14.8), solved for L_{sb} as $L_{sb} = \frac{q}{\omega}[-\ln(1 - T_E)] = \frac{0.1 \text{ s}}{0.002 \text{ m}} \frac{\text{m}^2}{\text{s}}[-\ln(1 - 0.9)] = 115$ m. Repeating the application to different basin widths shows that identical results are obtained as long as the settling-basin area $A_{sb} = L_{sb}W = 3{,}450$ m².

14.4 Bridge Scour

River engineers are concerned with bridge crossings in regard to: (1) the careful selection of the bridge site to minimize the total bridge costs and (2) the protection against possible structural failure from scour undermining the roadway embankments and piers. River crossings should preferably be located in straight river reaches or in stable bends without lateral migration. Sites with narrow floodplain width, rock outcrops, or high bluffs are good locations for bridge crossings. Protecting embankments and piers against scour requires consideration of: (1) general scour (Section 14.4.1); (2) contraction scour (Section 14.4.2); (3) abutment scour (Section 14.4.3); and (4) pier scour (Section 14.4.4). The total scour depth is usually obtained from the sum of all components. Selected methods are presented below, and a detailed treatment of scour at bridges can be found in Richardson and Davis (2001). It must be remembered that all scour equations serve as approximations and engineering judgment should be exercised.

14.4.1 General Scour

Progressive degradation or aggradation can be associated with changes in the river regime caused by natural processes or human activities on the stream or watershed. Factors that affect long-term bed changes are: (1) dams and reservoirs both upstream and downstream of the bridge site; (2) changes in watershed land use such as urbanization, deforestation, forest fires; (3) channel stabilization and rectification; (4) natural or artificial cutoff of a meander bend; (5) changes in the downstream base level of the bridge reach, including headcuts; (6) gravel mining from the streambed; (7) diversion of water into or out of the stream; (8) lateral migration of a river bend; and (9) thalweg shifting in a braided stream.

The engineer must assess the present state of the stream and watershed and consider planned future changes in the river system. Alluvial rivers can change cross-section geometry, and the location of the thalweg may also shift considerably with time. The bed level may also vary considerably as a consequence of traveling of sand waves, bedforms, alternate bars, and riffles and pools in relatively straight river reaches. Live-bed scour occurs when the upstream sediment transport can be transported into the scour hole. Conversely, clearwater scour refers to conditions without bed-sediment transport.

When general scour is expected to be significant, as sketched in Figure 14.13, it is often advisable to construct a grade-control structure downstream of the bridge crossing. The purpose of the structure is to control the bed elevation at the bridge site and protect the abutment and piers.

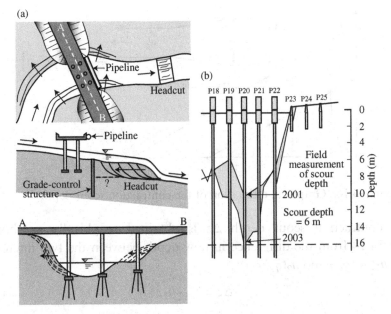

Figure 14.13. Alluvial channel deformation near bridge crossings (after Park et al., 2008). (a) Migration and headcut and (b) Nakdong River

Pipelines and gas lines should preferably be placed on the downstream side, so that they are not punctured by large floating debris during floods. It is also important to measure the scour depths at bridge crossings during floods, as shown for the Nakdong River in Figure 14.13b.

14.4.2 Contraction Scour

Contraction scour results from flow acceleration in river contractions (Nordin, 1977). The approach flow depth h_1 and width W_1, and the flow velocity V_1, give the flow rate $Q = W_1 q_1 = W_1 h_1 V_1$. To pass the same flow discharge in the approach and contracted sections requires

$$q_2 = \frac{W_1}{W_2} q_1, \tag{14.9}$$

where $q_1 = h_1 V_1$, and $q_2 = h_2 V_2$ uses subscript 2 for the contracted section.

The sediment-transport rate is $Q_s = W_1 q_{s1}$ where the unit sediment discharge q_{s1} is a function of flow depth and velocity, e.g. the Simons-Li-Fullerton method in Julien (2010). At equilibrium, the sediment-transport rate in the contracted section must be

$$q_{s2} = \frac{W_1}{W_2} q_{s1}. \tag{14.10}$$

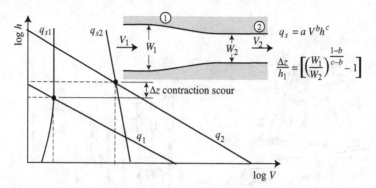

Figure 14.14. Graphical estimation of the contraction scour depth

The contraction scour depth Δz is obtained graphically as shown in Figure 14.14 from the difference in flow depth between the two interception points $h_1(q_1, q_{s1})$ and $h_2(q_2, q_{s2})$:

$$\Delta z = h_2 - h_1. \tag{14.11}$$

The contracted scour depth represents an average over the channel width and symmetry is assumed in the calculation. The analysis of the Rimac River by Kuroiwa et al. (2011) is a good example of engineering analysis of narrow rivers.

14.4.3 Abutment Scour

Abutments and spur dikes can have different shapes and angles to the flow. As sketched in Figure 14.15, the tip of an earth and rockfill abutment will generally have a spill-through shape when compared with sheet piles standing as vertical walls.

The abutment scour depth depends on the amount of flow intercepted by the bridge abutments. The equilibrium scour depth for local live-bed scour in sand is

$$\frac{\Delta z}{h_1} = 1.1 \left(\frac{L_a}{h_1}\right)^{0.4} \mathrm{Fr}_1^{0.33} \quad \text{for a spill – through abutment,} \tag{14.12a}$$

$$\frac{\Delta z}{h_1} = 4\,\mathrm{Fr}_1^{0.33} \quad \text{when } L_a/h_1 > 25. \tag{14.12b}$$

For vertical abutments, the scour depth nearly doubles (Liu et al., 1961; Gill, 1972).

$$\frac{\Delta z}{h_1} = 2.15 \left(\frac{L_a}{h_1}\right)^{0.4} \mathrm{Fr}_1^{0.33}, \tag{14.13}$$

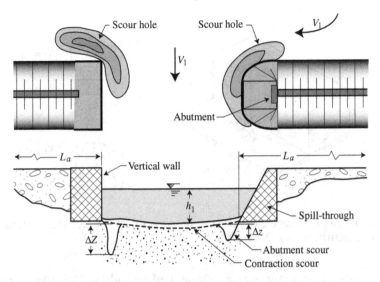

Figure 14.15. Abutment and contraction scour shape and location

where Δz is the abutment scour depth, h_1 is the average upstream flow depth in the main channel, $\mathrm{Fr}_1 = V_1/\sqrt{gh_1}$ is the upstream Froude number, and L_a is the abutment length normal to the river bank.

14.4.4 Pier Scour

Pier scour is induced by the swirling flow around the pier (Figure 14.16a). The flow moves downward on the front side of the pier and swirls around both sides of the base of the pier in the shape of a horseshoe. The horseshoe vortex removes bed material away from the base region in front of and along the side of the pier. The vortex strength decreases as the depth of scour increases. The equilibrium scour depth is reached when the transport rates entering and leaving the scour hole are equal.

The Colorado State University (CSU) equation calculates the pier scour as

$$\frac{\Delta z}{h_1} = 2.0 K_1 K_2 K_3 \left(\frac{a}{h_1}\right)^{0.65} \mathrm{Fr}_1^{0.43}, \qquad (14.14)$$

where Δz is the scour depth, h_1 is the flow depth just upstream of the pier, K_1 is the correction factor for pier shape from Figure 14.16b, K_2 is the correction factor for the flow orientation angle θ_p relative to the pier, K_3 is the correction factor for bed forms, a is the pier width, and $\mathrm{Fr}_1 = V_1/\sqrt{gh_1}$ is the upstream Froude number. The correction factor K_2 can be calculated from $K_2 = [\cos\theta_p + (L_p/a)\sin\theta_p]^{0.65}$, where L_p is the pier length. The factor

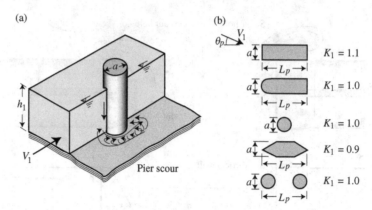

(a)   (b)

Pier scour

Figure 14.16. Pier scour: (a) Horseshoe vortex and (b) pier shape factor K_1 (after Richardson and Davis, 2001)

K_3 is 1.1 for clear-water scour, plane bed, antidunes, ripples, and small dunes. It increases to 1.2 when dunes are between 3 and 9 m in height.

The CSU equation has been slightly modified to account for: (1) scour in coarse bed material ($d_{50} \geq 2$ mm and $d_{95} \geq 20$ mm) and (2) scour in shallow flows around wide spread footings. Two additional scour-reduction factors, K_4 and K_5, have been developed for those conditions in Richardson and Davis (2001). While the factor K_4 is based on substantial research from Molinas et al. (1998), engineering judgment should be used in applying K_5 because it is based on limited data from flume experiments.

As a rule of thumb, the scour depth is approximately 2–3 times the pier width. In most cases, the pier scour depth remains less than three times the pier width. Exceptions include elongated piers that are not properly aligned with the flow direction. Debris-laden streams or ice jams can cause additional scour. The extra scour resulting from debris or ice accumulation against the pier must be given due consideration. Ice and debris can both produce static and dynamic pressure up to 200 Pa on bridge piers. These forces can largely exceed the dynamic forces exerted on the piers by the flowing water. Additional scour may be found under pressure flow which occurs when the flow depth reaches the bridge deck. Except for very wide bridges and shallow flow depths, the scour from pressure flow will most likely be observed downstream of the bridge crossing.

The extent to which a pier footing or pile cap affects local scour at a pier is not clearly determined. Under some circumstances the footing may serve to impede the horseshoe vortex and reduce the depth of scour. In other cases in which the footing extends above the streambed into the flow, it may increase the effective width of the pier, thereby increasing the local pier scour. In the calculations, the pier width can be used if the top of the pier footing is slightly above or below the streambed elevation. The footing width is used when the pier footing projects well above the streambed or when general scour is expected in the river reach.

Scour may also be reduced if riprap is placed around bridge piers. A riprap layer twice the design diameter thick should extend between 1.5 and 6 times the pier width. Ample additional information is available in the FHWA publications HEC18 (2012), HEC20 (2012), HEC23a, b (2009), and Melville and Coleman (2000). Example 14.5 illustrates how to calculate the local scour depth around a bridge pier.

Example 14.5 Abutment and pier scour depths

A 200-m-long bridge is to be constructed over a sand-bed channel with 300-m-long spill-through abutments $1V:2H$. Six rectangular bridge piers 1.5 m thick and 12 m long are aligned with the flow. Under a 100-year flood, the upstream flow velocity is 3.75 m/s and the flow depth is 2.8 m upstream of the piers. Estimate (a) the abutment scour depth and (b) the pier scour depth.

(a) Abutment scour

Step 1: The approach Froude number is

$$\text{Fr}_1 = \frac{V_1}{\sqrt{gh_1}} = \frac{3.75}{\sqrt{9.81 \times 2.8}} = 0.71.$$

Step 2: The ratio of abutment length to flow depth $L_a/h_1 = (300/2.8) = 107$.

Step 3: The abutment scour depth is calculated with Equation (14.12b) because $L_a/h_1 > 25$

$$\Delta z = 4h_1 \text{Fr}_1^{0.33} = 4 \times 2.8 \times 0.71^{0.33} = 10 \text{ m.}$$

Guidebanks in Section 9.2.3 should be considered to reduce abutment scour.

(b) Pier scour

Step 1: The Froude number is $\text{Fr}_1 = 0.71$.

Step 2: The rectangular pier shape corresponds to $K_1 = 1.1$ from Figure 14.15b.

Step 3: The pier length/width ratio $L_p/a = (12/1.5) = 8$ and $K_2 = 1.0$ as long as the pier is aligned with the flow, $\theta_p = 0$.

Step 4: The pier scour depth calculated from the CSU equation for small dunes $K_3 = 1.1$

$$\Delta z = 2h_1 K_1 K_2 K_3 \left(\frac{a}{h_1}\right)^{0.65} \text{Fr}_1^{0.43}$$

$$= 2 \times 2.8 \times 1.1 \times 1 \times 1.1 \times \left(\frac{1.5}{2.8}\right)^{0.65} \times 0.71^{0.43} = 3.9 \text{ m}$$

Type	Length (ft)	Breadth (ft)	Draft (ft)	Capacity (tons)
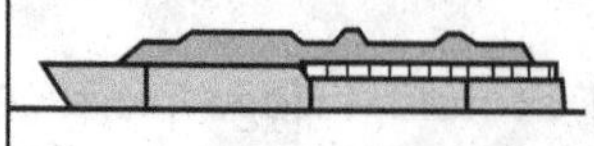 Open hopper barges Standard Jumbo Super jumbo	175 195 250–290	26 35 40–52	9 9 9	1,000 1,500 2,500–3,000
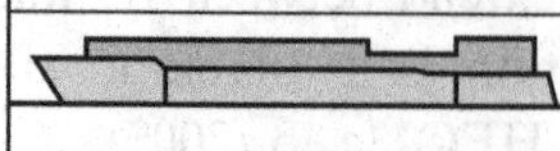 Covered hopper barges Standard Jumbo	175 195	26 35	9 9	1,000 1,500
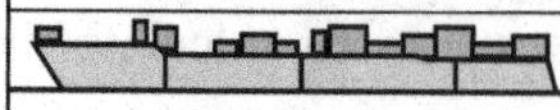 Integrated chemical and petroleum barges	150–300	50–54	9	1,900–3,000
 Towboats	65–160	25–50	5–9	300–11,000 hp

Note: 1 m = 3.28 ft; 1 ton = 2,000 lb = 8.96 kN

Figure 14.17. Barge and towboats characteristics (after Petersen, 1986)

14.5 Navigable Waterways

Rivers can be opened for navigation to generate benefits from commercial shipping. The reduced cost of moving commodities by means of waterways instead of other modes of transportation is compared with the costs of construction, operation, and maintenance. Navigation requirements are discussed in Section 14.5.1, followed with waterway alignment and cutoffs in Section 14.5.2, and locks and dams in Section 14.5.3.

14.5.1 Navigation Requirements

General requirements for channel depth, width, and lock dimensions for commercial navigation are governed by a number of factors, including the type and volume of future tonnage, and the type and size of barges and tows used on the waterways. Tows on the Lower Mississippi frequently have more than 40 barges and transport 50,000–60,000 tons (450–540 MN) of cargo (Figure 14.17). High-powered towboats have an average of ~3,000 hp (2.2 MW), with three 10-ft (3-m) diameter, five-bladed stainless-steel propellers housed in Kort nozzles.

Three barge types are common: (1) open-hopper barges for transporting coal, sand and gravel, and sulfur; (2) covered-hopper barges for grain and mixed cargo; and (3) tank barges for petroleum and chemicals (Figure 14.17). Barge sizes vary around a standard 35 ft (10.7 m) in width and 195 ft (60 m) in length.

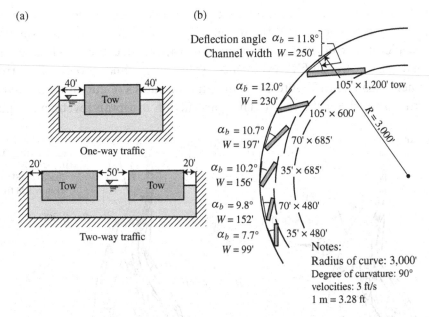

Figure 14.18. Navigation in straight and sinuous channels: (a) straight channel and (b) river bend

Pilots navigate towboats at the stern of the tow with control of the engine thrust and direction of the rudder. The navigation width depends on channel alignment, size of tow, and whether one-way or two-way traffic is planned. One-way traffic may be adequate when the traffic is light if the reach is relatively straight with good navigability and if passing lanes are provided. Multiple-way traffic permits heavy traffic to move faster, except when tows are meeting or passing. Figure 14.18a shows the recommended channel widths for commercial navigation in straight channels.

Wider navigation channels are required in bends. The crab angle α_b varies with: (1) the radius of curvature of the channel; (2) the speed, power, and design of the craft, and wind forces; (3) whether the tow is empty or loaded; (4) whether the traffic is going up or down the river; and (5) the flow pattern. The navigation-channel width is a direct function of the drift angle, which is larger for down-bound tows than for up-bound tows. For example, the drift angle for down-bound tows of various sizes in a bend with a 3,000-ft (914-m) radius is shown in Figure 14.18b.

14.5.2 Waterway Alignment and Cutoffs

Navigation is preferable in fairly straight channels or bends with long radii of curvature. In wide channels, the waterway alignment can be controlled with revetments, spur dikes, and longitudinal dikes. Once the radius of cur-vature has been determined, the design of river-alignment structures is done by considering the cost and performance of various structures including long-itudinal dikes, trench-fill revetments, spur dikes, and levees. The river-alignment example in Figure 14.19 maintains a radius of curvature of 8,000 ft (2.7 km) throughout the reach. The construction of riprap revetments, spurs,

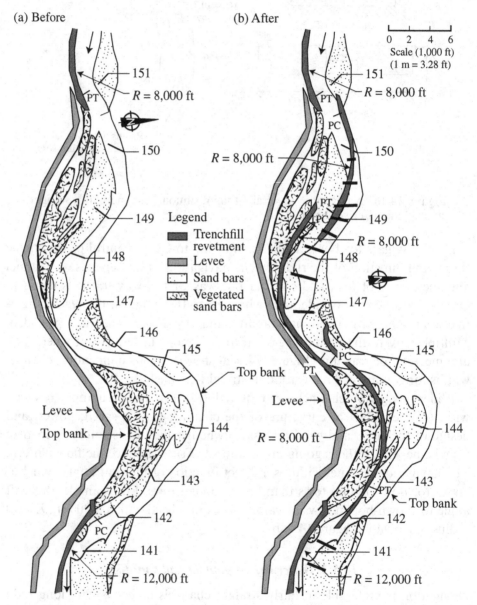

Figure 14.19. Example of river navigation control (from Petersen, 1986)

and longitudinal dikes provides flow control as long as dikes overlap and tie-backs prevent flanking of the structures. Note the increase in slope that is due to the reduced reach length.

Engineered neck cutoffs can improve river alignment for navigation, as shown in Figure 14.20. Pilot channels are excavated from the downstream end for the construction of artificial cutoffs. Pilot channels have $1V{:}3H$ side-slopes with bottom widths from 50 to 200 ft (15 to 60 m) and bottom elevations from 6 to 12 ft (2 to 4 m) below the low-water reference plane. When the meander length far exceeds the pilot-channel length, the slope ratio is favorable for the natural enlargement of the pilot channel. Wider and deeper pilot channels are required when there is little length or slope advantage. A parallel trench is excavated for the placement of a riprap revetment. Earth plugs near the upstream end of the excavated pilot channels are left in place to block low flows. Plugs are designed to be overtopped and washed out during floods after the revetment is completed.

Case Study 14.2 illustrates the morphological changes of the Rhine/Waal Rivers in The Netherlands. Case Study 14.3 reviews the engineered chute cutoffs of the Mississippi River.

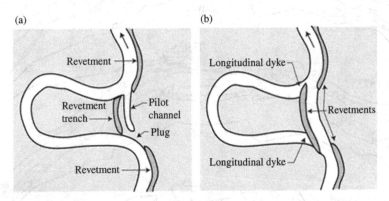

Figure 14.20. (a) Engineered pilot channel and (b) meander cutoff

---

Case Study 14.2 Control of the Rhine and the Waal Rivers, The Netherlands

Living communities along the Rhine River have coped with flood disasters for centuries RIZA (1999) and ten Brinke (2005). In winter, drifting ice formed ice jams that raised the water levels and increased the risk of levee overtopping. Levees on the floodplain and groynes or spur dykes along the river bank have been built since the eighteenth century to channelize the river, as shown in Figure CS14.2.1. The width of the main channel decreased from more than 500 to 260 m. Islands and sandbanks were removed, and the banks were protected from erosion with a series of groynes spaced approximately every 200 m on both sides of the river. As a result, the bed of the Rhine River has eroded considerably over the past 100 years. In the past 60 years, the Rhine's channel at Lobith has degraded by an average of 1–2 cm each year. The navigable channel of the Waal River is 150-m wide and a flow depth of at least 2.5 m is exceeded 95 percent of the time. The current river-engineering challenge is to increase the navigation width and depth without detrimental effects to the environment. Measures include the use of bottom vanes and bendway weirs to widen the bends and local extension of groynes and longitudinal dykes to increase the flow depth.

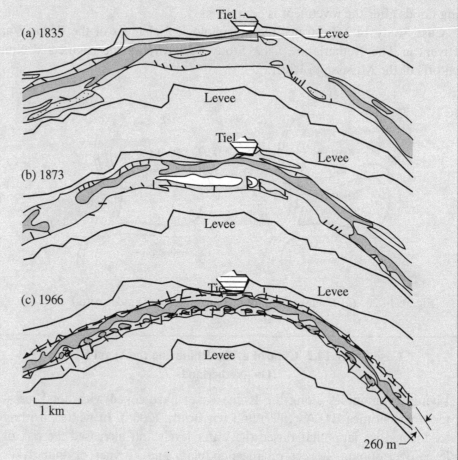

Figure CS14.2.1. The Waal River near Tiel in the Netherlands (after Jansen et al., 1979)

Case Study 14.3 Choctaw Bar on the Mississippi River, Mississippi, United States

The Choctaw Bar area is located on the Mississippi River between River Mile RM 557–565 AHP (Above Head of Passes), i.e. ~17 river miles upstream of Greenville, MS. Choctaw Bar is close to the mainline levee where crevasses have occurred several times. As shown in Figure CS14.3.1, Choctaw Bar formed a chute cutoff near the right bank of the river, and the Eutaw-Mounds revetment is located on the left bank (Combs, 1994). Stabilization of the left bank began in the 1940s and cost more than US $10 million. In addition, dikes in the vicinity of Chicot Landing were constructed in 1968 for ~$1.2 million.

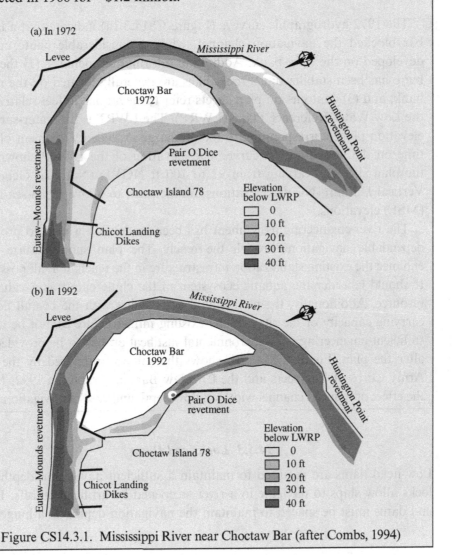

Figure CS14.3.1. Mississippi River near Choctaw Bar (after Combs, 1994)

River stabilization ensures flood protection and a 9-ft-deep navigable channel. Channel maintenance for navigation has become a significant problem within the reach. Table CS14.3.1 lists the dredging requirements for the years 1971–3.

Table CS14.3.1. *Dredging data (after Combs, USACE WES, 1998)*

Year	Days	Cubic yards
1971	15	404,000
1972	27	1,531,000
1973	71	4,121,000

The 1972 hydrographic surveys (Figure CS14.3.1a) indicate that a large bar blocked the navigation channel, while a considerable chute cutoff developed on the right bank. Additional information includes: (1) the left bank has been stabilized; (2) rock dikes are generally located on the right bank; and (3) contours on plan sheets refer to elevations in feet relative to the Low Water Reference Plane (LWRP). The LWRP is the water-surface elevation plane corresponding to the discharge exceeded 97 percent of the time on the flow-duration curve. The elevation of the LWRP shown on the plan sheets is varying from 97 to 100 ft NGVD (National Geodetic Vertical Datum). NGVD elevations are identical to above mean-sea-level (MSL) elevations.

The river-engineering assignment has been to develop a plan to provide dependable navigation through the reach. The plan should utilize and enhance the existing stabilization infrastructure in the reach, if at all possible. It should consider the aquatic ecosystem in the chute cutoff as a valuable resource. Additionally, the plan should not conflict with the overall flood-carrying capacity of the river. If the existing infrastructure cannot be used, sufficient engineering and environmental justification has to be provided to alter the plan. Figure CS14.3.1b shows the solution proposed by the US Army Corps of Engineers and the Choctaw Bar topography in 1992. Note the effect of reduced channel width on flow depth and ease of navigation.

14.5.3 Locks and Dams

Low-head dams are required to maintain a sufficient navigation depth, and locks allow ships to navigate in a river steepened by artificial cutoffs. Locks and dams must be spaced to maintain the navigation depth via a hinge-pool

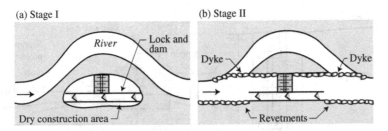

Figure 14.21. Lock and dam construction for navigable waterways

system. Spacing the dams farther apart may eliminate dams and reduce the initial cost of construction. However, maintenance dredging and channel contraction with dikes are also required for maintaining the sediment-transport capacity of the natural channel.

Locks and dams are constructed beside the river, as shown in Figure 14.21. An appropriate dry construction site is located next to the river in the first stage. When the construction is complete, the river is then diverted on the structure in the second stage.

Straight reaches are desirable locations for locks because it is easier to navigate them and to approach locks through them than bends. However, wide straight reaches are also prone to sedimentation and may require dredging to maintain sufficient flow depth. Strong currents during floods and adverse crosscurrents from spillway discharges may also affect traffic in the lock approaches. In river-navigation projects the lock is usually located near the bank at one end of the dam to minimize the adverse effects of spillway discharge on traffic. A typical lock and dam layout is illustrated in Figure 14.22.

The lock enables vessels to gain access to lower or higher water levels on either side of the dam. It is an open chamber with gates at both the upper and the lower pool. Locks fill the chamber from the upper pool and discharge to the lower pool, as illustrated in Figure 14.23:

(a) A downstream-bound tow approaches a lock, the emptying valves and lower lock gates are closed, and the water surface in the lock chamber is brought to the same elevation as that of the upper pool by opening the filling valves.

(b) The upper lock gates are then opened for the tow to move into the lock chamber.

(c) The upper lock gates and filling valves are then closed, and the emptying valves are opened to bring the water surface down to the level of the lower pool.

(d) The lower lock gates are then opened, and the tow moves out of the lock chamber into the lower pool.

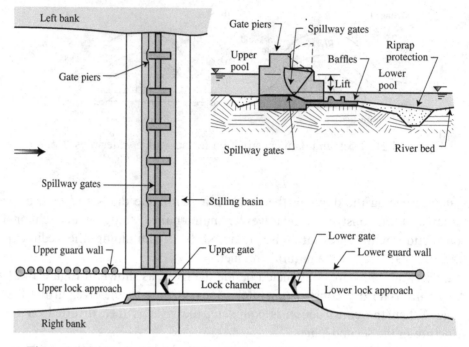

Figure 14.22. Typical lock and dam layout (after Petersen, 1986)

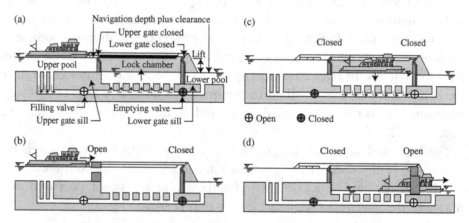

Figure 14.23. Lock and dam operations: (a) chamber filling, (b) entering vessel, (c) chamber emptying, and (d) leaving vessel

The procedure is reversed for a tow moving upstream. Undersized locks may force large tows to be very slow and cautious when entering a lock chamber. Large locks are more expensive and require longer filling/emptying times. Filling and emptying times should be as short as possible without excessive turbulence, surges, or crosscurrents in the lock chamber that might damage the tow or cause the tow to damage the lock.

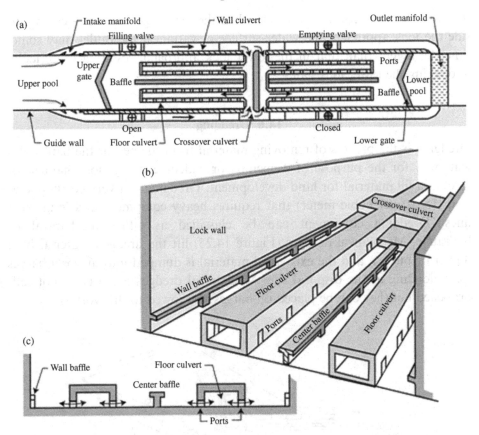

Figure 14.24. Lock chamber design: (a) planview, (b) 3-D perspective, and (c) cross-section (after Petersen, 1986)

Locks fill and empty through culverts and ports in or on the chamber floor or at the base of the lock walls, as shown in Figure 14.24. Deep locks provide cushions of water to dampen turbulence so that tows are not damaged and stresses in the hawsers (the cables that secure tows to the lock walls) are within acceptable limits. In the lock chamber, flow from the wall culverts passes into a crossover culvert across the center of the lock chamber. The splitter wall at the crossover culvert distributes flow equally into two longitudinal floor culverts with ports. Ports in the upper guard wall reduce crosscurrents by permitting the flow intercepted by the lock to pass through the wall to the baffles. The total cross-sectional area of port openings in the guard wall should be equivalent to the cross-sectional area of the approach channel. Upper guard-wall ports should be designed as low as possible to dissipate as much energy as possible in the deeper part of the lock chamber. Currents and velocities from a lock-emptying system can be dangerous to tows approaching the lower lock. The lock-emptying outlet should be outside the lower lock approach. However, when the discharge outlet is located outside the lock approach, the water-surface elevation at the outlet may sometimes be higher than that in the lower lock approach, resulting in difficulty in opening the lower lock gates.

14.6 Dredging

Dredging is the process of removing material from the bed or the banks of a waterway for the purpose of deepening or widening navigation channels or to obtain fill material for land development. Dredging is a very costly operation ($5–10 per cubic meter) that requires heavy equipment and long pipelines. Dredging equipment can be classified as either mechanical or hydraulic. Mechanical dredges (Figure 14.25) lift the dredged material by a dipper or bucket, and the excavated material is dumped into disposal barges for unloading at the disposal site. Mechanical dredges are usually not self-propelled but they can be placed on barges and towed to the work site.

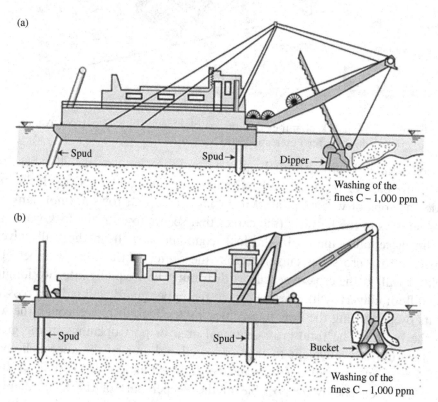

Figure 14.25. Mechanical dredges: (a) dipper dredge and (b) bucket dredge (after USACE, 1983)

Dipper dredges have considerable digging power to excavate hard compacted material and blasted-rock fragments. It can operate with very little maneuvering space and can be accurately controlled in the vicinity of bridges, harbors, and other structures.

Bucket dredges use interchangeable buckets (clamshell, orange peel, dragline) for different operational purposes. An open bucket digs into 12 yd^3 (9 m^3) of bed material, and then closes to be raised and emptied. A modified system in which buckets are fixed on a conveyor belt has also been used in Europe. Considerable fine material is lost from the bucket as it is raised from depths of up to 100 ft (30 m), and the maximum concentration of the suspended turbidity plume is typically less than 1,000 ppm.

Hydraulic suction dredges, shown in Figure 14.26, can be categorized by means of picking up the dredged material (plain suction, dustpan, and cutterhead dredges) and by means of disposal of the dredged material (hopper, pipeline, and sidecasting dredges). Hopper dredges, sketched in Figure 14.26a, are self-propelled seagoing vessels used primarily for maintenance dredging and progressive deepening by successive passes.

Hopper dredges draw concentrated material in contact with the channel bottom through suction pipes and store it in hoppers in the dredge. Sediment resuspension occurs near the suction lines at concentrations up to several tens of parts per thousand when there is overflow from the hoppers. Adverse water-quality effects are unlikely unless the dredged material is highly

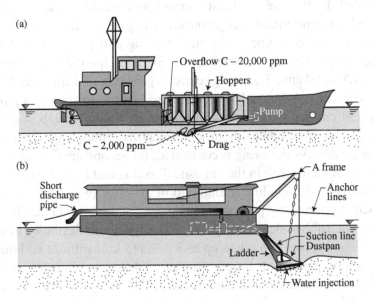

Figure 14.26. Hydraulic dredges: (a) Hopper dredge and (b) dustpan dredge (after USACE, 1983)

contaminated. Except during fish migration and spawning, turbidity in river waters is more likely to be an aesthetic problem than a biological problem. When fully loaded, the dredge moves to the disposal area under its own power. Hopper dredges are emptied by the opening of the bottom doors and dumping of the entire contents in a few seconds. A well-defined plume of dredged material entrains water as it settles to the bottom. Most of the material forms a mound on the bottom, and some spreads horizontally.

Sidecasting dredges are self-propelled seagoing vessels designed to remove material from shallow coastal harbors. A sidecasting dredge picks up bottom material through two suction pipes and pumps it directly overboard beside the dredge. The dredge operates back and forth across the bar, successively deepening the channel on each pass. Sidecasting dredges are not suitable for dredging contaminated material.

Hydraulic pipeline dredges loosen the bottom material with a cutterhead or with water jets (dustpan dredges). The slurry is then pumped through a floating discharge line to the disposal site. Dustpan dredges (Figure 14.26b) are self-propelled vessels suitable for dredging only noncohesive material in waters without significant wave action. Dustpan dredges are equipped with: (1) pressure water jets that loosen the bottom material and (2) a wide-flared and flat suction-line intake for sediment removal. It normally discharges into open water through a relatively short pipeline, up to 100-ft (30-m) long; a longer disposal line requires a booster pump.

Cutterhead dredges are the most efficient and versatile, and thus the most widely used. The cutterhead dredge shown in Figure 14.27a has a rotating cutter around the suction pipe intake and can dig and pump alluvial material including compacted clays and hardpans. Suction-pipe diameters range from 8 to 30 in. (20 to 90 cm). The dredge consists generally of a cutterhead, baskets, ladder, suction line, A-frame, H-frame, pumps, gantry, spuds, and a pipeline up to 30 in. (90 cm) in diameter. As illustrated in Figure 14.27b, a cutterhead dredge operates by circling about one anchored spud with the cutterhead describing an arc. As the swing is completed, the second spud is anchored and the first one raised to circle in the opposite direction and move forward.

The floating discharge line is made up of 30–50-ft (9–16-m)-long pipe sections, each supported by pontoons. They are connected together by flanges, ball joints, or rubber sleeves to give the dredge some flexibility in moving. Land pipeline disposal can range up to 3 miles (5 km) without additional sections of shore pipe. For longer transport distances, booster pumps are required on the discharge line. Slurries of 10–20 percent sediment concentration by weight are typically pumped in pipelines up to 30 in (90 cm) in diameter at velocities ranging from 15 to 20 ft/s (5 to 8 m/s).

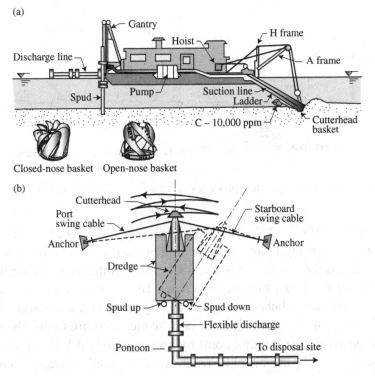

Figure 14.27. Cutterhead dredge: (a) elevation view, and (b) planview (after USACE, 1983)

The disposal of dredged material has received considerable environmental consideration since the 1960s. Not all dredged material is contaminated, and only a small percentage of the sites are highly contaminated (Petersen, 1986). The clay content and organic matter of dredged material tend to adsorb many contaminants, while sands easily release contaminants through mixing, resuspension, and transport. Confined land disposal is sometimes the best alternative for disposing highly polluted sediment. The dredge material is ponded until the suspended solids have settled out. The long-term storage capacity of land disposal sites depends on consolidation of the dredged material, compressibility of foundation soils, effectiveness in dewatering the dredged material, and management of the site.

14.7 River Ice

This section presents a brief introduction to river ice characteristics that cause engineering problems in cold regions. Ice floats as it is approximately 10 percent lighter than water. Very cold temperatures increase the density of ice. The mass density of ice in kg/m^3 is $\rho_{ice} \cong 916.5 - 0.133 \times T°\mathrm{C}$. Pure

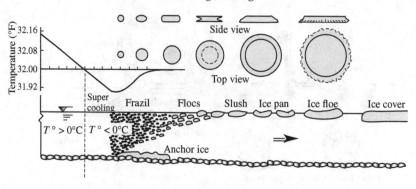

Figure 14.28. Ice formation process in rivers (after Michel, 1971)

water freezes at 0°C and the freezing process releases latent heat at a rate of 333.4 J/g. The freezing temperature and latent heat depend primarily on salinity and seawater freezes at −0.054°C. The compressive strength of ice increases with the strain rate and can reach 10–20 MPa at deformation rates ~10 s^{-1} (Carter and Michel, 1971). Ice is very weak in tension and will crack easily. Cooling of turbulent river flows in winter continues until the water is supercooled below the freezing point by approximately 0.1°C. Small ice crystals named "frazil" after an old French word, "fraisil" (meaning "cinders"), form in supercooled waters. Frazil tends to be very sticky during the supercooled phase, which can cause numerous river-engineering problems, such as clogging of the intakes of water treatment plants. Anchor ice also forms in supercooled waters of shallow streams. Frazil becomes passive once the water temperature returns to the freezing point and crystals grow and agglomerate into slush, and float to form ice pans and floes, as sketched in Figure 14.28. More ice mechanical properties are found in Michel (1978).

Different types of ice cover are depicted on Figure 14.29. Black ice will form when rivers and lakes freeze during very cold nights. Cracking of the ice sheet happens during very cold nights and/or snow accumulation adding sufficient weight to crack the sheet of ice and let the water permeate and freeze the lower part of the snowpack. Hummocked ice typically forms a mound with ice from different sources including ice blocks, snow, and underlying frazil and slush. Hanging dams are typically observed below rapids and falls. During cool nights large volumes of frazil are generated in the fast-flowing open areas and when the frazil piles up under the ice cover in the mild river reaches, the accumulations can become so significant that the flow is forced near the river bed. The reduced cross-section areas and higher flow velocities near the bed scour the fine bed material or transport the frazil farther downstream. Accumulations of frazil can reach 10–15 m in thickness

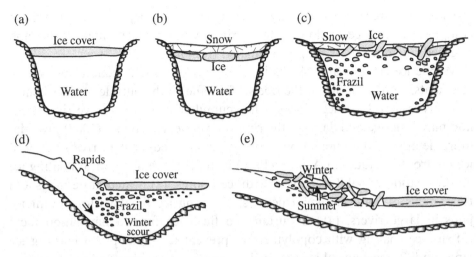

Figure 14.29. Different types of ice cover: (a) black ice, (b) snow cover, (c) hummocked ice, (d) hanging dam, and (e) ice jam

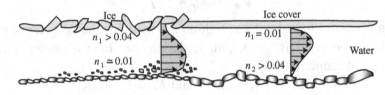

Figure 14.30. Velocity profiles with river-ice cover

over tens of river kilometers, e.g. La Grande River, as discussed in Michel and Drouin (1981). For instance, the lobated river planform features shown for the Matamek River in Figure 2.2e are attributed to the large-scale scour below river falls during the winter months.

Ice jams form when floating ice blocks accumulate in front of a solid ice cover (Beltaos, 1995). The increased roughness of the ice pack made of ice blocks pointing in many different directions decreases the flow velocity and induces an increase in flow depth that can cause winter flooding. The accumulation can sometimes be caused by natural obstructions such as narrow river reaches, river pools below river rapids, or structural obstructions such as bridge piers and spur dykes. The impact of the accumulation of ice blocks on riverbanks can remove large boulders from the riverbanks and can severely damage large riverbank protection measures. The breakup of ice jams causes pulses in river flows that can be very significant.

The presence of ice at the free surface of rivers increases resistance to flow. As shown in Figure 14.30, the stream-flow velocity near the free surface decreases and at a given discharge, the slower velocities will require a greater cross-section area, which slightly increases the flow depth. The analysis of composite resistance to flow n_0 for a river with river-ice roughness n_1 and riverbed roughness n_2 is simply calculated from $n_0 \cong [0.5(n_1^{3/2} + n_2^{3/2})]^{2/3}$.

Mobile ice crystals and blocks in rivers move in the direction opposite to sediments, i.e. they move to the free surface and to the outside of river bends. For this reason, large ice blocks will accumulate against concave riverbanks and may dislodge and damage the riprap revetments. Ettema (2008) provides more details on the interaction between ice and sediment in rivers. Typical ice-control techniques in Ashton (1986) include: (1) redirection of floating ice and formation of solid ice covers with ice booms; (2) heated grate inlets and trash racks of water-intake structures; (3) using ice breakers to prevent ice jams in large rivers; (4) air curtains to flush ice from a navigation lock; (5) surface coating with copolymers to prevent ice buildup; (6) blasting ice jams; and (7) mechanical ice removal.

♦♦ *Problem 14.1*

During a river closure by the end-dumping method, the flow depth in the critical section is 12 ft, or 3.7 m. Estimate the stone diameter and the stone weight required for closure.
[*Answer:* For $V_c = 20$ ft/s, $d_s = 2.6$ ft, and $F_w = 1,500$ lb.]

♦ *Problem 14.2*

Estimate the scour depth below a plunging jet at a velocity of 18 ft/s (5.5 m/s) at an angle of 60° from the horizontal. The jet thickness is 1 ft (30 cm) and the tailwater depth is 6 ft (1.8 m). Sand covers a thick layer of gravel ~5 ft (1.5 m) below the surface.
[*Answer:* $\Delta z \simeq 5.8$ ft (1.8 m).]

♦*Problem 14.3*

Estimate the scour depth below the sluice gate of Example 14.2 when the opening is 1 m. Determine the flow velocity, assuming conservation of energy on both sides of the gate.
[*Answer:* $V_1 \simeq \sqrt{2gh} = 9.9$ m/s, $q = Vh = 9.9$ m^2/s, and $V_2 \simeq 2$ m/s, $\Delta z \simeq$ 12 m.]

♦♦*Problem 14.4*

Estimate the scour depth below the grade-control structure in Example 14.3 when: (a) the discharge is 320 m^3/s and (b) for scour in medium sand ($Q = 160$ m^3/s).

◆◆*Problem 14.5*

Estimate the scour depth below a 2-ft-diameter culvert flowing full at a velocity of 15 ft/s.

[*Answer:* The discharge $Q = 47$ ft³/s (1.33 m³/s) and $\Delta z \simeq 5$ ft (1.5 m).]

Problem 14.6

Repeat the length calculations for the settling basin in Example 14.4 for a width of 60 m and show that the basin area is 3,450 m². Also determine the trap efficiency of a basin area of 2,000 m².

[*Answer*: $T_E = 74$ percent.]

◆◆*Problem 14.7*

Calculate the local scour depth at the end of a 100 m long spill-through abutment in a river flowing at a velocity of 3 m/s with a flow depth of 4 m. If the water spreads 200 m farther out onto the floodplain during floods, estimate the scour depth, considering that the Froude number does not change significantly.

[*Answer:* The scour depth is $\Delta z = 12.5$ m and does not change because $L_a/h_1 > 25$.]

◆◆*Problem 14.8*

Estimate the scour depth around a rectangular pier 1 m wide and 5 m long in a river that is 4 m deep. The flow velocity is 3 m/s at an angle of 30° from the pier alignment.

◆*Problem 14.9*

Reevaluate the pier scour depth from Example 14.5 should the flow alignment against the pier change to 20° in the coming years.

[*Answer:* With $L_p/a = 8$, $\theta_p = 20°$, and $K_2 \simeq 2.33$, the scour depth would more than double to $\Delta z \simeq 9$ m.]

◆*Problem 14.10*

In the case of long-contraction scour, consider that the unit sediment–transport rate is a power function of mean velocity V and flow depth h, like the Simons–Li–Fullerton approximation, i.e. $q_s = aV^b h^c$. Use Equations (14.9–14.11) to demonstrate that the scour depth is

$\frac{\Delta z}{h_1} = \left[\left(\frac{W_1}{W_2} \right)^{\frac{1-b}{c-b}} - 1 \right]$, which reduces to $\Delta z = h_1 \left[\left(\frac{W_1}{W_2} \right)^{0.75} - 1 \right]$ when $b = 4$ and $c = 0$.

◆◆*Problem 14.11*

Consider the field surveys near Cottage Bend of the Mississippi River in Figure P14.11.1. The elevations are in feet below the Low Water Reference Plane. Locate where navigation problems may occur. Sketch structures to redirect the flow and eliminate dredging problems. Give approximate information (height, length, angles, type of material, etc.) for a preliminary design.

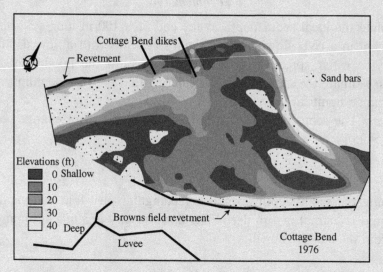

Figure P14.11.1. Mississippi River near Cottage Bend (after P. Combs, personal communication)

15

River Estuaries

This chapter relates to features observed in wide rivers and estuaries. Section 15.1 presents the theory of surface waves with applications to wind waves. Section 15.2 specifically deals with tides and saline wedges in river estuaries.

15.1 Surface Waves

This section focuses on the properties of surface waves, in Section 15.1.1, and wave celerity, in Section 15.1.2. The wave amplitude is determined from the displacement and velocity, in Section 15.1.3. For single waves, the energy, in Section 15.1.4, and propagation, in Section 15.1.5, precedes the analysis of power and celerity for a group of waves, in Section 15.1.6. Finally, applications to wind waves are found in Section 15.1.7.

15.1.1 Wave Properties

As sketched in Figure 15.1, a small-amplitude gravity wave of wavelength λ propagates in a wide canal of depth h without friction. The water-surface elevation $\tilde{\eta}$ changes with time t and the wave propagates at a celerity c.

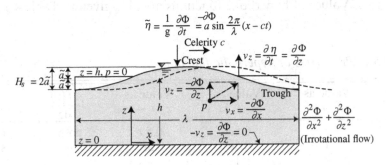

Figure 15.1. Surface-wave definition sketch

The velocity potential Φ is defined such that the velocity components are $v_x = -\partial\Phi/\partial x$ and $v_z = -\partial\Phi/\partial z$. The equation of continuity can be rewritten as

$$\frac{\partial^2 \Phi}{\partial x^2} + \frac{\partial^2 \Phi}{\partial z^2} = 0. \tag{15.1}$$

To solve the continuity equation, the variables are separated for a wave propagating at a celerity c in the x direction:

$$\Phi = f(z)\cos\frac{2\pi}{\lambda}(x - ct). \tag{15.2}$$

Substituting Equation (15.2) into Equation (15.1) yields an equation of the form

$$\frac{d^2 f(z)}{dz^2} - f(z) = 0. \tag{15.3}$$

The solution to this equation requires hyperbolic functions. The hyperbolic sine (sinh), hyperbolic cosine (cosh), and hyperbolic tangent (tanh) are defined as:

$$\sinh z = \frac{e^z - e^{-z}}{2}, \quad \cosh z = \frac{e^z + e^{-z}}{2}, \quad \tanh z = \frac{\sinh z}{\cosh z} = \frac{e^z - e^{-z}}{e^z + e^{-z}}. \tag{15.4}$$

These hyperbolic functions are analogous to the well-known circular functions

$$\sin z = \frac{e^{iz} - e^{-iz}}{2i}, \quad \cos z = \frac{e^{iz} + e^{-iz}}{2}, \quad \tan z = \frac{\sin z}{\cos z}, \tag{15.5}$$

where $i = \sqrt{-1}$ and $e^{iz} = \cos z + i\sin z$. Like their circular cousins, $\sinh z$ is an odd function, while $\cosh z$ is even. It follows that $\cosh^2 z - \sinh^2 z = 1$. By definition, $\operatorname{cosech} z = 1/\sinh z$, $\operatorname{sech} z = 1/\cosh z$, and $\coth z = 1/\tanh z$. Properties of hyperbolic functions are listed in Table 15.1 and sketched in Figure 15.2. Values of hyperbolic functions are also given in Table 4.1.

Table 15.1. *Properties of hyperbolic functions*

$f(z)$	Value as $z \to 0$	Value as $z \to \infty$	Value as $z \to -\infty$	$\dfrac{df(z)}{dz}$	$\int f(z)\,dz$
$\sinh z$	z	$0.5e^z$	$-0.5\,e^{-z}$	$\cosh z$	$\cosh z$
$\cosh z$	1	$0.5e^z$	$0.5e^{-z}$	$\sinh z$	$\sinh z$
$\tanh z$	z	1	-1		

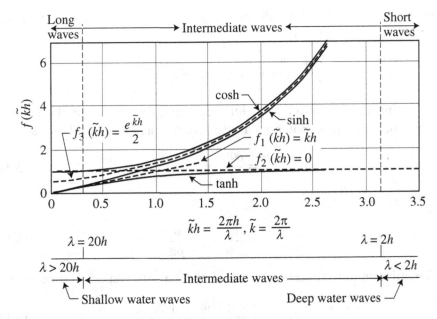

Figure 15.2. Hyperbolic functions for surface water waves

The complete solution for the potential function in Equation (15.2) is

$$\Phi = \left[\tilde{A} \cosh\left(\frac{2\pi z}{\lambda}\right) + \tilde{B} \sinh\left(\frac{2\pi z}{\lambda}\right) \right] \cos\left[\frac{2\pi}{\lambda}(x - ct)\right], \qquad (15.6)$$

where boundary conditions define the celerity c and coefficients $\tilde{A}$ and $\tilde{B}$. The first boundary condition is that the vertical velocity is zero at $z = 0$, or

$$v_z = \frac{\partial \Phi}{\partial z} = 0. \qquad (15.7)$$

Substituting Equation (15.6) into Equation (15.7) implies that $\tilde{B} = 0$. Two conditions remain and the celerity c is determined in Section 15.1.2, and the wave amplitude $\tilde{A}$ defines the flow velocity and displacement in Section 15.1.3.

15.1.2 Wave Celerity

The wave celerity is obtained from the boundary condition at the free surface. The expression emerges from the equation of motion without friction. For 2-D flow ($v_y = 0$), the equation of motion in the vertical z direction is

$$\frac{\partial v_z}{\partial t} + v_x \frac{\partial v_z}{\partial x} + v_z \frac{\partial v_z}{\partial z} = -g - \frac{1}{\rho}\frac{\partial p}{\partial z}. \qquad (15.8)$$

Considering irrotational flow, $\partial v_z/\partial x = \partial v_x/\partial z$, and $v_z = -\partial\Phi/\partial z$, Equation (15.8) defines the flow potential as

$$\frac{\partial}{\partial z}\left[\frac{-\partial\Phi}{\partial t} + \frac{1}{2}\left(v_x^2 + v_z^2\right) + gz + \frac{p}{\rho}\right] = 0. \tag{15.9}$$

Considering small flow velocity, thus neglecting the term v^2, and assuming a hydrostatic gravitation potential as $g(z+\tilde{\eta})$, the pressure term becomes

$$\frac{p}{\rho} = \frac{\partial\Phi}{\partial t} - g(z+\tilde{\eta}) \tag{15.10a}$$

or alternatively

$$\tilde{\eta} = \frac{1}{g}\frac{\partial\Phi}{\partial t} - z - \frac{p}{\rho g}. \tag{15.10b}$$

Differentiating this reduced equation of motion with respect to time gives

$$v_z = \frac{\partial\tilde{\eta}}{\partial t} = \frac{1}{g}\frac{\partial^2\Phi}{\partial t^2} - \frac{1}{\rho g}\frac{\partial p}{\partial t}. \tag{15.11a}$$

The atmospheric pressure p is constant and the velocity v_z finally reduces to

$$v_z = \underbrace{-\frac{\partial\Phi}{\partial z}}_{\text{velocity potential}} = \underbrace{\frac{1}{g}\frac{\partial^2\Phi}{\partial t^2}}_{\text{velocity potential}} \tag{15.11b}$$

After substituting the potential function from Equation (15.6) with $\tilde{B}=0$, into Equation (15.11b), we obtain the wave celerity at the free surface, $z = h$

$$c^2 = \frac{g\lambda}{2\pi}\tanh\frac{2\pi h}{\lambda}. \tag{15.12a} \blacklozenge\blacklozenge$$

This is the general relationship for wave celerity c derived by Airy.

Figure 15.2 shows two asymptotic cases for the wave celerity: (1) long waves or shallow-water waves when $\lambda > 20h$ and (2) short waves, also called deep-water waves, when $\lambda < 2h$. The celerity of shallow-water waves ($\tanh x = x$) depends solely on flow depth.

$$c^2 \simeq \frac{g\lambda}{2\pi}\frac{2\pi h}{\lambda} = gh. \tag{15.12b}$$

For deep-water waves, $\lambda < 2h$, the celerity depends on the wavelength λ as:

$$c^2 \simeq \frac{g\lambda}{2\pi}. \tag{15.12c}$$

The wavelength λ also relates to the wave period T and celerity c as $\lambda = cT$, and the following identities are obtained directly from Equation (15.12a–c):

$$T = \sqrt{\frac{2\pi\lambda}{g} \coth \frac{2\pi h}{\lambda}}, \tag{15.13a}\blacklozenge$$

$$\lambda = \frac{gT^2}{2\pi} \tanh \frac{2\pi h}{\lambda}. \tag{15.13b}$$

Figure 15.3 shows the relationship between celerity c, period T, and flow depth h. Table 15.2 also shows that deep-water waves do not solely depend on flow depth.

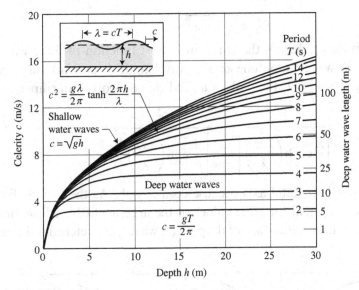

Figure 15.3. Wave celerity c as a function of flow depth h and period T

Table 15.2. *Examples of shallow- and deep-water waves*

h (m)	T (s)	c (m/s)	λ (m)	Wave type
15	20	$c = \sqrt{gh} = 12$	240	Shallow
10	6	$c = \sqrt{\dfrac{g\lambda}{2\pi} \tanh\dfrac{2\pi h}{\lambda}} = 8$	50	Intermediate
5	2	$c \simeq \sqrt{\dfrac{g\lambda}{2\pi}} = 3$	6	Deep

15.1.3 Displacement and Velocity

The velocity components are obtained from Equation (15.6) with $\tilde{B} = 0$:

$$v_x = \frac{\partial \tilde{\xi}}{\partial t} = -\frac{\partial \Phi}{\partial x} = \frac{2\pi \tilde{A}}{\lambda} \cosh \frac{2\pi z}{\lambda} \sin \frac{2\pi}{\lambda} (x - ct),$$ (15.14a)

$$v_z = \frac{\partial \tilde{\eta}}{\partial t} = -\frac{\partial \Phi}{\partial z} = -\frac{2\pi \tilde{A}}{\lambda} \sinh \frac{2\pi z}{\lambda} \cos \frac{2\pi}{\lambda} (x - ct)$$ (15.14b)

and integrating over time gives the horizontal $\tilde{\xi}$ and vertical $\tilde{\eta}$ displacements:

$$\tilde{\xi} = \frac{\tilde{A}}{c} \cosh \frac{2\pi z}{\lambda} \cos \frac{2\pi}{\lambda} (x - ct),$$ (15.15a)

$$\tilde{\eta} = \underbrace{\frac{\tilde{A}}{c} \sinh \left(\frac{2\pi z}{\lambda} \right)}_{\text{amplitude } \tilde{a}} \sin \left[\frac{2\pi}{\lambda} (x - ct) \right].$$ (15.15b)

This relationship defines the constant $\tilde{A}$ as a function of three wave characteristics: (a) wave amplitude $\tilde{a}$; (b) wavelength λ; and (c) wave celerity c. From $\cos^2 + \sin^2 = 1$, the path of a fluid element forms an ellipse:

$$\frac{\tilde{\xi}^2}{\left[\frac{\tilde{A}}{c} \cosh \left(\frac{2\pi z}{\lambda} \right) \right]^2} + \frac{\tilde{\eta}^2}{\left[\frac{\tilde{A}}{c} \sinh \left(\frac{2\pi z}{\lambda} \right) \right]^2} = 1$$ (15.15c)◆

with a major horizontal axis because cosh > sinh. As the hyperbolic sine and cosine become equal for large values of the argument, this ellipse turns into a circular path near the surface of deep-water waves, as sketched in Figure 15.4.

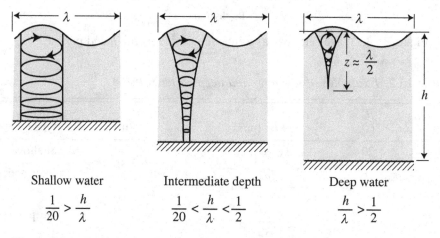

Shallow water	Intermediate depth	Deep water
$\frac{1}{20} > \frac{h}{\lambda}$	$\frac{1}{20} < \frac{h}{\lambda} < \frac{1}{2}$	$\frac{h}{\lambda} > \frac{1}{2}$

Figure 15.4. Particle path in water waves

The velocity near the bed can result in resuspension of the bed material. Indeed, from Equation (15.15b), the constant $\tilde{A} = \tilde{a}c/\sinh(2\pi h/\lambda)$ and therefore the maximum near-bed velocity from Equation (15.14a) is given by

$$v = \frac{2\pi\tilde{A}}{\lambda} = \frac{2\pi\tilde{a}}{T\sinh(2\pi h/\lambda)}. \qquad (15.15d)\blacklozenge\blacklozenge$$

This leads to the interesting concept that the bed material can be resuspended by shallow waves, but not by deep-water waves. The corresponding applied bed shear stress is estimated from

$$\tau_{0\text{Pa}} \cong 2.5v_{\text{m/s}}^2. \qquad (15.15e)\blacklozenge$$

The applied shear stress varies with the square of the wave height. Comparison with the critical shear stress in Table 2.2 gives information regarding whether the bed material will ($\tau > \tau_c$) or will not ($\tau < \tau_c$) be brought into suspension.

15.1.4 Wave Energy

The total energy of a wave is divided into two components: (1) potential energy (PE) and (2) kinetic energy (KE). Both components are generally expressed in terms of average energy over a complete wavelength per unit surface area.

Potential Energy

The PE that is due to the pressure wave formed on the free surface is obtained by subtracting the PE without the wave from the PE with the wave, as sketched in Figure 15.5a. The incremental PE per unit width, dPE_a, in a small column of water, is the height of the center of gravity times the mass increment dm.

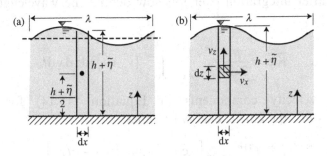

Figure 15.5. Definition sketch for (a) potential energy and (b) kinetic energy

In a water column $h + \tilde{\eta}$ high and dx long for water of specific weight γ, we obtain dPE_a = (height to center of gravity) $g\,dm = [0.5(h + \tilde{\eta})^2]\gamma\,dx$. The average PE per unit surface area is obtained from the integration of the incremental PE over one wavelength λ and one wave period T:

$$\overline{PE_a} = \frac{\gamma}{2\lambda T} \int_t^{t+T} \int_x^{x+\lambda} (h + \tilde{\eta})^2 \, dx \, dt. \tag{15.16a}$$

To simplify the notation, it is convenient to define the wave number $\tilde{k} = 2\pi / \lambda$ and the angular frequency $\tilde{\sigma} = \tilde{k}c = 2\pi / T$. Surface waves of amplitude $\tilde{a}$ are described as $\tilde{\eta} = \tilde{a}\sin(\tilde{k}x - \tilde{\sigma}t)$ from Equation (15.15b), and transform Equation (15.16a) into

$$\overline{PE_a} = \frac{\gamma}{2\lambda T} \int_t^{t+T} \int_x^{x+\lambda} [h^2 + 2\tilde{a}h\sin(\tilde{k}x - \tilde{\sigma}t) + \tilde{a}^2 \sin^2(\tilde{k}x - \tilde{\sigma}t)] \, dx \, dt \tag{15.16b}$$

or simply $\overline{PE_a} = \frac{\gamma h^2}{2} + \frac{\gamma \tilde{a}^2}{4}$, which is the average PE per unit surface area. Since the PE in the absence of a wave, as sketched in Figure 15.5a, is

$$\overline{PE_b} = \frac{\gamma}{2\lambda T} \int_t^{t+T} \int_x^{x+\lambda} h^2 \, dx \, dt = \frac{\gamma h^2}{2}, \tag{15.16c}$$

the average PE per unit area that is attributable to the wave is

$$PE = \overline{PE_a} - \overline{PE_b} = \frac{\gamma \tilde{a}^2}{4}. \tag{15.17}\blacklozenge$$

Kinetic Energy

From Figure 15.5b, the kinetic energy KE per unit width of a small element dx long and dz high with velocity components v_x and v_z is given by $d(KE) = \frac{1}{2}(v_x^2 + v_z^2)dm = \frac{1}{2}\rho(v_x^2 + v_z^2)dz\,dx$. The average KE per unit surface area is then given after integration over the flow depth, one wavelength, and one wave period:

$$\overline{KE} = \frac{\rho}{2\lambda T} \int_t^{t+T} \int_x^{x+\lambda} \int_0^{h+\tilde{\eta}} (v_x^2 + v_z^2)dz \, dx \, dt. \tag{15.18a}$$

From the velocity components in Equation (15.14) for the wave $\tilde{\eta} = \tilde{a}\sin(\tilde{k}x - \tilde{\sigma}t)$, the kinetic energy becomes

$$\overline{KE} = \frac{\rho}{2\lambda T} \int_t^{t+T} \int_x^{x+\lambda} \int_0^{h+\tilde{\eta}} \tilde{k}^2 \tilde{A}^2 \left[\begin{array}{c} \cosh^2 \tilde{k}z \sin^2(\tilde{k}x - \tilde{\sigma}t) \\ + \sinh^2 \tilde{k}z \cos^2(\tilde{k}x - \tilde{\sigma}t) \end{array} \right] dz \, dx \, dt. \tag{15.18b}$$

By using the following identities: (1) $\cosh^2 \tilde{k}z = \frac{1}{2}(1 + \cosh 2\tilde{k}z)$; (2) $\sinh^2 \tilde{k}z = -\frac{1}{2}(1 - \cosh 2\tilde{k}z)$; (3) $\cos^2(\tilde{k}x - \tilde{\sigma}t) - \sin^2(\tilde{k}x - \tilde{\sigma}t) = \cos 2(\tilde{k}x - \tilde{\sigma}t)$; and (4) $\cos^2(\tilde{k}x - \tilde{\sigma}t) + \sin^2(\tilde{k}x - \tilde{\sigma}t) = 1$, the average KE density becomes

$$\overline{\mathrm{KE}} = \frac{\rho \tilde{k}^2 \tilde{A}^2}{4\lambda T} \int_t^{t+T} \int_x^{x+\lambda} \int_0^{h+\tilde{\eta}} [\cosh 2\tilde{k}z - \cos 2(\tilde{k}x - \tilde{\sigma}t)] \mathrm{d}z \, \mathrm{d}x \, \mathrm{d}t = \frac{\rho \tilde{k} \tilde{A}^2}{8} \sinh 2\tilde{k}h.$$

This reduces further from: (1) $\sinh 2\tilde{k}h = 2\sinh \tilde{k}h \cosh \tilde{k}h$; (2) $\tilde{A} = \tilde{a}c / \sinh \tilde{k}h$ from Equation (15.15b); and (3) $c^2 = (g/\tilde{k})\tanh \tilde{k}h$ from relations (15.12a) to finally give

$$\overline{\mathrm{KE}} = \frac{\gamma \tilde{a}^2}{4}. \tag{15.19}\blacklozenge$$

Interestingly, $\overline{\mathrm{KE}} = \overline{\mathrm{PE}}$ from Equation (15.17) and the wave energy $\tilde{E}$ per unit area is

$$\tilde{E} = \overline{\mathrm{PE}} + \overline{\mathrm{KE}} = \frac{\gamma \tilde{a}^2}{2}. \tag{15.20}\blacklozenge\blacklozenge$$

15.1.5 Wave Propagation

The propagation of shallow-water waves in river estuaries is sketched in Figure 15.6. The law of conservation of energy provides a convenient way of expressing the effect of changes in channel width and flow depth on wave amplitude.

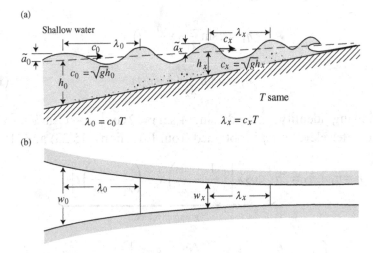

Figure 15.6. Shallow-water wave propagation in a river estuary: (a) elevation and (b) plan view

The total energy for the entire surface of the wave from Equation (15.20) is

$$W_0 \lambda_0 \tilde{E} = W_0 \lambda_0 \frac{1}{2} \gamma \tilde{a}_0^2 = W_x \lambda_x \frac{1}{2} \gamma \tilde{a}_x^2, \tag{15.21}$$

where W is the width of the river estuary, λ is the wavelength, and $\tilde{a}$ is the wave amplitude. The subscript $_0$ refers to the reference wave condition and the subscript x to any other location. Because the celerity c of shallow-water waves varies as $c = \sqrt{gh_x}$ while the period $T = \lambda/c$ is constant, the tide amplitude $\tilde{a}_x$ can be determined after substituting $\lambda_x = T\sqrt{gh_x}$ into Equation (15.21) to obtain

$$\frac{\tilde{a}_x}{\tilde{a}_0} = \left(\frac{W_0}{W_x}\right)^{1/2} \left(\frac{h_0}{h_x}\right)^{1/4}. \tag{15.22}\blacklozenge$$

This is known as Green's law, which clearly indicates that the wave amplitude will increase as the channel width and flow depth decrease.

15.1.6 *Wave Power and Group Celerity*

The superposition of several waves is examined by considering two waves as

$$\tilde{\eta} = \tilde{a}_1 \sin \tilde{\sigma}_1 t + \tilde{a}_2 \sin \tilde{\sigma}_2 t. \tag{15.23}$$

Two waves with different wavelengths cause local interference. The group celerity c_G is determined by adding two waves of equal amplitude and slightly different wavelength,

$$\tilde{\eta}_1 = \tilde{a} \sin \frac{2\pi}{\lambda_1} (x - c_1 t), \tag{15.24a}$$

$$\tilde{\eta}_2 = \tilde{a} \sin \frac{2\pi}{\lambda_2} (x - c_2 t). \tag{15.24b}$$

The following identity is used: $\sin x + \sin y = 2 \cos \frac{1}{2}(x - y) \sin \frac{1}{2}(x + y)$. The resultant water elevation $\tilde{\eta}$ is obtained from Equations (15.23) and (15.24):

$$\tilde{\eta} = \tilde{\eta}_1 + \tilde{\eta}_2 = 2\tilde{a} \underbrace{\left\{ \cos \pi \left[\left(\frac{1}{\lambda_1} - \frac{1}{\lambda_2}\right) x - \left(\frac{c_1}{\lambda_1} - \frac{c_2}{\lambda_2}\right) t \right] \right\}}_{\text{group envelope}}$$

$$\times \left\{ \sin \pi \left[\left(\frac{1}{\lambda_1} + \frac{1}{\lambda_2}\right) x - \left(\frac{c_1}{\lambda_1} + \frac{c_2}{\lambda_2}\right) t \right] \right\}. \tag{15.25}$$

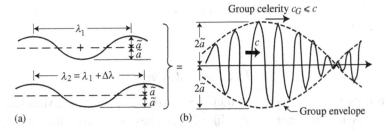

Figure 15.7. Definition sketches for (a) the sum of two waves and (b) the group celerity

The mean wavelength $\lambda = 0.5(\lambda_1 + \lambda_2)$ and celerity $c = 0.5(c_1 + c_2)$ give:

$$\tilde{\eta} \simeq 2\tilde{a}\underbrace{\left\{\cos\pi\left[d\left(\frac{1}{\lambda}\right)x - d\left(\frac{c}{\lambda}\right)t\right]\right\}}_{\text{group envelope}}\sin\frac{2\pi}{\lambda}(x - ct). \tag{15.26}$$

The result shown in Figure 15.7 is a short wave of celerity c traveling inside an envelope of amplitude $2\tilde{a}$ described by the term in braces of Equation (15.26).

The region of maximum amplitude of the envelope is progressing in the positive x direction with the group celerity c_G determined from

$$c_G = \frac{d\left(\frac{c}{\lambda}\right)}{d\left(\frac{1}{\lambda}\right)} = \frac{\frac{\lambda dc - cd\lambda}{\lambda^2}}{-\frac{d\lambda}{\lambda^2}} = c - \lambda\frac{dc}{d\lambda}. \tag{15.27}$$

In deep water, the wave celerity increases with λ such that individual short waves travel faster than the group wave. Over time, the short waves travel with each crest disappearing at the front of the group while a new crest grows from the rear.

The group celerity obtained from combining Equations (15.14) and (15.27) gives

$$c_G = \frac{c}{2}\left(1 + \frac{2\tilde{k}h}{\sinh 2\tilde{k}h}\right), \tag{15.28}\blacklozenge\blacklozenge$$

which varies between $c/2$ when $\tilde{k}h$ is large (deep-water waves) and c when $\tilde{k}h$ is small (shallow-water waves). In shallow water, the group and wave celerity are identical because wave propagation is independent of wavelength.

The power available in surface waves per unit surface area is given here without derivation as

$$\tilde{P} = \frac{\gamma\tilde{a}^2}{2}c_G. \tag{15.29}\blacklozenge$$

It is concluded that the wave energy from Equation (15.20) travels at the group celerity.

15.1.7 Wind Waves

The minimum wind speed producing gravity waves is approximately 6 m/s. Wind scales for extreme events like hurricanes and tornadoes are listed in Table 15.3. Tornadoes have extremely high wind speeds but they cover small areas for a short time period. There is more concern with hurricanes because high wind speeds can be sustained for long periods of time.

Wind waves generate a spectrum of wavelengths and amplitudes. The significant wave height $\tilde{H}_s$ represents the average height of the large waves in the upper third of the spectrum. The significant wave properties include the height, $\tilde{H}_s = 2\tilde{a}$, which is twice the amplitude $\tilde{a}$, and the significant wavelength is λ_s. Under a constant wind speed U_w, the wave height increases with time and downwind direction called the fetch length $\tilde{F}$ until equilibrium is obtained. The minimum windstorm duration $\tilde{t}_{s\,min}$ is obtained from the ratio of $\tilde{F}$ to the group celerity c_G. In shallow water, the storm duration must exceed $\tilde{t}_{s\,min} \approx \tilde{F}/\sqrt{gh}$.

The significant wave height $\tilde{H}_s$ depends on wind speed U_w, flow depth h, and fetch length $\tilde{F}$. At equilibrium, the work done $\tau\tilde{F}$ by the shear stress $\tau \sim \rho U_w^2$ exerted over the fetch length $\tilde{F}$ corresponds to the energy of the wave, or $\tau\tilde{F} \sim \rho g \tilde{H}_s^2$ from Equation (15.20). As a first approximation, the wave height $\tilde{H}_s$ then becomes $\tilde{H}_s \approx 0.003 U_w \sqrt{\tilde{F}/g}$. Accordingly, the wave height increases linearly with the wind speed and square root of the fetch length. The empirical formulas of Ijima and Tang (1966) can estimate the

Table 15.3. *Wind scales for hurricanes and tornadoes*

Type	Pressure (millibars)	Wind speed (mph)	Surge height (ft)	Type	Wind speed (mph)
Hurricanes (Saffir–Simpson scale)				Tornadoes (Fujita scale)	
Category 1	>979	74–95	4–5	F 0	40–73
Category 2	965–979	96–110	6–8	F 1	74–113
Category 3	945–964	111–130	9–12	F 2	114–158
Category 4	920–944	131–155	13–18	F 3	159–207
Category 5	<920	>155	>18	F 4	208–261
		1 m/s = 2.24 mph	1 m = 3.28 ft	F 5	262–319

significant wave height and period. For rivers, the following approximations are recommended:

$$\tilde{H}_s \cong 0.003 \left(\frac{U_w^2}{g}\right) \left(\frac{g\tilde{F}}{U_w^2}\right)^{0.42}, \tag{15.30a}$$

$$T \cong 0.58 \frac{U_w}{g} \left(\frac{g\tilde{F}}{U_w^2}\right)^{0.25}. \tag{15.30b}$$

The wavelength is estimated from either $\lambda_s \simeq 15\tilde{H}_s$ or Equation (15.13b). In rivers, wind waves are usually limited by the fetch length $\tilde{F}$. Short waves carry their energy near the surface, which causes riverbank erosion.

In shallow rivers under very strong winds, waves become limited by the flow depth and the maximum wave height is $\tilde{H}_s \simeq 0.35h$. Waves cannot develop further because the shear stress is transmitted to the river bottom, which causes significant resuspension of fine bed material. Example 15.1

Example 15.1 Wind waves and resuspension

Consider a wide and shallow river with a depth of 5 m and a 10-km fetch length. Estimate the wave height, length, and period under sustained winds at 45 mph (20 m/s). If the bed material is fine sand ($d_s = 0.2$ mm), will there be significant resuspension of the bed material? The significant wave characteristics are estimated in terms of:

(a) the storm duration must exceed $t > \frac{\tilde{F}}{\sqrt{gh}} = \frac{10,000}{\sqrt{9.81 \times 5}} = 1,430\text{s} \cong 25$ min,

(b) height $\tilde{H}_s \cong 0.003 \left(\frac{20^2}{9.81}\right) \left(\frac{9.81 \times 10,000}{20^2}\right)^{0.42} = 1.23$ m from Equation (15.30a), which is very close to a depth-limited wave because $\tilde{H}_s \cong 0.25h$,

(c) period $T \cong 0.58 \times \frac{20}{9.81} \left(\frac{9.81 \times 10,000}{20^2}\right)^{0.25} = 4.7$ from Equation (15.30b),

(d) wavelength $\lambda \cong \frac{9.81 \times 4.7^2}{2\pi} \tanh\left(\frac{2\pi \times 5}{15 \times 1.23}\right) = 32$ m from Equation (15.13b).

From Section 15.1.3, we can also examine resuspension of the bed material from the near-bed flow characteristics.

(e) the wave-induced flow velocity from Equation (15.15d) is

$$v_{m/s} = \frac{\pi \tilde{H}_s}{T \sinh(2\pi h/\lambda)} = \frac{\pi \times 1.23}{4.7 \sinh(2\pi \times 5/32)} = 0.73 \text{ m/s}$$

(f) from Equation (15.15e), the bed shear stress $\tau_0 \cong 2.5 \times 0.73^2 = 1.33$ Pa exceeds the critical shear stress for fine sand (0.15 Pa in Table 2.2). The waves will bring the bed material into suspension. Only particles coarser than 2 mm would not move.

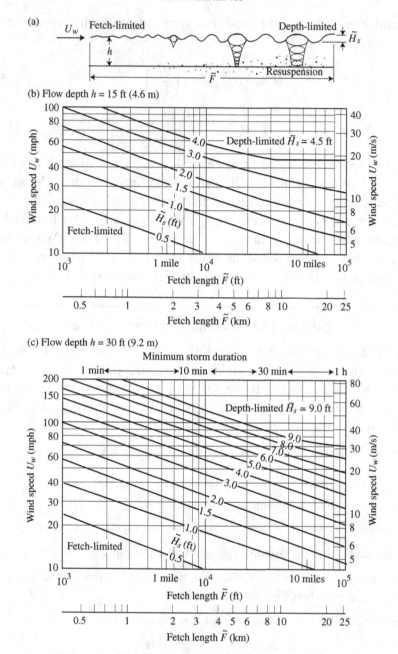

Figure 15.8. Amplitude of wind waves: (a) sketch, (b) shallow depth h = 15 ft (4.6 m), and (c) deep channel h = 30 ft

estimates the properties of wind waves and resuspension of bed material. Approximate wave heights are shown as a function of fetch length in Figure 15.8 for constant wind speeds at flow depths of 15 ft (4.6 m) and 30 ft (9.2 m), respectively.

15.2 River Estuaries

This section focuses on tides and saline wedges in river estuaries. The Coriolis force is discussed in Section 15.2.1, with the propagation of tides in Section 15.2.2. Section 15.2.3 briefly covers saline wedges in river estuaries.

15.2.1 Coriolis Acceleration

Considering the Earth rotation sketched in Figure 15.9, gravity is proportional to the universal gravitation constant $G_u = 6.673 \times 10^{-11}$ Nm^2/kg^2. Given that the mass of the Earth $m_E = 5.98 \times 10^{24}$ kg and radius $R_E = 6{,}371$ km, the gravitational acceleration is $g = G_u m_E / R_E^2 = 9.8$ m/s^2.

The centrifugal acceleration is proportional to the square of the angular velocity multiplied by the radius of rotation. As sketched in Figure 15.9, the centrifugal acceleration that is due to the Earth's rotation ($\omega_E = 7.27 \times 10^{-5}$ rad/s) is $a_{cE} = \omega_E^2 R_E \cos\phi = 0.034$ m/s$^2 \cos\phi$, where ϕ is the latitude. The centrifugal acceleration is zero at the poles but reaches 0.3 percent of the gravitational acceleration at the equator. This combined action varies with latitude and draws the water away from the poles toward the equator.

Moving water at the surface of the Earth is also subjected to the Coriolis acceleration. The Coriolis acceleration is a function of the angular-velocity vector of the Earth's rotation Ω directed toward the North Pole and the flow-velocity vector V at the surface of the Earth. The Coriolis acceleration vector due to Earth rotation $\vec{\Omega} = \hat{y}\omega_E \cos\phi + \hat{z}\omega_E \sin\phi$ is multiplied by the water-velocity vector $\vec{V}$ given v_x to the East, v_y to the North, and v_z vertical up to define

$$a_{\text{cor}} = -2\vec{\Omega} \times \vec{V}$$
$$= 2\omega_E [\hat{x}(v_y \sin\phi - v_z \cos\phi) - \hat{y}v_x \sin\phi + \hat{z}v_x \cos\phi]. \qquad (15.31)$$

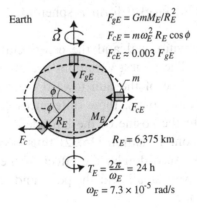

Figure 15.9. Earth rotation sketch

For horizontal flow at the surface of the Earth ($v_z = 0$), fluid motion is deflected to the right in the northern hemisphere and to the left in the southern hemisphere. The magnitude of the Coriolis acceleration is

$$|a_{cor}| \simeq 2 \, \omega_E V \sin \phi. \tag{15.32}$$

In a river, the ratio of the Coriolis acceleration to the pressure gradient gS is defined as $2 \, \omega_E V \sin \phi / gS$. The pressure gradient is normally balanced by the bed resistance, but in deep rivers with tides, the free surface slope can be much larger than the very flat riverbed slope. In a river bend of radius R_1, the Coriolis acceleration can be compared to the centrifugal acceleration (V^2/R_1) as $2\omega_E R_1 \sin \phi / V$. This ratio is negligible in all but fairly straight large rivers. Example 15.2 illustrates cases where the Coriolis acceleration is not negligible.

15.2.2 Tides in River Estuaries

Tides describe the oscillatory waves in large water masses. Tidal oscillations result from a balance between gravitational, centrifugal, and Coriolis forces exerted by the Earth, Moon, and Sun. Let us first consider the Earth and the Moon as a dual system, sketched in Figure 15.10. The Earth diameter and mass are, respectively, 3.7 and 81 times those of the Moon. The distance separating them is ~30 times the Earth diameters. The two large bodies have a common center of mass and the rotation about point C has a period of ~27.3 days. Accordingly, the primary tidal cycle will be slightly longer than a half-day period, i.e. 12 h 25 min.

The surface water is subjected to the gravitational accelerations of the Earth and the Moon and the centripetal acceleration. Locally at the Earth's surface, the centripetal acceleration about point C forces the water toward the Moon at point A, but away from the Moon at point B. Their resultant force causes the water surface to form a spheroid that is oblate along the Earth–Moon axis.

At the poles, the gravitational and the centripetal accelerations do not change during the day. The largest variations are obtained near the equator with the gravitational force of the Sun toward the East at sunrise and the West at sunset. On a daily basis, these differences generate flow velocities that are proportional to the cosine of the latitude ϕ. The magnitude of the Coriolis acceleration from Equation (15.32) thus becomes proportional to the product $\sin \phi \cos \phi$. Accordingly, the effect of the Coriolis acceleration on tides should be relatively small near the poles and equator, and at a maximum near the 45° latitude.

Example 15.2 Coriolis acceleration in the Lower Mississippi River

Consider a river bend with radius of curvature $R_1 = 8$ miles (12.9 km) of the lower Mississippi River at a latitude $\phi = 30°$. The downstream water-surface slope of the river is $S \approx 6 \times 10^{-5}$. At a flow depth of 60 ft (18 m) and Manning $n = 0.018$, the water moves horizontally to the South at 10 ft/s (3 m/s).

Step 1: The westward Coriolis acceleration due to the rotation of the Earth is $|a_{\text{cor}}| \simeq 2 \, \omega_E V \sin\phi = 2 \times 7.3 \times 10^{-5} \times 10 \times \sin 30° = 7.3 \times 10^{-4}$ ft/s^2 (2.2 $\times$ 10^{-4} m/s^2).

Step 2: The centripetal acceleration in the river bend is $a_{cent} = V^2/R_1 = 100$ ft^2/s^2 $\times$ 8 $\times$ 5,280 ft = 2.4 $\times$ 10^{-3} ft/s^2 (7.3 $\times$ 10^{-4} m/s^2).

Step 3: The downstream gravitational acceleration component is $gS = 32.2$ ft/s^2 $\times$ 6 $\times$ 10^{-5} = 1.9 $\times$ 10^{-3} ft/s^2 (5.8 $\times$ 10^{-4} m/s^2).

The Coriolis acceleration is ~ 30 percent of the centripetal acceleration in this river bend. The Coriolis acceleration is also ~ 38 percent of the downstream gravitational acceleration and should not be neglected. Velocity measurements from the Acoustic Doppler Current Profiler at the Old River Control Complex of the Mississippi River are shown in Figure E15.2.1.

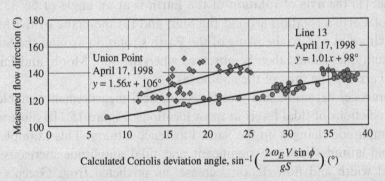

Figure E15.2.1. Measured and calculated velocity orientation angles from vertical profiles on the Lower Mississippi River (after Akalin, 2002)

In theory, the velocity orientation angle along vertical profiles should deviate clockwise in proportion to the magnitude of the velocity vector. While perfect agreement would be obtained when $y = x + ct$, these measurements are in very good agreement with the theoretical calculations, particularly when the velocity is high. However, in practice, on account of the presence of large dunes, many velocity profiles in the Mississippi River showed large-scale random turbulent velocity fluctuations masking the effects of the Coriolis acceleration. Winkley (1989) also concluded that the Coriolis force exerts an influence on the geometry of the Lower Mississippi River. The Coriolis effects become even more significant for oscillations of longer periods, such as tides.

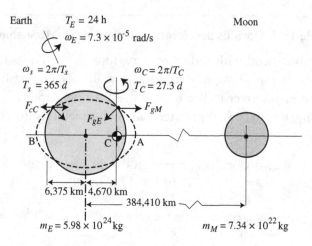

Figure 15.10. Earth–Moon influence on tides

A more accurate analysis of tides includes the effects of the Sun, whose gravitational effects are 2.17 times smaller than those of the Moon. The maximum tides should be obtained when the Earth, Moon, and Sun are directly aligned, i.e. at new and full Moons. Further complications arise because: (1) the axis of rotation of the Earth is at an angle of 66°33′ to the plane of Earth's rotation around the Sun; and (2) the plane of the Moon is also inclined at 5°9′ to the plane of the Earth's rotation around the Sun. The maximum tides should then be observed when the Sun, Moon, and Earth are aligned around the time of the equinoxes.

In practice, harmonic analyses based on field measurements enable accurate predictions of tidal levels at most locations. Figure 15.11a illustrates the tidal amplitude changes up the Saint Lawrence estuary. This example shows that mid-latitude tides are significant and tidal amplitude increases as the channel width and flow depth decrease, as predicted from Green's law in Section 15.1.5. Figure 15.11b also shows an example of tidal-wave propagation up the Saint Lawrence estuary on March 8, 2000. The reader will notice the tidal-wave amplification between Sept-Iles and Québec, and attenuation from fluvial friction between Québec and Trois-Rivières.

15.2.3 Saline Wedges in River Estuaries

Where a river enters a sea or an ocean, a saline wedge defines the interface between freshwater and saltwater. The mass density of saltwater ($\rho_{sea} \simeq 1,035$ kg/m^3) exceeds the mass density ρ_m of the river freshwater.

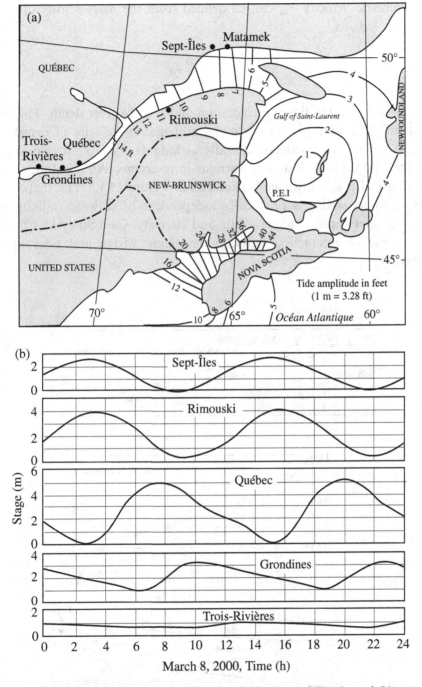

Figure 15.11. Tidal wave (a) amplification in the Bay of Fundy and (b) propagation up the Saint Lawrence estuary

A densimetric velocity V_Δ can be defined from the mass-density difference $\Delta\rho = \rho_{sea} - \rho_m$ as

$$V_\Delta = \sqrt{\frac{\Delta\rho}{\rho_m} gh}, \qquad (15.33)$$

where g is the gravitational acceleration and h is the river depth. The physical significance of V_Δ is linked to the propagation velocity of density currents. A saline wedge extends laterally as long as $V_\Delta > V$. It also properly describes the velocity of density currents in reservoirs (An and Julien, 2014). The shape of a saline wedge is sketched in Figure 15.12. The shape of the arrested saline wedge is practically independent of seawater salinity, river velocity, water depth, channel width, and viscosity. Case Study 15.1 provides an example of interaction between the saline wedge and tides in river estuaries.

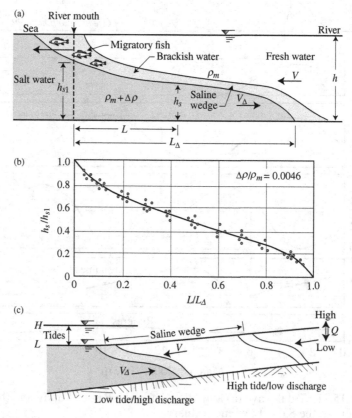

Figure 15.12. Brackish water zones in estuaries. (a) Brackish water zones in estuaries, (b) saline wedge profiles, and (c) influence of tides

Case Study 15.1 Smolt migration in the Matamek River estuary, Canada

The Matamek River is a small tributary of the Saint Lawrence River estuary located as shown on Figure 15.11a. When leaving rivers to enter the sea, a juvenile Atlantic salmon (*salmo salar*), or smolt, spends 20–30 days in the saline wedge of the river estuary to gradually adapt to saltwater. The capture-recapture method was used in the mid-1970s to study salmon migration and estimate the smolt population in the Matamek River estuary. Smolt captures with a seine were made for a period of 2 years at a fixed location in the estuary (point A in Figure 5.2) during the high tides of June and early July. The daily average number of smolts captured per seine haul is shown in Figure CS15.1.1. After the first year, the two peaks in fish migration intrigued the scientists at the Research Station. Hypotheses emerged to explain why the two different fish populations were migrating 2 weeks apart. Biologists debated that males and females could migrate at different times. When the moon cycle was plotted against the salmon capture chart, the peaks corresponded to the half-moons. It was then further argued that migrating smolts were sensitive to the Moon cycle. A systematic research program was set up for the subsequent year and the results of the detailed measurements were puzzling with a very long first migration and a very small second pulse. It is only during the second year that the high-water levels were plotted as a function of time in Figure CS15.1.1. It was then noticed that the low captures were observed during the very high tides. Based on tidal levels, engineers developed regression models to predict whether fishing would be good or not on any given day. Once biologists figured out that the mathematical model predictions were fairly accurate, further thinking was needed to physically understand what was going on.

With physical introspection, it dawned on us that the saline wedge moves upstream–downstream as a function of tidal level and flow discharge, as sketched in Figure 15.12c. During half-moons, when the tide is moderately high, fishing takes place in the saline wedge. At higher tides, the saline wedge has to move farther upstream. We waved goodbye to the hypotheses regarding: (1) the Moon's influence on fish migration and (2) the presence of two distinct salmon populations in the Matamek River. This example illustrates very well the importance of physical understanding in environmental and hydraulic studies. The complementarity of observation, physics, and mathematics has been the object of my Hunter Rouse Lecture (Julien, 2017). This brings us back to Figure 1.1 and I thank the reader for exploring the wonders of river mechanics with me.

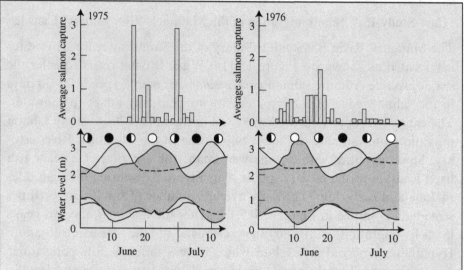

Figure CS15.1.1. Smolt capture and tide levels of the Matamek River estuary

Exercise 15.1

Substitute Equation (15.6) into Equation (15.7) to demonstrate that $\tilde{B} = 0$ at the boundary condition that the vertical velocity is zero at the bottom of the canal.

◆◆*Exercise 15.2*

Substitute Equation (15.6) into Equation (15.11b) to obtain wave-celerity Equation (15.12).

Exercise 15.3

From the definition of velocity in Equation (15.14), demonstrate that the convective-acceleration terms in Equation (15.9) are small compared to the local acceleration for small amplitude waves ($\tilde{a} \ll \lambda$). [*Hint:* Examine the ratio $(v_x^2 + v_z^2)/(\partial\Phi/\partial t)$.]

◆*Exercise 15.4*

Define the group–velocity relationship in Equation (15.28) by substituting the wave-celerity Equation (15.12) into Equation (15.27). Show that $c_G = c/2$ for deep-water waves.

◆*Exercise 15.5*

Calculate the densimetric velocity of the saline wedge for the Matamek River. The flow velocity is 1 m/s and the flow depth is ~4 m.
[*Answer:* $V_\Delta = 1.17$ m/s based on $\rho_{sea} = 1,035$ kg/m³.]

Exercise 15.6

Check the Beaufort scale for wind speeds and determine the equivalent wind speeds in m/s.
[*Answer:* $V = 8$–10 m/s for scale 5 and $V = 32$–37 m/s for scale 12.]

◆◆*Problem 15.1*

Estimate the average rate of sea-level rise from the water levels in Figure P15.1.

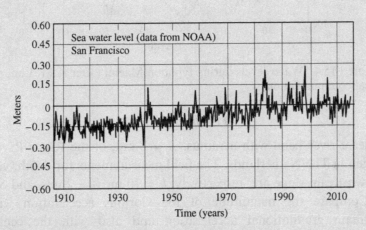

Figure P15.1. Water levels at San Francisco (after P. O'Brien, personal communication)

◆◆*Problem 15.2*

Consider a Category 1 hurricane with a wind speed of 75–90 mph (~40 m/s). Estimate the wave height, wavelength, and wave period in a river that is 10-m deep with a fetch length of 10 km.
[*Answer:* $\tilde{H}_s \simeq 2.8$m, $T \simeq 6.6$ s, and $\lambda_s \simeq 50$ m. The celerity $c = \lambda_s T \simeq 8$ m/s, and the waves are depth limited with resuspension of bed sediment.]

♦Problem 15.3

Estimate the wave height, length, and period in a river that is 4-m deep, with a fetch length of 1 km. For the wind speed of a Category 3 hurricane, how long would it take for the waves to fully develop?

♦Problem 15.4

The water-level measurements at the city of Masan in South Korea during typhoon Maemi are shown in Figure P15.4. If the measurements represent the maximum surge height, determine the category from the Saffir–Simpson scale.

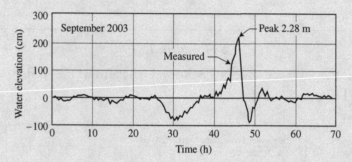

Figure P15.4. Water levels during typhoon Maemi (after K.M. Lim, personal communication)

♦Problem 15.5

Consider a river bend with a radius, $R = 2$ km, of the Rhine River near Nijmegen in The Netherlands. The field measurements for depth, velocity, and downstream slope are, respectively, $h = 10$ m, $V \simeq 2$ m/s, and $S \simeq 1 \times 10^{-4}$. Compare the magnitude of the Coriolis acceleration with the downstream gravitational acceleration and also with the centrifugal acceleration.

[*Answer:* $a_{cor} = 2 \times 10^{-4}$ m/s^2, $a_{cor}/gS \simeq 22$ percent, and $a_{cor}/a_{cent} \simeq 10$ percent.]

♦Problem 15.6

Consider a branch of the Mississippi River delta with a discharge of 270,000 ft^3/s, a channel width of 1,500 ft, and average depth of 45 ft. Estimate the densimetric velocity V_Δ of the saline wedge. Can the density current propagate upstream?

[*Answer:* From $\Delta\rho/\rho_m = 0.035$, $V = 4.0$ ft/s $= 1.2$ m/s, $V_\Delta = 7.11$ ft/s $= 2.17$ m/s. Yes.]

◆◆*Problem 15.7*

Consider the longitudinal profile of the Muda River estuary in Figure P15.7.

Determine the following when the river discharge is very low:

(1) How far could saltwater intrusion propagate in 1983 and 1994?
(2) If the river water is pumped into an irrigation canal located 30 km from the river mouth, describe what kind of problem could emerge.
(3) Two highway bridges were built 15 and 40 km upstream from the river mouth. What kind of problem do you expect?
(4) Which of the two bridges may experience future problems?
(5) What is the purpose of the estuary barrage?
(6) With reference to Julien et al. (2010), what may be the cause of riverbed degradation in the Muda River?

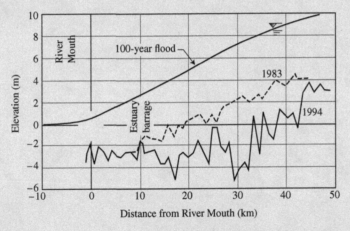

Figure P15.7. Longitudinal profile of the Muda River (after Julien et al., 2010)

Bibliography

AASHTO (1982). Hydraulic analyses for the location and design of bridges. In *Highway Drainage Guidelines, VII*. Washington, DC: American Association State Highway and Transportation Officials.

Abbott, M. B. and D. R. Basco (1989). *Computational Fluid Dynamics*. London: Addison-Wesley.

Abdullah, J. (2013). Distributed runoff simulation of extreme monsoon rainstorms in Malaysia using TREX. Ph.D. dissertation, Colorado State University, 226 p.

Abdullah, J. and P. Y. Julien (2014). Distributed flood simulations on a small tropical watershed with the TREX Model. *J. Flood Eng.*, 5(1–2), 17–37.

Abdullah, J., J. Kim and P. Y. Julien (2013). Hydrologic modeling of extreme events. In *Encyclopedia of Natural Resources*. New York, NY: Taylor & Francis.

Abdullah, J., S. N. Muhammad, P. Y. Julien, J. Ariffin and A. Shafie (2016). Flood flow simulations and return period calculation for the Kota Tinggi watershed, Malaysia. *J. Flood Risk Manag.*, DOI: 10.1111/jfr3.12256, 17.

Abes, L. M. (1991). Local scour around bridge piers in pressure flow. Ph.D. dissertation, Colorado State University, Fort Collins, CO, 303 p.

Ab. Ghani, A., N. A. Zakaria, R. Abdullah and S. S. Ahamad, eds. (2004). Rivers '04, Proc. 1st Intl. Conf. Manag. Rivers 21st Century: Issues and Chall. Penang, Malaysia, 667 p.

Ables, J. H., Jr. (1978). Filling and emptying system for Bay Springs Lock, Tennessee–Tombigbee Waterway, Mississippi. Tech. Rep. H-78-19, USACE, Waterways Exp. Station, Vicksburg, MS.

Abt, S. R. and T. L. Johnson (1991). Riprap design for overtopping flow. *J. Hyd. Eng., ASCE*, 117(8), 959–72.

Abt, S. R., M. S. Khattak, J. D. Nelson, et al. (1987). Development of riprap design criteria by riprap testing in flumes: phase I. NUREG/CR-4651 ORNL/TM-10100, Division of Waste Management, US Nuclear Regulatory Commission, Washington, DC, 111 p.

Abt, S. R., M. R. Peterson, C. C. Watson and S. A. Hogan (1992). Analysis of ARS low-drop grade-control structures. *J. Hyd. Eng., ASCE*, 118(10), 1424–34.

Abt, S. R., R. J. Wittler, J. F. Ruff, et al. (1988). Development of riprap design criteria by riprap testing in flumes: phase II. NUREG/CR-4651 ORNL/TM-10100/V2, Division of Waste Management, US Nuclear Regulatory Commission, Washington, DC, 84 p.

Ackers, P. and F. G. Charlton (1970). Dimensional analysis of alluvial channels with special reference to meander length. *J. Hyd. Res., IAHR*, 8, 287–314.

Aguirre-Pe, J. and R. Fuentes (1990). Resistance to flow in steep rough streams. *J. Hyd. Eng., ASCE*, 116(11), 1374–87.

Akalin, S. (2002). Water temperature effect on sand transport by size fraction in the Lower Mississippi River. Ph.D. dissertation, Colorado State University, Fort Collins, CO, 218 p.

Akashi, N. and T. Saitou (1986). Influence of water surface on scour from vertical jets. *J. Hydrosci. Hyd. Eng., Japan Soc. Civil Eng.*, 4 55–69.

Albert, J. (2004). Hydraulic analysis and double mass curves of the Middle Rio Grande from Cochiti to San Marcial, New Mexico. M.S. thesis, Colorado State University, 207 p.

Albertson, M. L., Y. B. Dai, R. A. Jensen and H. Rouse (1950). Diffusion of submerged jets. *Trans. ASCE*, 115, 639–97.

Al-Mattarneh, H., L. Mohd Sidek and M. Z. Yusoff, eds. (2008). Innovations in water resources and environmental engineering. In *Proc. Intl. Conf. Const. Build. Tech. 2008*. UPENA, UiTM, Shah Alam, Malaysia, 486 p.

Altinbilek, H. D. and Y. Basmaci (1973). Localized scour at the downstream of outlet structures. In *Proc. 11th Cong. Large Dams*. Intl. Comm. Large Dams, 105–21.

Ambrose, R. B., J. L. Martin and T. A. Wool (1993). WASP5, A hydrodynamic and water quality model – model theory, user's manual, and programmer's guide. US Environmental Protection Agency, Office of Research and Development, Environmental Research Laboratory, Athens, GA.

An, S. D. (2011). Interflow dynamics and three-dimensional modeling of turbid density currents in Imha Reservoir, South Korea. Ph.D. dissertation, Colorado State University, 151 p.

An, S. D. and P. Y. Julien (2014). Case-study: three-dimensional modeling of turbid density currents in Imha Reservoir, South Korea. *J. Hyd. Eng., ASCE*, 140(5), 15.

An, S. D., P. Y. Julien and S. K. Venayagamoorthy (2012). Numerical simulation of particle driven gravity currents. *J. Env. Fluid Mech.*, 12(6), 495–513.

An, S. D., H. Ku and P. Y. Julien (2015). Numerical modelling of local scour caused by submerged jets. *J. Sci. Tech., Maejo Intl.*, 9(3), 328–43.

Anctil, F., J. Rousselle and N. Lauzon (2012). *Hydrologie, Cheminement de l'Eau*, 2e Ed. Montreal: Presses Internationales Polytechnique, 391 p.

Anderson, A. G. (1967). On the development of stream meanders. In *Proc. 12th Congress*. IAHR, Delft, The Netherlands, 1, 370–8.

Annable, W. K., C. C. Watson and P. J. Thompson (2010). Quasi-equilibrium conditions of urban gravel-bed stream channels in Southern Ontario. *River Res. App.*, 26, 1–24.

Annandale, G. W. (2006). *Scour Technology*. New York, NY: McGraw-Hill, 430 p.

Anthony, D. J. (1987). Stage dependent channel adjustments in a meandering river, Fall River, Colorado. M.S. thesis, Earth Res., Colorado State University, Fort Collins, CO, 180 p.

—— (1992). Bedload transport and sorting in meander bends, Fall River, Rocky Mountain National Park, Colorado. Ph.D. dissertation, Earth Res., Colorado State University, Fort Collins, CO.

Anthony, D. J. and M. D. Harvey (1991). Stage-dependent cross-section adjustments in a meandering reach of Fall River, Colorado. *Geomorphology*, 4, 187–203.

Anthony, D. J., M. D. Harvey, J. B. Laronne and M. P. Mosley, eds. (2001). *Applying Geomorphology to Environmental Management.* Highlands Ranch: Water Resource Publications, 484 p.

Arcement, G. K. and V. R. Schneider (1984). Guide for selecting Manning's roughness coefficients for natural channels and flood plains. USGS Water Supply Paper 2339, US Geological Survey, Washington, DC.

ASCE Hydraulics (1998). Bank mechanics, and modeling of river width adjustment, river width adjustment. I: processes and mechanisms. *J. Hyd. Eng., ASCE,* 124(9), 881–902.

——— (1998). Bank mechanics, and modeling of river width adjustment, river width adjustment. II: modeling. *J. Hyd. Eng., ASCE,* 124(9), 903–17.

ASCE Manual of Practice 54 (1977). *Sedimentation Engineering.* Ed. V. A. Vanoni, New York, NY: ASCE.

ASCE Manual of Practice 110 (2008). Sedimentation engineering – processes, measurements, modeling and practice. In *Task Committee to Update Manual 54.* Ed. M. H. Garcia, New York, NY: ASCE, 1132 p.

ASCE Manual of Practice 124 (2013). Inland navigation, channel training works. In *Task Committee on Inland Navigation.* Ed. T. J. Pokrefke, New York, NY: ASCE, 186 p.

Ashmore, P. and G. Parker (1983). Confluence scour in coarse braided streams. *Wat. Res. Res., AGU,* 19, 392–402.

Ashton, G. D., ed. (1986). *River and Lake Ice Engineering.* Littleton: Water Resource Publications, 485 p.

Atlas, D. and C. Ulbrich (1977). Path and area-integrated rainfall measurement by micro-wave attenuation in the 1–3 cm band. *J. Appl. Meteorol.,* 16, 1322–31.

Bagnold, R. A. (1960). Some aspects of the shape of river meanders, physiographic and hydraulic studies of rivers. Professional Paper 282E, US Geological Survey, Washington, DC.

Baird, D. C., L. Fotherby, C. C. Klumpp and S. M. Scurlock (2015). Bank stabilization design guidelines. SRH-2015-25, US Bureau of Reclamation, Denver, 331 p.

Barkau, R. L. (1993). UNET – one dimensional unsteady flow through a full network of open channels. Rep. CPD-66, USACE, Hydrologic Engineering Center, Davis, CA.

Barry, J. M. (1997). *Rising Tide, The Great Mississippi Flood of 1927 and How It Changed America.* New York, NY: Simon and Schuster, 524 p.

Bastian, D. F. (1995). *Grant's Canal: The Union's Attempt to By-Pass Vicksburg.* Shippenburg, PA: Burd Street Press, 88 p.

Battjes, J. and R. J. Labeur (2017). *Unsteady Flow in Open Channels.* Cambridge: Cambridge University Press.

Bauer, T. R. (2000). Morphology of the Middle Rio Grande from Bernalillo Bridge to the San Acacia Diversion Dam, NM. M.S. thesis, Civil Engineering, Colorado State University, 308 p.

Bauer, T., C. Leon, G. Richard and P. Y. Julien (2000). Middle Rio Grande, Bernalillo Bridge to San Acacia, hydraulic geometry, discharge and sediment data base and report. Civil Engineering, Colorado State University, Fort Collins, CO.

Begin, Z. B., D. F. Meyer and S. A. Schumm (1981). Development of longitudinal profiles of alluvial channels in response to base level lowering. *E. Surf. Proc. Landf.,* 6, 49–68.

Beltaos, S., ed. (1995). *River Ice Jams*. Highlands Ranch, CO: Water Resource Publications, 372 p.

Bender, T. R. and P. Y. Julien (2012). Bosque reach overbank flow analysis 1962–2002. Colorado State University Report for USBR, Albuquerque, NM, 175 p.

Bernard, R. S. and M. L. Schneider (1992). Depth-averaged numerical modeling for curved channels. HL-92-9, USACE, Waterways Experiment Station, Vicksburg, MS.

Best, J. L. and P. J. Ashworth (1997). Scour in large braided rivers and the recognition of sequence stratigraphic boundaries. *Nature*, 387, 275–7.

Bestgen, K. and S. Platania (1991). Status and conservation of the Rio Grande silvery minnow, *Hibognathus amarus*. *Southwest. Nat.*, 36, 225–32.

Bethemont, J. (1994). Enfoncement de lits fluviaux: processus naturels et impacts des activités humaines. *Revue de Geographie de Lyon*, 69, 103.

Bhowmik, N., E. V. Richardson and P. Y. Julien (2008). Daryl B. Simons – hydraulic engineer, researcher and educator. *J. Hyd. Eng., ASCE*, 134(3), 287–94.

Biedenharn, D. S. (1995). Lower Mississippi River channel response: past present and future. Ph.D. dissertation, Civil Engineering, Colorado State University, Fort Collins, CO.

Biedenharn, D. S., P. G. Combs, G. J. Hill, C. F. Pinkard, Jr. and C. B. Pinkiston (1989). Relationship between the channel migration and radius of curvature on the Red River. In *Proc. Intl. Symp. Sed. Transp. Modeling*. New York, NY: ASCE, 536–41.

Biedenharn, D. S., C. M. Elliott and C. C. Watson (1997). The WES stream investigation and streambank stabilization handbook. USACE, Waterways Exp. St., Vicksburg, 436 p.

Biedenharn, D. S., C. R. Thorne, P. J. Soar, R. D. Hey and C. C. Watson (2001). Effective discharge calculation guide. *Intl. J. Sed. Res.*, 16(4), 445–59.

Blaisdell, F. W. and C. L. Anderson (1989). Scour at cantilevered outlets: plunge pool energy dissipator design criteria. ARS-76, USDA, Agricultural Research Service, Washington, DC.

Bledsoe, B. P., E. D. Stein, R. J. Hawley and D. Booth (2012). Framework and tool for rapid assessment of stream susceptibility to hydromodification. *J. AWRA*, 48(4), 788–808.

Blench, T. (1969). *Mobile-Bed Fluviology*. Edmonton: University of Alberta Press.

—— (1986). *Mechanics of Plains River*. Edmonton: Printing Service University of Alberta, 111 p.

Bley, P. A. and J. R. Moring (1988). Freshwater and ocean survival of Atlantic salmon and steelhead: a synopsis. *U.S. Fish and Wildlife Serv. Biol. Rep.* 88(9), 22.

Bogardi, J. L. (1974). *Sediment Transport in Alluvial Streams*. Budapest: Akademiai Kiado.

Bols, P. (1978). The iso-erodent map of Java and Madura. Belgian technical assistance project ATA105, Soil Research Institute, Bogor.

Bondurant, D. C. (1963). *Missouri River Division*. Omaha, NE: USACE.

Bormann, N. E. (1988). Equilibrium local scour depth downstream of grade-control structures. Ph.D. dissertation, Civil Engineering, Colorado State University, 214 p.

Bormann, N. E. and P. Y. Julien (1991). Scour downstream of grade-control structures. *J. Hyd. Eng., ASCE*, 117(5), 579–94.

Bos, M. G., ed. (1989). *Discharge Measurement Structures. Pub. 20*, Wageningen: International Institute for Land Reclamation and Improvement, 401 p.

Bounvilay, B. (2003). Transport velocities of bedload particles in rough open channel flows. Ph.D. dissertation, Civil Engineering, Colorado State University, 155 p.

Boyce, R. (1975). Sediment routing and sediment delivery ratios. In *Present and Prospective Technology for Predicting Sediment Yields and Sources*, USDA-ARS-S-40. US Department of Agriculture, Washington, DC, 61–5.

Bradley, J. B. (1984). Transition of a meandering river to a braided system due to high sediment concentration flows. In River Meandering. In *Proc. Conf. River '83*. ASCE, 89–100.

Bragg, M. (1990). *Historic Names and Places on the Lower Mississippi River*. Vicksburg, MS: Mississippi River Commission, USACE, 282 p.

Brahms, A. (1753). Anfangsgründe der Deich und Wasserbaukunst, Zurich.

Bray, D. I. (1979). Estimating average velocity in gravel bed rivers. *J. Hyd. Eng., ASCE*, 105, 1103–22.

—— (1980). Evaluation of effective boundary roughness for gravel-bed rivers. *Can. J. Civ. Eng.*, 7, 392–7.

—— (1982a). Flow resistance in gravel-bed rivers, Chap. 6. In *Gravel-Bed Rivers*. New York, NY: Wiley, 109–37.

—— (1982b). Flow resistance in gravel-bed rivers, Chap. 19. In *Gravel-Bed Rivers*. New York, NY: Wiley, 517–52.

—— (1991). Resistance to flow in gravel-bed rivers. Tech. Rep. HTD-91-1, CSCE Monographs: Desktop Series, 2. Canadian Society for Civil Engineering, Montreal, 95 p.

Breusers, H. N. C. (1967). Two-dimensional local scour in loose sediment. In *Closure of Estuarine Channels in Tidal Regions*. Delft: Delft Hydraulic Publications, 64.

Breusers, H. N. C., G. Nicolet and H. W. Shen (1977). Local scour around cylindrical piers. *J. Hyd. Res., IAHR*, 15, 211–52.

Breusers, H. N. C. and A. J. Raudkivi (1991). *Scouring*. Rotterdam: Balkema, 143 p.

Briaud, J. L., H. C. Chen, K. A. Chang, et al. (2007). Establish guidance for soil properties-based prediction of meander migration rate. FHWA/TX-07/0-4378-1, Texas Transportation Institute, Texas Department of Transportation, 315 p.

Brice, J. C. (1981). Stability of relocated stream channels. FHWA/RD-80/158. Federal Highway Administration, USDOT, Washington, DC.

—— (1982). Stream channel stability assessment. FHWA/RD-82/21, Federal Highway Administration, USDOT, Washington, DC.

—— (1984a). Assessment of channel stability at bridge sites. *Transp. Res. Record*, 2(950).

—— (1984b). Planform properties of meandering rivers. In *River Meandering: Proc. Conf. Rivers '83*. New York, NY: ASCE.

Brice, J. C. and J. C. Blodgett (1978). Countermeasures for hydraulic problems at bridges. FHWA/RD-78-162 & 163, Federal Highway Administration, USDOT, Washington, DC.

Brookes, A. and F. D. Shields, Jr., eds. (1996). *River Channel Restoration*. Chichester: Wiley, 433 p.

Brown, S. A. (1985a). Design of spur-type streambank stabilization structures. FHWA/RD-84-101, Federal Highway Administration, USDOT, Washington, DC.

—— (1985b). Streambank stabilization measures for highway stream crossings. FHWA-RD-80-160, Federal Highway Administration, USDOT, Washington, DC.

—— (1985c). Streambank stabilization measures for highway engineers. FHWA/ RD-84-100, Federal Highway Administration, USDOT, Washington, DC.

Brown, S. A. and E. S. Clyde (1989). Design of riprap revetment, Hydraulic Engineering Circular 11. FHWA/IP-89-016, Federal Highway Administration, USDOT, Washington, DC.

Brown, S. A., R. S. McQuivey and T. N. Keefer (1980). Stream channel degradation and aggradation: analysis of impacts to highway crossings. FHWA/RD-80-159, Federal Highway Administration, USDOT, Washington, DC.

Bruens, A. (2003). Entraining mud suspensions. Communications on Hydraulic and Geotechnical Engineering, Rep. 03-1, ISSN 0169-6548, TUD, Delft University of Technology, The Netherlands, 137 p.

Bunte, K. (1996). Analyses of the temporal variation of coarse bedload transport and its grain size distribution, Squaw Creek Montana. Gen. Tech. Rep. RM-GTR-288, USDA, Rocky Mountain Forest and Range Experiment Station, Fort Collins, 124 p.

Bunte, K., K. W. Swingle and S. R. Abt (2007). Guidelines for using bedload traps in coarse-bedded mountain streams: construction, installation, operation, and sample processing. Gen. Tech. Rep. RMRS-GTR-191, USDA, Forest Service, Rocky Mountain Research Station, Fort Collins, CO, 91 p.

Burke, T. K. (1994). Datalink, a real-time data acquisition system for hydrologic modeling. M.S. thesis, Civil Engineering, Colorado State University, 257 p.

Burrows, R. L., W. W. Emmett and B. Parks (1981). Sediment transport in the Tanana River near Fairbanks, Alaska, 1977–79. USGS Water Resources Investigations 81-20, Anchorage, AK.

Buska, J. S., E. F. Chacho, C. M. Collins and L. W. Gatto (1984). Overview of Tanana River monitoring and research studies near Fairbanks, Alaska. USACE Cold Regions Research and Engineering Laboratory, Hanover, NH.

Busnelli, M. M. (2001). Numerical simulation of free surface flows with steep gradients. Communications on Hydraulic and Geotechnical Engineering, Rep. 01-3, ISSN 0169-6548,TUD, Delft University of Technology, The Netherlands, 180 p.

Bussi, G., F. Frances, J. J. Montoya and P. Y. Julien (2014). Distributed sediment yield modelling: importance of initial sediment conditions. *J. Env. Model. Software*, 58, 58–70.

Calder, I. R. (1992). Hydrologic effect of land-use change, Chap. 13. In *Handbook of Hydrology*. New York, NY: McGraw-Hill, 50 p.

Callander, R. A. (1969). Instability and river channels. *J. Fluid Mech., CUP*, 36, 465–80.

—— (1978). River meandering. *Ann. Rev. Fluid Mech.*, 10, 129–58.

Campbell, C. S. and M. Ogden (1999). *Constructed Wetlands in the Sustainable Landscape*. New York, NY: Wiley, 270 p.

Carstens, M. R. (1966). Similarity laws for localized scour. *J. Hyd. Eng., ASCE*, 92, 13–36.

Carter, D. and B. Michel (1971). Lois et mécanismes de l'apparente fracture fragile de la glace de rivière et de lac. Rep. S-22, Génie Civil, Université Laval, Québec.

Caruso, B. S., T. J. Cox, R. L. Runkel, et al. (2008). Metals fate and transport modelling in streams and watersheds: state-of-the-science and US-EPA workshop review. *J. Hydrol. Proc.*, 22, 4011–21.

Cecen, K., M. Bayazit and M. Sumer (1969). Distribution of suspended matter and similarity criteria in settling basins. *Proc. 13th Congress of the IAHR*, Kyoto, Japan, Vol. 4, 215–25.

Cha, Y. K., J. S. Lee, D. C. Lee, et al. (2003). *Coastal Engineering*. Seoul: Saeron, 407 p.

Chabert, J. and P. Engeldinger (1956). *Étude des affouillements autour des piles de ponts*. Chatou: Lab. National d'Hydraulique.

Chabert, J., M. Remillieux and I. Spitz (1962). Correction des rivières par panneaux de fond. *Bull. Centre Rech. D'Essais de Chatou, France, Sér. A*, 1, 49–63.

Chang, H. H. (1979a). Geometry of rivers in regime. *J. Hyd. Eng., ASCE*, 105, 691–706.

—— (1979b). Minimum stream power and river channel patterns. *J. Hydrol.*, 41, 303.

—— (1980). Stable alluvial canal design. *J. Hyd. Eng., ASCE*, 106, 873–91.

—— (1984). Analysis of river meanders. *J. Hyd. Eng., ASCE*, 110, 37–50.

—— (1988). *Fluvial Processes in River Engineering*. New York, NY: Wiley.

—— (2006). *Generalized Computer Program FLUVIAL-12 Mathematical Model for Erodible Channels – User's Manual*. San Diego, CA: Chang Consultants, 66 p.

Chapra, S. C. (1997). *Surface Water-Quality Modeling*. New York, NY: McGraw-Hill, 844 p.

Charlton, F. G. (1982). River stabilization and training in gravel-bed rivers, Chap. 23. In *Gravel-Bed Rivers*, New York, NY: Wiley, 635–57.

Chaudhry, M. H. (2008). *Open-Channel Flow*, 2nd ed. New York, NY: Springer, 523 p.

Chauvin, J. L. (1962). Similitude des modèles de cours d'eau à fond mobile. *Bull. Centre Rech. d'Essais de Chatou, France, Sér. A*, 1, 64–91.

Chee, S. P. and E. M. Yuen (1985). Erosion of unconsolidated gravel beds. *Can. J. Civ. Eng.*, 12, 559–66.

Chen, C. L. (1976). Flow resistance in broad shallow grassed channels. *J. Hyd. Eng., ASCE*, 102, 307–22.

Chen, Y. H. and C. F. Nordin (1976). Temperature effects in the transition from dunes to plane bed. Missouri River Division Sediment Series 14, USACE, Omaha, NE, 37 p.

Cheng, N. S. (2015). Resistance coefficients for artificial and natural coarse-bed channels: alternative approach for large scale roughness. *J. Hyd. Eng., ASCE*, 141(2), 515–29.

Chien, N. (1954). Meyer–Peter formula for bedload transport and Einstein bedload function. Rep. 7, Institute of Engineering Research, University of California, Berkeley, CA.

—— (1957). A concept of the regime theory. *Trans. ASCE*, 122, 785–93.

Chiew, Y. M. (1990). Mechanics of local scour depth at submarine pipelines. *J. Hyd. Eng., ASCE*, 116, 515–29.

—— (1991). Prediction of maximum scour depth at submarine pipelines. *J. Hyd. Eng., ASCE*, 117, 452–66.

Chinnarasri, C., D. Kositgittiwong and P. Y. Julien (2012). Model of flow over spill-ways by computational fluid dynamics. In *Proc. Inst. Civil Eng.*, Paper 1200034, 12 p.

Chitale, S. V. (1970). River channel patterns. *J. Hyd. Eng., ASCE*, 96, 201–21.

—— (1973). Theory and relationship of river channel patterns. *J. Hydrol.*, 19, 285–308.

Choi, G. W. (1991). Hydrodynamic network simulation through channel junctions. Ph.D. dissertation, Civil Engineering, Colorado State University, 220 p.

Chow, V. T. (1959). *Open-Channel Hydraulics.* New York, NY: McGraw-Hill, 680 p.

Churchill, M. A., H. L. Elmore and R. A. Buckingham (1962). Prediction of stream reaeration rates. *J. Sanitary Eng. Div., ASCE*, 88(SA4), 1–46, Paper 3199.

Clements, W. (2004). Small-scale experiments support causal relationships between metal contamination and macroinvertebrate community responses. *Ecol. Appl.*, 14(3), 954–67.

Cline, T. J. (1988). Development of a watershed information system for HEC-1 with applications to Macks Creek, Idaho. M.S. thesis, Civil Engineering, Colorado State University, 238 p.

Cline, T. J., A. Molinas and P. Y. Julien (1989). An Auto-CAD-based watershed information system for the hydrologic model HEC-1. *Wat. Res. Bull., AWRA*, 25(3), 641–52.

Combs, P. G. (1994). Prediction of the loop rating curve in alluvial rivers. Ph.D. dissertation, Civil Engineering, Colorado State University, Fort Collins, CO.

Cooper, K. (2011). Evaluation of the relationship between the RUSLE R-factor and the mean annual precipitation. M.S. technical report, Civil Engineering, Colorado State University, 37 p.

Copeland, R. R. and W. A. Thomas (1989). Corte Madera Creek sedimentation study: numerical model investigation. Tech. Rep. HL-89-6, USACE, Waterways Exp. Station, Vicksburg, MS.

Cowley, D. E., P. D. Shirey and M. D. Hatch (2006). Ecology of the Rio Grande silvery minnow (Cyprinidae: *Hybognathus amarus*) inferred from specimens collected in 1874. *Rev. Fisheries Science*, 14, 111–25.

Cox, A. L., C. I. Thornton and S. R. Abt (2014). Articulated concrete block stability assessment for embankment-overtopping conditions. *J. Hyd. Eng., ASCE*, 140(5), 7.

Creager, W. P., J. D. Justin and J. Hinds (1945). *General Design*, Vol. 1. Engineering Dams Series, New York, NY: Wiley.

Cunge, J. A. (1969). On the subject of a flood propagation computation method (Muskingum method). *J. Hyd. Res., IAHR*, 7, 205–30.

CURTAW (1991). *Guide for the Design of River Dikes, Vol. 1 – Upper River Area.* ISBN 90 376 00 11 5, Gouda, The Netherlands, 208 p.

CWPRS (2008). Guidelines for operation of desilting basins. Ministry of Water Resources, Government of India, Pune, 22 p.

D'Aoust, S. G. and R. G. Millar (2000). Stability of ballasted woody debris habitat structures. *J. Hyd. Eng., ASCE*, 126(11), 810–17.

Dawod, A. M. (1986). Modeling of soil erosion using rainfall volume (for engineering and planning application). Ph.D. dissertation, Colorado State University, 135 p.

Dawod, A. M. and P. Y. Julien (1987). On predicting upland erosion losses from rainfall depth Part 2: field applications in Iraq. *J. Stoch. Hydrol. Hyd.*, 1, 135–40.

Delft Hydraulics (1996). Case: gravity based structure. Rep. Q 2225, Delft Hydraulics, Delft, The Netherlands.

Derbyshire, K. (2006). Fisheries guidelines for fish-friendly structures. Fish Habitat Department of Primary Industries and Fisheries, Guideline FHG 006, Queensland Government, 64 p.

Derrick, D. L. (1997). Harland Creek Bendway Weir/Willow post bank stabilization demonstration project. In *Proc. Conf. Manag. Landscapes Disturbed by Channel Incision*. Oxford, MS.

Derrick, D. L., T. J. Pokrefke, M. B. Boyd, J. P. Crutchfield and R. R. Henderson (1994). Design and development of bendway weirs for the Dogtooth Bend Reach, Mississippi River. HL-94-10, USACE, Waterways Experiment Station, Vicksburg, MS, 104 p.

Derruau, M. (1972). *Les Formes du Relief Terrestre*. Paris: Masson et Cie, 120 p.

de Vriend, H. J. (1976). A mathematical model of steady flow in curved open shallow channels. Rep. 76-1, Civil Engineering, University of Technology, Delft, The Netherlands.

—— (1977). A mathematical model of steady flow in curved shallow channels. *J. Hyd. Res., IAHR*, 15, 37–53.

de Vries, M. (1973). Application of physical and mathematical models for river problems. DHL Pub. 112, Delft Hydraulics, The Netherlands.

DID (2009). River Management, DID Manual. Vol. 2, Department of Irrigation and Drainage, Government of Malaysia, Kulala Lumpur, Malaysia, 612 p.

Di Toro, D. M. (2001). *Sediment Flux Modeling*. New York, NY: Wiley, 624 p.

Doe, W. W., III (1992). Simulation of the spatial and temporal effects of army maneuvers on watershed response. Ph.D. dissertation, Civil Engineering, Colorado State University, 301 p.

Doe, W., B. Saghafian and P. Y. Julien, III (1996). Land use impact on watershed response: the integration of two dimensional hydrologic modelling and Geographic Information Systems. *J. Hydrol. Proc.*, 10, 1503–11.

Doehring, F. K. and S. R. Abt (1994). Drop height influence on outlet scour. *J. Hyd. Eng., ASCE*, 120, 1470–6.

Duan, J. G. (2001). Simulation of streambank erosion processes with a two-dimensional numerical model, Chap. 13. In *Landscape Erosion and Evolution Modeling*, New York, NY: Kluwer Academic/Plenum Publishers, 389–427.

Duan, J. G. and P. Y. Julien (2005). Numerical simulation of the inception of channel meandering. *Earth Surf. Proc. Landforms*, 30, 1093–110.

—— (2010). Numerical simulation of meandering evolution. *J. Hydrol.*, 391, 34–46.

Dudley, R. K. and S. P. Platania (1997). Habitat use of Rio Grande silvery minnow. Report to the New Mexico Department of Game and Fish, Santa Fe, and USBR, Albuquerque, NM, 96 p.

Einstein, A. (1926). Die Ursache der Mäanderbildung der Flussläufe und des sogenannten Baerschen Gesetzes, Naturwissenschaften, 11.

Einstein, H. A. and N. Chien (1955). Effect of heavy sediment concentration near the bed on velocity and sediment distribution. MRD Series No. 8. University of California Institute of Engineering Research and USACE Missouri River Div., Omaha, Nebraska.

Elliott, C. M., ed. (1984). *River Meandering Proceedings of the Conference Rivers '83*. New York, NY: ASCE.

Engelund, F. (1970). Instability of erodible beds. *J. Fluid Mech., CUP*, 42, 225–44.

—— (1974). Flow and bed topography in channel bends. *J. Hyd. Div., ASCE*, 100, 1631–48.

Engelund, F. and O. Skovgaard (1973). On the origin of meandering and braiding in alluvial streams. *J. Fluid Mech., CUP*, 57, 289–302.

England, J. F., Jr. (2006). Frequency analysis and two-dimensional simulations of extreme floods on a large watershed. Ph.D. dissertation, Civil Engineering, Colorado State University, 237 p.

England, J. F., Jr., J. E. Godaire, R. E. Klinger, T. R. Bauer and P. Y. Julien (2010). Paleohydrologic bounds and extreme flood frequency of the Upper Arkansas River, Colorado, USA. *J. Geomorphol.*, 124, 1–16.

England, J. F., P. Y. Julien and M. L. Velleux (2014). Physically-based extreme flood frequency analysis using stochastic storm transposition and paleoflood data. *J. Hydrol.*, 510, 228–45.

England, J. F., Jr., M. L. Velleux and P. Y. Julien (2007). Two-dimensional simulations of extreme floods on a large watershed. *J. Hydrol.*, 347(1–2), 229–41.

EPA (1997). EPA's National hardrock mining framework. EPA Rep. 833-B-97-003, Washington, DC.

EPA and USACE (1998). Evaluation of dredged material proposed for discharge in waters of the U.S. – testing manual. EPA-823-B-98-004, Washington, DC, 87 p.

Ettema, R. (1980). Scour at bridge piers. Rep. 216, The University of Auckland, New Zealand.

—— (2008). Ice effects on sediment transport in rivers, Chap. 13. In *ASCE MOP 110*, 613–48.

Fahlbusch, F. E. (1994). Scour in rock riverbeds downstream of large dams. *Intl. J. Hydropower Dams*, 1(4), 30–2.

FAP24 (1996). Floodplain levels and bankfull discharge, river survey project, special reports 6 and 7. Government of the People's Republic of Bangladesh, 36 p. and 40 p.

Falconer, R. A., B. Lin, E. L. Harris and C. A. M. E. Wilson, eds. (2002). Hydroinformatics 2002. In *Proc. 5th Intl. Conf. on Hydroinformatics*. Cardiff, UK, 794 p.

Farhoudi, J. and K. V. H. Smith (1985). Local scour profiles downstream of hydraulic jumps. *J. Hyd. Res., IAHR*, 23, 343–58.

Farias, H. D., J. D. Brea and C. M. Garcia, eds. (2011). Hidraulica Fluvial: Procesos de Erosion y Sedimentacion, Obras de Control y Gestion de Rios. ISBN: 978-987-1780-05-1, Univ. Nac. deSantiago del Estero, Argentina, 122 p.

Federal Interagency Stream Restoration Working Group (1998). *Stream Corridor Restoration, Principles, Processes and Practices*. Washington, DC: US Federal Government.

FEMA (1986). Design manual for retrofitting flood-prone residential structures. Rep. FEMA 114, Federal Emergency Management Agency, 265 p.

Fennema, R. J. and M. H. Chaudhry (1986). Explicit numerical schemes for unsteady free-surface flows with shocks. *Wat. Res. Res., AGU*, 22, 1923–30.

—— (1990). Numerical solution of two-dimensional transient free-surface flows. *J. Hyd. Eng., ASCE*, 116, 1013–34.

Ferguson, R. I. (1984). The threshold between meandering and braiding, channels, and channel control structures. In *Proceedings 1st International Conference on Hydraulic Design in Water Resource Engineering*. Ed. K. V. H. Smith, New York, NY: Springer-Verlag, 6-15–6.30.

Ferguson, R. (2007). Flow resistance equations for gravel – and boulder bed streams. *Wat. Res. Res., AGU*, 43(5), W05427.

FHWA (1984). Guide for selecting Manning's roughness coefficients for natural channels and flood plains. FHWA-TS-84-204, Turner-Fairbank Highway Research Center, McLean, VA, 62 p.

—— (1989). Design of riprap revetments. HEC-11, Turner-Fairbank Highway Research Center, VA.

FHWA-HEC18 (2012). Evaluating Scour at Bridges, 5th ed. Hydraulic Engineering Circular 18, FHWA-HIF-12-003, Federal Highway Administration, USDOT, Washington, DC, 340 p.

FHWA-HEC20 (2012). Stream Stability at Highway Structures, 4th ed. Hydraulic Engineering Circular 20, FHWA-HIF-12-004, Federal Highway Administration, USDOT, Washington, DC, 328 p.

FHWA-HEC23a (2009). Bridge Scour and Stream Instability Countermeasures: Experience, Selection, and Design Guidance, Vol. 1, 3rd ed. Hydraulic Engineering Circular 23, FHWA-NHI-09-111, Federal Highway Administration, USDOT, Washington, DC, 256 p.

FHWA-HEC23b (2009). Bridge Scour and Stream Instability Countermeasures: Experience, Selection, and Design Guidance, Vol. 2, 3rd ed. Hydraulic Engineering Circular 23, FHWA-NHI-09-112, Federal Highway Administration, USDOT, Washington, DC, 376 p.

FHWA-HRE (2001). River engineering for highway encroachments – highways in the river environment. FHWA NHI 01-004, Hydraulic Design Series No. 6, Federal Highway Administration, USDOT, Washington, DC, 646 p.

Fischenich, C. (2001). Stability thresholds for stream restoration materials. EMRRP Tech. Notes Coll. (ERDC TN-EMRRP-SR-29), USACE, ERDC, Vicksburg, MS.

Fischenich, C. and H. Allen (2000). Stream management. Water operations special rep. ERDC/EL SRW-00-1, USACE, ERDC, Vicksburg, MS.

Fischenich, C. and R. Seal (2000). Boulder clusters. EMRRP Tech. Notes Coll., ERDC TN-EMRRP-SR-11, USACE, ERDC, Vicksburg, MS.

Follum, M. L., C. W. Downer, J. D. Niemann, S. M. Roylance and C. M. Vuyovich (2015). A radiation-derived temperature index snow routine for the GSSHA hydrologic model. *J. Hydrol.*, 529, 723–36.

Fortier, S. and F. C. Scobey (1926). Permissible canal velocities. *Trans. ASCE*, 89, Paper 1588, 940–84.

Fotherby, L. M. (2009). Valley confinement as a factor of braided river pattern for the Platte River. *Geomorphology*, 103(4), 562–76.

Fournier, F. (1969). Transports solides effectués par les cours d'eau. *Bull. IAHS*, 14, 7–49.

Franco, J. J. and C. D. McKellar (1968). Navigation conditions at Lock and Dam No. 3, Arkansas River, Arkansas and Oklahoma. H-68-8, USACE, Waterways Experiment Station, Vicksburg, MS.

Fredsøe, J. (1978). Meandering and braiding of rivers. *J. Fluid Mech., CUP*, 84, 609–24.

—— (1979). Unsteady flow in straight alluvial streams: modification of individual dunes. *J. Fluid Mech., CUP*, 91, 497–512.

Fredsøe, J. and R. Deigaard (1992). *Mechanics of Coastal Sediment Transport*. Singapore: World Scientific, 369 p.

Frenette, M., M. Caron, P. Y. Julien and R. J. Gibson (1984). Interaction between salmon parr population and discharge in the Matamec River. *Can. J. Fish. Aquat. Sci.*, 41, 954–64.

Frenette, M. and B. Harvey (1971). *Hydraulique Fluviale I*. Québec: Les Presses de l'Université Laval, 157 p.

Frenette, M. and P. Y. Julien (1980). Rapport synthèse sur les caractéristiques hydro-physiques du bassin de la rivière Matamec. Rep. CENTREAU-80-06, Laval Univ., Québec, Canada, 283 p.

Frenette, M. and P. Julien (1981). Etude hydrodynamique du site d'installation d'une barrière à saumons sur la rivière Mitis, Hydrotech Rep., Québec, Canada, 21 p.

Frenette, M. and P. Y. Julien (1986). LAVSED-I – Un modèle pour prédire l'érosion des bassins et le transfert de sédiments fins dans les cours d'eau nordiques. *Can. J. Civ. Eng., CSCE*, 13, 150–61.

—— (1987). Computer modelling of soil erosion and sediment yield from large watersheds. *Intl. J. Sed. Res.*, 1, 39–68.

Frenette, M., J. Llamas and M. Larinier (1974). Modèle de simulation du transport en suspension des rivières Châteauguay et Chaudière. Rapport CRE 74/05, CENTREAU, Univ. Laval, Québec, Canada, 33 p.

Friedman, D., J. Schechter, B. Baker, C. Mueller, G. Villarini and K. D. White (2016). US Army Corps of Engineers nonstationarity detection tool user guide. USACE, Climate Preparedness and Resilience Community of Practice, Washington, DC, 57 p.

Friesen, N. (2007). Amplification of supercritical surface waves in steep open channels near Las Vegas, NV. M.S. thesis, Civil Engineering, Colorado State University, 140 p.

Frings, R. (2007). From gravel to sand – downstream fining of bed sediments in the lower river Rhine. Netherlands Geographical studies 368, ISBN 0169-4839, Utrecht, 219 p.

Froehlich, D. C. (1988). Analysis of on site measurements of scour at piers. In *Proc. National Hyd. Eng. Conf.* New York, NY: ASCE.

Fujita, Y. and Y. Muramoto (1982a). Experimental study on stream channel processes in alluvial rivers. *Bull. Disaster Prev. Res. Inst., Kyoto Univ.*, 32, 49–96.

—— (1982b). The widening process of straight stream channels in alluvial rivers. *Bull. Disaster Prev. Res. Inst., Kyoto Univ.*, 32, 115–41.

—— (1985). Studies on the process and development of alternate bars. *Bull. Disaster Prev. Res. Inst., Kyoto Univ.*, 35, 55–86.

Galay, V. J. (1983). Causes of river bed degradation. *Water Res. Res., AGU*, 19, 1057–90.

Galay, V. (1987). *Erosion and Sedimentation in the Nepal Himalaya. An Assessment of River Processes*. Singapore: Kefford Press.

Garcia de Jalon, D. and M. Gonzalez del Tanago (1988). *Rios Y Riberas, Enciclop. Naturaleza de Espana*. Madrid: Borja Cardelus, 128 p.

Garde, R. J. (1989). *3rd International Workshop on Alluvial River Problems*. Rotterdam: Balkema, 329 p.

Gatto, L. W. (1984). Tanana River monitoring and research program: relationships among bank recession, vegetation, soils, sediments, and permafrost on the Tanana River near Fairbanks, Alaska. Rep. 84-21, USACE, CRREL, Hanover, NH.

Gessler, D., B. Hall, M. Spasojevic, et al. (1999). Application of 3D mobile bed hydrodynamic model. *J. Hyd. Eng., ASCE*, 125(7), 737–49.

Gibson, R. J. (1993). The Atlantic salmon in fresh water: spawning, rearing and production. *Reviews in Fish Biology and Fisheries*, 3, 39–73.

Gibson, R. J. and R. E. Cutting, eds. (1993). The production of juvenile Atlantic salmon, Salmo salar, in natural waters. Can. Spec. Publ. Fish. Aquat. Sci., National Research Council, Ottawa, Canada, 262 p.

Gill, M. A. (1972). Erosion of sand beds around spur dikes. *J. Hyd. Div., ASCE*, 98, 1587–602.

Gole, C. V. and S. V. Chitale (1966). Inland delta building activity of the Kosi River. *J. Hyd. Eng., ASCE*, 92(2), 111–26.

Goode, J. (2009). Substrate influences on bedrock channel forms and processes. Ph.D. dissertation, Earth Res., Colorado State University, 158 p.

Gordon, N. D., T. A. McMahon and B. L. Finlayson (1992). *Stream Hydrology an Introduction for Ecologists*. New York, NY: Wiley.

Graf, W. H. (1984). *Hydraulics of Sediment Transport.* New York, NY: McGraw-Hill.

Graf, W. H. and M. S. Altinakar (1993). *Hydraulique Fluviale,* Vol. 16, Tome 1 and 2. Lausanne: Presses Polytechniques et Universitaires Romandes, 259, 378 p.

Gray, D. M. and T. D. Prowse (1992). Snow and floating ice, Chap. 7. In *Handbook of Hydrology.* New York, NY: McGraw-Hill, 58 p.

Gray, D. H. and R. B. Sotir (1996). *Biotechnical and Soil Bioengineering: A Practical Guide for Erosion Control.* New York, NY: Wiley.

Greco, M., A. Carravetta and R. Della Morte, eds. (2004). *River Flow 2004.* Leiden: Balkema, 1455 p.

Green, W. H. and G. A. Ampt (1911). Studies of soil physics, 1, the flow of air and water through soils. *J. Agric. Sci.,* 4, 1–24.

Gregory, K. J. (1977). *River Channel Changes.* New York, NY: Wiley-Interscience.

Griffiths, G. A. (1981). Stable-channel design in gravel-bed rivers. *J. Hydrol.,* 52, 291–305.

Grozier, R. V., J. F. McCain, L. F. Lang and D. R. Merriman (1976). The Big Thompson River flood of July 31–August 1, 1976, Larimer County, Colorado. Flood Information Report, US Geological Survey and Colorado Water Conservation Board.

Guo, J. (1998). Turbulent velocity profiles in clear water and sediment-laden flows. Ph.D. dissertation, Civil Engineering, Colorado State University, 237 p.

Guo, J., ed. (2002). Advances in hydraulics and water engineering. In *Proc. 13th IAHR-APD Congress.* Singapore, 1098 p.

Guo, J. and P. Y. Julien (2001). Turbulent velocity profiles in sediment-laden flows. *J. Hyd. Res., IAHR,* 39(1), 11–23.

—— (2003). Modified log-wake law for turbulent flow in smooth pipes. *J. Hyd. Res., IAHR,* 41(5), 493–501.

—— (2004). Efficient algorithm for computing Einstein integrals. *J. Hyd. Eng., ASCE,* 130(12), 30–7.

—— (2005a). Application of the modified log-wake law in open channels. *J. Appl. Fluid Mech.,* 1(2), 17–23.

—— (2005b). Shear stress in smooth rectangular open-channel flows. *J. Hyd. Eng., ASCE,* 131(1), 1198–201.

Guo, J., P. Y. Julien and R. N. Meroney (2005). Modified log-wake law for zero-pressure-gradient turbulent boundary layers. *J. Hyd. Res., IAHR,* 43(4), 421–30.

Haan, C. T., B. J. Barfield and J. C. Hayes (1993). *Design Hydrology and Sedimentology for Small Catchment.* San Diego, CA: Academic Press, 588 p.

Hager, W. H. and A. J. Schleiss (2009). *Constructions Hydrauliques. 15.* Lausanne: Presses Polytech. et Univ. Romandes, 597 p.

Hagerty, D. J. (1992). Identification of piping and sapping erosion of streambanks, HL-92-1, University of Louisville, KY, for USACE, Vicksburg, MS.

Hagerty, D. J., M. F. Spoor and J. F. Kennedy (1986). Interactive mechanisms of alluvial-stream bank erosion. In *River Sedimentation, Vol. III Proc. 3rd Intl. Symp. River Sed.* Oxford, MS: University of Mississippi, 1160–8.

Halgren, J. (2012). TREX-SMA: A multi-event hybrid hydrological model applied at California Gulch Colorado. Ph.D. dissertation, Civil Engineering, Colorado State University, 300 p.

Harrison, J. S. and W. W. Doe, III (1997). Erosion modeling in Pinon Canyon maneuver site using the Universal Soil Loss Equation in GRASS. CEMML TPS

97-21, Center for Ecological Management of Military Lands, Colorado State University, 47 p.

Hartley, D. (1990). Boundary shear stress induced by raindrop impact. Ph.D. dissertation, Civil Engineering, Colorado State University, 193 p.

Hartley, D. M. and P. Y. Julien (1992). Boundary shear stress induced by raindrop impact. *J. Hyd. Res.*, 30, 341–59.

Hasan, A. J. (2005). Permodelan Hidrodinamik Sungai. River hydrodynamic modelling – the practical approach. ISBN 983-42905-0-0, NAHRIM, Malaysia, 155 p.

Hayashi, T. (1970a). The formation of meanders in rivers. *Trans. Japan Soc. Civ. Eng.*, 2, 180.

—— (1970b). On the cause of the initiations of meandering of rivers. *Trans. Japan Soc. Civ. Eng.*, 2, 235–9.

Henderson, F. M. (1961). Stability of alluvial channels. *J. Hyd. Div., ASCE*, 87, 109–38.

—— (1963). Stability of alluvial channels. *Trans. ASCE*, 128(3440), 657–86.

Herbich, J. B. (1992). *Handbook of Dredging Engineering*. New York, NY: McGraw-Hill.

Hey, R. D. (1978). Determinate hydraulic geometry of river channels. *J. Hyd. Eng., ASCE*, 104, 869–85.

Hey, R. D., J. C. Bathurst and C. R. Thorne, eds. (1982). *Gravel-Bed Rivers: Fluvial Processes, Engineering and Management*. New York, NY: Wiley.

Hey, R. D. and C. R. Thorne (1986). Stable channels with mobile gravel beds. *J. Hyd. Eng., ASCE*, 112, 671–89.

Hibma, A. (2004). Morphodynamic modeling of estuarine channel-shoal systems. Communications on Hydraulic and Geotechnical Engineering 04-3, ISSN 0169-6548, TUD, Delft University of Technology, 122 p.

Hickin, E. J. and G. C. Nanson (1975). The character of channel migration on the Beatton River, northeast B.C., Canada. *Geol. Soc. Am. Bull.*, 86, 487–494.

Hirano, M. (1973). River-bed variation with bank erosion. *Proc. Japan Soc. Civ. Eng.*, 210, 13–20.

Hite, J. E. (1992). Vortex formation and flow separation at hydraulic intakes. Ph.D. dissertation, Washington State University, Pullman, WA, 195 p.

Hoffmans, G. J. C. M. (1994a). Scour due to plunging jets. Rep. W-DWW-94-302, Ministry of Transport, Public Works and Water Management, Road and Hydraulic Engineering Division, Delft, The Netherlands.

—— (1994b). Scour due to submerged jets. Rep. W-DWW-94-303, Ministry of Transport, Public Works and Water Management, Road and Hydraulic Engineering Division, Delft, The Netherlands.

—— (1995). Ontgrondingen rondom brugpijlers en aan de kop van kribben. Rep. W-DWW-94-312, Ministry of Transport, Public Works and Water Management, Road and Hydraulic Engineering Division, Delft, The Netherlands.

Hoffmans, G. J. C. M. and R. Booij (1993). Two-dimensional mathematical modeling of local-scour holes. *J. Hyd. Res., IAHR*, 31, 615–34.

Hoffmans, G. J. C. M. and K. W. Pilarczyk (1995). Local scour downstream of hydraulics structures. *J. Hyd. Eng., ASCE*, 121, 326–40.

Hoffmans, G. J. C. M. and H. J. Verheij (1997). *Scour Manual*. Rotterdam: Balkema.

Hofland, B. (2005). Rock & roll turbulence-induced damage to granular bed protections. Communications on Hydraulic and Geotechnical Engineering Report 05-4, ISSN 0169-6548, TUD, Delft University of Technology, The Netherlands, 221 p.

Holmes, R. R. (1996). Sediment transport in the Lower Missouri and the Central Mississippi Rivers, June 26 through September 14, 1993. Geological Survey Circular 1120-I, US Department of the Interior, 23 p.

Hooke, R. L. (1975). Distribution of sediment transport and shear stress in a meander bend. *J. Geol.*, 83, 543–60.

Hooke, J. M. (1979). An analysis of the processes of river bank erosion. *J. Hydrol.*, 42, 39–62.

—— (1980). Magnitude and distribution of rates of river bank erosion. *Earth Surf. Proc.*, 5, 143–57.

Horner, C. L. (2016). Middle Rio Grande habitat suitability criteria. M.S. technical report, Civil Engineering, Colorado State University, 61 p.

Hussain, H. (1999). Analysis of different models to predict the mean flow velocity in hyperconcentrations, mudflows and debris flows. M.S. thesis, Civil Engineering, Colorado State University, 170 p.

Hussein, K. and V. H. Smith (1986). Flow and bed deviation angle in curved open channels. *J. Hyd. Res., IAHR*, 24, 93–108.

Hutchings, J. A. and M. E. B. Jones (1998). Life history variation and growth rate thresholds for maturity in Atlantic salmon, *Salmo salar. Can. J. Fish. Aquat. Sci.*, 55, 22–47.

Ijima, T. and F. L. W. Tang (1966). Numerical calculations of wind waves in shallow waters. In *Proc. 10th Coastal Eng. Conf.* Tokyo, Japan, 38–45.

Ikeda, S. (1982). Lateral bedload on side slopes. *J. Hyd. Div., ASCE*, 108, 1369–73.

—— (1984a). Flow and bed topography in channels with alternate bars. In *River Meandering, Proc. Conf. Rivers '83*. New York, NY: ASCE, 733–46.

—— (1984b). Prediction of alternate bar wavelength and height. *J. Hyd. Eng., ASCE*, 110, 371–86.

—— (1987). Bed topography and sorting in bends. *J. Hyd. Eng., ASCE*, 113(2), 190–206.

Ikeda, S., G. Parker and K. Sawai (1981). Bend theory of river meanders, part 1. linear development. *J. Fluid Mech., CUP*, 112, 363–77.

Ikeda, H. and A. Ohta (1986). On the formation of stationary bars in a straight flume. Annual Report of the Institute of Geoscience, 12, Tsukuba University, Japan, 42–6.

Inglis, C. C. (1947). Meanders and the bearing on river training. Institution of Civil Engineers, Maritime and Waterways, Paper 7.

—— (1949). The effect of variations in charge and grade on the slopes and shapes of channels. In *Proc. 3rd Congress*. IAHR, Delft, The Netherlands, II.1.1–II.1.10.

Isbash, S. V. (1935). Construction of dams by dumping stones in flowing water, translated by A. Dorijikov, US Army Engineer District, Eastport, ME.

Ivicsics, L. (1975). *Hydraulic Models*. Fort Collins, CO: Water Resource Publications, 310 p.

Jacobson, R. B. and K. A. Oberg (1993). Geomorphic changes on the Mississippi River flood plain at Miller City, Illinois, as a result of the flood of 1993. Geological Survey Circular 1120-J, US Department of the Interior, Washington, DC, 22 p.

Jain, S. C. and E. E. Fischer (1980). Scour around bridge piers at high velocities. *J. Hyd. Eng., ASCE*, 106, 1827–42.

Jansen, P., L. van Bendegom, J. van den Berg, M. de Vries and A. Zanen (1979). *Principles of River Engineering: The Non-Tidal Alluvial River*. San Francisco, CA: Pitman.

Jarrett, R. D. (1985). Determination of roughness coefficients for streams in Colorado, USGS Water Resources Investigations Rep. 85-4004, Lakewood, CO, 54 p.

Jarrett, R. D. and J. M. Boyle (1986). Pilot study for collection of bridge-scour data. USGS Water Resources Investigations Rep. 86-4030, Denver, 46 p.

Jarrett, R. D. and J. E. Costa (1986). Hydrology, geomorphology and dam-break modeling of the July 15, 1982 Lawn Lake dam and Cascade Lake dam failures, Larimer County, CO, USGS Open File Rep. 84-612, and also USGS Professional Paper 1369, US Geological Survey, Washington, DC, 78 p.

—— (1988). Evaluation of the flood hydrology in the Colorado front range using precipitation, streamflow, and paleoflood data for the Big Thompson River basin. USGS Water Resources Investigations Rep. 87-4177, Denver, CO, 37 p.

Ji, U. (2006). Numerical model for sediment flushing at the Nakdong River Estuary Barrage. Ph.D. dissertation, Civil Engineering, Colorado State University, 195 p.

Ji, U., P. Y. Julien and S. K. Park (2011). Sediment flushing at the Nakdong River Estuary Barrage. *J. Hyd. Eng., ASCE*, 137(11), 1522–35.

Ji, U., P. Y. Julien, S. K. Park and B. Kim (2008). Numerical modeling for sedimentation characteristics of the Lower Nakdong River and sediment dredging effects at the Nakdong River Estuary Barrage. *Korean J. Civ. Eng.*, 28(4B), 405–11.

Ji, U., M. Velleux, P. Y. Julien and M. Hwang (2014). Risk assessment of watershed erosion at Naesung Stream, South Korea. *J. Environ. Manag.*, 136, 16–26.

Jia, Y., S. Scott and S. S. Wang (2001). 3D Numerical model validation using field data and simulation of flow in Mississippi River. World Water Forum, ASCE.

Johnson, P. A. (1992). Reliability-based pier scour engineering. *J. Hyd. Eng., ASCE*, 118, 1344–58.

—— (1995). Comparison of pier-scour equations using field data. *J. Hyd. Eng., ASCE*, 121, 626–9.

Johnson, B. E. (1997). Development of a storm event based two-dimensional upland erosion model. Ph.D. dissertation, Civil Engineering, Colorado State University, 254 p.

Johnson, T. L. (2002). Design of erosion protection for long-term stabilization. Manual NUREG-1623, Office of Nuclear Material Safety and Safeguards, US Nuclear Regulatory Commission, Washington, DC, 161 p.

Johnson, W. W. and M. T. Finley (1980). Handbook of acute toxicity of chemicals to fish and aquatic invertebrates. Fish and Wildlife Services, Resources Publication 137, US Department of the Interior, Washington, DC, 106 p.

Johnson, B. E., P. Y. Julien, D. K. Molnar and C. C. Watson (2000). The two-dimensional upland erosion model CASC2D-SED. *J. AWRA*, 36(1), 31–42.

Johnson, A. W. and J. M. Stypula, eds. (1993). Guidelines for bank stabilization projects in riverine environments of King County. Surface Water Management Division, King County Department of Public Works, Seattle, WA.

Johnson, B., Z. Zhang, M. Velleux and P. Y. Julien (2011). Development of a distributed watershed Contaminant Transport, Transformation, and Fate (CTT&F) Sub-model. *J. Soil Sed. Contam.*, 20(6), 702–21.

Jones, J. S. (1983). Comparison of prediction equations for bridge pier and abutment scour. TRB Record 950, *2nd Bridge Eng. Conf.*, 2, Transportation Research Board, Washington, DC.

—— (1989). Laboratory studies of the effect of footings and pile groups on bridge pier scour. In *Proc. 1989 Bridge Scour Symp.* Washington, DC: Federal Highway Administration.

Jordan, B. (2008). An urban geomorphic assessment of the Berryessa and upper Penitencia Creek watersheds in San Jose, California. Ph.D. dissertation, Civil Engineering, Colorado State University, 146 p.

Jorgeson, J. D. (1999). Peak flow analysis using a two-dimensional watershed model with radar precipitation data. Ph.D. dissertation, Civil Engineering, Colorado State University, 192 p.

Jorgeson, J. and P. Y. Julien (2005). Peak flow forecasting with CASC2D and radar data. *Water International, IWRA*, 30(1), 40–9.

Julien, P. Y. (1979). Erosion de bassin et apport solide en suspension dans les cours d'eau nordiques. M.Sc. thesis, Civil Engineering, Laval University, Québec, Canada, 186 p.

—— (1982). Prédiction d'apport solide pluvial et nival dans les cours d'eau nordiques à partir du ruissellement superficiel. Ph.D. dissertation, Civil Engineering, Laval University, Québec, Canada, 240 p.

—— (1985). Planform geometry of meandering alluvial channels, Report CER84-85PYJ5, Civil Engineering, Colorado State University, Fort Collins, CO, 49 p.

—— (1986). Concentration of fine sediment particles in a vortex. *J. Hyd. Res., IAHR*, 24(4), 255–64.

—— (1988). Downstream hydraulic geometry of noncohesive alluvial channels. In *International Conference on River Regime*, New York, NY: Wiley, 9–16.

—— (1989). Géométrie hydraulique des cours d'eau à lit alluvial. In *Proc. IAHR Conf.* Natural Resources Council, Ottawa, Canada, B9–16.

—— (1996). Transforms for runoff and sediment transport. *J. Hydrol. Eng., ASCE*, 1(3), 114–22.

Julien, P.Y. (2002). *River Mechanics*, First Ed. Cambridge: Cambridge University Press, 434 p.

Julien, P. Y. (2009). Fluvial transport of suspended solids. In *Encyclopedia of Inland Waters*. Amsterdam: Elsevier.

—— (2010). *Erosion and Sedimentation*, 2nd ed. Cambridge: Cambridge University Press, 371 p.

—— (2017). Our hydraulic engineering profession, 2015 Hunter Rouse Lecture. In 60th Anniversary State-of-the-Art Reviews, *J. Hyd. Eng., ASCE*, ISSN 0733-9429.

Julien, P. Y., A. Ab. Ghani, N. A. Zakaria, R. Abdullah and C. K. Chang (2010). Flood mitigation of the Muda River, Malaysia. *J. Hyd. Eng., ASCE*, 136(4), 251–61.

Julien, P. Y. and D. Anthony (2002). Bedload motion by size fractions in meander beds. *J. Hyd. Res., IAHR*, 40(2), 125–33.

Julien, P. Y. and B. Bounvilay (2013). Velocity of rolling bedload particles. *J. Hyd. Eng., ASCE*, 139(2), 1344–59.

Julien, P. Y. and A. M. Dawod (1987). On predicting upland erosion losses from rainfall depth part 1: probabilistic approach. *J. Stoch. Hydrol. and Hyd.*, 1, 127–34.

Julien, P. Y. and M. Frenette (1985). Modeling of rainfall erosion. *J. Hyd. Eng., ASCE*, 111, 1344–59.

—— (1986). LAVSED II – a model for predicting suspended load in northern streams. *Can. J. Civ. Eng., CSCE*, 13, 162–70.

—— (1987). Macroscale analysis of upland erosion. *Hydrol. Sci. J., IAHS*, 32, 347–58.

Julien, P. Y., N. Friesen, J. G. Duan and R. Eykholt (2010). Celerity and amplification of supercritical surface waves. *J. Hyd. Eng., ASCE*, 136(9), 656–61.

Julien, P. Y. and M. Gonzalez del Tanago (1991). Spatially-varied soil erosion under different climates. *Hydrol. Sci. J.*, 36(6), 511–24.

Julien, P. Y. and J. S. Halgren (2014). Hybrid hydrologic modelling, Chap. 17. In *Handbook of Engineering Hydrology*. New York, NY: Taylor & Francis, 22 p.

Julien, P. Y. and D. M. Hartley (1986). Formation of roll waves in laminar sheet flow. *J. Hyd. Res., IAHR*, 24(1), 5–17.

Julien, P. Y. and G. J. Klaassen (1995). Sand–dune geometry of large rivers during floods. *J. Hyd. Eng., ASCE*, 121(9), 657–63.

Julien, P. Y., G. J. Klaassen, W. T. M. ten Brinke and A. W. E. Wilbers (2002). Case study: bed resistance of the Rhine River during the 1998 flood. *J. Hyd. Eng., ASCE*, 128(12), 1042–50.

Julien, P. Y. and Y. Q. Lan (1991). Rheology of hyperconcentrations. *J. Hyd. Eng., ASCE*, 117(3), 346–53.

Julien, P. Y. and G. E. Moglen (1990). Similarity and length scale for spatially-varied overland flow. *Wat. Res. Res., AGU*, 26(8), 1819–32.

Julien, P. Y. and A. Paris (2010). Mean velocity of mudflows and debris flows, *J. Hyd. Eng., ASCE*, 136(9), 676–9.

Julien, P. Y. and Y. Raslan (1998). Upper regime plane bed. *J. Hyd. Eng., ASCE*, 124(11), 1086–96.

Julien, P. Y., G. A. Richard and J. Albert (2005). Stream restoration and environmental river mechanics. *Intl. J. River Basin Manag., IAHR & INBO*, 3(3), 191–202.

Julien, P. Y. and R. Rojas (2002). Upland erosion modeling with CASC2D-SED. *Intl. J. Sed. Res.*, 17(4), 265–74.

Julien, P. Y., B. Saghafian and F. Ogden (1995). Raster-based hydrologic modeling of spatially-varied surface runoff. *Water Res. Bull., AWRA*, 31(3), 523–36.

Julien, P. Y. and D. B. Simons (1984). Analysis of hydraulic geometry relationships in alluvial channels. CER83-84PYJ-DBS45, Civil Engineering, Colorado State University, Fort Collins, CO.

—— (1985). Sediment transport capacity of overland flow. *Trans. ASAE*, 28(3), 755–62.

Julien, P. Y., M. L. Velleux, U. Ji and J. Kim (2014). Upland erosion modelling, Chap. 9. In *Handbook of Environmental Engineering*. New York, NY: Springer Science, 437–65.

Julien, P. Y. and J. Wargadalam (1995). Alluvial channel geometry: theory and applications. *J. Hyd. Eng., ASCE*, 121(4), 312–25.

Jun, B. H., S. I. Lee, I. W. Seo and G. W. Choi, eds. (2005). Abstracts of the 32rd IAHR Congress, COEX, Seoul, Korea, 1466 p.

Junid, S. (1992). *Rivers of Malaysia*. Selangor Darul Ehsan: Design Dimensions, 184 p.

Kalkwijk, J. P. and H. J. deVriend (1980). Computations of the flow in shallow river bends. *J. Hyd. Res., IAHR*, 18, 327–42.

Kamphuis, J. W. (2010). *Introduction to Coastal Engineering and Management.* Hackensack, NJ: World Scientific, 525 p.

Kane, B. (2003). Specific degradation as a function of watershed characteristics and climatic parameters. Ph.D. dissertation, Civil Engineering, Colorado State University, 213 p.

Kane, B. and P. Y. Julien (2007). Specific degradation of watersheds. *Intl. J. Sed. Res.*, 22(2), 114–9.

Kang, D. H. (2005). Distributed snowmelt modeling with GIS and CASC2D at California Gulch, Colorado. M.S. thesis, Civil Engineering, Colorado State University, 208 p.

Katopodis, C. (1992). Introduction to fishway design. Freshwater Institute, Central and Arctic Region, Department of Fisheries and Oceans, Winnipeg, Canada, 71 p.

Kawai, S. and P. Y. Julien (1996). Point bar deposits in narrow sharp bends. *J. Hyd. Res., IAHR*, 34(2), 205–18.

Keefer, T. N., R. S. McQuivey and D. B. Simons (1980). Interim report – stream channel degradation and aggradation: causes and consequences to highways. FHWA/RD-80/038, Federal Highway Administration, USDOT, Washington, DC.

Kellerhals, R. (1967). Stable channels with gravel-paved beds. *J. Waterways Harbors Div., ASCE*, 93, 63–84.

Kellerhals, R. and M. Church (1989). The morphology of large rivers: characterization and management. In *Proc. Intl. Large River Symp. 1986*, Dept. of Fisheries and Oceans, Ottawa, Canada, 31–48.

Kennedy, R. G. (1895). The prevention of silting in irrigation canals. *Minutes Proc. Inst. Civ. Eng., London*, 119, 281–90.

Keown, M. P. (1983). *Streambank Protection Guidelines for Landowners and Local Governments.* Vicksburg, MS: USACE, Waterways Experiment Station.

Keulegan, G. H. (1938). Laws of turbulent flow in open channels. *J. Res. Natl. Bur. Stand.*, 21, 707–41; see also USBR, Washington, DC, Research Paper 1151.

Kilinc, M. (1972). Mechanics of soil erosion from overland flow generated by simulated rainfall. Ph.D. dissertation, Civil Engineering, Colorado State University, Fort Collins, CO.

Kilinc, M. S. and E. V. Richardson (1973). Mechanics of soil erosion from overland flow generated by simulated rainfall. Hydrol. Pap. 63, Colorado State University, 54 p.

Kim, H. S. (2006). Soil erosion modeling using RUSLE and GIS on the IMHA watershed, South Korea. M.S. thesis, Civil Engineering, Colorado State University, 118 p.

Kim, J. (2012). Hazard area mapping during extreme rainstorms in South Korean mountains. Ph.D. dissertation, Civil Engineering, Colorado State University, 146 p.

Kim, H. Y. (2016). Optimization of Sangju weir operations to mitigate sedimentation problems. Ph.D. dissertation, Civil Engineering, Colorado State University, Fort Collins, CO, 393 p.

Kim, H. S. and P. Y. Julien (2006). Soil erosion modeling using RUSLE and GIS on the Imha Watershed, South Korea. *Water Eng. Res. J., KWRA*, 7(1), 29–41.

Kim, J., P. Y. Julien, U. Ji and J. Kang (2011). Restoration modeling analysis of abandoned channels of the Mangyeong River. *J. Env. Sci.*, 555–64.

Kim, Y., H. Shin and D. Park (2011). Variation of shear strength of unsaturated weathered granite soil with degree of saturation and disturbance at Dongrae, Inje and Yeonki sites. *Korea Soc. Hazard Mitig.*, 11(4), 157–62.

Klaassen, G. J. (1990). On the scaling of Braided Sand-Bed rivers. In *Mobile Bed Physical Models*. Boston, MA: Kluwer Academic, 56–71.

—— (1992). Experience from a physical model for a bridge across a braided river with fine sand as bed material. In *Proc. 5th Intl. Symp.* Karlsruhe: River Sed., 509–20.

Klaassen, G. J., E. Mosselman, G. Masselink, et al. (1993). Planform changes in large braided sand-bed rivers. Delft Hydraul. Pub. No. 480, Delft, The Netherlands.

Klaassen, G. J. R. and J. J. van der Zwaard (1974). Roughness coefficients of vegetated flood plains. *J. Hyd. Res., IAHR*, 12, 43–63.

Klaassen, G. J. R. and B. H. J. van Zarter (1990). On cutoff ratios of curved channels. Delft Hydraul. Pub. No. 444, Delft, The Netherlands.

Klaassen, G. J. R. and K. Vermeer (1988). Confluence scour in large braided rivers with fine bed material. In *Proc. Intl. Conf. Fluvial Hydraulics*. Budapest, Hungary.

Kleinhans, M. G. (2002). Sorting out sand and gravel: sediment transport and deposition in sand-gravel bed rivers, Netherlands Geographic Studies 293, Faculty of Geographic Science, Utrecht University, 317 p.

Kleitz, M. (1877). Note sur la théorie du movement non-permanent des liquides et sur l'application de la propagation des crues des rivières. *Ann. Ponts Chaussées*, 5(16), 133–96.

Klimek, K. E. (1997). Spherical particle velocities on rough dry surfaces. M.S. thesis, Civil Engineering, Colorado State University, 89 p.

Knauss, J., ed. (1987). *Swirling Flow Problems at Intakes*. Rotterdam: Balkema, 165 p.

Knighton, D. (1998). *Fluvial Forms and Processes*. Baltimore: Arnold, 383 p.

Knuuti, K. and D. McComas (2003). Assessment of changes in channel morphology and bed elevation in Mad River, California, 1971–2000. ERDC/CHL TR-03-16, USACE, ERDC, Vicksburg, MS, 54 p.

Kobus, H., P. Leister and B. Westrich (1979). Flow field and scouring effects of steady and pulsating jets impinging on a movable bed. *J. Hyd. Res., IAHR*, 17, 175–92.

Koch, F. G. and C. Flokstra (1980). Bed level computations for curved alluvial channels. In *Proc. XIX Congress IAHR*. Delft, The Netherlands, 2, 357.

Kositgittiwong, D., C. Chinnarasri and P. Y. Julien (2012a). Numerical simulation of flow velocity profiles along a stepped spillway. *J. Process. Mech. Eng.*, 14, 1–9.

—— (2012b). Two-phase flow over stepped and smooth spillways: numerical and physical models. *Ovidius Univ. Ann. Ser.: Civ. Eng.*, 14, 147–54.

Koutsunis, N. A. (2015). Impact of climatic changes on downstream hydraulic geometry and its influence on flood hydrograph routing-applied to the Bluestone Dam watersheds. M.S. technical report, Civil Engineering, Colorado State University, 60 p.

KOWACO (2004). *Water, Nature & People, Exploring Five Major Rivers in Korea*. Daejon: Korea Water Resources Corporation, 172 p.

Kuhnle, R. A. (1993). Incipient motion of sand gravel sediment mixtures. *J. Hyd. Eng., ASCE*, 119, 1400–15.

Kuroiwa, J. M., A. J. Mansen, F. M. Romero, L. F. Castro and R. Vega (2011). Narrowing of the Rímac River due to anthropogenic causes – partial engineering solutions. In *World Environmental and Water Resources Congress*. Palm Springs, CA: ASCE – EWRI.

Lacey, G. (1929–30). Stable channels in alluvium. *Proc. Inst. Civ. Eng., London*, 229, 259–92.

Lagasse, P. F., P. E. Clopper, L. W. Zevenbergen and J. F. Ruff (2006). Riprap design criteria, recommended specifications, and quality control. NCHRP Rep. 568, National Cooperative Highway Research Program, Transportation Research Board, Washington, DC, 226 p.

Lagasse, P. F., J. D. Schall, F. Johnson, E. V. Richardson and F. M. Chang (1995). Stream stability at highway structures. Hydraulic Engineering Circular No. 20, FHWA-IP-90-014, Federal Highway Administration, USDOT, Washington, DC.

Lagrange, J. L. de (1788). *Mécanique analytique*. 2-II(2), 192.

Lai, A. T. (1998). Bedforms in the Waal River, characterization and hydraulic roughness. M.Sc. thesis, The International Institute for Infrastructure, Hydraulic and Environment Engineering, IHE Delft, The Netherlands.

Lai, Y. G. (2008). SRH-2D version 2: theory and user's manual sedimentation and river hydraulics – two-dimensional river flow modeling. US Department of Interior, USBR Technical Service Center, Denver, CO.

Lai, K. X. O. (2016). Impact of the Smart tunnel outflow on the hydraulics of the Kerayong River, Malaysia. M.S. technical report, Civil Engineering, Colorado State University, 139 p.

Lai, Y. G. and C. T. Yang (2004). Development of a numerical model to predict erosion and sediment delivery to river systems, progress report no. 2: sub-model development and an expanded review. US Department of Interior, USBR Technical Service Center, Denver, CO.

Lambe, T. W. and R. V. Whitman (1969). *Soil Mechanics*. New York, NY: Wiley, 553 p.

Lan, Y. (1990). Dynamic modeling of meandering alluvial channels. Ph.D. dissertation, Civil Engineering, Colorado State University, 248 p.

Lane, E. W. (1953). Progress report on studies on the design of stable channels of the Bureau of Reclamation. *Proc. ASCE*, 79 (280).

—— (1955a). Design of stable channels. *Trans. ASCE*, 120, 1234–79.

—— (1955b). The importance of fluvial geomorphology in hydraulic engineering. *Proc. ASCE*, 81, 1–17.

—— (1957). A study of the shape of channels formed by natural streams flowing in erodible material. MRD No. 9, USACE, Missouri River Div., Omaha, Nebraska.

Langbein, W. B. and S. A. Schumm (1958). Yield of sediment in relation to mean annual precipitation. *Trans. AGU*, 39, 1076–84.

Langbein, W. B. and L. B. Leopold (1966). River meander – theory of minimum variance. USGS Prof. Paper 422-H, US Geol. Survey, Washington, DC.

Langhaar, H. L. (1956). *Dimensional Analysis and Theory of Models*. New York, NY: Wiley.

Larras, J. (1963). Profondeurs maximales d'érosion des fonds mobiles autour des piles en rivière. *Ann. Ponts Chaussées*, 133, 410–24.

Larsen, A. K. (2007). Hydraulic modeling analysis of the Middle Rio Grande – Escondida Reach, New Mexico. M.S. thesis, Civil Engineering, Colorado State University, 209 p.

Laursen, E. M. (1960). Scour at bridge crossings. *Proc. ASCE*, 86, 39–54.

—— (1963). An analysis of relief bridge scour. *Proc. ASCE*, 89, 93–118.

—— (1980). Predicting scour at bridge piers and abutments. *Gen. Rep. No. 3*, Arizona Department of Transportation, Phoenix, AZ.

Laursen, E. M. and M. W. Flick (1983). Scour at sill structures. FHWA/AZ83/184, Arizona Department of Transportation, Arizona Transportation Traffic Institute, Tempe, AZ.

Lee, J. S. (2002). *River Engineering*. Seoul: Saeron, 591 p.

—— (2007). *Hydrology*. Seoul: Saeron, 484 p.

Lee, J. K. and D. C. Froehlich (1989). Two dimensional finite element modeling of bridge crossings. FHWA-RD-88-149, Federal Highway Administration, USDOT, Washington, DC.

Lee, J. S. and P. Y. Julien (2006a). Electromagnetic wave surface velocimeter. *J. Hyd. Eng., ASCE*, 132(2), 146–53.

—— (2006b). Downstream hydraulic geometry of alluvial channels. *J. Hyd. Eng., ASCE*, 132(12), 1347–52.

—— (2012a). Resistance factors and relationships for measurements in fluvial rivers. *J. Korea Contents Association, JKCA*, 12(7), 445–52.

—— (2012b). Utilizing the concept of vegetation freeboard equivalence in river restoration. *Intl. J. Contents, KCA*, 8(3), 34–41.

Lee, J. H. and P. Y. Julien (2016a). ENSO impacts on temperature over South Korea. *Intl. J. Climatology, Roy. Met. Soc.*, 13.

—— (2016b). Teleconnections of the ENSO and South Korean precipitation patterns. *J. Hydrol.*, 534, 237–50.

—— (2017a). Influence of El Nino/Southern Oscillation on South Korean Streamflow Variability. *J. Hydrol. Proc.*, 1–17.

Lee, J. S. and P. Y. Julien (2017b). Composite Flow Resistance. *J. Flood Eng.*, Vol. 8 No. 2, ISSN: 0976-6219. July–December 2017, pp. 55–75.

Lee, J. S., P. Y. Julien, J. Kim and T. W. Lee (2012). Derivation of roughness coefficient relationships using field data in vegetated rivers. *J. Kor. Wat. Res. Ass.*, 45, 137–49.

Lee, J. H. W. and K. M. Lam, eds. (2004). Environmental hydraulics and sustainable water management. *Proc. 4th Intl. Symp. Env. Hyd. and 14th IAHR-APD Cong.* Hong Kong, 2319 p.

Leliavsky, S. (1961). *Précis d'Hydraulique Fluviale*. Paris: Dunod, 256 p.

—— (1966). *An Introduction to Fluvial Hydraulics*. New York, NY: Dover, 257 p.

Leon, C. (2003). Analysis of equivalent widths of alluvial channels and application for instream habitat in the Rio Grande. Ph.D. dissertation, Colorado State University, 283 p.

Leon, C., P. Y. Julien and D. C. Baird (2009). Case study: equivalent widths of the Middle Rio Grande, New Mexico. *J. Hyd. Eng., ASCE*, 135(4), 306–15.

Leon, C., G. Richard, T. Bauer and P. Y. Julien (1999). Middle Rio Grande Cochiti to Bernalillo Bridge, hydraulic geometry, discharge and sediment data base and report. Vols. I–III, Civil Engineering, Colorado State University Fort Collins, CO.

Leonard, B. P. (1979). A stable and accurate convective modeling procedure based on quadratic upstream interpolation. *Comp. Meth. Appl. Mech. Eng.*, 19, 59–98.

Leon Salazar, C. (1998). Morphology of the Middle Rio Grande from Bernalillo Bridge to the San Acacia Diversion Dam, New Mexico. M.S. thesis, Colorado State University, 209 p.

Leopold, L. B., R. A. Bagnold, R. G. Wolman and L. M. Brush (1960). Flow resistance in sinuous or irregular channels. USGS Prof. Paper 282-D, Washington, DC, 111–34.

Leopold, L. B. and T. Maddock, Jr. (1953). The hydraulic geometry of stream channels and some physiographic implications. USGS Prof. Paper 252, Washington, DC.

Leopold, L. B. and T. Maddock (1954). *The Flood Control Controversy.* New York, NY: Ronald Press.

Leopold, L. B. and M. G. Wolman (1957). River channel patterns: braided, meandering and straight. USGS Prof. Paper 282-B, Washington, DC.

—— (1960). River meanders. *Geol. Soc. Am. Bull.,* 71, 769–94.

Leopold, L. B., M. G. Wolman and J. P. Miller (1964). *Fluvial Processes in Geomorphology.* San Francisco: Freeman.

Li, R. M. and H. W. Shen (1973). Effect of tall vegetation on flow and sediment. *J. Hyd. Div., ASCE,* 99(5), 793–814.

Liggett, J. A. and J. A. Cunge (1975). Numerical methods of solution of the unsteady flow equations. In *Unsteady Flow in Open Channels.* Fort Collins, CO: Water Resources Publication.

Liggett, J. A. and D. A. Woolhiser (1967). Difference solutions of shallow-water equations. *Proc. ASCE,* 93, 39–71.

Limerinos, J. T. (1970). Determination of the Manning's coefficient for measured bed roughness in natural channels. USGS Water Supp. Pap. 1891-B, Washington, DC.

Lindley, E. S. (1919). *Regime Channels.* Punjab: Punjab Engineering Congress.

Liu, H. K., F. M. Chang and M. M. Skinner (1961). Effect of bridge construction on scour and backwater. Rep. CER60-HKL22, Civil Engineering, Colorado State University, Fort Collins, CO.

Lowe, J., III. and W. V. Binger (1982). 2nd Annual USCOLD Lecture, Tarbela Dam Project, US Commission on Large Dams, Atlanta, GA, 103 p.

Lu, Y. and H. Allen (2001). Partitioning of copper onto suspended particulate matter in river waters. *Sci. Total Environ.,* 277(1–3), 119–32.

Lu, N. and J. W. Godt (2013). *Hillslope Hydrology and Stability.* Cambridge: Cambridge University Press, 437 p.

MacCormack, R. W. (1969). The effect of viscosity in hypervelocity impact cratering. American Institute of Aeronautics and Astronautics, Paper 69–354.

MacDonald, T. E. (1991). Inventory and analysis of stream meander problems in Minnesota. M.S. thesis, University of Minnesota, Minneapolis, MN.

Maddock, T. (1970). Indeterminate hydraulics of alluvial channels. *J. Hyd. Div., ASCE,* 96, 2309–23.

Mahoney, H. A., E. D. Andrews, W. W. Emmett, et al. (1976). Data for calibrating unsteady-flow sediment-transport models, East Fork River, Wyoming, 1975. USGS Open File Rep. 76-22, Washington, DC, 293 p.

Marcus, K. B. (1991). Two-dimensional finite element modeling of surface runoff from moving storms on small watersheds. Ph.D. dissertation, Colorado State University, 299 p.

Marinier, G., W. L. Chadwick and R. B. Peck (1981). 1st Annual USCOLD Lecture, James Bay Hydro Development, US Commission on Large Dams.

Martins, R., ed. (1989). *Recent Advances in Hydraulic Physical Modelling.* Dordrecht: Kluwer, 627 p.

Martin Vide, J. P. (2006). *Ingenieria de Rios.* ISBN 978-84-8301-900-9, Barcelona: Universidad Politecnica de Catalunya, 381 p.

Martyusheva, O. (2014). Smart water grid. M.S. technical report, Civil Engineering, Colorado State University, 80 p.

Mason, P. J. and K. Arumugam (1985). Free jet scour below dams and flip buckets. *J. Hyd. Eng., ASCE,* 111, 220–35.

May, D. R. (1993). The space-time correlation structure of convective rainstorms in the Lagrangian reference frame. Ph.D. dissertation, Civil Engineering, Colorado State University, 206 p.

May, D. R. and P. Y. Julien (1998). Eulerian and Lagrangian correlation structures of convective rainstorms. *Water Res. Res., AGU,* 34(10), 2671–83.

Mayerle, R., C. Nalluri and P. Novak (1991). Sediment transport in rigid bed conveyances. *J. Hyd. Res., IAHR,* 29, 475–96.

Maynord, S. T. (1988). Stable riprap size for open channel flows. Tech. Rep. HL-88-4, USACE, Waterways Experiment Station, Vicksburg, MS.

—— (1992). Riprap stability: studies in near-prototype size laboratory channel. Tech. Rep. HL-92-5, USACE, Waterways Experiment Station, Vicksburg, MS.

—— (1995). *Corps Riprap Design Guidance for Channel Protection in River, Coastal and Shoreline Protection/Erosion Control using Riprap and Armourstone.* New York, NY: Wiley, 4–53.

McCain, J. F., L. R. Hoxit, R. A. Maddox, et al. (1979). Storm and flood of July 31–August 1, 1976, in the Big Thompson River and Cache la Poudre River basin, Larimer and Weld Counties, CO. USGS Professional Paper 1115, US Geological Survey, Washington, DC.

McCarley, R. W., J. J. Ingram, B. J. Brown and A. J. Reese (1990). Flood-control channel national inventory. Miscellaneous Paper HL-90-10, USACE, Waterways Experiment Station, Vicksburg, MS.

McCuen, R. H. (2016). *Hydrologic Analysis and Design,* 4th ed. Boston, CA: Pearson, 790 p.

McCuen, R. H., P. A. Johnson and R. M. Ragan (1995). Hydrologic design of highways. Hydraulic Design Series No. 2, Federal Highway Administration USDOT, Washington, DC.

McCullah, J. and D. Gray (2005). Environmentally sensitive channel- and bank-protection measures. NCHRP Rep. 544, Transportation Research Board, Washington, DC, 59 p.

McKee, E. D. (1989). Sedimentary structures and textures of Rio Orinoco channel sands, Venezuela and Columbia, Water Supply Paper 2326, US Geological Survey, Washington, DC, 23 p.

Meade, R. H., ed. (1995). Contaminants in the Mississippi River 1987–92. Geol. Survey Circ. 1133, US Department of the Interior, Denver Federal Center, Denver, CO, 140 p.

Meade, R. H. and J. A. Moody (2010). Causes for the decline of suspended-sediment discharge in the Mississippi River system, 1940–2007. *Hydrol. Proc.,* 24, 35–49.

Meier, C. I. (1995). Transport velocities of single bed-load grains in hydraulically smooth open-channel flow. M.S. thesis, Civil Engineering, Colorado State University, 93 p.

Melone, A. M., E. V. Richardson and D. B. Simons (1975), Exclusion and ejection of sediment from canals. Unpublished Civil Engineering Report, Colorado State University, 192 p.

Melville, B. W. and S. E. Coleman (2000). *Bridge Scour*. Highlands Ranch: Water Resources Publications, 550 p.

Melville, B. W. and D. M. Dongol (1992). Bridge pier scour with debris accumulation. *J. Hyd. Div., ASCE*, 118.

Melville, B. W. and A. J. Sutherland (1988). Design method for local scour at bridge piers. *J. Hyd. Div., ASCE*, 114.

Michel, B. (1971). Winter regime of rivers and lakes. CRREL Monog. III-B1a, Hanover, NH, 139 p.

—— (1978). *Ice Mechanics*. Québec: Presses Université Laval, 499 p.

Michel, B. and M. Drouin (1981). Backwater curves under ice cover of the La Grande River. *Can. J. Civ. Eng.*, 8(3), 351–63.

Michelot, J. L. (1995). Gestion patrimoniale des milieu naturels fluviaux. Guide technique, L'atelier Technique des Espaces Naturels, Montpellier, France, 67 p.

Middleton, G. V. (1965). Primary sedimentary structures and their hydrodynamic interpretation. *Soc. Econ. Paleontol. Mineral. Spec. Pub.*, 12, 265.

Mihelcic, J. R. (1999). *Fundamentals of Environmental Engineering*. New York, NY: Wiley, 335 p.

Milliman, J. D. and J. P. M. Syvitski (2017). Geomorphic/tectonic control of sediment discharge to the ocean: the importance of small mountainous rivers. *J. Geol.*, 100, 525–44.

Ministerio de Obras Publicas (1972). Mediciones en Rios Grandes, Direccion general de recursos hidraulicos, Caracas, Venezuela, 93 p.

Minor, H. E. and W. H. Hager, eds. (2004). River engineering in Switzerland. *Soc. Art of Civil Eng.*, 6, 140.

Mirtskhoulava, Ts. Ye. (1988). *Basic Physics and Mechanics of Channel Erosion*. Leningrad: Gidrometeoizdat.

—— (1991). Scouring by flowing water of cohesive and noncohesive beds. *J. Hyd. Res., IAHR*, 29.

MOCT (2007a). Cheongmi-Cheon basic improvement plan. Vol. 4, Ministry of Construction Technical Report, Seoul Province Office, South Korea, 533 p.

—— (2007b). Cheongmi-Cheon basic improvement plan, pre-environmental impact report, Ministry of Construction Technical Report, Seoul Province Office, South Korea, 533 p.

Moglen, G. E. (1989). The effects of spatial variability of overland flow parameters on runoff hydrographs. M.S. thesis, Civil Engineering, Colorado State University, 157 p.

Molinas, A. (2000). User's manual for BRI-STARS (BRIdge and Stream Tube model for Alluvial River Simulation). FHWA-RD-99-190, Transportation Research Board, NTIS, 238 p.

Molinas, A., M. I. Abdou, H. M. Noshi, et al. (1998). Effects of gradation and cohesion on bridge scour. Lab. Studies, Vol. 1 to 6, FHWA, Reston, VA, and Colorado State University, Fort Collins, CO.

Mollars, J. D. and J. R. Jones (1984). *Airphoto Interpretation and the Canadian Landscape*. Ottawa: Deptartment of Energy, Mines and Resources.

Molnar, D. K. (1997). Grid size selection for 2-D hydrologic modeling of large watersheds. Ph.D. dissertation, Civil Engineering, Colorado State University, 201 p.

Molnar, D. K. and P. Y. Julien (1998). Estimation of upland erosion using GIS. *J. Comp. Geosci.*, 24(2), 183–92.

—— (2000). Grid-size effects on surface runoff modeling. *J. Hydrol. Eng., ASCE*, 5(1), 8–16.

Montoya Monsalve, J. J. (2008). Desarrollo de un modelo conceptual de produccion, transporte y deposito de sedimentos. Ph.D. dissertation, University Politec. Valencia, Spain, 236 p.

Moody, J. A. (1995). Propagation and composition of the flood wave on the Upper Mississippi River, 1993. Geological Survey Circular 1120-F, US Department of the Interior, Washington, DC, 21 p.

Mooney, D. M., C. L. Holmquist-Johnson and S. Broderick (2007). Rock Ramps design guidelines, USBR, Technical Service Center, Denver, CO, 110 p.

Moriasi, D. N., J. G. Arnold, M. W. Van-Liew, et al. (2007). Model evaluation guidelines for systematic quantification of accuracy in watershed simulations. *J. Am. Soc. Agric. Biol. Eng., ASABE*, 50(3), 885–900.

Mossa, M., Y. Yasuda and H. Chanson, eds. (2004). *Fluvial, Environmental and Coastal Developments in Hydraulic Engineering*. London: Balkema, 235 p.

Mosselman, E. (1989). Theoretical investigation on discharge-induced river-bank erosion. Communications on Hydraulic and Geotechnical Engineering, 89-3, TUD, Delft University of Technology, 56 p.

—— (1992). Mathematical modelling of morphological processes in rivers with erodible cohesive banks. Communications on Hydraulic and Geotechnical Engineering, 92-3, TUD, Delft University of Technology, 134 p.

Muhammad, N. S. (2013). Probability structure and return period calculations for multi-day monsoon rainfall events at Subang, Malaysia. Ph.D. dissertation, Civil Engineering, Colorado State University, 177 p.

Muhammad, N. S., P. Y. Julien and J. D. Salas (2015). Probability structure and return period of multiday monsoon rainfall. *J. Hydrol. Eng., ASCE*, 11.

Murillo-Muñoz, R. E. (1998). Downstream fining of sediments in the Meuse River. M.Sc. thesis, International Institute of Infrastructure Hydraulic and Environment Engineering, Delft, The Netherlands.

Mussetter, R. A. (1989). Dynamics of mountain streams. Ph.D. dissertation, Civil Engineering, Colorado State University, 174 p.

Nanson, G. C. and E. J. Hickin (1983). Channel migration and incision on the Beatton River. *J. Hyd. Eng., ASCE*, 109, 327–37.

—— (1986). A statistical examination of bank erosion and channel migration in Western Canada. *Bull. Geol. Soc. Am.*, 97, 497–504.

Naudascher, E. and D. Rockwell, (1994). *Flow-Induced Vibrations: An Engineering Guide*. Rotterdam: Balkema, 413 p.

NCASI (1999). Scale considerations and the detectability of sedimentary cumulative watershed effects. Tech. Bull. 776, National Council for Air and Stream Improvement, Research Triangle Park, NC, 327 p.

Ndolo Goy, P. (2015). GIS-based soil erosion modeling and sediment yield of the N'Djili River basin, Democratic Republic of Congo. M.S. thesis, Civil Engineering, Colorado State University, 220 p.

Neill, C. R., ed. (1973). Guide to Bridge Hydraulics. Roads and Transportation Association of Canada, University of Toronto Press, Canada.

Neill, C., D. Hotopp and B. Hunter (2013). Some hydrotechnical features of Padma River, Bangladesh. In *Proc. 21st Can. Hydrotech. Conf.* CSCE, Banff, 11 p.

Neill, C. R. and E. K. Yaremko (1988). Regime aspects of flood control channelization. In *Proc. Intl. Conf. River Regime*. New York, NY: Wiley.

Nelson, P. A., A. K. Brew and J. A. Morgan (2015). Morphodynamic response of a variable-width channel to changes in sediment supply. *Wat. Res. Res., AGU*, 51, 18.

Nelson, J. M., R. R. McDonald and P. J. Kinzel (2006). Morphologic evolution in the USGS surface-water modeling system. In *Proc. 8th Fed. Interagency Sed. Conf.* Reno, 8 p.

Nelson, P. A., R. R. McDonald, J. M. Nelson and W. E. Dietrich (2015). Coevolution of bed surface patchiness and channel morphology: 1. Mechanisms of forced patch formation, and 2. Numerical experiments. *J. Geophys. Res. Earth Surf., AGU*, 37.

Nelson, P. A. and G. Seminara (2011). Modeling the evolution of bedrock channel shape with erosion from saltating bed load. *Geophys. Res. Lett., AGU*, 38, 5.

Nelson, J. M. and J. D. Smith (1989). Flow in meandering channels with natural topography. In *River Meandering*. Washington, DC: AGU Water Resource Monitor, 69–102.

Nezu, I. and H. Nakagawa (1993). *Turbulence in Open-Channel Flows*. Rotterdam: Balkema.

Nicollet, G. (1975). Affouillement au pied des piles de pont en milieu cohésif. In *Proc. 16th Cong.* IAHR, Brazil, Paper B60, 478–84.

Nordin, C. F., Jr. (1964). Aspects of low resistance and sediment transport Rio Grande near Bernalillo, New Mexico. USGS Water Supp. Paper 1498-H, Washington, DC.

—— (1977). Graphical aids for determining scour depth in long contractions. In USGS Open File Rep. 77-837, Denver, CO, 12 p.

Nordin, C. F. and D. Perez-Hernandez (1989). Sand waves, bars, and wind-blown sands of the Rio Orinoco, Venezuela and Colombia. USGS Water Supply Paper 2326-A, 74 p.

Nordin, C. F. and G. V. Sabol (1974). Empirical data on longitudinal dispersion in rivers. USGS Wat. Res Invest. 20-74, Denver Fed. Center, Lakewood, CO, 332 p.

Novak, S. J. (2006). Hydraulic modeling analysis of the Middle Rio Grande from Cochiti Dam to Galisteo Creek, New Mexico. M.S. thesis, Civil Engineering, Colorado State University, 158 p.

Novak, P., A. I. B. Moffat, C. Nalluri and R. Narayanan (2001). *Hydraulic Structures*, 3rd ed. London: Spon Press, 666 p.

Novotny, V. (2003). *Water Quality – Diffuse Pollution and Watershed Management*, 2nd ed. Hoboken, NJ: Wiley, 864 p.

NRCS (2007). Stream restoration design. In *National Engineering Handbook*, Part 654. Washington, DC: USDA.

Nunnally, N. R. and F. D. Shields (1985). Incorporation of environmental features in flood control channel projects. Tech. Rep. E-85-3, USACE, Waterways Experiment Station, Vicksburg, MS.

O'Brien, J. S. and P. Y. Julien (1985). Laboratory analysis of mudflow properties. *J. Hyd. Eng., ASCE*, 114(8), 877–87.

O'Brien, J. S., P. Y. Julien and W. T. Fullerton (1993). Two-dimensional water flood and mud flow simulation. *J. Hyd. Eng., ASCE*, 119(2), 244–61.

O'Connor, D. J. and W. E. Dobbins (1958). Mechanism of reaeration of natural streams. *Trans. ASCE*, 123, 641–66.

Odgaard, A. J. (1981). Transverse slope in alluvial channel bends. *J. Hyd. Eng., ASCE*, 107, 1677–94.

—— (1982). Bed characteristics in alluvial channel bends. *J. Hyd. Div., ASCE*, 108, 1268–81.

Ogden, F. L. (1992). Two-dimensional runoff modeling with weather radar data. Ph.D. dissertation, Civil Engineering, Colorado State University, 211 p.

Ogden, F. L. and P. Y. Julien (1993). Runoff sensitivity to temporal and spatial rainfall variability at runoff plane and small basin scales. *Wat. Res. Res., AGU*, 29(8), 2589–97.

—— (1994). Runoff model sensitivity to radar rainfall resolution. *J. Hydrol.*, 158, 1–18.

Ogden, F. L., J. R. Richardson and P. Y. Julien (1995). Similarity in catchment response 2. Moving rainstorms. *Wat. Res. Res., AGU*, 31(6), 1543–7.

Olsen, N. R. B. (1991). A three-dimensional numerical model for simulation of sediment movements in water intakes. Ph.D. dissertation, Norwegian Institute of Technology, University of Trondheim, Norway, 106 p.

Orechwa, A. (2015). Soil contaminant mapping and prediction of sediment yield at an abandoned uranium mine. M.S. technical report, Civil Engineering, Colorado State University, 75 p.

Osman, A. M. and C. R. Thorne (1988). Riverbank stability analysis I: theory. *J. Hyd. Eng., ASCE*, 114, 134–50.

Ouellet, Y. (1972). *Compléments d'Hydraulique*. Québec: Presses de l'Univ. Laval, 236 p.

Owen, T. E. (2012). Geomorphic analysis of the Middle Rio Grande – Elephant Butte Reach, New Mexico. M.S. thesis, Civil Engineering, Colorado State University, 186 p.

Pagliara, S., A. Radecki-Pawlik, M. Palermo and K. Plesenski (2016). Block ramps in curved rivers: morphology analysis and prototype data supported design criteria for mild bed slopes. *River Res. Appl.*, DOI: 10.1002/rra.3083, 11.

Parilkova, J. and J. Vesely (2009). Laboratory of water management research of the department of water structures, ISBN 978-82-214-3889-7, Brno University of Technology, Brno: VITIUM, 92 p.

Park, K. (2013). Mechanics of sediment plug formation in the Middle Rio Grande, New Mexico. Ph.D. dissertation, Civil Engineering, Colorado State University, 199 p.

Park, S. K., P. Y. Julien, U. Ji and J. F. Ruff (2008). Case-study: retrofitting large bridge piers on the Nakdong River, South Korea. *J. Hyd. Eng., ASCE*, 134(11), 1639–50.

Parker, G. (1976). On the cause and characteristic scales of meandering and braiding in rivers. *J. Fluid Mech., CUP*, 76, 457–80.

Parker, G. and E. D. Andrews (1985). Sorting of bedload sediments by flow in meander bends. *Wat. Res. Res., AGU*, 21, 1361–73.

Parker, G. and M. H. Garcia, eds. (2005). River, Coastal and Estuarine Morphodynamics. In *Proc. 4th RCEM Symp.* London: Taylor & Francis, 1246 p.

Pemberton, E. L. and J. M. Lara (1984). Computing degradation and local scour. In *Technical Guidelines for Bureau of Reclamation*. Denver, CO: USBR, Eng. Res. Center.

Pemberton, E. L. and R. I. Strand (2005). Whitney M. Borland and the Bureau of Reclamation, 1930–1972. *J. Hyd. Eng., ASCE*, 131(5), 339–46.

Perry, C. A. (1994). Effect of reservoirs on flood discharges in the Kansas and the Missouri River basins, 1993. Geological Survey Circular 1120-E, US Department of the Interior, 20 p.

Peters, J. J. (1988). Etudes récentes de la navigabilité. In *Proc. Symp. L'accès maritime du Zaire*. Acad. Royale Sc. Outre-Mer, Bruxelles, Belgium, 89–110.

—— (1993). Problèmes de navigation fluviale dans les bassins de l'Amazone et de l'Orénoque dans les pays andins. *Bull. Séanc. Acad. r. Sci. Outre-Mer*, 38, 505–24.

—— (1994). *Manejo de Rios en la Cuenca de Pirai*. Santa Cruz: SEARPI, 141 p.

—— (1998). Amélioration du transport fluvial en amazonie bolivienne. *Bull. Séanc. Acad. r. Sci. Outre-Mer*, 44, 463–82.

Petersen, M. S. (1986). *River Engineering*. Englewood Cliffs, NJ: Prentice Hall.

PIANC (1992). Guidelines for the design and construction of flexible revetments incorporating geotextiles in marine environment. Suppl. PIANC Bull. 78/79, Brussels, Belgium.

Pilarczyk, K. W. (1995). *Design Tools Related to Revetments Including Riprap, River, Coastal and Shoreline Protection: Erosion Control Using Riprap and Armour Stone*. New York, NY: Wiley.

Pilarczyk, K. W. and R. B. Zeidler (1996). *Offshore Breakwaters and Shore Evolution Control*. Rotterdam: Balkema.

Pitlick, J. C. (1985). The effect of a major sediment influx on Fall River, Colorado. M.S. thesis, Earth Res., Colorado State University, Fort Collins, CO.

Platania, S. P. and C. S. Altenbach (1998). Reproductive strategies and egg types of seven Rio Grande basin cyprinids. *Copeia*, 3, 559–69.

Portland Cement Association (1984). Soil-cement slope protection for embankments: planning and design. PCA Pub. IS173.02W, Skokie, IL.

Ports, M. A., ed. (1989). Hydraulic engineering. In *Proc. Natl. Conf. Hyd. Eng.* New York, NY: ASCE.

Posada Garcia, L. (1995). Transport of sands in deep rivers. Ph.D. dissertation, Civil Engineering, Colorado State University, Fort Collins, CO, 158 p.

Preissmann, A. (1961). Propagation des intumescences dans les canaux et rivières. Congress of the French Assoc. for Computation, France.

—— (1971). Modèles pour le calcul de la propagation des crues. *La Houille Blanche*, 26, 219–24.

Preissmann, A. and J. A. Cunge (1961). Calcul du mascaret sur machine électronique. *La Houille Blanche*, 5, 588–96.

Przedwojski, B., R. Blazejewski and K. W. Pilarczyk (1995). *River Training Techniques: Fundamentals, Design, and Applications*. Rotterdam: Balkema.

Queen, S. (1994). Changes in bed material along the Lower Mississippi, 1932–89. M.S. thesis, Civil Engineering, Colorado State University, 84 p.

Rainwater, J. (2013). Review of sediment plug factors, Middle Rio Grande, NM. M.S. technical report, Civil Engineering, Colorado State University, 55 p.

Rajaratnam, N. (1981). Erosion by plane turbulent jets. *J. Hyd. Res., IAHR*, 19, 339–58.

—— (1982). Erosion by unsubmerged plane water jets. In *Applying Research to Hydraulic Practice*. New York, NY: ASCE, 280–8.

Rajaratnam, N. and R. K. MacDougall (1983). Erosion by plane wall jets with minimum tailwater. *J. Hyd. Eng., ASCE*, 109, 1061–4.

Rajaratnam, N. and B. A. Nwachukwu (1983). Erosion near groyne-like structures. *J. Hyd. Res., IAHR*, 21, 277–87.

Raslan, Y. M. (1994). Resistance to flow in the upper regime plane bed. Ph.D. dissertation, Civil Engineering, Colorado State University, 125 p.

—— (2000). Geometrical properties of dunes. M.S. thesis, Civil Engineering, Colorado State University, 127 p.

Raudkivi, A. J. (1976). *Loose Boundary Hydraulics*, 2nd ed. Oxford: Pergamon.

—— (1986). Functional trends of scour at bridge piers. *J. Hyd. Div., ASCE*, 112(1), 1–13.

—— (1993). *Sedimentation: Exclusion and Removal of Sediment from Diverted Water*. Rotterdam: Balkema.

Raudkivi, A. J. and R. Ettema (1985). Scour at cylindrical bridge piers in armoured beds. *J. Hyd. Eng., ASCE*, 111, 713–31.

Rawls, W. J., D. J. Brakensiek and N. Miller (1983). Green–Ampt infiltration parameters from soils data. *J. Hyd. Eng., ASCE*, 109, 62–70.

Reclamation (1974). *Design of Small Canal Structures*. Washington, DC: US Department of the Interior.

—— (1976). *Design of Gravity Dams*. Washington, DC: US Department of Interior.

—— (1977). *Design of Small Dams*. Washington, DC: US Department of Interior.

—— (2007). *Rock Ramp Design Guidelines*. Denver, CO: US Department of Interior, USBR Technical Service Center, 110 p.

—— (2015). Bank Stabilization Design Guidelines. Rep. SRH 2015-25, US Department of the Interior, USBR Technical Service Center, Denver, CO.

Reclamation and USACE (2015). Large wood national manual: assessment, planning, design, and maintenance of large wood in fluvial ecosystems: restoring process, function, and structure. US Government, Washington, DC, 628 pages + App.

Renard, K. G., G. R. Foster, G. A. Weesies, D. K. McCool and D. C. Yoder (1997). Predicting soil erosion by water: a guide to conservation planning with the Revised Universal Soil Loss Equation (RUSLE). In *Agricultural Handbook No. 703*. Washington, DC: USDA, 407.

Renard, K. G. and J. R. Freimund (1994). Using monthly precipitation data to estimate the R-factor of the revised USLE. *J. Hydrol.*, 157, 287–306.

Richard, G. A. (2001). Quantification and prediction of lateral channel adjustments downstream from Cochiti Dam, Rio Grande, NM. Ph.D. dissertation, Civil Engineering, Colorado State University, 276 p.

Richard, G. A. and P. Y. Julien (2003). Dam impacts and restoration of an alluvial river – Rio Grande, New Mexico. *Intl. J. Sed. Res.*, 18(2), 89–96.

Richard, G. A., P. Y. Julien and D. C. Baird (2005a). Case study: modeling the lateral mobility of the Rio Grande below Cochiti Dam, New Mexico. *J. Hyd. Eng., ASCE*, 131(11), 931–41.

—— (2005b). Statistical analysis of lateral migration of the Rio Grande. *Geomorphology*, 71, 139–55.

Richard, G., C. Leon and P. Y. Julien (2000). Bernardo Reach geomorphic analysis – Middle Rio Grande, New Mexico. Civil Engineering, Colorado State University Report for USBR, Albuquerque, NM.

Richardson, J. R. (2003). The effect of moving rainstorms on overland flow using one-dimensional finite elements. Ph.D. dissertation, Civil Engineering, Colorado State University, 238 p.

Richardson, E. V. and S. R. Davis (2001). Evaluating scour at bridges, 4th ed. Hydraulic Engineering Circle 18, RFHWA NHI 01-001, Federal Highway Administration, USDOT, Washington, DC.

Richardson, J. R. and P. Y. Julien (1994). Suitability of simplified overland flow equations. *Wat. Res. Res., AGU*, 30(3), 665–71.

Richardson, E. V. and P. F. Lagasse (1996). Stream stability and scour at highway bridges. *Water Intl.*, 21, 108–18.

Richardson, E. V. and P. F. Lagasse, eds. (1999). Stream stability and scour at highway bridges: compendium of papers ASCE water resources engineering conferences 1991–98. ASCE, ISBN 0-7844-0407-0, 1040 p.

Richardson, E. V. and D. B. Simons (1984). Use of spurs and guidebanks for highway crossings. TRB Record 950, 2nd Bridge Eng. Conf., Vol. 2, Transportation Research Board, Washington, DC.

RIZA (1999). *Twice a river, Rhine and Meuse in the Netherlands*. Arnhem: RIZA 99.003. ISBN 90 369 52 239. 127 p.

Roberson, J. A., J. J. Cassidy and M. H. Chaudhry (1997). *Hydraulic Engineering*, 2nd ed. New York, NY: Wiley, 653 p.

Rojas, R., P. Y. Julien, M. Velleux and B. E. Johnson (2008). Grid size effect on watershed soil erosion models. *J. Hydrol. Eng., ASCE*, 134(9), 793–802.

Rojas Sanchez, R. (2002). GIS-based upland erosion modeling, geovisualization and grid size effects on erosion simulations with CASC2D-SED. Ph.D. dissertation, Civil Engineering, Colorado State University, 140 p.

Rosgen, D. (1996). *Applied River Morphology*. Pagosa Springs: Wildland Hydrology.

Rousar, L., Z. Zachoval and P. Julien (2016). Incipient motion of coarse uniform gravel. *J. Hyd. Res, IAHR*, 17.

Rousseau, C. (1979). Analyse des caractéristiques hydrologiques d'une couverture nivale. M.S. thesis, Civil Engineering, Laval University, Québec, Canada.

Rozovskii, I. L. (1957). Flow of water in bends of open channels. Academy of Sciences of the Ukrainian SSR., Institute of Hydrology and Hydraulic Engineering, Kiev., Transl. by Y. Prushansky, 1961 Israel Prog. Scient. Transl., S. Monson, Jerusalem, PST Cat. 363.

Ruff, J. F., S. R. Abt, C. Mendoza, A. Shaikh and R. Kloberdanz (1982). Scour at culvert outlets in mixed bed materials. FHWA/RD-82/011, Washington, DC.

Ruff, J. F., A. Shaikh, S. R. Abt and E. V. Richardson (1987). Riprap stability in side-sloped channels. Civil Engineering, Unpublished Report, 34 p.

Saghafian, B. (1992). Hydrologic analysis of watershed response to spatially varied infiltration. Ph.D. dissertation, Civil Engineering, Colorado State University, 215 p.

Saghafian, B. and P. Y. Julien (1995). Time to equilibrium for spatially variable watersheds. *J. Hydrol.*, 172, 231–45.

Saghafian, B., P. Y. Julien and F. L. Ogden (1995). Similarity in catchment response 1. Stationary storms. *Wat. Res. Res., AGU*, 31(6), 1533–41.

Saghafian, B., P. Y. Julien and H. Rajae (2002). Runoff hydrograph simulation based on time variable isochrone technique. *J. Hydrol.*, 261, 193–203.

Sahaar, A. S. (2013). Erosion mapping and sediment yield of the Kabul River basin, Afghanistan. M.S. thesis, Civil Engineering, Colorado State University, 151 p.

Saint-Venant, J. C. B. (1870). Démonstration élémentaire de la formule de propagation d'une onde ou d'une intumescence dans un canal prismatique, et remarques sur les propagations du son et de la lumière, sur les ressauts, ainsi que sur la distinction des rivières et des torrents. *C. R. Acad. Sci., Paris*, 71, 186–95.

—— (1871). Théorie du mouvement non-permanent des eaux avec application aux crues des rivières et à l'introduction des marées dans leur lit. *C. R. Acad. Sci., Paris*, 73(148–54), 237–40.

Salas, J. D., G. Gavilan, F. R. Salas, P. Y. Julien and J. Abdullah (2014). Uncertainty of the PMP and PMF, Chap. 28. In *Handbook of Engineering Hydrology, Modeling Climate Change and Variability*. Hoboken, NJ: Taylor & Francis, 575–603.

Santoro, V. C. (1989). Experimental study on scour and velocity field around bridge piers. M.S. thesis, Civil Engineering, Colorado State University, 142 p.

Santoro, V. C., P. Y. Julien, E. V. Richardson and S. R. Abt (1991). Velocity profiles and scour depth measurements around bridge piers. *Proc. Ann. Meeting Transp. Res. Board*. FHWA, paper # 910874.

Sasal, M. (1992). Hydraulic design of channel constrictions, CER92-93MS100, Civil Engineering, Colorado State University, Fort Collins, CO, 83 p.

Sauveé, S. F., W. Hendershot and H. E. Allen (2000). Solid-solution partitioning of metals in contaminated soils: dependence on pH, total metal burden, and organic matter. *Environ. Sci. Technol.*, 34(7), 1125–31.

Sauveé, S. F., S. Manna, M. C. Turmel, A. G. Roy and F. Courchesne (2003). Solid-solution partitioning of Cd, Cu, Ni, Pb, and Zn in the organic horizons of a forest soil. *Environ. Sci. Technol.*, 37(22), 5191–6.

Schumm, S. A. (1963). Sinuosity of alluvial rivers on the Great Plains. *Bull. Geol. Soc. Am.*, 74, 1089–100.

Schumm, S. (1969). River metamorphosis. *J. Hyd. Eng., ASCE*, 95, 255–73.

Schumm, S. A. (1972). *River Morphology. Benchmark Papers in Geology, Colorado State University*. Fort Collins, CO: Dowden, Hutchinson & Ross, 429 p.

—— (1977). *The Fluvial System*. New York, NY: Wiley.

—— (1991). *To Interpret the Earth, Ten Ways to be Wrong*. Cambridge: Cambridge University Press.

Schumm, S. A., M. D. Harvey and C. C. Watson (1984). *Incised Channels: Morphology, Dynamics and Control*. Littleton, CO: Water Resources Publication.

Schumm, S. A., M. P. Mosley and W. E. Weaver (1987). *Experimental Fluvial Geomorphology*. New York, NY: Wiley.

Schumm, S. A., C. C. Watson and A. W. Burnett (1982). Investigation of neotectonic activity within the Lower Mississippi Valley Division. USACE Lower Miss. Valley Div. Potamology program (P-I), 2, 158.

Schumm, S. A. and B. R. Winkley, eds. (1994). *The Variability of Large Alluvial Rivers*. New York, NY: ASCE, 469 p.

Schwab, G. O., R. K. Frevert, T. W. Edminster and K. K. Barnes (1981). *Soil Water Conservation Engineering*, 3rd ed. New York, NY: Wiley.

Scruton, D. A. and R. J. Gibson (1993). The development of habitat suitability curves for juvenile Atlantic salmon (*Salmo salar*) in riverine habitat insular Newfoundland. Canada, 149–61.

Seddon, J. A. (1900). River Hydraulics. *Trans. ASCE*, 43, 179–229.

Shafie, A. (2009). Extreme flood event: a case study on floods of 2006 and 2007 in Johor, Malaysia. M.S. technical report, Civil Engineering, Colorado State University, 82 p.

Shah, S. C. (2006). Variability in total sediment load using BORAMEP on the Rio Grande Low Flow Conveyance Channel. M.S. thesis, Civil Engineering, Colorado State University, 195 p.

Shah-Fairbank, S. C. (2009). Series expansion of the modified Einstein procedure. Ph.D. dissertation, Civil Engineering, Colorado State University, 238 p.

Shah-Fairbank, S. C. and P. Y. Julien (2015). Sediment load calculations from point measurements in sand-bed rivers. *Intl. J. Sed. Res.*, 30, 1–12.

Shah-Fairbank, S., P. Y. Julien and D. C. Baird (2011). Total sediment load from SEMEP using depth-integrated concentration measurements. *J. Hyd. Eng., ASCE*, 137(12), 1606–14.

Sharma, A. (2000). Two-dimensional subsurface flow modeling for watersheds under spatially and temporally variable rainfall. PhD dissertation, Civil Engineering, Colorado State University, Fort Collins, CO, 154 p.

Sharp, J. J. (1981). *Hydraulic Modelling*. Boston, MA: Butterworths, 242 p.

Shen, H. W., ed. (1971a). River Mechanics, Vols. I and II. P.O. Box 606, Fort Collins, CO.

Shen, H. W. (1971b). Scour near piers. In *River Mechanics*. Fort Collins, CO: Colorado State University, Vol. II, 23, 1–25.

Shen, H. W. and P. Y. Julien (1992). Erosion and sediment transport, Chap. 12. In *Handbook of Hydrology*. New York, NY: McGraw-Hill.

Shen, H. W., V. R. Schneider and S. S. Karaki (1969). Local scour around bridge piers. *Proc. ASCE*, 95, 1919–40.

Shields, A. (1936). Anwendung der Aehnlichkeitsmechanik und der Turbulenzforschung auf die Geschiebebewegung, Mitteilungen der preussischen Versuchanstalt für Wasserbau und Schiffbau, Berlin.

Shields, F. D., Jr., S. R. Pezeshki, G. V. Wilson, W. Wu and S. M. Dabney (2008). Rehabilitation of an incised stream with plant materials: the dominance of geomorphic processes. *Ecology and Society* 13(2), 54.

Shih, H. M. (2007). Estimating overland flow soil transport capacity and surface erosion rate using unit stream power. Ph.D. dissertation, Civil Engineering, Colorado State University, 172 p.

Shimizu, Y., M. W. Schmeeckle and J. M. Nelson (2000). Three-dimensional calculation of flow over two-dimensional dunes. *Ann. J. Hyd. Eng., Japan Soc. Civ. Eng.*, 43, 623–8.

Shin, Y. H. (2007). Channel changes downstream of the Hapcheon re-regulation Dam in South Korea. Ph.D. dissertation, Civil Engineering, Colorado State University, 217 p.

Shin, Y. H. and P. Y. Julien (2010). Changes in hydraulic geometry of the Hwang River below the Hapcheon re-regulation Dam, South Korea. *Intl. J. Riv. Basin Manag., IAHR*, 8(2), 139–50.

—— (2011). Effect of flow pulses on degradation downstream of Hapcheon Dam, South Korea. *J. Hyd. Eng., ASCE*, 137(1), 100–11.

Siddique, M. (1991). A non-equilibrium model for reservoir sedimentation. Ph.D. dissertation, Civil Engineering, Colorado State University, 330 p.

Sieben, J. (1997). Modelling of hydraulics and morphology in mountain rivers. Communications on Hydraulic and Geotechnical Engineering, ISSN 0169-6548, TUD, Delft University of Technology, 222 p.

Simons, D. B. (1957). Theory and design of stable channels in alluvial material. Ph. D. dissertation, Civil Engineering, Colorado State University, Fort Collins, CO.

Simons, D. B. and M. L. Albertson (1963). Uniform water conveyance channels in alluvial material. *Trans. ASCE*, 128, 65–167.

Simons, L. and Assoc. and Louis Berger Intl. (1984). Geomorphic Analysis of the Niger River Basin. Report prepared for the Niger Basin Authority, 240 p. + App.

Simons, D. B., Y. H. Chen and L. J. Swenson (1984). Hydraulic test to develop design criteria for the use of Reno mattresses, Simons and Li, Fort Collins, CO: Colorado State University.

Simons, D. B. and P. Y. Julien (1983). Engineering analysis of river meandering. In *River Meandering, Proc. Conf. Rivers '83*. New York, NY: ASCE, 530–44.

Simons, D. B. and F. Sentürk (1992). *Sediment Transport Technology, Water and Sediment Dynamics*. Littleton, CO: Water Resources Pub.

Sin, K. S. (2010). Methodology for calculating shear stress in a meandering channel. M.S. thesis, Civil Engineering, Colorado State University, Fort Collins, CO.

—— (2014). Three-dimensional computational modeling of curved channel flow. Ph.D. dissertation, Civil Engineering, Colorado State University, 112 p.

Sistermans, P. G. J. (2002). Graded sediment transport by non-breaking waves and a current. Communications on Hydraulic and Geotechnical Engineering, Report 02-2, ISSN 0169-6548, TUD, Delft University of Technology, 205 p.

Sixta, M. J. (2000). Hydraulic modeling and meander migration of the Middle Rio Grande, New Mexico. M.S. thesis, Civil Engineering, Colorado State University, 260 p.

Skempton A. W. (1964). Long-term stability of clay slopes. *Geotech.*, 14(2), 77–101.

Skinner, M. M. and M. D. Stone (1983). Identification of instream hazards to trout habitat quality in Wyoming, FWS/OBS-83/13, US Department of Interior, Fish and Wildlife Services, 69 p.

Sloff, C. J. (1994). Modelling turbidity currents in reservoirs. Communications on Hydraulic and Geotechnical Engineering, ISSN 0169-6548, TUD, Delft University of Technology, 141 p.

Smith, M. (1988). Plan morphology and channel alignment of Rio Apure, Venezuela. M.S. thesis, Civil Engineering, Colorado State University, Fort Collins, CO, 149 p.

Smith, J. A. (1992). Precipitation, Chap. 3. In *Handbook of Hydrology*. New York, NY: McGraw-Hill, 47 p.

Smith, C. E. (1998). Modeling high sinuosity meanders in a small flume. *Geomorphology*, 25, 19–30.

Smith, M. E. and C. F. Nordin (1988). Alignment characteristics of Rio Apure, Rep. CER99-89MES-CFN3, Civil Engineering, Colorado State University, 137 p.

Soar, P. J. and C. R. Thorne (2001). Channel restoration design for meandering rivers. Rep. ERDC/CHL CR-01-1, USACE, ERDC, Vicksburg, MS, 437 p.

Sotir, R. B. and J. C. Fischenich (2001). Live and inert fascine streambank erosion control. EMRRP Tech. Notes Coll., ERDC TN-EMRRP-SR-31, USACE, ERDC, Vicksburg, MS, 8 p.

Spah, J. A. (2000). Rainfall runoff and the effects of initial soil moisture associated with the Little Washita River Watershed, Oklahoma. M.S. thesis, Civil Engineering, Colorado State University, 140 p.

Stedinger, J. R., R. M. Vogel and E. Foufoula-Georgiou (1992). Frequency analysis of extreme events, Chap. 18. In *Handbook of Hydrology*. New York, NY: McGraw-Hill, 66 p.

Stedinger, J. R., D. C. Heath and K. Thompson (1996). Risk analysis for dam safety evaluation: hydrologic risk. IWR Rep. 96-R-13, Cornell University, 81 p.

Stein, O. R. (1990). Mechanics of headcut migration in rills. Ph.D. dissertation, Civil Engineering, Colorado State University, 215 p.

Stein, O. R. and P. Y. Julien (1993). Criterion delineating the mode of headcut migration. *J. Hyd. Eng., ASCE*, 119(1), 37–50.

—— (1994). Sediment concentration below free overfall. *J. Hyd. Eng., ASCE*, 120(9), 1043–59.

Stein, O. R., P. Y. Julien and C. V. Alonso (1993). Mechanics of jet scour downstream of a headcut. *J. Hyd. Res., IAHR*, 31(6), 723–38.

Steininger, A. (2014). Dam overtopping and flood routing with the TREX watershed model. M.S. thesis, Civil Engineering, Colorado State University, Fort Collins, CO, 82 p.

Sternberg, H. (1875). Untersuchungen über das Lagen-und Querprofil geschiebeführender Flüsse. *Z. Bauwesen*, 25, 483–506.

Stevens, M. A. and C. F. Nordin, Jr. (1987). Critique of the regime theory for alluvial channels. *J. Hyd. Eng., ASCE*, 113, 1359–80.

Stevens, M. A. and D. B. Simons (1971). Stability analysis for course granular material on slopes, Chap. 17. In *River Mechanics*. Fort Collins, CO: Water Resource Publication.

—— (1976). Safety factors for riprap protection. *J. Hyd. Div., ASCE*, 102, 637–55.

Storage (2004). *First International Conference on Service Reservoirs*. Geneva: Services Industriels de Genève, 229 p.

Straub, T. D. (2007). Erosion dynamic of a stepwise small dam removal, Brewster Creek Dam at St. Charles, Illinois. Ph.D. dissertation, Civil Engineering, Colorado State University, 174 p.

Strickler, A. (1923). Beiträge zur Frage der Geschwindichkeit – Formel und der Rauhigkeitszahlen für Ströme, Kanäle un Geschlossene Leitungen. Mitteilungen des Eidgenössischen Amtes für Wasserwirtschaft, Bern, Switzerland, 16g.

Struiksma, N., K. W. Olesen, C. Flokstra and H. J. de Vriend (1985). Bed deformation in curved alluvial channels. *J. Hyd. Res., IAHR*, 21, 57–79.

Sundborg, A. (1956). The river Klaralven: a study of fluvial processes. *Geografis. Ann. Stockholm*, XXXVIII, 127–316.

Swain, R. E., J. F. England, Jr., K. L. Bullard and D. A. Raff (2006). Guidelines for evaluating hydrologic hazards, US Department of the Interior, USBR, Denver, CO, 83 p.

Sylte, T. L. and J. C. Fishenich (2000). Rootwad composites for streambank stabilization and habitat enhancement. EMRRP Tech. Notes Coll., ERDC TN-EMRRP-SR-21, USACE, ERDC, Vicksburg, MS.

Talmon, A. M., M. C. L. M. van Mierlo and N. Struiksma (1995). Laboratory measurements of the direction of sediment transport on transverse alluvial-bed slopes. *J. Hyd. Res., IAHR*, 22, 495–517.

Teh, S. H. (2011). Soil erosion modeling using RUSLE and GIS on Cameron Highlands, Malaysia for hydropower development. M.S. thesis, School Renew. Energy Sc., Akureyri, Iceland, 76 p.

Teh, S. H., L. Mohd Sidek, P. Y. Julien and J. Luis (2014). GIS-based upland erosion mapping, Chap. 16. In *Handbook of Engineering Hydrology*. New York, NY: Taylor & Francis, 18 p.

ten Brinke, W. (2005). The Dutch Rhine a Restrained River. Veen magazines, B.V., Diemen, The Netherlands, 228 p.

Thomas, W. A. and W. H. McAnally (1985). Users manual for the generalized computer program system: open channel flow and sedimentation, TABS-2. USACE, Waterways Experiment Station, Vicksburg, MS.

Thorne, C. R., J. C. Bathurst and R. D. Hey, eds. (1987). *Sediment Transport in Gravel-Bed Rivers*. New York, NY: Wiley.

Thorne, C. R. and A. M. Osman (1988). Riverbank stability analysis II: applications. *J. Hyd. Eng., ASCE*, 114, 151–72.

Thorne, C. R., S. Rais, L. W. Zevenbergen, J. B. Bradley and P. Y. Julien (1983). Measurements of bend flow hydraulics on the Fall River at low stage. Rep. WRFSL 83-9p, Civil Engineering, Colorado State University, 48 p.

Tsai, C. W. and B. C. Yen (2004). Shallow water wave propagation in convectively accelerating open-channel flow induced by the tailwater effect. *J. Eng. Mech., ASCE*, 130(3), 320–36.

Turner, T. M. (1996). *Fundamentals of Hydraulic Dredging*, 2nd ed. ISBN 0-7844-0147-0. Reston: ASCE, 258 p.

USACE and USBR (2015). Best practices in Dam and Levee safety risk – Part IX – risk assessment/management. Version 4.0, Document prepared for internal use in support of performing risk analysis for dam and levee safety projects.

US Army Corps of Engineers (1981). The streambank erosion control evaluation and demonstration act of 1974, Final Report to Congress, Executive Summary and Conclusions, Washington, DC.

—— (1983). *Dredging and Dredged Material Disposal*. EM 1110-2-502. Washington, DC: US Government Printing Office.

—— (1987). Land cover and aquatic habitat maps of the Mississippi River, Cairo Illinois to head of Passes, Louisiana, Mississippi River Commission, 58 maps.

—— (1988). Flood control and navigation maps of the Mississippi River, Cairo Illinois to Gulf of Mexico, 56th ed. Lower Miss. Valley Div. 68 maps.

—— (1991). Hydraulic design of flood control channels. EM 1110-2-1601, US Government Printing Office, Washington, DC.

—— (1994a). Channel stability assessment for flood control projects. Engineer Manual EM 1110-2-1218, US Government Printing Office, Washington, DC.

—— (1994b). Design of development of bendway weirs for the Dogtooth Bend reach, Mississippi River. HL-94-10, US Government Printing Office, Washington, DC.

—— (1995). River analysis system, HEC-RAS User's Manual Version 1.0. Hydrologic Engineering Center, Davis, CA.

—— (1999). Lower Mississippi River sediment study. Two compact discs with report and data files.

US Bureau of Reclamation (1976). *Design of Gravity Dams*. US Department of the Interior, Washington, DC: US Government Printing Office, 553 p.

—— (1977). *Design of Small Dams*, 2nd ed. US Department of the Interior, Washington, DC: US Government Printing Office, 816 p.

—— (1983). *Design of Small Canal Structures*. US Department of the Interior, Washington, DC: US Government Printing Office, 435 p.

—— (1997). *Water Measurement Manual*, 3rd ed. US Department of the Interior, Washington, DC: US Government Printing Office.

—— (2006). Fish protection at water diversions – a guide for planning and designing fish exclusion facilities, US Department of Interior, Denver, CO, 480 p.

USBR and USACE (2015). Large Wood National Manual, assessment, planning, design and maintenance of large wood in fluvial ecosystems: restoring process, function and structure, www.usbr.gov/pn/, 628 p.

US Department of Agriculture (2007). Stream restoration design, Part 654 National Engineering Handbook, Washington, DC, 660 p.

US Department of Transportation, Federal Highway Administration (1988a). Scour at bridges. Tech. Advisory T5140.20, updated by Technical Advisory T514.23, 1991, Evaluating scour at bridges, Washington, DC.

—— (1988b). Interim procedures for evaluating scour at bridges, Off. Eng. Bridge Div., Washington, DC.

US Fish and Wildlife Service (2007). Rio Grande Silvery Minnow (*Hybognathus amarus*) Recovery Plan. Albuquerque, NM, xiii + 175 p.

Vaill, J. E. (1995). Application of a sediment-transport model to estimate bridge scour at selected sites in Colorado, 1991–93. Wat. Res. Invest. Rep. 95-4179, US Geological Survey, Denver, 37 p.

van Ledden, M. (2003). Sand-mud segregation in estuaries and tidal basins. Communications on Hydraulic and Geotechnical Engineering, Rep. 03-2, ISSN 0169-6548,TUD, Delft University of Technology, 218 p.

van Rijn, L. C. (1984). Sediment transport, part II, suspended load transport. *J. Hyd. Div., ASCE*, 110, 1613–41.

—— (1993). *Principles of Sediment Transport in Rivers, Estuaries and Coastal Seas.* Amsterdam: Aqua Publications.

van Vuren, S. (2005). Stochastic modelling of river morphodynamics. Communications on Hydraulic and Geotechnical Engineering, Rep. 05-2, ISSN 0169-6548,TUD, Delft University of Technology, 275 p.

Vanoni V. A. (1946). Transportation of suspended sediment by water. *Trans. ASCE*, 3, pap. 2267, 67–133.

Velleux, M. (2005). Spatially distributed model to assess watershed contaminant transport and fate. Ph.D. dissertation, Civil Engineering, Colorado State University, 261 p.

Velleux, M., J. England and P. Y. Julien (2008). TREX: Spatially distributed model to assess watershed contaminant transport and fate. *Sci. Total Environ.*, 404(1), 113–28.

Velleux, M., P. Y. Julien, R. Rojas-Sanchez, W. Clements and J. England (2006). Simulation of metals transport and toxicity at a mine-impacted watershed: California Gulch, Colorado. *Env. Sci. Technol.*, 40(22), 6996–7004.

Velleux, M., A. Redman, P. Paquin, et al. (2012). Exposure assessment for potential risks from antimicrobial copper in urbanized areas. *Env. Sci. Technol.*, 46, 6723–32.

Velleux, M., S. Westenbroek, J. Ruppel, M. Settles and D. Endicott (2001). A user's guide to IPX, the in-place pollutant export water quality modeling framework. V. 2.7.4., EPA/600/R-01/079, US Environmental Protection Agency, Office of Research and Development, National Health and Environmental Effects Research Laboratory, Grosse Ile, Michigan. 179 p.

Vensel, C. W. (2005). Review of sedimentation issues on the Mississippi River. M.S. technical report, Civil Engineering, Colorado State University, 62 p.

Vischer, D. and W. H. Hager (1992). Hochwasser-rueckhaltebecken. *Verlag der Fachvereine Zurich, Schweitzerische Hochschule und Technik AG*, Zurich, Switzerland, 211 p.

Vischer, D. and W. H. Hager, eds. (1995). *Energy Dissipators.* Rotterdam: Balkema, 201 p.

Vischer, D. and R. Sinniger (1999). Hydropower in Switzerland. *Soc. Art Civil Eng.*, 4, 131.

Visser, P. J. (1995). Application of sediment transport formulae to sand-dike breach erosion. Communications on Hydraulic and Geotechnical Engineering, ISSN 0169-6548, TUD, Delft University of Technology, 78 p.

Voicu, R., D. Banaduc, E. Kay, E. Schneider-Binder and A. Curtean-Banaduc (2017). Improvement of lateral connectivity in a sector of River Hartibaciu (Olt/Danube basin). *Transylv. Rev. Syst. Res.*, 19(2), 53–68.

Vreugdenhill, C. G. (1972). Mathematical methods for flood waves. DHL Res. Rep. 89-IV, Delft Hydraulics, Delft, The Netherlands.

Wargadalam, J. (1993). Hydraulic geometry equations of alluvial channels. Ph.D. dissertation, Civil Engineering, Colorado State University, Fort Collins, CO, 203 p.

Warren, S. D., H. Mitasova, M. R. Jourdan, et al. (2000). Digital terrain modeling and distributed soil erosion simulation/measurement for minimizing environmental impacts of military training (CS-752). CEMML TPS 00-2, Forest Sc., Colorado State University, 65 p.

Water Environment Federation (WEF) (2001). *Natural Systems for Wastewater Treatment*, 2nd ed. Alexandria, VA: Task force of the WEF, 326 p.

Water Resources Council, Hydrology Committee (1981). *Guidelines for Determining Flood Frequency*. Bull. 17B, U.S. Washington, DC: Water Resources Council.

Watson, C. C., Biedenharn, D. S. and S. H. Scott (1999). *Channel Rehabilitation: Processes, Design, and Implementation*. Vicksburg, MS: USACE, ERDC, 312 p.

Watson, C. C., M. D. Harvey, D. S. Biedenharn and P. Combs (1988). Geotechnical and hydraulic stability numbers for channel rehabilitation: Part I, the approach, and Part II, Application. In *Proc. Hyd. Div. Natl. Conf.* Colorado Springs, CO: ASCE, 120–31.

Weinhold, M. R. (2001). Application of a site-calibrated Parker-Klingeman bedload transport model, Little Granite Creek, Wyoming. M.S. thesis, Colorado State University, 79 p.

Wemelsfelder, P. (1947). Hoogwatergolf doorbrak moehnetalsperre. *Het Ingenieur*, 42, 103–5.

White, W. R., ed. (1987). Topics in fluvial hydraulics. XXII IAHR Cong., Lausanne, 385 p.

—— (1988). *International Conference of River Regime*. Wallingford: Wiley.

White, W. R., R. Bettess and E. Paris (1982). Analytical approach to river regime. *J. Hyd. Div., ASCE*, 108, 1179–93.

Whiting, P. J. and W. E. Dietrich (1990). Boundary shear stress and roughness over mobile alluvial beds. *J. Hyd. Div., ASCE*, 116, 1495–511.

Wiberg, P. and J. D. Smith (1987). Calculations of the critical shear stress for motion of uniform and heterogeneous sediment. *Wat. Res. Res., AGU*, 23, 1471–80.

Williams, D. T. (1995). Selection and predictability of sand transport relations based upon a numerical index. Ph.D. dissertation, Civil Engineering, Colorado State University, 152 p.

Williams, G. P. (1978). Bank-full discharge of rivers. *Wat. Res. Res., AGU*, 14, 1141–54.

Williamsm, D. T. and P. Y. Julien (1989). On the selection of sediment transport equations. *J. Hyd. Eng., ASCE*, 115(11), 1578–81.

Williams, G. P. and M. G. Wolman (1984). Downstream effects of dams on alluvial rivers. USGS Prof. Paper 1286, US Government Printing Office, Washington, DC.

Winkley, B. (1977). Man-made cutoffs on the Lower Mississippi River, conception, construction, and river response. Potamology Invest. Rep. 300-2, USACE, Vicksburg District, MS, 209 p.

—— (1989). The Lower Mississippi River and the Coriolis force. In *Proc. Natl. Conf. Hyd. Eng.* New York, NY: ASCE, 1114–19.

Wischmeier, W. H. and D. D. Smith (1978). Predicting rainfall erosion losses: a guide to conservation planning. In *USDA Agriculture Handbook* 53–7, Washington, DC.

Wohl, E. (2000). Mountain Rivers. Water Res. Monog. 14, AGU, Washington, DC.

Wolman, M. G. and L. B. Leopold (1957). River floodplains: some observations on their formation. USGS Prof. Paper 282-C, US Geological Survey, Washington, DC.

Woo, H. S. (2002). *River Hydraulics*, 1st ed. Seoul: Cheong Mun Gak, South Korea, 844 p.

Woo, H. S., ed. (2006). River Restoration Casebook. River Restoration Comm., Paju, Geonggi Do: Cheong Mun Gak, South Korea, 227 p.

Woo, H. S. and P. Y. Julien (1990). Turbulent shear stress in heterogeneous sediment-laden flows. *J. Hyd. Eng., ASCE*, 116(11), 1416–21.

Woo, H. S., P. Y. Julien and E. V. Richardson (1986). Washload and fine sediment load. *J. Hyd. Eng., ASCE*, 112(6), 541–5.

—— (1987). Transport of bed sediment in clay suspensions. *J. Hyd. Eng., ASCE*, 113(8), 1061–6.

—— (1988). Suspension of large concentrations of sands. *J. Hyd. Eng., ASCE*, 114, 888–98.

Woo, H. S., W. Kim and U. Ji (2015). *River Hydraulics*, 2nd ed. Paju, Geonggi Do: Cheong Mun Gak, South Korea, 688 p.

Woolhiser, C. A. (1975). Simulation of unsteady overland flow, Chap. 12. In *Unsteady Flow in Open Channels*, Washington, DC: Water Resource Publications, 485–507.

World Meteorological Organization (1986). Manual for estimation of probable maximum precipitation, 2nd ed. Oper. Hydrol. Rep. 1, WMO 332, Geneva, 252.

—— (2010). Manual on stream gauging, Vol. I Fieldwork. And Vol. 2, WMO1044, Geneva, Switzerland.

Wu, B., A. Molinas and P. Y. Julien (2004). Bed-material load computations for nonuniform sediments. *J. Hyd. Eng., ASCE*, 130(10), 1002–12.

Wu, B., Z. Y. Wang, G. Wang, et al., eds. (2002). *Flood Defence '2002*. Beijing: Science Press, 1733 p.

Yalin, M. S. (1971). *Theory of Hydraulic Models*. London: Macmillan.

—— (1992). *River Mechanics*. New York, NY: Pergamon, 220 p.

Yalin, M. S. and E. Karahan (1979). Inception of sediment transport. *J. Hyd. Div., ASCE*, 105, 1433–43.

Yang, C. T. (1976). Minimum unit stream power and fluvial hydraulics. *J. Hyd. Div., ASCE*, 102, 919–34.

Yang, C. T. and F. Simoes (2000). User's manual for GSTARS 2.1. USBR, Technical Service Center, Denver, CO, 94 p.

Yanmaz, A. M. and H. D. Altinbilek (1991). Study of time-dependent local scour around bridge piers. *J. Hyd. Eng., ASCE*, 117, 1247–68.

Yen, C. L. (1970). Bed topography effect on flow in a meander. *J. Hyd. Div., ASCE*, 96, 57–73.

Yen, C. L. and K. T. Lee (1995). Bed topography and sediment sorting in channel
      bend with unsteady flow. *J. Hyd. Eng., ASCE*, 121, 591–9.
Yu, B. and C. J. Rosewell (1996). A robust estimator of the R-Factor for the
      Universal Soil Loss Equation. *Trans. ASAE*, 39, 559–61.
Zeller, J. (1967a). Flussmorphologische Studie zum Mäanderproblem. *Geogr. Helv.*,
      22(2), 57–95.
——— (1967b). Meandering channels in Switzerland. In *Proc. Symp. River Morph.*
      IAHR.

Index

493